The Dynamic Genome and Mental Health

The Dynamic Genome and Mental Health

The Role of Genes and Environments

in Youth Development

Edited by
Kenneth S. Kendler
Sara R. Jaffee
Daniel Romer

THE ANNENBERG
PUBLIC POLICY CENTER
OF THE UNIVERSITY OF PENNSYLVANIA

OXFORD
UNIVERSITY PRESS

OXFORD
UNIVERSITY PRESS

Library of Congress Cataloging-in-Publication Data

The dynamic genome and mental health : the role of genes and environments in youth development /
edited by Kenneth S. Kendler, Sara R. Jaffee, Daniel Romer.
    p. cm.
  Includes bibliographical references and index.
  ISBN 978-0-19-973796-3 (pbk. : alk. paper)
  1. Youth—Mental health. 2. Youth development. 3. Genetic psychology. 4. Mental illness—Genetic
aspects. 5. Mental illness—Environmental aspects. I. Kendler, Kenneth S., 1950- II. Jaffee, Sara R.,
III. Romer, Daniel, 1946-
  RJ503.D96 2011
  618.92'89—dc22   2010029640

978-0-19-973796-3
1 3 5 7 9 10 8 6 4 2

Typeset in Berling
Printed on acid-free paper
Printed in the United States of America

Preface

This volume represents the work of researchers studying some of the most exciting advances in behavioral science: the role of genes and environments in the development of traits and syndromes related to mental health. What was once thought to be a static division between unalterable effects of genes versus more malleable influences of the environment is now recognized as a dynamic interplay that continuously unfolds during development. As a result, it is no longer possible to assume that a high heritability estimate for a human trait, such as intelligence, or an illness, such as schizophrenia, implies that the environment has no ability to influence a person's fate. In regard to a wide range of behavioral and mental phenotypes, heritability is not destiny.

Because of the exciting developments in this new understanding of gene–environment relationships, the Annenberg Public Policy Center at the University of Pennsylvania, in collaboration with Oxford University Press, in April 2009, convened a group of researchers who are at the forefront of these emerging disciplines to present their work and to collaborate on the volume that is presented here. The lead editors, Kenneth S. Kendler and Sara R. Jaffee, deserve much of the credit for identifying the themes of the meeting and the participants who were invited. The plan of the conference was to invite presenters who could provide overviews of the major approaches to studying the interplay between genes and environment, particularly as they relate to psychiatric disorders and complex traits such as intelligence. These presentations are contained in the first half of the book. Other researchers were asked to provide examples of these approaches as applied to the development of common mental or behavioral health problems that emerge in childhood and adolescence. These presentations are found in the second half of the volume.

Kendler leads off the volume with a discussion of how the boundary between genes and environments can no longer be drawn with assurance regarding phenotypes related to behavioral and mental health. The effects of genes extend beyond "the body," producing what are called gene–environment correlations (rGE). He introduces the distinction between traditional quantitative twin studies that estimate aggregate effects of genes summed over the entire genome versus the molecular approach that has become available in recent years for estimating effects of specific genetic variants. Although the former approach tends to find powerful effects of genes in the aggregate, the latter has failed to find robust relations at the level of individual variants. This pattern presents a puzzle that several authors attempt to explain in subsequent chapters. Kendler discusses one potential explanation for the puzzle: the controversial topic of gene–environment interaction (G × E), and provides some illustrations of the difficulties of identifying this form of interplay.

Rudolf Uher follows with a spirited defense of the G × E approach, especially as it applies to the recent controversy surrounding the discovery by Caspi and colleagues of an interaction between the serotonin transporter (SERT) gene and early experience of maltreatment in the development of depression. The idea that genes only affect development in the context of certain environments is an appealing explanation for the absence of robust evidence for the effects of specific genetic variants on mental health. Hence, he argues that the concept will continue to inspire research that can explain the effects of genomic variation in the development of mental health conditions such as depression. Michael J. Shanahan and Shawn Bauldry then discuss the contribution of life-course sociology to our understanding of the dynamic nature of environments. They note the difficulty of characterizing environments and the implications this has for studying G × E effects, especially as applied to research involving the SERT gene. Their chapter underscores the reality that environments are at least as complex and difficult to measure as the genome and that to fully understand environmental and genetic effects will require careful study of development in context.

Sara R. Jaffee follows with an overview of gene–environment correlations, including the differences between active (persons choose their environments), evocative (persons evoke reactions from others), and passive correlations (parent and child behavior are correlated due to shared genotypes). Each of these forms of correlation poses unique challenges to studying the influence of environments because environments are influenced by the characteristics of the people who inhabit them (and hence their genetic makeup). Nevertheless, Jaffee provides an overview of the major approaches to disentangling these interrelated influences. In a related vein, William Dickens, Eric Turkheimer, and Christopher Beam follow with a discussion of the reciprocal effects model of intelligence, which uses the concepts of active and evocative rGE to explain the development of skills such as intelligence. Their model indicates that environmental influences can play a large part in the development of skills, even when heritability is high. That is, environmental exposures can amplify initially modest genetic effects into large phenotypic differences. As a result, their model suggests that the standard twin design cannot provide a pure measure of genetic influence because of the continuous interplay between genes and environments during development.

Briana Horwitz, Kristine Marceau, and Jenae Neiderhiser follow with a discussion of how genes and environments influence the quality of family relationships, including those between parents and their children, as well as those between spouses. Families are complex environments in which parents share genes with their children, and each family member's behavior can be influenced by all of the forms of rGE described earlier by Jaffee. Nevertheless, the authors show how novel twin and adoption designs are being developed to disentangle these interrelated influences. Furthermore, there is evidence that different parenting styles may be effective for preventing disorders in children, depending on the genetic propensities of the children.

One of the exciting advances in the study of environments is the role that experience can play in the subsequent expression of genes during development. Jonathan Mill describes how "epigenetic" effects produce "reversible regulation of gene expression... independently of DNA sequence." Although these effects are most easily observed in animal models, studies with humans are increasingly showing the same effects. He describes how the "epigenome" intervenes dynamically between genotype and phenotype, and how this form of influence may help to provide a mechanism that underlies G × E effects, as well as discordance in phenotypes despite high heritability estimates. In a related vein, Maria Glymour, Wim Veling, and Ezra Susser describe a research strategy borrowed from economics that can be applied to the study of epigenetic effects on development. Using the effects of exogenous stressors, such as famines, on the development of schizophrenia as inspiration, they describe the use of Mendelian randomization as a method for identifying the effects of biomarkers that may induce epigenetic effects on neurodevelopment. Their discussion illustrates how genetic variants can be used to understand such effects on healthy neurodevelopment in children.

We conclude the first part of the volume with an examination of another exciting stream of research concerning the short-term effects of social environments on gene transcription. Steve W. Cole shows how socioenvironmental effects on the resulting "transciptome" can have dramatic implications for our understanding of health and illness. Indeed, it was Cole who suggested the notion of the dynamic genome as a title for the volume. He illustrates these arguments with examples from his own research on the effects of loneliness on genetic regulation of the immune response. Furthermore, he suggests that the transcriptome may be a more fruitful target than the genome for studying genetic effects on health because it integrates the influence of genome, epigenome, and the environment. He lays out an ambitious research agenda for the new field of social genomics and its implications for prevention of illness.

In the second half of the volume, authors discuss the study of gene–environment interplay as it applies to more specific behavioral and mental health problems. In the first section, presenters summarize research regarding the affective disorders.

Thalia C. Eley provides an overview of research on the interplay between genes and environments in the development of anxiety and depression in young people. She examines both quantitative and molecular approaches. She notes that, although both conditions exhibit considerable genetic overlap, the environments that trigger each condition tend to be more specific, with anxiety more related to threatening events and depression more linked to loss. Furthermore, different environmental conditions appear to initiate changes in the symptoms of each condition. Perhaps most important for the aims of this volume, she notes the extensive evidence for the role of both rGE and G × E processes in the emergence of these two related conditions.

BJ Casey, Erika Ruberry, and Victoria Libby discuss their latest research on the amygdala's response to threat cues and the role of the prefrontal cortex in exerting control over this response. They show that individual differences in this circuit are particularly important in adolescent response to threats. In addition, variation in the brain-derived neurotrophic factor (BDNF) gene may underlie these individual differences. They describe a study in which this hypothesis was supported in both a human and mouse model, and they discuss how the use of imaging technology may be integrated into the study of molecular and behavioral genetic research, with potential implications for treatment of anxiety conditions.

Francheska Perepletchikova and Joan Kaufman end this section with a discussion of their studies of maltreated children who were placed under the care of child protective services in the state of Connecticut. Consistent with the G × E research of Caspi and colleagues, they find that, compared with non-maltreated children, those with the critical polymorphisms of the

SERT as well as the *BDNF* gene exhibit greater risk for depressive symptoms. In addition, they find that children who have greater sources of social support are less susceptible to the effects of both maltreatment and high-risk genotypes. These findings suggest that interventions to prevent the effects of maltreatment may be feasible despite the increased risks posed by certain genotypes.

The next section focuses on the emergence of problems in conduct and other externalizing conditions, including substance use and misuse during childhood and adolescence. Janet Audrain-McGovern and Kenneth P. Tercyak review the latest findings regarding the role of genes in susceptibility to nicotine addiction. They present a biobehavioral model of adolescent smoking that includes the influence of genes that can affect both the attraction to tobacco use (e.g., dopamine function) as well as the reaction to the ingestion of tobacco ingredients (e.g., metabolism of nicotine). In addition, the comorbidity between tobacco use and disorders such as depression may have a genetic underpinning. At the same time, certain forms of experience may also influence these underlying processes, thereby altering the likelihood that tobacco use will be maintained. Continued study of these effects holds promise for the future identification of effective interventions to reduce the uptake of and dependence on tobacco use.

Danielle Dick discusses the influence of genes in the development of alcohol use and dependence. She notes that alcohol dependence is highly comorbid with other externalizing conditions, such as conduct disorder and the use and misuse of other drugs. Not surprisingly, genetic influences on alcohol dependence share considerable variation with these other conditions. In addition, the influence of genes on initiation and use of alcohol is weak during pre- and early adolescence but increases with age, suggesting that environments exert greater influence on initial exposure to and use of alcohol. Furthermore, environmental variation can directly alter the contribution of genes to the emergence of alcohol use and dependence. This has been shown in both quantitative studies using twin designs and studies examining specific genetic variants previously identified as related to alcohol use.

Brian M. D'Onofrio, Paul J. Rathouz, and Benjamin B. Lahey discuss the importance of identifying different forms of *r*GE in understanding the etiology of antisocial behavior. If a presumed environmental risk is actually a result of a shared genetic liability (e.g., between parent and child), then the implications for theory and intervention will be very different from that of a direct environmental influence. They illustrate the operation of such shared genetic influence in the relation between maternal smoking during pregnancy and antisocial behavior in offspring. Nevertheless, they also note the direct effects of various environmental influences, such as parental divorce and family poverty, on child outcomes. They provide some excellent examples of analytic techniques and quasi-experimental designs that can disentangle these potentially confounded influences.

In the final chapter in this section, Kendler discusses evidence for both rGE and G × E influences on adolescent conduct disorder and alcohol use. Intensive analyses of a large database of male twins who reported on the risk behavior of themselves and their peers from ages 8 to 25 provide a window into the genetic and environmental correlates of such behavior. The analyses showed that youth affiliated with deviant peers who matched their own genetic risk for deviance (i.e., active rGE). At the same time, peer group deviance moderated the influence of genetic risk on alcohol consumption, providing evidence of G × E for this risk behavior. Furthermore, the interaction was much more evident among preadolescents than among older youth, suggesting age dependence for this form of G × E. Kendler notes that his findings illustrate the operation of all of the processes that are featured in this volume and especially the dynamic nature of the interplay between genes and environments across development.

A final topic in this section concerns the interplay between genes and environment in one of the most debilitating conditions, schizophrenia. Elaine F. Walker, Hanan D. Trotman, Joy Brasfield, Michelle Estgerberg, and Molly Larsen discuss a variety of hormonal changes that occur during adolescence that might introduce risks for psychosis, especially among genetically predisposed individuals. Of particular importance are effects of glucocorticoid responses to stress that appear to increase during adolescence. They also describe prenatal hormonal influences associated with maternal stress responses that may introduce epigenetic effects in the newborn that pose risks for later emergence of psychosis.

Valerie Mondelli, Marta DiForti, Carmine M. Pariante, and Robin M. Murray end this section with a discussion of various environmental influences on the emergence of psychosis. They focus on the renewed interest in stress as a precipitant to psychosis and the various environmental conditions that can contribute to stressful experience. It is also likely that certain genetic variants increase the risk for psychosis when stress is experienced, but success in isolating such genotypes has been difficult. Nevertheless, a focus on how the stress response might interact with relevant genotypes provides direction for future research on the etiology of schizophrenia and psychosis.

We end the volume with a concluding statement provided by the editors. They discuss the scientific, ethical, and policy implications of the research presented in this book. They also suggest areas of further study to enable greater use of the new insights provided by the exciting field of gene-by-environment interplay in mental and behavioral health.

In conclusion, we thank the authors of this volume for their patience in responding to requests for clarification of their presentations. We strove to make the writing as accessible as possible, in the hope that it will stimulate readers to become better acquainted with the issues raised by this exciting field of research. We also thank the Annenberg Foundation, which provided support for the Adolescent Health Communication Institute of the Annenberg Public Policy Center and to Kathleen Hall Jamieson, Director of the Center. Without this support, it would be difficult to organize and complete a project of this scope. We also thank Joan Bossert of Oxford University Press for her steadfast support and collaboration in the project. Finally, no small amount of gratitude is owed the two lead editors of the volume, Kenneth S. Kendler and Sara R. Jaffee, for their tireless efforts to bring this group of scientists together and to make the work come alive in this volume.

Daniel Romer
Annenberg Public Policy Center
University of Pennsylvania

CONTENTS

Schizophrenia and Stress

Part III Implications for the Future

CONTRIBUTORS

Janet Audrain-McGovern, PhD
Department of Psychiatry
University of Pennsylvania
Philadelphia, PA

Shawn Bauldry, MA
Department of Sociology
University of North Carolina at
   Chapel Hill
Chapel Hill, NC

Christopher Beam, MA
Department of Psychology
University of Virginia
Charlottesville, VA

Joy Brasfield, MA
Department of Psychology
Emory University
Atlanta, GA

BJ Casey, PhD
Sackler Institute for Developmental
   Psychobiology
Departments of Psychiatry, Neurology
   and Neuroscience
Weill Medical College of Cornell
   University
New York, NY

Steve W. Cole, PhD
Division of Hematology-Oncology
UCLA School of Medicine
Los Angeles, CA

Brian M. D'Onofrio, PhD
Department of Psychological and
   Brain Sciences
Indiana University
Bloomington, IN

Danielle M. Dick, PhD
Virginia Institute for Psychiatric and
   Behavioral Genetics
Department of Psychiatry
Virginia Commonwealth University
   Richmond, VA

William Dickens, PhD
Department of Economics
Northeastern University
Boston, MA

Marta DiForti, MD, MRCPsych
Department of Psychosis Studies
Institute of Psychiatry
King's College London
London, UK

Thalia C. Eley, PhD
Social, Genetic and Developmental
  Psychiatry Centre
Institute of Psychiatry
King's College London
London, UK

Michelle Esterberg, PhD
Department of Psychology
Emory University
Atlanta, GA

M. Maria Glymour, ScD
Department of Society, Human
  Development, and Health
Harvard School of Public Health
Boston, MA

Briana Horwitz, PhD
Department of Psychology
Penn State University
University Park, PA

Sara R. Jaffee, PhD
Institute of Psychiatry
King's College London
London, UK

Joan Kaufman, PhD
Child and Adolescent Research and
  Education (CARE) Program
Department of Psychiatry
Yale School of Medicine
New Haven, CT

Kenneth S. Kendler, MD
Virginia Institute for Psychiatric and
  Behavioral Genetics
Department of Psychiatry
Virginia Commonwealth University
Richmond, VA

Benjamin B. Lahey, PhD
Department of Health Studies
  University of Chicago
Chicago, IL

Molly Larsen, MA
Department of Psychology
Emory University
Atlanta, GA

Victoria Libby, BA
Sackler Institute for Developmental
  Psychobiology
Department of Psychiatry
Weill Medical College of Cornell
  University
New York, NY

Kristine Marceau, PhD
Department of Psychology
Penn State University
University Park, PA

Jonathan Mill, PhD
Social, Genetic and Developmental
  Psychiatry Centre
Institute of Psychiatry
King's College London
London, UK

Valerie Mondelli, MD, PhD
Sections of Perinatal Psychiatry &
  Stress, Psychiatry and Immunology
Department of Psychological Medicine
  Institute of Psychiatry
King's College London
London, UK

Robin M. Murray, FRS, MD, DSc, FRCP,
  FRCPsych, FMedSci
Department of Psychosis Studies
Institute of Psychiatry Kings College
  London
London, UK

Jenae Neiderhiser, PhD
Department of Psychology
Penn State University
University Park, PA

Carmine M. Pariante, MD, MRCPsych, PhD
Sections of Perinatal Psychiatry & Stress, Psychiatry and Immunology
Department of Psychological Medicine
Institute of Psychiatry
King's College London
London, UK

Francheska Perepletchikova, PhD
Department of Psychiatry
Yale School of Medicine
New Haven, CT

Paul J. Rathouz, PhD
Department of Health Studies
University of Chicago
Chicago, IL

Daniel Romer, PhD
Adolescent and Health Communication Institutes
Annenberg Public Policy Center
University of Pennsylvania
Philadelphia, PA

Erika Ruberry, BA
Sackler Institute for Developmental Psychobiology
Department of Psychiatry
Weill Medical College of Cornell University
New York, NY

Michael J. Shanahan, PhD
Department of Sociology
University of North Carolina
Chapel Hill, NC

Ezra Susser, MD
Mailman School of Public Health
Columbia University
New York State Psychiatric Institute
New York, NY

Kenneth P. Tercyak, PhD
Division of Health Outcomes & Health Behaviors
Department of Oncology
Georgetown University Medical Center
Washington, DC

Hanan D. Trotman, PhD
Department of Psychology
Emory University
Atlanta, GA

Eric Turkheimer, PhD
Department of Psychology
University of Virginia
Charlottesville, VA

Rudolf Uher, PhD
Social, Genetic and Developmental Psychiatry Centre
Institute of Psychiatry
King's College London
London, UK

Wim Veling, MD, PhD
Parnassia Psychiatric Institute
Rotterdam, The Netherlands

Elaine Walker, PhD
Department of Psychology
Emory University
Atlanta, GA

The Dynamic Genome and Mental Health

Part I

Overviews of Research on
Gene–Environment Interplay

1

A Conceptual Overview of Gene–Environment Interaction and Correlation in a Developmental Context

KENNETH S. KENDLER

I want to begin this introductory chapter by setting this volume in its historical and conceptual context. The first suggestions that hereditary factors play an important role in psychiatric disorders and other important mental functions date back over a century ago. We now stand at a historical cusp in the development of the closely related fields of psychiatric and behavioral genetics. The last 20 years has seen a growth in the size and sophistication of traditional psychiatric genetic studies—largely in twins and families, but also utilizing adoptions and other extended and complex designs—that have demonstrated beyond a reasonable doubt that genetic factors play an important etiologic role in these conditions. The current data are sufficiently robust that no etiologic theory for the major psychiatric disorders (or psychological traits like personality or intelligence) that ignores genetic factors could hope to be comprehensive.

Two other major developments have been noteworthy. First, the field of psychiatric genetics has been moving from static to dynamic models. Older conceptualizations viewed genetic and environmental influences on illness as easily separable, passive components of an etiologic model. It was sufficient to articulate the magnitude of genetic risk, the magnitude of environment risk and assume they simply added together. As will be amply demonstrated in this book, this approach is beginning to be abandoned and in its place are models that assume that the genetic and environmental risk factors of psychiatric illness dynamically interrelate with each other over time. Furthermore, genes in general and DNA specifically have long been conceptualized as fixed and static. After all, we inherit our genes from our parents and, aside from the rare mutation we might pick up along the way, pass them along to our children. However, as I hope to demonstrate in this chapter, and in more detail in other parts of this book, this simplistic conceptualization is deeply flawed. Although DNA may be fixed, we know from many studies, both of basic cell biology and development, that gene expression is highly dynamic and variable over time. Put cells in culture and add specific nutrients they do not normally see and new sets of genes are turned on to metabolize that nutrient. We know, too, that key developmental milestones in human development—the age at onset of menses and menopause in women, the development of bald spots and stomach paunches in men—are under strong genetic control. So, it is, fortunately, becoming increasingly untenable to develop etiologic models for psychiatric and key psychological traits that do not include genes, environment, and development.

Second, our conference has occurred in the middle of another phase of the technology-driven molecular genetic revolution. Gene-finding methods for complex traits like psychiatric disorders have gone through several paradigm shifts, from linkage and positional cloning to candidate genes to genome-wide association studies. Here is not the place to try to chart the history of gene-finding methods for psychiatric disorders and psychological traits. Only a few comments are in order. Our best lessons to date have been negative ones. We have clearly learned that very few genes of large effect exist for any of these traits that we care about. Otherwise, we would have detected them and reliably replicated them by now. Rather, those risk genes that do exist are certain to have small overall effect sizes that make replication difficult. Furthermore, the pathway from gene to phenotype is likely to be extraordinarily complex and involve features such as genetic background and environmental experiences. These will add to the problems with obtaining similar results across samples.

However, the molecular revolution is upon us in full force. Genotyping is becoming faster and cheaper. More and more researchers—again, well-demonstrated in the chapters making up this volume—are supplementing the more traditional genetic studies with molecular tools.

Before getting into the body of this chapter, it is finally important to articulate the two very different ways of thinking about and measuring genetic effects that will be well-demonstrated in this volume. The first involves latent variable modeling, one common tool of genetic epidemiology. This approach does *not* involve the actual direct measurement of genes. No DNA is involved. No lab is needed. Instead, we infer the effect of genes from the examination of special relationships in these "experiments of nature."

The one we will confront most commonly in this book will be twins. Adoptees are also very useful subjects of study, and both approaches can be expanded (with a commensurate increased ability to ask subtler questions); for example, by examining twins and their parents, or twins and their children.

This approach has one critical strength. It permits us to assess the impact of all genetic variation all at once. We get a single aggregate measure averaged over our entire genome. Think of this as sitting in an airplane, viewing a landscape from 40,000 feet. It is a very good way to get the general lay of the land.

Such studies have one parallel deep weakness. They tell us absolutely nothing about the biology of the process—of which particular genes might be involved, how they are acting, and in what particular ways. Molecular genetic studies present us with the mirror-image of strengths and limitations. Instead of the bird's eye view from the airplane, you are a scout on the ground, looking at each boulder and tree. A lot of detail is involved, but it doesn't help at all with the big picture. Molecular genetics proposes more specific hypotheses; in that sense, according to Karl Popper, they reflect better science. Compared to latent variable modeling, molecular genetic methods are more mechanistic and more reductionist. But because the effect of individual genes is small (and often *very* small), getting reliable results can be very hard, especially when we want to know not only whether particular genetic variants impact on our trait of interest, but also want to ask even more subtle questions, like whether those gene variants interact with environmental exposures.

Our record with replication for these two kinds of studies has differed widely. Although some exceptions exist, heritability estimates from twin studies have agreed surprisingly well with one another for the major psychiatric syndromes such as schizophrenia (Sullivan, Kendler, & Neale, 2003), major depression (Sullivan, Neale, & Kendler, 2000), and alcohol dependence. By contrast, candidate gene association studies and linkage studies have been notoriously difficult to replicate. This has been equally or even more true for subtler effects that we might care about even more, such as gene–environment interactions (Risch et al., 2009). Although this is not the place to go into detailed arguments about α levels and problems of multiple testing, the main point can be easily summarized: In aggregate, the impact of genetic risk factors on psychiatric and substance-use disorders is large, thereby generally producing robust results. By contrast, the effect that any individual molecular genetic variant may have on disease risk, we are discovering, is quite small, hard to detect, and harder still to replicate.

We are now ready to move to the main points of this introductory chapter, beginning with ideas about how we define the "environment."

THE ENVIRONMENT

We have in our field two different concepts of what "the environment" represents. For convenience, I will label them the "everything except" and the "specified" definitions. Geneticists typically take the "everything except" definition. That is, they define the environment as everything except that which is genetic. For those used to quantitative genetic terminology, we use the symbols E for individual specific environmental effects, and C for shared or common environment. One useful way to define E is "anything that makes identical twins reared together different." This is a commonsensical definition since identical twins share all their genes and their shared environment.

Traditionally, social scientists, by contrast, utilize a "specified" definition of the environment. This definition might run something like this: the environment reflects some "outside the skin" processes that can be meaningfully specified, measured, potentially manipulated, and that impact on behaviors of interest.

The "everything except" and "specified" definitions of the environment differ in important ways that, if not properly understood, can lead to confusion and sterile debates. The geneticists' "everything except" definition of the environment is broader than the "specified" definition of the social scientists.' A geneticist's environment will typically include measurement error, stochastic developmental processes, effects of somatic mutations, and epigenetic modifications, none of which would be considered environmental from a traditional social science perspective. Note that I am not claiming that one definition is more real than the other. My simple point is that we must understand the differences and not to be confused by them.

As a clinician researcher, I need to make one additional point here. The dominating research paradigm in which we all work has been termed *nomothetic*. That is, we look for general laws that we can apply across individuals within a group. However, I am concerned that, when studying human beings, we sometimes confront aspects of the environment that are inherently understandable only from an idiographic perspective; that is, only from the perspective of one individual taken as a single, unique case study. We human beings live our lives "drenched in meaning," and sometimes this meaning arises in ways that are difficult to capture using our standard research methodology. I will provide one disguised case history to illustrate this point. One of our interviewers in our large twin study complained to me that our quite detailed stressful life event inventory missed a key event described by a twin whom she had recently interviewed. The woman had developed her depressive episode late in adult life and had stated that it was the result of the death of her pet parakeet. (We had decided at the beginning of the interview project to not count deaths of pets.) The story that emerged was that this single woman, now in her late 40s, had lived with her widowed mother for the last 20 years. They had a close, mutually caring relationship, and our subject had cared lovingly for her mother through a long terminal illness. One year before her mother's death, her mother had given her a parakeet as a birthday present. It did not take psychoanalytic training to suggest that the death of the parakeet was depressogenic for our subject because of its association with her mother. That is, the parakeet functioned as a living link between this woman and her deceased mother. On hearing this story, I despaired at our inability to assess systematically these kinds of important but highly idiosyncratic environmental events.

HOW GENES AND ENVIRONMENTS INTERRELATE

The major focus of this book is to explore the interrelationship between genetic and environmental effects on psychopathology and key psychological traits. Just as we can work with two definitions of the environment, our job is further complicated by two contrasting ideas of what it means to have genes and environment "interact." The first approach—and the one that is most commonly represented in the chapters of this book—is a statistical perspective dominated by the great statistical geneticist, Ronald Fisher, who developed the analysis of variance and founded the school of what is commonly called *biometrical* or *statistical genetics*. In this approach, we develop and test statistical models in which the genes and environment are sometimes measured only indirectly (so-called *latent variable modeling*), and sometimes directly. But we don't go into the biology of the process. Rather, we observe processes from afar and model them.

The other approach is a biological or molecular perspective. Since the early work by Jacob and Monod on the operon model of gene regulation, for which they won the 1965 Nobel Prize, we have known that environmental effects can profoundly influence gene expression. If you switch the source of food for bacteria—say, from glucose to lactose—a new set of genes that metabolize the lactose molecule get "turned on." This, too, is a way of thinking about how genes and environment "interact," but one that differs rather dramatically from the Fisherian statistical viewpoint. From this perspective, when we say "interact," we mean in a biological way, getting "down and dirty" and measuring environmental exposures in biologically meaningful ways and looking at processes like gene expression.

Again, our job is to realize that these are complementary ways of thinking about how genes and environment interrelate and to avoid getting confused between these different conceptual and terminological frameworks. In this chapter, I will emphasize the statistical approach.

GENE–ENVIRONMENT INTERACTION

Here, I want to introduce the two major ways in which we have been thinking about how genes and environment "interrelate" with each other: *gene–environment interaction* and *gene–environment correlation*. As will become clear, these are two quite different kinds of things. Contrary to what is commonly conceived, gene–environment correlation is much "firmer" and biological. It merely reflects an expansion of the boundaries of gene action beyond the skin. By contrast, gene–environment interaction can be more ephemeral and statistical by nature.

Our whole idea of statistical interactions is bound up irrevocably with the Fisherian worldview best expressed in the development of the analysis of variance. In this highly influential statistical technique, Fisher posited an approach that first took into account main effects. For example, if we studied the height of a particular plant 10 weeks after planting, we could examine the impact of the two different plant strains (reflecting genes) and the two different fertilizers (reflecting the environment). We would get a main effect for each of these. (That is, one plant strain would grow taller, on average, than the other, regardless of which fertilizer was used, and the plants nourished with one type of fertilizer would grow taller than plants nourished with the other fertilizer, regardless of plant strain.) Then, we would look for a gene–environment (or, more technically, a strain × fertilizer) interaction. This interaction would reflect any explanatory power left over *after you took account of the main effects*.

In many cases, no significant interaction is detected. In this case, we have effects of genes on the phenotype, effect of environment on the phenotype, and no interaction. This is what statisticians call an *additive model*—one in which the effects of genes and environment "just" add together.

What does it mean if the additive model fails, if we do detect a significant gene × environment interaction? What we mean is that their impact on the phenotype (here, plant height) is not independent of one another. The impact of genes depends on environmental exposure, and the impact of the environment depends on the effect of genes. Note that these two statements are conceptually equivalent. There is no inherent order in an interaction, so it is as correct to say "genes modify environmental effect" and "the environment modifies genetic effects." Expressed in yet another way, the central concept of genotype by environment interaction is that of *dependence* or *conditionality*. You cannot understand how genes are acting without taking the environment into account, and vice versa.

But let me digress here briefly to outline yet another source of terminological confusion in this area. In the lay language, the term *interact* sometimes only means "to act together." This is perfectly consistent with the technical concept of an additive model, in which we simply have main effects of genes and environment. So, someone without statistical sophistication might take results in which we find only main effects of genes and main effects of environment, and say that we have a gene–environment interaction, when what they really mean is that genes and environment are "acting together" in an additive manner.

Operating under the assumption that a picture is worth a thousand words, let us now review four examples of gene–environment effects. For the sake of simplicity, let's assume that we have five groups differing in level of genetic liability for a particular trait Y, from low to high. The group with the lowest liability is at the bottom of the figure and is marked by little diamonds, and the highest liability group is at the top of the figure and marked by asterisks (*). It might help to think of Y as a personality trait associated with risk for psychiatric disorders, like neuroticism or sensation seeking. Furthermore, we assume that the environment also comes in five increasing categories. Level 1 reflects a very benign environment that conveys no increase at all on trait Y. As the environment becomes more pathogenic—from levels 2 to 5—it has a progressively greater and greater impact on trait Y.

Figure 1.1 depicts an additive model. The key feature here is that the lines are all parallel with one another. Put another way, this means that, as you go from a low-risk to a high-risk environment (from environment 1 to environment 5), the increase in the level of Y is the same across all five genotypes. Genes and environment act independently of one another.

Figure 1.2 depicts the first of three different kinds of interaction—what I have called a "fan-shaped" interaction (and my guess is that these kinds of interactions are particularly common in nature). Note that, compared to Figure 1.1, the impact of genes is dependent on the environment, and vice versa. The key characteristic of a fan-shaped interaction is that, in benign environments, the difference in level of our outcome variable (Y) as a function of level of genetic liability is quite modest. That is, genes are not doing that much in a protective environment. However, with increasingly severe environmental exposures, the difference between genotypes increases. Literally, the curve fans out. Genes have a much more potent impact on the phenotype in a stressful environment.

Another useful way to conceptualize such fan-shaped interactions is to see that genes in this context do two different things. First, they set the mean level of genetic liability. That is, in each environment, the low-risk group always has the lowest level of Y, and the high-risk group the highest. However, the genes are also doing something different. They are impacting on the sensitivity of the individual to the impact of the environment. This is depicted by the steeper slope of those at high genetic liability than in those with low genetic risk.

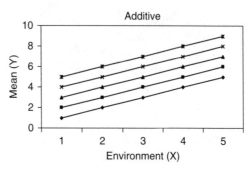

Figure 1.1. **An additive model of genetic and environmental effects.** For this and the next three figures, we are predicting the mean level of a quantitative trait Y from the effect of genes and environment. The lines depict five different genotypes with varying levels of liability to trait Y. The very-high-risk genotype is depicted by an asterisk, the high-risk by a cross, a mean level of risk is depicted by a triangle, a low level by a square, and the lowest level of risk by a diamond. The environmental level of risk is depicted on the x-axis and ranges from level 1 (very low risk) to 5 (very high risk). The key feature of this additive model is that the lines are all parallel—that is, the increase in the level of trait Y associated with a more adverse environment is the same for all genotypes.

Let's now contrast the fan-shaped interaction with that depicted in Figure 1.3—a type 1 crossover interaction. Here, the order of genetic effects changes as a function of the environment—something we did not see with the fan-shaped interactions. Those at lowest risk in environment 1, because of their genotype, are at highest risk in environment 5. Note here that you would expect the environment on average to have an impact on the phenotype because the average level of individuals in environment 5 (the highest-risk environment) will be substantially greater than those in the most benign environment (environment 1). However, in general, you would not expect to see any main effect of genotype in this situation. This is because a balancing would generally occur of the risk-decreasing effects in benign environments and the risk-increasing effects in the malignant environments. (The actual details of this have to do with the frequency of environmental exposures among other things, but this level of detail need not concern us here.)

Finally, what might be considered to be the most extreme kind of crossover is illustrated in Figure 1.4: the type 2 crossover interaction. What is particularly notable about this model is that it generally predicts that there will be no main effect of genes or of environment. Given that the genotypes and environmental influences are of approximately equal frequency, you can see clearly from the diagram that the mean level of outcome measure Y

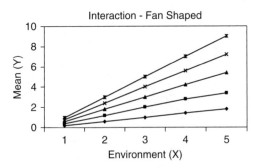

Figure 1.2. **A fan-shaped interaction of genetic and environmental effects on trait *Y*.** See Figure 1.1 and the text for details.

does not change as a function of the environment. Similarly, genotypes will not tend to have any average effect because they will be balanced with opposite effects in benign versus malignant environments.

There are other important differences between these classes of interactions. The fan-shaped interaction is probably the most common, but also the most difficult to interpret. This is because a statistical transformation of the scale of measurement can make an interaction go away. That is, if you examine the raw scale scores for a particular trait, you may find significant evidence for a fan-shaped interaction. However, if you take a log or square-root transformation of the scale scores, you will often find the interaction has disappeared.

Does this mean the interaction is not real? This is a profound issue that we cannot hope to decide here. Part of the answer has to do with the degree of "grounding" of the particular scale of measurement that one is examining. Often, in the kinds of research we do, the particular measures are relatively arbitrary and might reflect the number of endorsed items on a personality scale, or the number of *Diagnostic and Statistical Manual of Mental Disorders, Fourth Edition* (DSM-IV) criteria met. It is in that case difficult to strongly argue that the number of DSM criteria is inherently more real than the square root of those numbers. It would be somewhat different if we were measuring things that were more "real," such as physiological measurement or weight. This adds an extra interpretational difficulty to many analyses of genotype–environment interaction that do not carefully explore the degree to which transformations of the scale of measurement can make the interactions go away.

A related problem is the common use of logistic regression in the analyses of genotype–environment interaction. Logistic regression is a convenient statistical tool when the dependent measure is dichotomous, such as

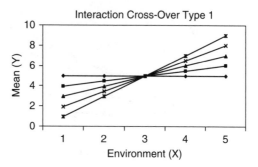

Figure 1.3. **A crossover type 1 interaction of genetic and environmental effects on trait Y.** See Figure 1.1 and the text for details.

whether the subject does or does not have a particular disorder. However, if you look in your statistics textbooks, you will discover that logistic regression involves a logarithmic transformation of the probability of being affected. Most of us recall from high school algebra that taking a logarithm profoundly changes the nature of relationships between variables. In particular, two variables that multiply as regular numbers will add together when you take logarithms. This renders relatively treacherous the interpretation of interactions that rely solely on logistic regression. The interpretation of these results depends in part on a long argument in the epidemiological literature about whether the additive or the multiplicative model of risk is the more appropriate.

This issue is illustrated in Figure 1.5. I here have depicted results from a simple design in which both genes and environment are dichotomized. All subjects can be put into one of four groups, in which those at low risk are indicated by a minus sign (-) and those at high risk indicated by a plus sign (+). The risk of illness is set at 1% for the lowest-risk group, those at both low genetic and environmental risk (G- E-). Those with high genetic risk and low environmental risk (G+ E-), and low genetic but high environmental risk (G- E+) both have a three-fold increased risk of illness (to 3%). The critical question then is, what happens to those who are at excess risk for both genetic and environmental reasons? Example 1 would be expected if the impact of genes and environment added together on the scale of probabilities. Acting alone, each increases the risk of illness by 2% (from 1% to 3%). Summing those produces an increase of 4%. So, an additive model would predict a risk for the G+E+ group of 1% + 4% = 5%. What happens if you examined this question using logistic regression? Although we do not need to go into the detail here (in which you are looking at what is additive on the scale of the log of odds ratios), the answer is ~8.2%. When you graph

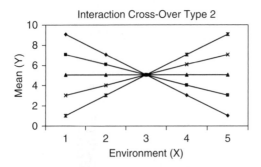

Figure 1.4. **A crossover type 2 interaction of genetic and environmental effects on trait Y.** See Figure 1.1 and the text for details.

this in example 2, it visually looks as though genes and environment are positively interacting and producing particularly high risk for those doubly exposed individuals. However, when analyzed by a logistic regression analysis, within statistical error, this would come out as a perfectly additive model. Only in a result like that depicted in example 3, in which the highest-risk group has a risk of 15%, would this come out as a statistical interaction using logistic regression.

An important paper in this area was published in 2006, by my colleague, Lindon Eaves (Eaves, 2006). It should be compulsory reading for those interested in the problems of genotype and environment interaction. He simulated the effect of candidate genes and specific environmental factors in predicting a normally distributed continuous variable using a purely additive model (just as in Figure 1.1). Then, the resulting continuous results were dichotomized at a particular threshold value, and the dichotomized data were analyzed by logistic regression. Depending on the nature of the simulation and whether random or selected samples were used, genotype–environment interaction was detected (spuriously) in 70% to 100% percent of the simulations. These sobering results indicate that there should be a warning label saying, "buyer beware" for genotype–environment studies that detect interactions using logistic regression for dichotomous dependent measures. It is quite challenging in such studies to determine whether this is a true result or an artifact of the statistical measures employed.

If these weren't difficulties enough, there is an added problem in the interpretation of gene–environment interactions. This is the question of where you might be on the dose-response curve. Look at Figure 1.6. Here, it looks as if we have a nice, clear additive effect depicted as a cumulative logit function. The risk for an illness is depicted on the y axis and environmental risk is depicted on the x axis. We have two populations, with

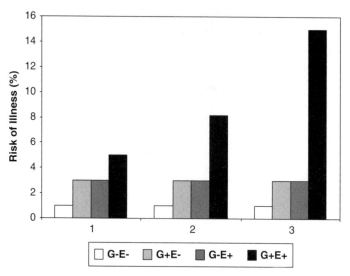

Figure 1.5. **A demonstration of the possible problems in the interpretation of gene–environment interactions.** This figure depicts a design in which we study four groups of individuals: those with (i) no genetic or environmental risk factors (G–E-), (ii) with genetic risk factors but no environmental risk factors (G+E-), (iii) with environmental risk factors but no genetic risk factors (G-E+), and (iv) those with both genetic and environmental risk factors (G+E+). We assume that, acting independently, the genetic and environmental risk factors each increase the risk of illness three-fold. The main question then is, what will they do to risk of illness when both are present? Model 1 (left-hand side of figure) depicts results that are additive on the scale of probabilities. Model 2 (middle of the picture) depicts results that are additive on the scale obtained using logistic regression (taking the log of odds). Model 3 (right-hand side of picture) depicts results that would be seen as a positive multiplicative interaction on the scale both of raw probabilities and by logistic regression.

low- and high-risk genotypes. The high-risk genotype is "left-shifted" with respect to the low-risk genotype, producing higher risks of illness with the same level of environmental risk.

If we happened to be examining environmental risks in our population from somewhere around –0.5 to about +1.5 in our figure, we'll get a nice clear additive relationship because this is where the curves for the two genotypes are roughly parallel to one another. However, what if we happen to be assessing environmental risk in the range of -2 to -1 in our figure? As an inspection of the figure indicates, you will see a substantial shift in risk with the high-risk genotype, but the low-risk genotype has a floor effect, such that increasing the environmental exposure translates into almost no change in risk. It will look like an interaction. Similarly, if we examine subjects with quite high environmental risk, that is, above 2 in our figure, we get the reverse effect. We see substantial changes in risk in our low-risk genotypes as the environmental risk increases. However, in subjects with the high-risk genotype, because of a ceiling effect, we will see almost no change in risk for the disorder with increasing severity of the environment. How often might we be interpreting interactions when we are in one small part of a larger dose-response curve of which we are largely ignorant?

ANOTHER SOURCE OF CONCERN

Let me close this discussion on genotype–environment interaction by looking at two real-world examples that might put our previous theoretical comments in context. The first, Figure 1.7, represents 12 clones of green algae (*Chlamydomonas reinhardtii*) grown in eight different environments, varying in all those things that algae like, including nitrogen, phosphorous, and carbon (Bell, 1991). The *y* axis reflects the log of the rate of growth when the initial area of the growth occurred at low density, and the *x* axis is a summary measure of the quality of the environment. What do we see? Mostly, we see a fan-shaped effect. On average, at low environmental qualities, at the left-hand of the picture, the different clones of green algae tend to cluster, all having a

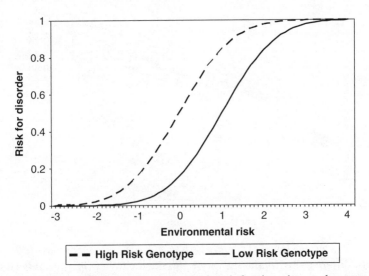

Figure 1.6. **Two cumulative risk curves** (with risk for disorder on the *y* axis) for those at low and high genetic risk across increasing levels of environmental risk depicted on the *x* axis.

fairly low growth rate. However, as the environment improves, the difference between the various clones increases. But that does not describe the entire picture. For example, clone 9 actually grows more poorly at better environments than worse, suggesting aspects of a crossover effect. Note how strikingly different the slopes of clones 2 and 3 are in terms of how rapidly their growth improves with better environments. Clearly, clone 3 is much more sensitive to the quality of the environment in its growth media than is clone 2. The real world does not map in a pretty or entirely clear fashion onto our three prior examples of gene–environment interactions.

Our second real-world example comes from the biomedical literature. This study examines 30 professional boxers who underwent neurological and behavioral assessment in conjunction with ApoE genotyping. The environment measure that was assessed was exposure to head injury, here indexed by the number of professional bouts (< or ≥12 fights). The outcome was a 10-point clinical rating scale (0–9) that assessed neurological damage. The high-risk genotype is the ApoE4 allele, the one that increases risk for Alzheimer disease. The low-risk genotype is simple: the absence of that

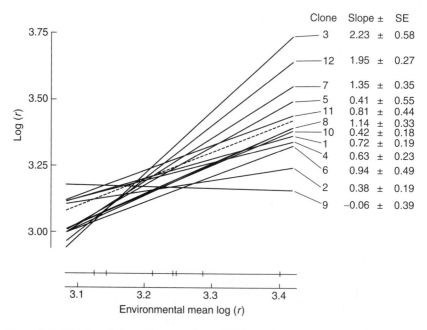

Figure 1.7. **Twelve clones of green algae (*Chlamydomonas reinhardtii*) grown in eight different environments, varying in the quality of the nutrients (Bell, 1991).** The *y* axis reflects the log of the rate of growth when the initial area of the growth occurred at low density, and the *x* axis is a summary measure of the quality of the environment.

allele. What we see is that, in low exposure conditions (i.e., boxers with few professional bouts), the mean brain injury scores are very low and don't differ by genotype. The impact of the ApoE4 allele is only seen in those who have had especially adverse environmental experiences (here, getting their head knocked around a lot; see Fig. 1.8). This result also resembles (in a more truncated way because the authors dichotomized their environment exposures) the fan-shaped interaction depicted in Figure 1.2.

GENOTYPE–ENVIRONMENT CORRELATION

Gene–environment correlation is a "very different sort of thing" from G × E interaction and has a long tradition in the field of evolutionary biology. The idea that our skin represents the boundary of gene expression may make sense from a physiological perspective but is hardly appropriate in an evolutionary context. Take the classical example of within-species sexual selection. The male of the species, under selective pressure, develops external characteristics (coloration, a large tail, etc.) that appeal to the sensory organs of the female of the species, with the goal of appearing to be robust and hence a good sexual partner. So, genes in the male are trying to influence perceptual and judgmental processes within the sensory apparatus and brain of the female. Across species, something rather similar happens in predator–prey relationships. Camouflage has evolved repeatedly in potential prey.

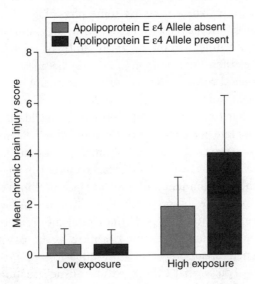

Figure 1.8. **Chronic brain injury scores of boxers.** As a function of their ApoE genotype and their exposure to head injury.

What is the target of that camouflage? The obvious answer is the sensory system of the predator. To be more explicit, the wing feathers of the owl have evolved to avoid detection by the auditory system of the mouse.

Finally, an array of animals substantially modify their environments as a result of their genetic programming in ways that are adaptive. Beavers build dams, bower-birds build elaborate nests, and termites in Africa can produce quite substantial termite mounds stretching over 10 meters long. For those who would like to learn more about this process, I recommend the book *The Extended Organism* by J. Scott Turner, which provides many examples of animals who markedly shape their environments (Turner, 2000). Another good book that builds on this concept is *The Extended Phenotype* by Richard Dawkins (Dawkins, 1982). If there were any questions that there might be genes within a species that impact on construction of the environment, Carol Lynch performed an elegant selection study for nest construction in female mice (Lynch, 1980) and showed, after 15 generations, a ten-fold divergence in the quality of the nest and a realized heritability of 28%.

From a reductionist physiological perspective, it might make sense to think that gene expression stops at the skin. However, when viewed in a broader evolutionary framework, this is an untenably restrictive view about what genes do. For no organism is this truer than for humans.

Let me illustrate this important point another way. Look at the top part of Figure 1.9, which presents the traditional etiological model in medical epidemiology. Susceptibility to disease is seen as arising from an internal or inside-the-skin genetic pathway (in which DNA impacts on RNA, which influences protein, which impacts on some physiological substrate of susceptibility) and from the environment, with the direction of causality entirely from environment to the organism. In this model, humans are passive recipients of their environment. This may be an accurate picture for some environmental risks, but it cannot begin to capture the full picture, especially for behavioral and psychiatric disorders. The lower part of this figure depicts what I have previously termed an outside-the-skin pathway (Kendler, 2001). Here, we conceive of the brain as a behavioral transducing device. As one wag referred to it, for us humans, "our brain has feet." That is, our genes influence our brain, which impacts on our behavior, which then leaps out into the environment and does all kinds of things to influence our environment in behaviorally and psychologically important ways. It is critical to note that we have now "flipped" the causal interrelationship between organism and environment, from the organism as the passive recipient to active "environment creator."

Figure 1.10 depicts what I think is the realistic situation, especially for psychiatric disorders and psychological traits. It suggests that, in tracing pathways from genes to disease susceptibility, we have to think about *two*

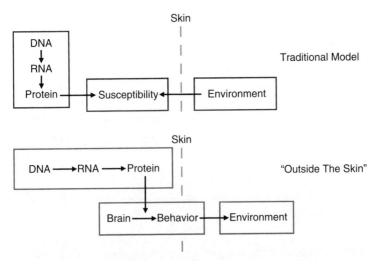

Figure 1.9. **Models of environment and disease etiology.** The top model depicts the traditional model, in which disease susceptibility results from within-the-skin genetic pathways and the environment, with the direction of effect being entirely from the environment to the organism. The bottom model depicts an outside-the-skin pathway, in which genetic factors influence the brain and behavior, so that the organism goes out and alters the environment.

paths. We have to be concerned both about the traditional, reductionist, physiological within-the-skin pathways, as well as the outside-the-skin pathways that arise as a result of gene–environment correlation.

How strong is the empirical evidence that this outside-the-skin pathway impacts on those phenotypes that we in psychiatry and psychology care about? The best summary of which I know is a 2007 review paper that I conducted with a talented graduate student, Jessica Baker (Kendler & Baker, 2007). Jessica and I identified 55 independent studies that examined the heritability of variables traditionally considered to be "environmental." We organized these reports into seven categories, examining general and specific stressful life events, parenting reported by child, parenting reported by parent, family environment, social support, peer interactions, and marital quality. We found 35 different environmental measures in these categories that were examined by at least two studies. Weighted heritability estimates for them ranged between 7% and 39%, with most falling between 15% and 35%. This indicates, as we might expect, that genetic influences on these environmental factors are of overall moderate importance. The weighted heritability for all environmental measures in these studies was at least 27%. Of note, substantial heritability was seen when either informant report or direct or videotape observations were used. We concluded "genetic influences

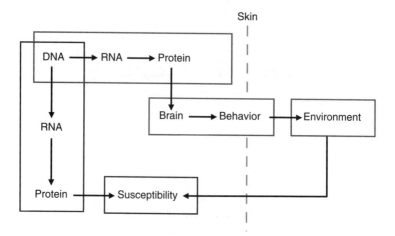

Figure 1.10. **A combined disease pathway including both within- and outside-the-skin genetic pathways.** See Figure 1.9 and text for a further description of this model.

on measures of the environment are persuasive in extent and modest to moderate in impact. These findings largely reflect actual behavior rather than only perceptions" (Kendler & Baker, 2007).

One particularly striking example of outside-the-skin pathways has emerged recently in the area of molecular genetics. In a recent short article (Berrettini, 2008), Wade Berrettini reviewed the unusually consistent results obtained in associating variants in the α_3 and α_5 nicotinic receptors on chromosome 15q25 with the risk for heavy smoking or nicotine dependence. Berrettini reviewed six studies and, rather miraculously for psychiatric genetics, all six of them reported congruent and significant findings. Of particular interest is that several groups also reported on associations between these variants and risk for adenocarcinoma of the lung. Although there remains a slight possibility that nicotinic receptors could impact directly on lung function, by far the most plausible causal explanation for this story is that genetic variance in these nicotinic receptor genes, either by causing greater rewarding effects for nicotine or reducing nicotine-associated side effects, increased the risk for the development of nicotine dependence. The dependence state itself induces affected individuals to seek cigarettes in their environment and to expose themselves to large amounts of cigarette smoke by inhaling potential carcinogens deeply into their lungs. This results in an increased risk of lung carcinoma. Although you could say that these nicotinic receptors are oncogenes, these are strange oncogenes indeed that have to loop out into the environment to cause their effect. It would be hard to devise a clearer example of a critical outside-the-skin pathway for human disease.

What other kinds of methodological difficulties arise in the analyses of genotype–environment correlation? Note that none of the thorny issues about scales of measurement and transformation, which are so critical in the interpretation of genotype by environment interaction, are important here. Instead, we have conceptual questions and measurement effects. Epidemiologists have divided life stressors into those that are possibly dependent on the individual's behavior and those that are independent or "fateful" (Brown, Sklair, Harris, & Birley, 1973). In this context, fateful means "truly random" or "unrelated to the subject's behavior." Examples of such events might be a subject becoming unemployed because the factory at which he works closes down to relocate to Asia, or a person who drives at a safe speed behind a truck when a piece of heavy equipment falls off, causing a motor vehicle accident. We ought to be surprised if such events, which actually constitute a small minority of the stressors typically experienced, are much influenced by genetic factors. Two twin studies have found that, reassuringly, the genetic contribution to such independent stressful live events are absent (Kendler, Karkowski, & Prescott, 1999) or very modest (Plomin et al., 1990).

I alluded to another methodologic concern above, one about measurement of environmental variables. If you rely on the respondent for reports about his or her environmental exposures, there is always the concern about reporter or projection bias. To what extent do ratings of an environmental variable actually reflect the behavior of the individual, rather than merely perception? That is, you might find depressed subjects reporting poorer work or social relations because they see the world through self-deprecatory "lenses" (Beck, 1967). One obvious way to get around this problem is to use multiple raters. As noted above, in our literature review on the role of genetic factors in the environment (Kendler & Baker, 2007), we noted that the weighted mean heritability of the environment, when based on other-report, was only trivially smaller than when based on self-report. So, although this bias is probably there for a number of our measures, its aggregate effect is likely modest.

In this section, I have emphasized what has been termed "active" gene–environment correlation, ones in which the organism goes out and changes the environment. As will be discussed in more detail in Chapter 4 (by Sara R. Jaffee), two other kinds of gene–environment correlation (passive and evocative) also merit consideration.

One last methodologic point needs to be made: genotype–environment correlation can confound the interpretation of genotype–environment interaction. Imagine that we have two research subjects: Mr. Highstrung and Mr. Lowkey. Mr. Highstrung has a high genetic risk for depression in his family, and Mr. Lowkey does not. Partly as a result of this genetic loading, Mr. Highstrung has a difficult, conflict-prone personality with interpersonal problems at work and conflict-filled love relationships. Mr. Lowkey, by contrast, is rather easy-going and rarely gets into personal problems. In the first year of your study, Mr. Lowkey had a bad skiing accident (a high-stress event) but coped with it fairly well and remained mentally healthy. Mr. Highstrung, by contrast, experienced a rough and messy romantic breakup (also a high-stress event), after which he developed a major depression. You might want to leap to the conclusion that you have demonstrated genotype–environment interaction. After all, both subjects experienced a bad event, and only the one (Mr. Highstrung) with a strong family history developed depression. But, there is another interpretation. It might be that Mr. Highstrung's high genetic loading for depression both caused him to develop depression and caused him to get into a difficult relationship that resulted in the breakup. As illustrated in Figure 1.11, this could be an example not of genotype–environment interaction but rather of genotype–environment correlation.

This potential confound is a real problem. Theoretically, for genotype–environment interaction, you want a "pure" environmental risk factor that is uncorrelated with the subject's genotype. The dilemma is that finding such effects in humans are actually relatively difficult. That is, gene–environment

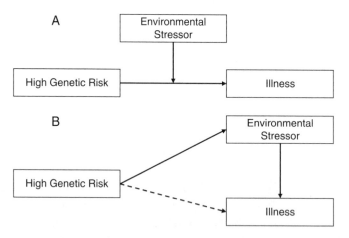

Figure 1.11. **Gene–environment interaction and gene–environment correlation.** As described in the text, this figure illustrates how an apparent gene–environment interaction, (A) in which an individual at high genetic risk becomes ill after exposure to an environmental stressor, could in fact result in part from gene–environment correlation, (B) in which the genetic risk factors for disease also increased the probability of exposure to the environmental stressor.

correlations are likely widespread among many areas of human behavior. Finding "pure" environmental factors that are entirely free of genetic influences can be problematic.

Things are not hopeless. Good statistical minds in the field (e.g., Shaun Purcell and Lindon Eaves) have developed methods to unconfound these two phenomena. The technical details need not concern us here, but this is a problem that deserves our continued attention.

DEVELOPMENT

The final subject of this overview chapter is development. As I noted in my introduction, too often, researchers in genetics have assumed that genes are static, fixed in their effects upon phenotypes. Indeed, in the popular imagination, to say that something is "genetic" implies stability, permanence, and unchangeability. This stereotype is in marked contrast to our increasing knowledge of the speed and complexity of the response of our genomes to environmental perturbations. Starting with the famous studies of Monod and Jacob, alluded to above, merely changing the kind of sugar in the solution surrounding a bacterium produces rapid and dramatic changes in patterns of gene expression. Technological developments in expression arrays have shown us, in a range of tissues, including brain, how rapidly gene expression changes in the setting of stress or in response to various common pharmacologic compounds including nicotine and alcohol.

More traditional genetic observations have shown for many decades that key developmental milestones in the human lifespan are under strong genetic control. The time of menarche in women is highly heritable (Treloar & Martin, 1990), as is the time of onset of menopause (Murabito et al., 2005). More anecdotally, anyone who has observed family members will see substantial correlations of the age of onset of the whitening of hair, the development of bald spots, and the development of the abdominal paunch so common in middle-aged American males these days.

In recent years, we have seen a number of developmental twin studies that have begun to clarify the dynamic nature of the human genome, as manifest in development (e.g., Boomsma, van Beijsterveldt, & Hudziak, 2005; Kendler, Gardner, & Lichtenstein, 2008; Silberg, Rutter, & Eaves, 2001). These studies have suggested that we need to add at least four (somewhat) new terms in our conceptual vocabulary. The first of these is *genetic innovation*. Genetic innovation is detected when, on following individuals over time and measuring the same phenotype, you see evidence for new genetic variation that impacts on that phenotype at a later developmental period, which was not evident at earlier time points. Not surprisingly, studies point to puberty as a developmental watershed, suggesting that genetic risk factors for some important psychiatric phenotypes, such as anxiety and depression, may shift dramatically pre- and post-puberty. But the phenomenon of genetic innovation may not be restricted to puberty, and several studies have suggested that it continues well into late adolescence and even early adulthood.

The reverse phenomenon, labeled by our second new term *genetic attenuation*, has also been demonstrated. This refers to the process whereby genetic effects important at one developmental period become considerably less important at later time points. That is, for example, genetic effects on anxiety prepubertally are much less potent postpubertally, as new genetic influences arise. It is important to note that these processes—genetic innovation and attenuation—are by no means unique to behavioral phenotypes. Very similar effects have been seen on plasma lipids, for example (Middelberg, Martin, & Whitfield, 2007).

Our third and fourth new terms are *developmental homotypy* and *heterotypy*. If the same set of genes influence the same phenotype over time, then we have demonstrated developmental homotypy. However, if the same set of genetic risk factors influence phenotype[1] at age[1] and phenotype[2] at age[2], then developmental heterotypy is at work. An example of the latter is seen in work by Danielle Dick, showing that variants in the *GABRA2* gene impact the risk for conduct disorder in childhood and adolescence, but the risk for alcohol dependence in adulthood (Dick et al., 2006). In developmental heterotypy, the phenotype that is affected by a particular set of genetic factors changes over time.

These and other findings speak powerfully against the static models that have dominated both molecular genetics and genetic epidemiology. Fortunately, we have available to us in genetic epidemiology an increasingly sophisticated set of statistical tools to address these questions. Trajectory analyses can allow us to clarify and distinguish different developmental patterns of symptomatology (or substance use) over time. Growth mixture models provide an alternative way of examining the dynamic nature of developmental data.

CONCLUSION

In this introductory chapter, I have introduced the three key themes of this book: gene–environment interaction, gene–environment correlation, and development. The underlying concept that runs through these themes, and that echoes throughout the chapters of this book, is the dynamic and interactive nature of the etiologic pathways from genes to behavioral phenotypes. Our old static model of genes and environment—which assumes, to a first approximation, that they simply "add together" in their impact on risk of illness—is often incorrect and sometimes dramatically so. Genes frequently shape the way we respond to environmental adversity. The boundary between genes and environment that many of us take for granted is far more porous and less clear than we might have thought. Genetic effects frequently "seep" into our environmental measures. This is not a question of inappropriate methodology. Rather, many of the key environments with which we humans surround ourselves are significantly shaped by our behavior, which in turn is importantly influenced by our genes. Add to this the necessity of looking at how these processes unfold throughout developmental time and you have an accurate picture of the complexity of how genes and environment impact on human behavior, which is the subject of this book.

REFERENCES

Beck, A. T. (1967). *Depression: Clinical, experimental and theoretical aspects.* New York: Hoeber.

Bell, G. (1991). The ecology and genetics of fitness in Chlamydomonas: III. Genotype-by-environment interaction within strains. *Evolution, 45,* 668–679.

Berrettini, W. (2008). Nicotine addiction. *American Journal of Psychiatry, 165,* 1089–1092.

Boomsma, D. I., van Beijsterveldt, C. E., & Hudziak, J. J. (2005). Genetic and environmental influences on anxious/depression during childhood: a study from the Netherlands Twin Register. *Genes Brain and Behavior, 4,* 466–481.

Brown, G. W., Sklair, F., Harris, T. O., & Birley, J. L. (1973). Life-events and psychiatric disorders. 1. Some methodological issues. *Psychological Medicine, 3,* 74–87.

Dawkins, R. (1982). *The extended phenotype: The gene as the unit of selection.* Oxford: Oxford University Press.

Dick, D. M., Bierut, L., Hinrichs, A., Fox, L., Bucholz, K. K., Kramer, J. et al. (2006). The role of GABRA2 in risk for conduct disorder and alcohol and drug dependence across developmental stages. *Behavior Genetics, 36,* 577–590.

Eaves, L. J. (2006). Genotype x environment interaction in psychopathology: fact or artifact? *Twin Research and Human Genetics, 9,* 1–8.

Kendler, K. S. (2001). Twin studies of psychiatric illness: an update. *Archives of General Psychiatry, 58,* 1005–1014.

Kendler, K. S., & Baker, J. H. (2007). Genetic influences on measures of the environment: A systematic review. *Psychological Medicine, 37,* 615–626.

Kendler, K. S., Gardner, C. O., & Lichtenstein, P. (2008). A developmental twin study of symptoms of anxiety and depression: evidence for genetic innovation and attenuation. *Psychological Medicine, 38,* 1567–1575.

Kendler, K. S., Karkowski, L. M., & Prescott, C. A. (1999). The assessment of dependence in the study of stressful life events: validation using a twin design. *Psychological Medicine, 29,* 1455–1460.

Lynch, C. B. (1980). Response to divergent selection for nesting behavior in *Mus musculus*. *Genetics, 96*, 757–765.

Middelberg, R. P., Martin, N. G., & Whitfield, J. B. (2007). A longitudinal genetic study of plasma lipids in adolescent twins. *Twin Research and Human Genetics, 10*, 127–135.

Murabito, J. M., Yang, Q., Fox, C., Wilson, P. W., & Cupples, L. A. (2005). Heritability of age at natural menopause in the Framingham Heart Study. *Journal of Clinical Endocrinology and Metabolism, 90*, 3427–3430.

Plomin, R., Lichtenstein, P., Pedersen, N. L., McClearn, G. E., & Nesselroade, J. R. (1990). Genetic influence on life events during the last half of the life span. *Psychology and Aging, 5*, 25–30.

Risch, N., Herrell, R., Lehner, T., Liang, K. Y., Eaves, L., Hoh, J., et al. (2009). Interaction between the serotonin transporter gene (5-HTTLPR), stressful life events, and risk of depression: A meta-analysis. *Journal of the American Medical Association, 301*, 2462–2471.

Silberg, J. L., Rutter, M., & Eaves, L. (2001). Genetic and environmental influences on the temporal association between earlier anxiety and later depression in girls. *Biological Psychiatry, 49*, 1040–1049.

Sullivan, P. F., Kendler, K. S., & Neale, M. C. (2003). Schizophrenia as a complex trait: evidence from a meta-analysis of twin studies. *Archives of General Psychiatry, 60*, 1187–1192.

Sullivan, P. F., Neale, M. C., & Kendler, K. S. (2000). Genetic epidemiology of major depression: review and meta-analysis. *American Journal of Psychiatry, 157*, 1552–1562.

Treloar, S. A., & Martin, N. G. (1990). Age at menarche as a fitness trait: Nonadditive genetic variance detected in a large twin sample. *American Journal of Human Genetics, 47*, 137–148.

Turner, J. S. (2000). *The extended organism: The physiology of animal-built structures* (1st ed.) Cambridge, MA: Harvard University Press.

2

Gene–Environment Interactions

RUDOLF UHER

For decades, research on the determinants of mental health and illness has been divided between nature and nurture. Social scientists and epidemiologists claimed that most cases of mental illness can be explained by socioecological exposures, such as poverty, parenting, and child abuse. These findings were quickly followed by proposals that manipulations of social environment can prevent and treat mental illness. At the same time, genetic analyses of twin and adoption studies demonstrated that genetic variation explains a large proportion of vulnerability to mental illness. Claims followed that specific genetic risk variants, once identified, will point to the molecular mechanisms involved in the causation of mental illness. Both approaches are near to reaching their limits and appear unlikely to deliver on their promises. Interventions aimed at modifying known environmental exposures have failed to substantially reduce the incidence of mental illness (e.g., Leventhal, Fauth, & Brooks-Gunn, 2005). Comprehensive measurement of variation across the whole human genome has led to the identification of only a few genetic variants that are weakly predictive of mental illness and explain only a tiny proportion of the presumably large heritability (Craddock, O'Donovan, & Owen, 2008; Plomin & Davis, 2009). These failures point to a need to go beyond researching one factor at a time.

In recent past, several paradigm-breaking studies raised the promise that joint consideration of genetic variants and environmental exposures can elucidate the complex causal pathways to mental illness and lead to more effective prevention and therapy (Caspi et al., 2002, 2003). However exciting and promising, this new wave of gene–environment interaction (G × E) studies is not without controversy: testing of complex causal models has met with some formidable methodological challenges, and concerns have been raised about the replicability of these findings. In this chapter, I review the rationale for studying genes and environments jointly rather than separately, scrutinize the concept of gene–environment interaction, give an overview of the current state of evidence, and examine the methodology and controversies surrounding this field. I conclude with some tentative suggestions for future directions in this emerging field of research.

WHY GENE-ENVIRONMENT INTERACTIONS?

The principal impetus for studying gene–environment interactions in humans was the ubiquitous observation that individuals differ in their vulnerability to nearly all environmental pathogens. For example, although many children are adversely affected by growing up in depriving institutions, the majority follow healthy developmental trajectories in spite of deprivation (Croft et al., 2007; Rutter, 2006).

However, individual differences in the response to environmental stressors are not limited to humans, and the general principle that effects of environmental exposures depend on genetic predisposition can also be deduced from experimental data on biological systems. Biologists have repeatedly observed that change in a single genetic or environmental factor rarely results in significant behavioral change in experimental animals and that effects of genetic modifications depend on laboratory conditions (Crabbe, Wahlsten, & Dudek, 1999; Edelman & Gally, 2001). This may be because *biological degeneracy*, denoting the availability of multiple alternative processes to ensure important functions, has provided evolutionary advantage in the context of changing environment and thus it is a general property of complex biological systems (Edelman & Gally, 2001; Uher, 2008b). It follows that multiple environmental and genetic insults are needed to perturb important biological and social functions, as seen in severe mental illness.

It is also the case that, from an evolutionary perspective, the existence of common heritable mental disorders that are associated with increased mortality and reproductive disadvantage presents a formidable puzzle. Any common genetic variant that is even weakly directly associated with a deadly and disabling disorder like schizophrenia should have been pruned out through natural selection (Keller & Miller, 2006). Balanced gene–environment interactions with no net effect of genetic variants averaged across ancestral environments provide a possible explanation for this apparent paradox (Uher, 2009).

Finally, the concept of gene–environment interaction is needed to reconcile the epidemiological findings of strong environmental and social risk factors for mental disorders with high heritability estimates for the same disorders: Because most identified risk factors are neither necessary nor sufficient causes of disease, the fractions of mental disorders attributable to genetic and environmental factors can sum to more than 100% (Rothman & Greenland, 2005). In other words, an epidemiologist saying that 60% of mental illness can be attributed to social disadvantage, and a geneticist saying that mental illness is 60% heritable can both be right, as long as gene–environment interactions operate (see next paragraph: Conceptualization).

In summary, biological and epidemiological data concur with evolutionary theory to indicate that gene–environment interactions are important and ubiquitous in the causation of mental illness. To quote Sir Michael Rutter: "It is highly unlikely that sensitivity to environment would be the one characteristic of living organisms that is uniquely outside genetic influence" (Rutter, 2008, page 3).

CONCEPTUALIZATION: BIOLOGICAL AND STATISTICAL INTERACTIONS

Now, let's turn to the question of what exactly we mean when we talk about gene–environment interactions. Although the word "interaction" is intuitively understood as a reciprocal influence or interdependence, pinning down the exact biological mechanisms and patterns in observational and experimental data that are consistent with the presence or absence of interaction has proven to be something of a challenge. To clarify the concept of gene–environment interactions, the distinction has to be made between biological interaction and statistical interaction. *Biological interaction* refers to a causal mechanism in which two causal factors (e.g., a genetic variant and environmental exposure) contribute to an outcome (e.g., mental illness) in the same individual, but neither of them is a sufficient cause in itself (Rothman, Greenland, & Walker, 1980; Rothman, Greenland, & Lash, 2008; Fig. 2.1). If this definition is satisfied, the following statements will be true. First, the effect of one of these factors (e.g., environmental exposure) on health or illness depends on the presence or absence of the other factor (genetic variant). Second, removal of either of the interacting factors can prevent the occurrence of cases that are due to this causal mechanism (Rothman & Greenland, 2005).

Establishing biological interactions in the causation of health and illness has major implications for basic research and for public health (Uher, 2008a). In terms of basic research, discovery of an interaction between two unrelated factors provides a much more precise clue to the causative mechanism than does discovering a single risk factor. From the public health perspective, the presence of biological gene–environment interaction means

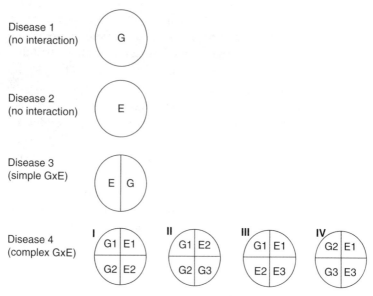

Figure 2.1. **The causal pie model of biological interaction** (according to Rothman and Greenland). The whole circle represents a sufficient cause of a disease. Disease 1 is purely genetically caused, and there is no interaction involved (all such diseases are rare; an example would be Lesch–Nyhan syndrome). Disease 2 is purely environmentally caused, with no interaction involved (most such conditions would be injuries rather than diseases). Disease 3 is caused by a simple biological gene–environment interaction including one specific genetic factor and one environmental factor: each of them is necessary, but neither is sufficient to cause the condition (such diseases are also rare; an example is phenylketonuria). Disease 4 is an example of a moderately complex disorder in which various genetic and environmental factors combine into multiple causal mechanisms: as a result, neither factor alone is sufficient or necessary for the causation of the disorder (all common types of mental illness fall into this category, and most will have more complex causation than what is shown in this schema).

that a preventative or therapeutic intervention that can remove or neutralize the environmental factor can be targeted to the carriers of a specific genetic variant and has the potential to remove all cases of illness due to this causal mechanism, even if a strong genetic contribution is present. However, biological interaction cannot be unambiguously determined from observational epidemiological data (Greenland, 2009). Determination of biological interactions requires experimental research, which is often not practical or ethically permissible in humans. Therefore, researchers rely on observation to infer the likelihood of biological interaction from patterns in observational data. Such inference is usually done through a test of *statistical interaction*.

In statistical terms, gene–environment interaction is defined as a different effect of an environmental exposure on disease risk in persons with different genotypes (Ottman, 1996). Statistical interaction is usually tested in a multiple regression framework as a significant contribution of a product term (genotype multiplied by environment) to the prediction of outcome, over and above the first-order or "main" effects of the explanatory variables (genes and environment). In many instances, the presence of significant statistical interaction coincides with a biological interaction in the causal process. But, in other instances, biological interaction will be present in the absence of statistical interaction or vice versa (Rothman et al., 2008; Rutter, 2008; Uher, 2008c; VanderWeele, 2009b). This mismatch is due to the fact that biological interaction is universal but statistical interaction is contingent on the scale of measurement, statistical model applied, distribution of genetic variants and exposures in the test population, and statistical power (Uher, 2008c; VanderWeele, 2009a,b; Kendler, this volume Chapter 1). For example, mental retardation in phenylketonuria is caused by a biological interaction between two copies of a deficient phenylalanine hydroxylase gene and exposure to phenylalanine in the diet. Both the mutation of phenylalanine hydroxylase gene and dietary phenylalanine are necessary for the mental impairment to occur, and neither of them is a sufficient cause on its own. The knowledge of this gene–environment interaction has enabled effective prevention of the severe outcome by excluding phenylalanine-containing food from the diet of individuals identified through screening. However, as phenylalanine is a ubiquitous component of the normal human diet, and there are virtually no individuals who are not exposed to it other than those who are advised to avoid phenylalanine based on screening, no statistical interaction could have been detected. Generally, biological interactions that involve rare (or ubiquitous) genetic variants or rare (or ubiquitous) environmental exposures are unlikely to be detected as statistical interactions (Uher, 2008c). However, even if substantial variability exists in both genetic variants and the environmental exposures in the test population, testing statistical interactions remains a challenge due to scale- or model-dependence and limited statistical power.

SCALE DEPENDENCE OF STATISTICAL INTERACTIONS

A further potential for discrepancy between biological and statistical interactions stems from the common use of logistic regression models for testing the effects of risk factors (and interactions between them) on categorical outcomes, such as the presence or absence of a disease. It has been demonstrated that the presence of a biological interaction is likely to cause a departure from additivity of risk: if 2 out of 100 individuals exposed to factor A and 10 out of 100 individuals exposed to factor B develop an illness, compared to 1 out of 100 among those not exposed to either A or B, then in the absence of any biological interaction between factor A and factor B, 11 out of 100 individuals exposed to both factor A and factor B would be expected to develop the illness, and a rate of illness exceeding 11/100 should be considered as evidence for synergistic biological interaction (Rothman et al., 2005, 2008; VanderWeele, 2009b). However, logistic regression tests departures from the multiplication of risks and would only detect a positive interaction if more than 20 out of 100 individuals exposed to both factor A and factor B develop the illness (Knol, van der Tweel, Grobbee, Numans, & Geerlings, 2007). Therefore, the use of logistic regression will, in most cases, underestimate the effect of interaction (see also Kendler, Chapter 1].

It has also been shown that, in some cases (e.g., when the outcome is constructed as a product of two very different distributions), logistic regression may falsely identify departures from multiplication of risks (Eaves, 2006; see also Kendler, Chapter 1). Fortunately, this scale- and model-dependence of statistical interactions is only a problem if both factors have a direct effect. If, for example, there is no difference in disease rates between individuals exposed to only one factor and those not exposed to either factor, then additive and multiplicative models will give equivalent results (adding a 0 or multiplying by 1 leaves the estimates unchanged). As most detected gene–environment interactions in the causation of mental illness are not accompanied by a direct effect of genes, their detection is essentially independent of which statistical model is used. In addition, methods have been made available that enable testing of statistical interactions as departures from risk additivity even for categorical outcomes (Knol et al., 2007; Richardson & Kaufman, 2009).

POWER TO DETECT STATISTICAL INTERACTIONS

Although statistical interactions have often been detected and reported in experimental studies, it has proven difficult to find significant statistical interactions in observational data, except when the effects are very strong and samples are large (McClelland & Judd, 1993). The multiple reasons for why significant statistical interactions in observational data are hard to come by relate to the asymmetrical approach to testing interactions and first-order effects in the Fisherian tradition[1]: Although first-order effects are tested in a

---

[1] Sir Ronald Aylmer Fisher (1890-1962), a British scientist, is one of the founders of statistics and of statistical genetics in particular. He introduced a framework, where interactions are always tested

model with no interaction terms, the contribution of interaction is always evaluated in a model that contains all lower-order effects. This asymmetry means that any covariance between the first-order predictors and the interaction term contributes to the test of first-order (main) effects but not to the test of interactions (McClelland & Judd., 1993). This approach protects against falsely detecting an interaction between correlated predictors, simply as a function of their nonindependence. However, with categorical and non-normally distributed data, a degree of correlation/covariance between the first-order predictors and their product (the interaction term) is inevitable due to asymmetry of their distributions. Unfortunately, genetic data are almost always asymmetrically distributed (with the exception of an allele frequency of exactly 0.5). With decreasing minor allele frequency, the amount of information in the gray zone of covariance between first-order effects and their products steeply increases, and the power to test statistical interaction decreases equally rapidly (Caspi et al 2010; Fig. 2.2). As a result, the power to test statistical interactions strongly depends on the distributions of the genotypes and exposures in the sample.

Generally, biological interactions that involve rare genetic variants or rare environmental exposures are unlikely to be detected as significant statistical interactions (Uher, 2008c). Even very large samples (around 10,000 individuals) are only powered to detect moderately strong gene—environment interactions (explaining 2%–5% of variance in outcome) between genotypes with a minor allele frequency of at least 0.1 and environmental factors to which at least 10% of the population has been exposed. The asymmetry of testing first-order effects and interactions in a multiple regression framework means that the effects of gene—environment interactions involving very common genetic variants (frequency >0.8) and common or ubiquitous exposures (>80% population exposed) are likely to be misattributed to direct effects of environment and genes, respectively, missing up on the opportunity for establishing the true causal mechanism (Uher, 2008c; Caspi et al 2010). Although large samples with good assessment of environmental exposures are being collected, which will allow reasonably powered tests of interactions between common genotypes and common exposures, the methods to test the roles of the rarer genotypes and the rarer exposures need to be developed to overcome the problem of statistical power. It is likely that prior biological and epidemiological knowledge will have to be incorporated to allow directed joint testing of multiple, functionally related genetic variants and multiple environmental exposures in a single model.

---

with first-order effects retained in the regression equation but are dropped from the equation if they are not significant.

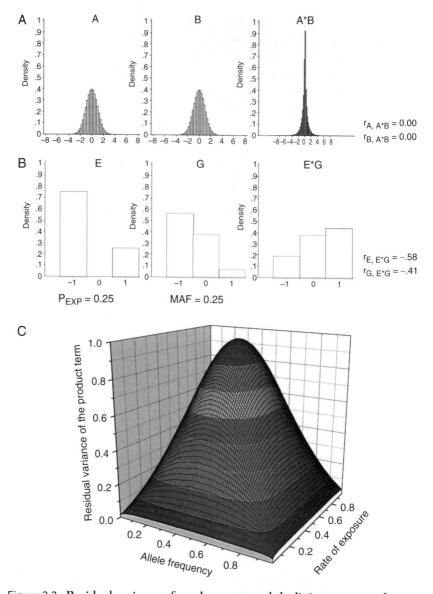

Figure 2.2. **Residual variance of product term and declining power to detect interactions with unevenly distributed genotypes and exposures.** Panels A and B demonstrate a key difference between interactions involving normally distributed continuous variables and those including asymmetrically distributed categorical ones, including most genotypes. If the product term (equivalent to the interaction term in a multiple regression) is calculated from two normally distributed symmetrical variables, it has a restricted variance (leptokurtic distribution), but is uncorrelated with the first-order predictors

Figure 2.2. (*Continued*)

(**A**). However, a product term "E*G" of two categorical variables (categorical exposure "E" of 25% population and a genotype "G" with a minor allele frequency of 25%) is significantly correlated with the first-order predictors (**B**). As a result, the residual variance of the product term (after factoring out first-order predictors) and the power to detect interactions declines rapidly with the rates of exposure and minor allele frequency departing from 50% (**C**). The full power for testing interactions between categorical variables is only preserved in the special case of an exposure rate of 50% and a minor allele frequency equal to 50% (the top segment of **C**). Modifed from Caspi, A., Hariri, A.R., Holmes, A., Uher, R. & Moffitt, T.E. (2010) Genetic sensitivity to the environment: the case of the serotonin transporter gene and its implications for studying complex diseases and traits. *American Journal of Psychiatry, 167,* 509–527.

EXAMPLES AND SYNTHESIS OF KNOWN INTERACTIONS BETWEEN MEASURED GENES AND ENVIRONMENT

Evidence for gene–environment interactions in the pathogenesis of mental illness has been accumulating over several decades from twin and adoption studies (Cadoret, Yates, Troughton, Woodworth, & Stewart, 1995; Kendler et al., 1995; Tienari et al., 2004). However, the progress of this research had been hindered by the limited ability to separate gene–environment interactions from gene–environment correlations when all genetic factors are considered jointly (Eley, Chapter 10; Rutter, 2008). The advent of large-scale molecular genetic studies has revolutionized the field by allowing direct testing and control for gene–environment correlations and has opened a universe of opportunities for translation and interface with biological sciences (Caspi & Moffitt, 2006). Since the first interaction between a measured genetic variant and environmental exposure on mental illness was published (Caspi et al., 2002), there has been a surge of interest in the topic, and a number of reports on measured gene–environment interactions followed (Table 2.1). Importantly, several of these reports have been followed by a number of replications and a wave of experimental studies that have brought strong supportive evidence and made steps toward elucidating some of the biological mechanisms underlying the gene–environment interactions. In the following paragraphs, I will focus on the first two reported interactions between measured genes and environments that have attracted the largest number of replication attempts and follow-up studies.

Table 2.1. Gene–Environment Interactions on Mental Health Outcomes Reported in the Literature

Gene	Exposure	Outcome	Direct first-order effects		Reference	Replicated	Reviewed
			Genotype	Exposure			
MAOA	Child abuse	Antisocial behavior	No	Yes	Caspi et al., 2002	Yes	Taylor, 2007
SERT	Child abuse/life events	Depression	No	Yes	Caspi, 2003	Yes	Uher et al., 2008 Brown et al., 2008 Munafo et al., 2009 Risch et al., 2009
CRHR1	Child abuse	Depression	No	Yes	Bradley et al.,2008	Yes	
FKBP5	Child abuse	Posttraumatic stress disorder	No	Yes	Binder et al.,2008		
COMT	Cannabis	Psychosis	No	Yes	Caspi et al.,2005		
DRD4	Season of birth	Attention-deficit hyperactivity disorder (ADHD)	No	Yes	Seeger et al.,2004		
DRD4	Parenting	ADHD	No	Yes	Sheese et al.,2007	Yes	Thapar et al., 2007
DRD2	Parenting (rules)	Alcohol use	No	Yes	van der Zwaluw, 2009		
DAT1	Mother alcohol use in pregnancy	ADHD	No	Yes	Brookes, 2006		

MAOA = monoamine oxidaze A, SERT = serotonin transporter gene (also known as SLC6A4); CRHR1 = corticotropin receptor 1 gene, FKBP5 = gene coding a glucocorticoid receptor co-chaperone, COMT = catechol-o-methyl transferase gene, DRD4 = dopamine receptor 4 gene, DRD2 = dopamine receptor 2 gene, DAT1 = dopamine transporter gene

The first report of measured gene–environment interaction focused on the relationship between childhood maltreatment and antisocial behavior. Although it is known that childhood abuse and neglect increase the risk of conduct disorder and antisocial behavior, many individuals who suffered maltreatment in their childhood grow up to be law-abiding and compassionate adults. A team of researchers working on the Dunedin Multidisciplinary Health and Development Study tested the hypothesis that a common functional variant of the monoamine oxidase A gene (*MAOA*) may be responsible for this marked individual variation in the consequences of childhood maltreatment. The *MAOA* was selected as a biologically highly plausible candidate, as it has been linked to aggression in both experimental animals and humans, and a rare mutation causing complete absence of functional *MAOA* was shown to be associated with aggression and antisocial behavior in affected men (Brunner, Nelen, Breakefield, Ropers, & van Oost, 1993). As direct associations between a functional variant of this gene and antisocial behavior varied across human populations, it was hypothesized that, rather than directly causing aggression, this genetic variant makes individuals vulnerable to early environmental adversity, such as childhood maltreatment.

As *MAOA* is located on the X chromosome, and antisocial behavior is much more common in men, this interaction hypothesis was primarily tested in male subjects. Among the 442 male members of the population-based Dunedin cohort, the effect of childhood maltreatment on antisocial behavior was found to be dependent on the MAOA genotype. Among subjects carrying the low-activity *MAOA* variant, there was a stronger positive relationship between childhood maltreatment and antisocial behavior (Fig. 2.3A). In comparison, carriers of the more common high-activity *MAOA* variant appeared to be relatively protected from the effects of childhood maltreatment on antisocial behavior. Interestingly, among those not exposed to any childhood maltreatment (the majority of subjects), the low-activity *MAOA* variant was actually associated with a somewhat lower level of antisocial behavior, indicating some pay-off between resilience in the face of adversity and a potential for benefit in favorable context, and resulting in no net effect of the genetic variant on outcome.

Within 2 years after publication of the first report, this remarkable relationship between *MAOA* and sensitivity to early adversity was replicated in a community-based sample of male twins with remarkably similar findings (Foley et al., 2004). Other replications followed. In 2007, eight studies were published of this gene–environment interaction, and six of them found a significant interaction in the expected direction. A meta-analysis of these eight studies concluded that this gene–environment interaction is proven beyond reasonable doubt (Taylor & Kim-Cohen, 2007; Fig. 2.3A).

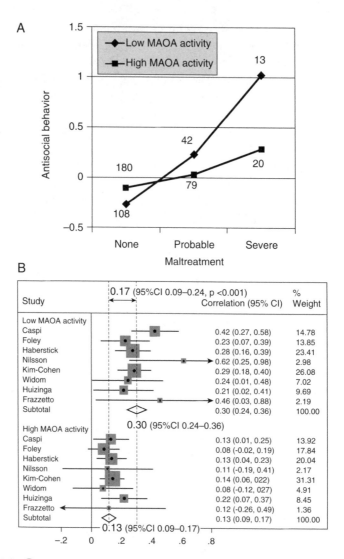

A

B

Figure 2.3. **Gene–environment interaction between *MAOA* and childhood maltreatment results in antisocial behavior**. (A) The relationship between childhood maltreatment and antisocial behavior is much stronger among male carriers of the low-activity *MAOA* genotype than among those carrying the high-activity genotype. Modified from Caspi, A., McClay, J., Moffitt, T. E., Mill, J., Martin, J., Craig, I. W., et al. (2002). Role of genotype in the cycle of violence in maltreated children. *Science*, 297, 851–854, with permission. (B) This relationship holds across most replication studies. Modified from Taylor, A., & Kim-Cohen, J. (2007). Meta-analysis of gene-environment interactions in developmental psychopathology. *Developmental Psychopathology*, 19, 1029–1037, with permission.

However, any nonreplications merit attention as they may hint to specifications of the mechanism. The two studies that did not find a gene–environment interaction in the expected direction stood out by having samples with higher proportions of African ancestry. It may be worth exploring whether measures of exposures or outcomes show differential functioning in African American populations, or whether a different genetic background is responsible for the differences.

The second reported interaction between measured genes and environmental adversity concerned depression. It is well known that the risk of depression is higher in individuals who have had more adverse experiences throughout their lives and that depressive episodes often have an onset following a severe stressful life event (Brown, Bifulco, & Harris, 1987). However, once again, huge individual variability exists in the sensitivity to stress, and many individuals do not become depressed even after enduring severe stress. Researchers working on the Dunedin Multidisciplinary Health and Development Study hypothesized that these individual differences in reactivity to stress could be mediated by a functional length polymorphism in the serotonin transporter gene (5-HTTLPR). Once again, this was a strong candidate, with high biological plausibility but no consistent evidence for a direct association with depression in human populations.

The serotonin transporter gene encodes the target of the most commonly used group of antidepressant drugs (the selective serotonin reuptake inhibitors, such as Prozac) and had been shown to mediate stress reactivity in mice (Holmes, Murphy, & Crawley, 2003), monkeys (Barr et al., 2004). and humans (Hariri et al., 2002; Caspi et al., 2010). Taken together, these findings suggested the hypothesis that 5-HTTLPR may modulate response to major stress and thus be involved in the causation of depression in a conditional rather than a direct manner. This hypothesis has been tested in 847 Caucasian members of the Dunedin cohort. It was shown that the depressogenic effects of both early adversity (childhood maltreatment) and stress in adulthood (stressful life events) increased with the number of 5-HTTLPR short alleles in a dose-dependent manner (Caspi et al., 2003).

Of the gene–environment interactions reported in the literature to date, this one has received the most attention, in terms of both enthusiasm and criticism. It was soon followed by replications (Kaufman et al., 2004; Kendler, Kuhn, Vittum, Prescott, & Riley, 2005); supported by observations from natural experiments, including exposure to a large-scale natural disaster (Kilpatrick et al., 2007); and extended through a number of good-quality human experimental studies demonstrating the dependence of stress-reactivity on 5-HTTLPR (Alexander et al., 2009; Fox, Ridgewell, & Ashwin, 2009; Gotlib, Joormann, Minor, & Hallmayer, 2008). However, it has also been criticized on both theoretical and empirical grounds. One theoretically contentious aspect of this interaction is the time period for stress measurement. Although previous studies clearly demonstrated that depressive episodes tend to follow within months of stressful life events, and that a single severe life event is sufficient to trigger depression (Alexander et al., 2009; Brown et al., 1987; Surtees & Wainwright, 1999), the original Dunedin study measured stressful life events across a 5-year period and showed a gradient of the risk of depression with increasing number of life events (Fig. 2.4). In one study, the expected gene–environment interaction was found for life events across a 5-year period, but not for life events that occurred in the year preceding the depression episode's onset (Wilhelm et al., 2006). It was suggested that such a long-term relationship may reflect an event-generating effect of early adversity and that gene–environment interaction may primarily involve childhood maltreatment rather than stressful life events in adulthood (Brown & Harris, 2008). This proposal is consistent with a generally good replicability of this gene–environment interaction in studies measuring childhood abuse and neglect and with the known role of serotonin transport in the early postnatal developmental period in animals (Ansorge, Zhou, Lira, Hen, & Gingrich, 2004; Ansorge, Morelli, & Gingrich, 2008).

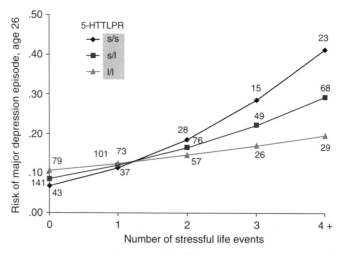

Figure 2.4. **Gene–environment interaction between the serotonin transporter gene and stressful life events results in depression**. The depressogenic effect of stressful life events increases with the number of short alleles of *5-HTTLPR*. Modified from Caspi, A., Sugden, K., Moffitt, T. E., Taylor, A., Craig, I. W., Harrington, H., et al. (2003). Influence of life stress on depression: moderation by a polymorphism in the 5-HTT gene. *Science*, 301, 386–389, with permission.

Empirically, there have been some intriguing nonreplications of this gene–environment interaction in several large studies (Gillespie, Whitfield, Williams, Heath, & Martin, 2005; Surtees et al., 2006). In a systematic review of 18 observational studies published up to August 2007, we concluded that the discrepancies between studies can be explained by differences in the methods used to assess environmental adversity (Uher & McGuffin, 2008). Studies that used an objective indicator or a structured interview to assess environmental exposures in the context and put them on a similar scale independent of subjective attributions reported positive results, but studies that relied on brief self-report instruments for the measurement of environmental exposures often did not replicate the gene–environment interaction (Uher & McGuffin, 2008). Others have also emphasized the importance of contextual measurement of environmental adversity and have expressed concerns about the standards of measuring exposures in large genetic studies (Brown et al., 2008; Monroe & Reid, 2008; see also Shanahan and Bauldry, Chapter 3).

The difference between objective interview and self-report measures of adversity is a question of specificity and objective validity, as simple reliability problems with increased error of measurement would have been overcome by the larger numbers of subjects included in some studies. Accordingly, it has been found that a brief self-report checklist records different types of events than do detailed interviews, both over-rating the trivial and omitting the essential (Monroe 2008). For example, some responders feel compelled to report some stressful life events, because that is what the researchers are interested in and their life would appear boring otherwise. On the other hand, some severe life events are missed because of the avoidance of traumatic memories or the literal interpretation of the limited accompanying instructions. For example, a participant who became depressed while his young child was in the hospital with a brain tumor reported no stressful life events on a checklist because the tumor was diagnosed before the reporting period.

Structured interviews, such as the Life-Events and Difficulties Schedule (Brown & Harris 1978), provide a much more accurate assessment because the interviewers are trained to probe for major types of events and evaluate the stressful nature of life events in context, in order to reduce the biases associated with self-report. In the example above, an interviewer would have inquired about and counted as life events each of the repeated admissions of the participant's son into the hospital, especially as these events fitted with the patient's vulnerability, stemming from the traumatic memory of being himself alone in a hospital as a small child. Another way of obtaining objective information about adverse events and circumstances is to access records of events that were collected independently of the researchers: for example, social services records of child abuse and neglect, court decisions, medical diagnoses of physical illness, or major well-documented events, such as hurricanes. These types of records are associated with negligible measurement errors compared to self-report checklists, and studies that use such events have consistently replicated the gene–environment interaction between 5-HTTLPR and adversity (Uher & McGuffin 2010).

An additional factor explaining some of the differences may be the age of the sample, as results from adolescents (especially male adolescents) are less consistent, possibly due to a failure to measure the types of environmental adversity that are relevant to male teenagers or to the occurrence of alternative adverse outcomes other than depression (Uher & McGuffin 2008).

More studies have been published since and, updating our review with studies published up to March 2009, we found 34 observational human studies that reported a test of interaction between 5-HTTLPR and environmental adversity on depression, of which 17 replicated the gene–environment interaction in the expected direction; eight were partial replications, finding an interaction in the expected direction only in females or only for one of several measures of adversity; and nine studies were nonreplications, finding either no interaction or an interaction in the opposite direction (Table 2.2; Uher & McGuffin 2010). The relationship between the method used to assess environmental adversity and results held almost perfectly: of the five studies with an objective measure of adversity, there were four replications, one partial replication (in adolescents), and no nonreplications; of the eight studies that used an interview to assess adversity, there were five replications, three partial replications, and no nonreplication. All studies that failed to replicate the gene–environment interaction used a questionnaire to assess adversity (Table 2.2; Uher & McGuffin 2010).

Overall, the relationship between method used to assess environmental adversity and study outcome was statistically significant ($x^2(1)$ for trend = 5.30, $p = 0.02$). The preponderance of nonreplications and partial replications in adolescent samples also held for the updated list of 34 studies ($x^2(1)$ for trend = 5.30, $p = 0.02$; Table 2.2). Thus, it can be concluded that (a) the serotonin transporter gene–environment interaction appears to involve exposure to objective stressors rather than subjects' memories or attribution; (b) the gene–environment interaction appears to be developmentally sensitive, and more research on adolescents is needed to determine whether the lack of replications in this age group is due to problems with measuring relevant environmental factors, alternative adverse outcomes or a developmental-stage–specific resilience; and (c) the heterogeneity in sampling and methodology between studies precludes a meaningful single analysis. In other words, both the positive and negative studies are too important to be ignored, and differences between them should be explored to yield further insights into the pathogenesis of depression.

META-ANALYSES OF G X E

It has been increasingly recognized that a count of positive and negative studies is not the best way of synthesizing results and, with some controversy (Egger, Schneider, & Davey, 1998; Stroup et al., 2000), the method of meta-analysis has started to be applied to observational studies, including those of gene–environment interactions. As mentioned earlier, a meta-analysis of eight studies on the interaction between *MAOA* and childhood maltreatment has been reported, and it confirmed that the gene–environment interaction is replicable across samples (Taylor & Kim-Cohen, 2007).

In 2009, two meta-analyses of published studies on the interaction between *5-HTTLPR* and stressful life events were reported, based on five and 14 studies, respectively, and concluding that there is insufficient evidence to support such a gene–environment interaction (Risch et al., 2009; Munafo, Durrant, Lewis, & Flint, 2009). The fact that only a minority of relevant studies were included in these meta-analyses reflects the challenges in synthesizing tests of interaction across heterogeneous literature. First, studies differed in the genetic model that was used: although the original Dunedin study reported findings consistent with an additive genetic model (depressogenic effect of environmental adversity gradually increasing with the number of short alleles: none, one, or two alleles), other studies collapsed the genotypic information into a binary variable, using either a dominant or a recessive genetic model. Any meta-analysis of such studies has to make a choice on the genetic model and will rely on the authors providing additional data not available in the printed articles.

A review of laboratory and epidemiological studies shows that available evidence does not unequivocally support one genetic model over another (Uher & McGuffin 2008; Caspi et al., 2010). Reporting findings for each of the three possible genotypes has been recommended, as such reporting contains the most information and facilitates data synthesis. However, in studies with categorical outcomes, an additive genetic model may lead to very small numbers of cases in some genotypes and thus complicate statistical analysis. Note that the genetic model is not an issue with X-linked markers, such as a *MAOA*, making a meta-analysis of gene–environment interactions involving such markers much more straightforward (Taylor & Kim-Cohen, 2007).

A second issue of major importance in gene–environment interactions concerns heterogeneity in those aspects of environmental adversity that are measured. Although *5-HTTLPR* appears to interact with objectively measured stressors, other genotypes may moderate the impact of emotional memories that may be better detected with self-reported exposures (Polanczyk et al., 2009). In both cases, it appears inappropriate to collapse data across studies using very different methods of assessing the environment

Table 2.2. Published Studies on Gene–Environment Interactions Involving *5-HTTLPR*, Stress, and Depression, Ordered by Method of Assessment of Environmental Stress. Studies are divided into replications, showing a significant effect in the same direction as the original Caspi et al. report; part-replications, showing a significant effect in the same direction only in some analyses (e.g., only one gender or only for one type of stress); and nonreplications, including nonsignificant findings and effects in the opposite direction. Objective measures of stress include records that are collected independently of participants' report and of researchers (e.g., social services record of child abuse, natural disaster, or physical illness established by objective examination). Interview measures involve an interviewer making a judgment on the significance of life stress in context, whereas questionnaire leaves the measurement of stress to the direct report of participants. Studies of adolescent samples (mean age 10–19) are marked with an "A." Studies included in the Risch et al. meta-analysis are marked with an "R," and studies included in the Munafo et al. meta-analysis are marked with an "M."* An unpublished study by Middledorp et al. was included in the meta-analysis by Risch et al. Reprinted from Uher, R., & McGuffin, P. (2010). The moderation by the serotonin transporter gene of environmental adversity in the etiology of depression: 2009 update. *Molecular Psychiatry*, 15: 18–22, with permission.

	REPLICATIONS	Adol	Meta	PART-REPLICATIONS	Adol	Meta	NON-REPLICATIONS	Adol	Meta
Objective	Kaufman et al., 2004			Cicchetti et al., 2007	A				
	Nakatani et al., 2005								
	Kilpatrick et al., 2007								
	Otte et al., 2007								
	Kohen et al., 2008								
	Nobile et al., 2009	A							
Interview	Caspi et al., 2003		R,M	Grabe et al., 2005		R			
	Kendler et al., 2005			Wilhelm et al., 2006		R			
	Mandelli et al., 2007			Sjoberg et al., 2006	A				
	Zalsman et al., 2006								
	Drachmann Bukh, 2009								

Self-report							
Taylor et al., 2006	R	Eley et al., 2004	A	R,M	Gillespie et al., 2005		R
Jacobs et al., 2006	R,M	Scheid et al., 2007		M	Surtees et al., 2006		R,M
Cervilla et al., 2007	R	Brummett et al., 2008			Covault et al., 2007	A	
Kim et al., 2007	R,M	Aguilera et al., 2009			Chipman et al., 2007	A	R
Dick et al., 2007					Chorbov et al., 2007	A	R
Lazary et al., 2008					Wichers et al., 2008		
					Power et al., 2008		R
					Laucht et al., 2009	A	R
					Zhang et al., 2009		
					Middeldorp*		R

(Brown et al., 2008; Monroe et al., 2006; Uher & McGuffin, 2008). Although the impact of method could be assessed in a meta-analysis using a test of heterogeneity, such tests tend to be underpowered and are only meaningful if a large number of studies are included.

Third, synthesis across studies using categorical outcomes (e.g., diagnosis) and continuous outcomes (e.g., a symptom score) is problematic as the two types of outcomes are traditionally tested using different regression models (logistic and linear, respectively) that differ substantially in how statistical interaction is conceptualized (see the section, Scale Dependence of Statistical Interactions).

The two reported meta-analyses of the interaction between *5-HTTLPR* and stressful life events attempted to achieve homogeneity by excluding a number of studies and by categorizing continuous outcomes (Munafo et al., 2009; Risch et al., 2009). Unfortunately, the result was a biased selection of studies, with more negative studies included and more positive studies omitted. For example, the meta-analysis by Risch et al. included 7/10 negative studies, 3/8 part-replications, and only 4/17 positive studies, and in addition, included one unpublished negative study. The result was a statistically significant selection bias toward including more nonreplications and fewer replications ($x^2(1)$ for trend = 5.48, $p = 0.02$; Table 2.2). A meta-analysis based on such a biased selection of evidence can hardly be expected to provide conclusive information on the phenomenon under study. Another sources of concern is the confounding of sample size and other aspects of methodology. Meta-analyses give more weight to studies with large sample sizes that often employ weaker measurement methodology. In addition, use of logistic regression with categorically recoded data has led to testing of statistical interaction as a departure from multiplication of risk, which fits poorly with the concept of biological interactions (Rothman et al., 2008). In conclusion, the methods currently used to synthesize information across studies struggle with the heterogeneity of observational data, and a more valid approach is needed to give consideration to the relevant aspects of methodology.

IMPLICATIONS OF GENE–ENVIRONMENT INTERACTIONS AND DIRECTIONS FOR FUTURE RESEARCH

The gene–environment interactions summarized in Table 2.1 share common features that allow some generalizations. All of them have resulted from strong hypothesis-driven research and included a known environmental exposure with a strong, direct, pathogenic effect and a functionally plausible candidate gene (Moffitt, Caspi, & Rutter, 2005). Notably, no direct effect of genotype on outcome occurred in any of these interactions. Compared to the quest for direct genotype–outcome associations, the gene–environment research has been remarkably successful. Four gene–environment interactions have been identified in the Dunedin cohort alone (Caspi et al., 2002, 2003, 2005, 2007), all based on specific hypotheses and with virtually no exploration (results on all but one candidate gene polymorphism genotyped in the Dunedin cohort up to 2007 have been published). Moreover, four specific gene–environment interactions have already been replicated (Table 2.1). This track record of gene–environment research indicates that interactions between genes and environment are common and important.

So far, research on gene–environment interplay has been limited to environmental exposures with strong direct effects. It is a question of time when researchers start to investigate the high-risk/high-gain territory of environmental factors that do not have strong direct effects but may play important roles in individuals with different susceptibility: for example, liberal versus controlling parenting, safe-but-boring versus risky-but-stimulating environments, etc. Similarly, it is unlikely that all important genes can be guessed, based on our limited knowledge of biology. Thus, in the coming years, we will see attempts to systematically search the genome for genetic variants that moderate individually variable responses to environmental exposures. Finally, biology not only tells us that the effects of genes and environments are unlikely to be independent, it also strongly indicates that causal pathways are unlikely to only include a single gene (Shao et al., 2008). The vast opportunities for systematic genome-wide searches and higher-order interactions are put into perspective by the problems with statistical power and the debates about replicability of the few gene–environment interactions known to date. A paradigm change in research methodology may already be on the way, with a shift from single-hypothesis testing to an exploration of multiple relationships and the exploitation of distributions of statistics from a large number of tests that will eventually probe the complex etiology of mental health and illness (Glymour, 2004).

In most reported gene–environment studies, no significant correlation was found between the measured gene and environmental exposures. Thus, the findings indeed reflect interactions and not indirect effects of the measured genes acting through an outside-the-skin route (see Kendler, Chapter 1). However, the role of genes that are not measured in determining the environmental exposures cannot be excluded in the available observational studies. If such genetic effects were important, the reported gene–environment interactions may contain a contribution from gene–gene interactions (between a gene that is measured and a gene that is not measured but that influences the exposure to an environmental factor). Experimental studies in which exposure can be manipulated are needed to separate true gene–environment interactions from any gene–environment correlations and epistasis (Caspi et al., 2006; Moffitt et al., 2005).

The experimental studies available to date indicate that, at least for the serotonin transporter gene, a true gene–environment interaction occurs, as the genotype moderates responses to randomly manipulated stressful exposures (Alexander et al., 2009; Gotlib et al., 2008). Similar conclusions can be drawn from natural experiments involving exposures, such as natural disasters, that are highly unlikely to be genetically determined (Kilpatrick et al., 2007). The gene–environment interaction between the COMT gene and cannabis in the pathophysiology of psychosis has also been supported in an experimental study (Henquet et al., 2006). More experimental studies and natural experiments are needed to probe other gene–environment interactions.

A salient feature of the reported gene–environment interactions is that there are no direct effects of genotypes on outcomes. This general observation is likely due to the fact that the same genotypes that make individuals vulnerable to the pathogenic effects of environmental adversity also afford an advantage in the absence of such adversity (Uher, 2008b), and this has been variously formulated as "biological sensitivity to context" (Boyce & Ellis, 2005) or "differential susceptibility" and "plasticity genes" (Belsky et al., 2009). One implication of this general principle is that there have to be multiple pathogenic mechanisms for a specific mental illness to occur. If a specific environmental exposure has a more pathogenic effect on carriers of a specific genotype, the absence of such exposure is not sufficient to explain the excess risk of mental illness in those not carrying this genotype in the absence of this exposure; thus, another pathogenic mechanism must exist that is responsible for these excess cases (Uher, 2008b). Such heterogeneity of mental illness causation may prove a fertile basis for targeted prevention and personalized medicine. Some of the genes involved in etiological gene–environment interactions, most notably the serotonin transporter, have already been shown to moderate the outcomes of pharmacological and psychosocial treatments (Brody, Beach, Philibert, Chen, & Murry, 2009; Huezo-Diaz et al., 2009; Serretti, Kato, De, & Kinoshita, 2007). Such prediction could be greatly enhanced by taking into account the history of environmental exposures, and a case should be made for collecting both genotypes and psychosocial histories in treatment and prevention trials.

CONCLUSION

In conclusion, gene–environment research has brought a new perspective to the causation of mental illness that is consistent with biological and evolutionary principles. Collaboration across disciplines and methodological developments are needed to translate this emerging knowledge into more comprehensive models that could be applied to the aim of reducing the burden of mental illness.

REFERENCES

Aguilera, M., Arias, B., Wichers, M., Barrantes-Vidal, N., Moya, J., Villa, H., et al. (2009). Early adversity and 5-HTT/BDNF genes: new evidence of gene-environment interactions on depressive symptoms in a general population. *Psychological Medicine, 39*, 1425–1432.

Alexander, N., Kuepper, Y., Schmitz, A., Osinsky, R., Kozyra, E., & Hennig, J. (2009). Gene-environment interactions predict cortisol responses after acute stress: Implications for the etiology of depression. *Psychoneuroendocrinology, 34*, 1294–1303.

Ansorge, M. S., Morelli, E., & Gingrich, J. A. (2008). Inhibition of serotonin but not norepinephrine transport during development produces delayed, persistent perturbations of emotional behaviors in mice. *Journal of Neuroscience, 28*, 199–207.

Ansorge, M. S., Zhou, M., Lira, A., Hen, R., & Gingrich, J. A. (2004). Early-life blockade of the 5-HT transporter alters emotional behavior in adult mice. *Science, 306*, 879–881.

Barr, C. S., Newman, T. K., Shannon, C., Parker, C., Dvoskin, R. L., Becker, M. L., et al. (2004). Rearing condition and rh5-HTTLPR interact to influence limbic-hypothalamic-pituitary-adrenal axis response to stress in infant macaques. *Biological Psychiatry, 55*, 733–738.

Belsky, J., Jonassaint, C., Pluess, M., Stanton, M., Brummett, B., & Williams, R. (2009). Vulnerability genes or plasticity genes? *Molecular Psychiatry, 14*, 746–754.

Binder, E. B., Bradley, R. G., Liu, W., Epstein, M. P., Deveau, T. C., Mercer, K. B., et al. (2008). Association of FKBP5 polymorphisms and childhood abuse with risk of posttraumatic stress disorder symptoms in adults. *Journal of the American Medical Association, 299*, 1291–1305.

Boyce, W. T., & Ellis, B. J. (2005). Biological sensitivity to context: I. An evolutionary-developmental theory of the origins and functions of stress reactivity. *Developmental Psychopathology, 17*, 271–301.

Bradley, R. G., Binder, E. B., Epstein, M. P., Tang, Y., Nair, H. P., Liu, W., et al. (2008). Influence of child abuse on adult depression: moderation by the corticotropin-releasing hormone receptor gene. *Archives of General Psychiatry, 65*, 190–200.

Brody, G. H., Beach, S. R., Philibert, R. A., Chen, Y. F., & Murry, V. M. (2009). Prevention effects moderate the association of 5-HTTLPR and youth risk behavior initiation: gene x environment hypotheses tested via a randomized prevention design. *Child Development, 80*, 645–661.

Brookes, K. J., Mill, J., Guindalini, C., Curran, S., Xu, X., Knight, J., et al. (2006). A common haplotype of the dopamine transporter gene associated with attention-deficit/hyperactivity disorder and interacting with maternal use of alcohol during pregnancy. *Archives of General Psychiatry, 63*, 74–81.

Brown, G. W., Bifulco, A., & Harris, T. O. (1987). Life events, vulnerability and onset of depression: some refinements. *British Journal of Psychiatry, 150,* 30–42.

Brown, G.W., & Harris, T.O. (1978). *Social origins of depression: A study of psychiatric disorder in women.* London: Tavistock.

Brown, G. W., & Harris, T. O. (2008). Depression and the serotonin transporter 5-HTTLPR polymorphism: a review and a hypothesis concerning gene-environment interaction. *Journal of Affective Disorders, 111,* 1–12.

Brummett, B. H., Boyle, S. H., Siegler, I. C., Kuhn, C. M., Ashley-Koch, A., Jonassaint, C. R., et al. (2008). Effects of environmental stress and gender on associations among symptoms of depression and the serotonin transporter gene linked polymorphic region (5-HTTLPR). *Behavior Genetics, 38,* 34–43.

Brunner, H. G., Nelen, M., Breakefield, X. O., Ropers, H. H., & van Oost, B. A. (1993). Abnormal behavior associated with a point mutation in the structural gene for monoamine oxidase A. *Science, 262,* 578–580.

Cadoret, R. J., Yates, W. R., Troughton, E., Woodworth, G., & Stewart, M. A. (1995). Genetic-environmental interaction in the genesis of aggressivity and conduct disorders. *Archives of General Psychiatry, 52,* 916–924.

Caspi, A., McClay, J., Moffitt, T. E., Mill, J., Martin, J., Craig, I. W., et al. (2002). Role of genotype in the cycle of violence in maltreated children. *Science, 297,* 851–854.

Caspi, A., & Moffitt, T. E. (2006). Gene-environment interactions in psychiatry: joining forces with neuroscience. *Nature Reviews Neuroscience, 7,* 583–590.

Caspi, A., Moffitt, T. E., Cannon, M., McClay, J., Murray, R., Harrington, H., et al. (2005). Moderation of the effect of adolescent-onset cannabis use on adult psychosis by a functional polymorphism in the catechol-O-methyltransferase gene: longitudinal evidence of a gene X environment interaction. *Biological Psychiatry, 57,* 1117–1127.

Caspi, A., Sugden, K., Moffitt, T. E., Taylor, A., Craig, I. W., Harrington, H., et al. (2003). Influence of life stress on depression: moderation by a polymorphism in the 5-HTT gene. *Science, 301,* 386–389.

Caspi, A., Williams, B., Kim-Cohen, J., Craig, I. W., Milne, B. J., Poulton, R., et al. (2007). Moderation of breastfeeding effects on the IQ by genetic variation in fatty acid metabolism. Proceedings of the National Academy of Science, USA, *104,* 18860–18865.

Caspi, A., Hariri, A.R., Holmes, A., Uher, R. & Moffitt, T.E. (2010) Genetic sensitivity to the environment: the case of the serotonin transporter gene and its implications for studying complex diseases and traits. *American Journal of Psychiatry, 167.* 509–527.

Cervilla, J. A., Molina, E., Rivera, M., Torres-Gonzalez, F., Bellon, J. A., Moreno, B., et al. (2007). The risk for depression conferred by stressful life events is modified by variation at the serotonin transporter 5HTTLPR genotype: evidence from the Spanish PREDICT-Gene cohort. *Molecular Psychiatry, 12,* 748–755.

Chipman, P., Jorm, A. F., Prior, M., Sanson, A., Smart, D., Tan, X., et al. (2007). No interaction between the serotonin transporter polymorphism (5-HTTLPR) and childhood adversity or recent stressful life events on symptoms of depression:

results from two community surveys. *American Journal of Medicine Genetics B: Neuropsychiatric Genetics, 144B,* 561–565.

Chorbov, V. M., Lobos, E. A., Todorov, A. A., Heath, A. C., Botteron, K. N., & Todd, R. D. (2007). Relationship of 5-HTTLPR genotypes and depression risk in the presence of trauma in a female twin sample. *American Journal of Medicine Genetics B: Neuropsychiatric Genetics, 144B,* 830–833.

Cicchetti, D., Rogosch, F. A., & Sturge-Apple, M. L. (2007). Interactions of child maltreatment and serotonin transporter and monoamine oxidase A polymorphisms: depressive symptomatology among adolescents from low socioeconomic status backgrounds. *Developmental Psychopathology, 19,* 1161–1180.

Covault, J., Tennen, H., Armeli, S., Conner, T. S., Herman, A. I., Cillessen, A. H., et al. (2007). Interactive effects of the serotonin transporter 5-HTTLPR polymorphism and stressful life events on college student drinking and drug use. *Biological Psychiatry, 61,* 609–616.

Crabbe, J. C., Wahlsten, D., & Dudek, B. C. (1999). Genetics of mouse behavior: interactions with laboratory environment. *Science, 284,* 1670–1672.

Craddock, N., O'Donovan, M. C., & Owen, M. J. (2008). Genome-wide association studies in psychiatry: lessons from early studies of non-psychiatric and psychiatric phenotypes. *Molecular Psychiatry, 13,* 649–653.

Croft, C., Beckett, C., Rutter, M., Castle, J., Colvert, E., Groothues, C., et al. (2007). Early adolescent outcomes of institutionally-deprived and non-deprived adoptees. II: language as a protective factor and a vulnerable outcome. *Journal of Child Psychology and Psychiatry, 48,* 31–44.

Dick, D. M., Plunkett, J., Hamlin, D., Nurnberger, J., Jr., Kuperman, S., Schuckit, M., et al. (2007). Association analyses of the serotonin transporter gene with lifetime depression and alcohol dependence in the Collaborative Study on the Genetics of Alcoholism (COGA) sample. *Psychiatric Genetics, 17,* 35–38.

Drachmann, B. J., Bock, C., Vinberg, M., Werge, T., Gether, U., & Vedel, K. L. (2009). Interaction between genetic polymorphisms and stressful life events in first episode depression. *Journal of Affective Disorders, 119:* 107–115.

Eaves, L. J. (2006). Genotype x Environment interaction in psychopathology: fact or artifact? *Twin Research and Human Genetics, 9,* 1–8.

Edelman, G. M., & Gally, J. A. (2001). Degeneracy and complexity in biological systems. *Proceedings of the National Academy of Science U.S.A, 98,* 13763–12768.

Egger, M., Schneider, M., & Davey, S. G. (1998). Spurious precision? Meta-analysis of observational studies. *BMJ, 316,* 140–144.

Eley, T. C., Sugden, K., Corsico, A., Gregory, A. M., Sham, P., McGuffin, P., et al. (2004). Gene-environment interaction analysis of serotonin system markers with adolescent depression. *Molecular Psychiatry, 9,* 908–915.

Foley, D. L., Eaves, L. J., Wormley, B., Silberg, J. L., Maes, H. H., Kuhn, J., et al. (2004). Childhood adversity, monoamine oxidase A genotype, and risk for conduct disorder. *Archives of General Psychiatry, 61,* 738–744.

Fox, E., Ridgewell, A., & Ashwin, C. (2009). Looking on the bright side: biased attention and the human serotonin transporter gene. *Proceedings Biological Sciences/ Royal Society, 276,* 1747–1751.

Gillespie, N. A., Whitfield, J. B., Williams, B., Heath, A. C., & Martin, N. G. (2005). The relationship between stressful life events, the serotonin transporter (5-HT-TLPR) genotype and major depression. *Psychological Medicine, 35,* 101–111.

Glymour, C. (2004). The automation of discovery. *Daedalus, 133,* 69–77.

Gotlib, I. H., Joormann, J., Minor, K. L., & Hallmayer, J. (2008). HPA axis reactivity: a mechanism underlying the associations among 5-HTTLPR, stress, and depression. *Biological Psychiatry, 63,* 847–851.

Grabe, H. J., Lange, M., Wolff, B., Volzke, H., Lucht, M., Freyberger, H. J., et al. (2005). Mental and physical distress is modulated by a polymorphism in the 5-HT transporter gene interacting with social stressors and chronic disease burden. *Molecular Psychiatry, 10,* 220–224.

Greenland, S. (2009). Interactions in epidemiology: relevance, identification, and estimation. *Epidemiology, 20,* 14–17.

Hariri, A. R., Mattay, V. S., Tessitore, A., Kolachana, B., Fera, F., Goldman, D., et al. (2002). Serotonin transporter genetic variation and the response of the human amygdala. *Science, 297,* 400–403.

Henquet, C., Rosa, A., Krabbendam, L., Papiol, S., Fananas, L., Drukker, M., et al. (2006). An experimental study of catechol-o-methyltransferase Val158Met moderation of delta-9-tetrahydrocannabinol-induced effects on psychosis and cognition. *Neuropsychopharmacology, 31,* 2748–2757.

Holmes, A., Murphy, D. L., & Crawley, J. N. (2003). Abnormal behavioral phenotypes of serotonin transporter knockout mice: parallels with human anxiety and depression. *Biological Psychiatry, 54,* 953–959.

Huezo-Diaz, P., Uher, R., Smith, R., Rietschel, M., Henigsberg, N., Marusic, A., et al. (2009). Moderation of antidepressant response by the serotonin transporter gene. *British Journal of Psychiatry, 195,* 30–38.

Jacobs, N., Kenis, G., Peeters, F., Derom, C., Vlietinck, R., & van Os, J. (2006). Stress-related negative affectivity and genetically altered serotonin transporter function: evidence of synergism in shaping risk of depression. *Archives of General Psychiatry, 63,* 989–996.

Kaufman, J., Yang, B. Z., Douglas-Palumberi, H., Houshyar, S., Lipschitz, D., Krystal, J. H., et al. (2004). Social supports and serotonin transporter gene moderate depression in maltreated children. *Proceedings of the National Acadmey of Science, U.S.A, 101,* 17316–17321.

Keller, M. C., & Miller, G. (2006). Resolving the paradox of common, harmful, heritable mental disorders: which evolutionary genetic models work best? *Behavior and Brain Science, 29,* 385–404.

Kendler, K. S., Kessler, R. C., Walters, E. E., MacLean, C., Neale, M. C., Heath, A. C., et al. (1995). Stressful life events, genetic liability, and onset of an episode of major depression in women. *American Journal of Psychiatry, 152,* 833–842.

Kendler, K. S., Kuhn, J. W., Vittum, J., Prescott, C. A., & Riley, B. (2005). The interaction of stressful life events and a serotonin transporter polymorphism in the prediction of episodes of major depression: a replication. *Archives of General Psychiatry, 62,* 529–535.

Kilpatrick, D. G., Koenen, K. C., Ruggiero, K. J., Acierno, R., Galea, S., Resnick, H. S., et al. (2007). The serotonin transporter genotype and social support and moderation of posttraumatic stress disorder and depression in hurricane-exposed adults. *American Journal of Psychiatry, 164*, 1693–1699.

Kim, J. M., Stewart, R., Kim, S. W., Yang, S. J., Shin, I. S., Kim, Y. H., et al. (2007). Interactions between life stressors and susceptibility genes (5-HTTLPR and BDNF) on depression in Korean elders. *Biological Psychiatry, 62*, 423–428.

Knol, M. J., van der Tweel, I., Grobbee, D. E., Numans, M. E., & Geerlings, M. I. (2007). Estimating interaction on an additive scale between continuous determinants in a logistic regression model. *International Journal of Epidemiology, 36*, 1111–1118.

Kohen, R., Cain, K. C., Mitchell, P. H., Becker, K., Buzaitis, A., Millard, S. P., et al. (2008). Association of serotonin transporter gene polymorphisms with post-stroke depression. *Archives of General Psychiatry, 65*, 1296–1302.

Laucht, M., Treutlein, J., Blomeyer, D., Buchmann, A. F., Schmid, B., Becker, K., et al. (2009). Interaction between the 5-HTTLPR serotonin transporter polymorphism and environmental adversity for mood and anxiety psychopathology: evidence from a high-risk community sample of young adults. *International Journal of Neuropsychopharmacology, 12*, 737–747.

Lazary, J., Lazary, A., Gonda, X., Benko, A., Molnar, E., Juhasz, G., et al. (2008). New evidence for the association of the serotonin transporter gene (SLC6A4) haplotypes, threatening life events, and depressive phenotype. *Biological Psychiatry, 64*, 498–504.

Leventhal, T., Fauth, R. C., & Brooks-Gunn, J. (2005). Neighborhood poverty and public policy: a 5-year follow-up of children's educational outcomes in the New York City moving to opportunity demonstration. *Developmental. Psychology, 41*, 933–952.

Mandelli, L., Serretti, A., Marino, E., Pirovano, A., Calati, R., & Colombo, C. (2007). Interaction between serotonin transporter gene, catechol-O-methyltransferase gene and stressful life events in mood disorders. *International Journal of Neuropsychopharmacology, 10*, 437–447.

McClelland, G. H., & Judd, C. M. (1993). Statistical difficulties of detecting interactions and moderator effects. *Psychological Bulletin, 114*, 376–390.

Moffitt, T. E., Caspi, A., & Rutter, M. (2005). Strategy for investigating interactions between measured genes and measured environments. *Archives of General Psychiatry, 62*, 473–481.

Monroe, S.M. (2008). Modern approaches to conceptualizing and measuring human life stress. *Annual Review of Clinical Psychology, 4*, 33–52.

Monroe, S. M., & Reid, M. W. (2008). Gene-environment interactions in depression research: genetic polymorphisms and life-stress polyprocedures. *Psychological Science, 19*, 947–956.

Monroe, S. M., Torres, L. D., Guillaumot, J., Harkness, K. L., Roberts, J. E., Frank, E., et al. (2006). Life stress and the long-term treatment course of recurrent depression: III. Nonsevere life events predict recurrence for medicated patients over 3 years. *Journal of Consulting Clinical Psychology, 74*, 112–120.

Munafo, M. R., Durrant, C., Lewis, G., & Flint, J. (2009). Gene X environment interactions at the serotonin transporter locus. *Biological Psychiatry, 65,* 211–219.

Nakatani, D., Sato, H., Sakata, Y., Shiotani, I., Kinjo, K., Mizuno, H., et al. (2005). Influence of serotonin transporter gene polymorphism on depressive symptoms and new cardiac events after acute myocardial infarction. *American Heart Journal, 150,* 652–658.

Nobile, M., Rusconi, M., Bellina, M., Marino, C., Giorda, R., Carlet, O., et al. (2009). The influence of family structure, the TPH2 G-703T and the 5-HTTLPR serotonergic genes upon affective problems in children aged 10-14 years. *Journal of Child Psychology and Psychiatry, 50,* 317–325.

Otte, C., McCaffery, J., Ali, S., & Whooley, M. A. (2007). Association of a serotonin transporter polymorphism (5-HTTLPR) with depression, perceived stress, and norepinephrine in patients with coronary disease: the Heart and Soul Study. *American Journal of Psychiatry, 164,* 1379–1384.

Ottman, R. (1996). Gene-environment interaction: definitions and study designs. *Preventive Medicine., 25,* 764–770.

Plomin, R., & Davis, O. S. (2009). The future of genetics in psychology and psychiatry: microarrays, genome-wide association, and non-coding RNA. *Journal of Child Psychology and Psychiatry, 50,* 63–71.

Polanczyk, G., Caspi, A., Williams, B., Price, T. S., Danese, A., Sugden, K., et al. (2009). Protective effect of CRHR1 gene variants on the development of adult depression following childhood maltreatment: replication and extension. *Archives of General Psychiatry., 66,* 978–985.

Power, T., Stewart, R., Ancelin, M. L., Jaussent, I., Malafosse, A., & Ritchie, K. (2008). 5-HTTLPR genotype, stressful life events and late-life depression: No evidence of interaction in a French population. *Neurobiology of Aging, 31:* 886–887.

Richardson, D. B., & Kaufman, J. S. (2009). Estimation of the relative excess risk due to interaction and associated confidence bounds. *American Journal of Epidemiology, 169,* 756–760.

Risch, N., Herrell, R., Lehner, T., Liang, K. Y., Eaves, L., Hoh, J., et al. (2009). Interaction between the serotonin transporter gene (5-HTTLPR), stressful life events, and risk of depression: a meta-analysis. *Journal of the American Medical Association, 301,* 2462–2471.

Rothman, K. J., & Greenland, S. (2005). Causation and causal inference in epidemiology. *American Journal of Public Health, 95 Suppl 1,* S144–150.

Rothman, K. J., Greenland, S., & Lash, T. L. (2008). *Modern epidemiology.* (3rd ed.) Philadelphia: Wolter Kluwer Health; Lippincott Williams & Wilkins.

Rothman, K. J., Greenland, S., & Walker, A. M. (1980). Concepts of interaction. *American Journal of Epidemiology, 112,* 467–470.

Rutter, M. (2006). Implications of resilience concepts for scientific understanding. *Annals of the New York Academy of Science., 1094,* 1–12.

Rutter, M. (2008). Introduction: whither gene-environment interactions? *Novartis Foundation Symposium, 293,* 1–12.

Scheid, J. M., Holzman, C. B., Jones, N., Friderici, K. H., Nummy, K. A., Symonds, L. L., et al. (2007). Depressive symptoms in mid-pregnancy, lifetime stressors and the 5-HTTLPR genotype. *Genes Brain and Behavior, 6,* 453–464.

Seeger, G., Schloss, P., Schmidt, M. H., Ruter-Jungfleisch, A., & Henn, F. A. (2004). Gene-environment interaction in hyperkinetic conduct disorder (HD + CD) as indicated by season of birth variations in dopamine receptor (DRD4) gene polymorphism. *Neuroscience. Letters, 366,* 282–286.

Serretti, A., Kato, M., De, R. D., & Kinoshita, T. (2007). Meta-analysis of serotonin transporter gene promoter polymorphism (5-HTTLPR) association with selective serotonin reuptake inhibitor efficacy in depressed patients. *Molecular Psychiatry, 12,* 247–257.

Shao, H., Burrage, L. C., Sinasac, D. S., Hill, A. E., Ernest, S. R., O'Brien, W., et al. (2008). Genetic architecture of complex traits: large phenotypic effects and pervasive epistasis. *Proceedings of the National Academy of Science U.S.A, 105,* 19910–19914.

Sheese, B. E., Voelker, P. M., Rothbart, M. K., & Posner, M. I. (2007). Parenting quality interacts with genetic variation in dopamine receptor D4 to influence temperament in early childhood. *Developmental Psychopathology, 19,* 1039–1046.

Sjoberg, R. L., Nilsson, K. W., Nordquist, N., Ohrvik, J., Leppert, J., Lindstrom, L., et al. (2006). Development of depression: sex and the interaction between environment and a promoter polymorphism of the serotonin transporter gene. *International Journal of Neuropsychopharmacology, 9,* 443–449.

Stroup, D. F., Berlin, J. A., Morton, S. C., Olkin, I., Williamson, G. D., Rennie, D., et al. (2000). Meta-analysis of observational studies in epidemiology: a proposal for reporting. Meta-analysis Of Observational Studies in Epidemiology (MOOSE) group. *Journal of the American Medical Association, 283,* 2008–2012.

Surtees, P. G., & Wainwright, N. W. (1999). Surviving adversity: event decay, vulnerability and the onset of anxiety and depressive disorder. *European Archives of Psychiatry and Clinical Neuroscience, 249,* 86–95.

Surtees, P. G., Wainwright, N. W., Willis-Owen, S. A., Luben, R., Day, N. E., & Flint, J. (2006). Social adversity, the serotonin transporter (5-HTTLPR) polymorphism and major depressive disorder. *Biological Psychiatry, 59,* 224–229.

Taylor, A., & Kim-Cohen, J. (2007). Meta-analysis of gene-environment interactions in developmental psychopathology. *Developmental Psychopathology, 19,* 1029–1037.

Taylor, S. E., Way, B. M., Welch, W. T., Hilmert, C. J., Lehman, B. J., & Eisenberger, N. I. (2006). Early family environment, current adversity, the serotonin transporter promoter polymorphism, and depressive symptomatology. *Biological Psychiatry, 60,* 671–676.

Thapar, A., Langley, K., Owen, M. J., & O'Donovan, M. C. (2007). Advances in genetic findings on attention deficit hyperactivity disorder. *Psychological Medicine, 37,* 1681–1692.

Tienari, P., Wynne, L. C., Sorri, A., Lahti, I., Laksy, K., Moring, J., et al. (2004). Genotype-environment interaction in schizophrenia-spectrum disorder. Long-term follow-up study of Finnish adoptees. *British Journal of Psychiatry, 184,* 216–222.

Uher, R. (2009). The role of genetic variation in the causation of mental illness: An evolution-informed framework. *Molecular Psychiatry, 14,* 1072–1082.

Uher, R. (2008a). Forum: The case for gene-environment interactions in psychiatry. *Current Opinions in Psychiatry, 21,* 318–321.

Uher, R. (2008c). Gene-environment interaction: overcoming methodological challenges. *Novartis Foundation Symposiums, 293,* 13–26.

Uher, R. (2008b). The implications of gene-environment interactions in depression: will cause inform cure? *Mol. Psychiatry, 13,* 1070–1078.

Uher, R., & McGuffin, P. (2008). The moderation by the serotonin transporter gene of environmental adversity in the aetiology of mental illness: review and methodological analysis. *Molecular Psychiatry, 13,* 131–146.

Uher, R., & McGuffin, P. (2010). The moderation by the serotonin transporter gene of environmental adversity in the etiology of depression: 2009 update. *Molecular Psychiatry, 15:* 18–22.

van der Zwaluw, C. S., Engels, R. C., Vermulst, A. A., Franke, B., Buitelaar, J., Verkes, R. J., et al. (2009). Interaction between dopamine D2 receptor genotype and parental rule-setting in adolescent alcohol use: evidence for a gene–parenting interaction. *Molecular Psychiatry, 15:* 727–735.

VanderWeele, T. J. (2009a). Marginal structural models for the estimation of direct and indirect effects. *Epidemiology, 20,* 18–26.

VanderWeele, T. J. (2009b). Sufficient cause interactions and statistical interactions. *Epidemiology, 20,* 6–13.

Wichers, M., Kenis, G., Jacobs, N., Mengelers, R., Derom, C., Vlietinck, R., et al. (2008). The BDNF Val(66)Met x 5-HTTLPR x child adversity interaction and depressive symptoms: An attempt at replication. *American Journal of Medicine Genetics. B: Neuropsychiatric Genetics, 147B,* 120–123.

Wilhelm, K., Mitchell, P. B., Niven, H., Finch, A., Wedgwood, L., Scimone, A., et al. (2006). Life events, first depression onset and the serotonin transporter gene. *British Journal of Psychiatry, 188,* 210–215.

Zalsman, G., Huang, Y. Y., Oquendo, M. A., Burke, A. K., Hu, X. Z., Brent, D. A., et al. (2006). Association of a triallelic serotonin transporter gene promoter region (5-HTTLPR) polymorphism with stressful life events and severity of depression. *American Journal of Psychiatry, 163,* 1588–1593.

Zhang, K., Xu, Q., Xu, Y., Yang, H., Luo, J., Sun, Y., et al. (2009). The combined effects of the 5-HTTLPR and 5-HTR1A genes modulates the relationship between negative life events and major depressive disorder in a Chinese population. *Journal of Affective Disorders, 114,* 224–231.

3

Improving Environmental Markers in Gene–Environment Research

Insights from Life-Course Sociology

MICHAEL J. SHANAHAN AND SHAWN BAULDRY

Most research that examines gene–environment interactions and correlations relies on *environmental markers*, simplified measures of social context that are assumed to capture (or summarize) the complexity of environmental processes that produce variation in behavioral development. Findings from life-course research, however, suggest that, in many cases, this assumption deserves reconsideration. Two themes in life-course sociology are especially noteworthy: (1) a systems view positing that context reflects many environmental candidates and how they are organized with respect to one another; and (2) growing recognition that static assessments of environmental features often fail to capture the substantial heterogeneity in people's actual experiences. Together, these two themes suggest promising strategies to enhance the validity of environmental measures used in gene–environment research.

This chapter provides a brief sketch of the systemic and dynamic properties of environments. We focus on life events as a specific case study because they are perhaps the most well-studied environmental candidates to date. The first section of the chapter introduces life-course sociology and its systemic and dynamic themes. The second section applies these themes to the specific environmental candidate of life events. Our review is not intended to exhaustively review relevant literatures but rather to highlight possibilities for improving measures of life events based on the systemic, dynamic view promoted by life-course research. The third section then presents a simulation study that examines the life events × serotonin transporter gene interaction (LE*HTT) in the prediction of depression (see also Chapters 2 and 12, by Uher and, Perepletchikova and Kaufman) with these complexities taken into account. That is, we consider the potential importance of improved measures of environmental candidates by way of a specific case study.

This simulation illustrates how the magnitude of gene–environment interactions involving stressors may well depend on the population-specific distribution of the severity of stressors (which reflects their systemic and dynamic properties). In turn, efforts to identify *the* gene–environment inter-action effect may be misguided. Our conceptual argument and simulation underscore the need for valid measures of environmental candidates that reflect their systemic and dynamic properties and the interpretation of gene–environment interactions and correlations in terms of well-described populations.

LIFE-COURSE SOCIOLOGY AND ENVIRONMENTAL MARKERS

Life-course sociology refers to a paradigm of interrelated concepts, hypotheses, and methods that collectively direct attention to the social and temporal complexities of environments and behavioral development (Elder & Shanahan, 2006; Mortimer & Shanahan, 2003). The foundations of the paradigm were largely induced from a series of empirical studies of children during the Great Depression and their subsequent life patterns (Elder, 1979/1999). Those studies suggested that the meaning of economic downturn was highly contingent, depending on prior circumstances, the timing of the event in the children's lives, the role configurations in the family, the family's efforts to gain control of the situation, and the subsequent circumstances. In other words, the meaning of the macro social change could not be understood without viewing it in terms of the individual's ongoing, multifaceted experiences. In fact, a simple summary of household income loss poorly predicted the children's adjustment. Following these studies, the central thrust of life-course sociology has been the identification of social mechanisms that explain why people react differently to the same transitions.

Two themes have been especially prominent in these subsequent efforts. First is a systems view, which posits that features of the environment acquire their significance because of how they are organized with respect to one another. In the case of the Great Depression studies, for example, income loss, age of the children, father's pre-Depression psychological profile, and role responsibilities during the crisis all came together to define environments of risk or resilience. Collectively, these factors greatly shaped how children made sense of their circumstances.

With respect to gene–environment studies, the recommendation for a systems view seems especially promising since many environmental candidates are forms of psychosocial risk that are known to be highly correlated with other risks, possibly forming "networks of risks." For example, childhood maltreatment and *MAOA* are known to interactively predict antisocial behaviors (Caspi et al., 2002, and this finding has been replicated many times (e.g., Kim-Cohen et al., 2006). From a systems view, maltreatment is highly correlated with poverty, family conflict, neighborhood disorganization, parental substance abuse, sibling hostility and lack of warmth, geographic mobility, income instability, and parental psychopathology—all factors that are, in turn, related to externalizing symptoms and antisocial behaviors in children. A systems view of maltreatment refocuses from the conventional task of isolating its unique predictive effect (both as a main, additive effect and as part of an interactive effect) to describing how maltreatment is organized with other environmental candidates and how such networks are related to antisocial behavior.

Indeed, given these networks of associations, it may be difficult to establish unique, additive, and interactive roles of maltreatment per se. Put differently: Is maltreatment necessary and sufficient (i.e., the sole causal factor implicated in the main and interactive effects predicting antisocial behavior), unnecessary but sufficient (meaning that one or more of these other correlated candidates may substitute for maltreatment as a causal factor), or unnecessary and insufficient (meaning that one or more correlated candidates may substitute and maltreatment can, but need not, be part of the causal process)?

Childhood maltreatment is one example of this potential complexity, but others can be identified, including health risks (e.g., Alamanian & Paradis, 2008; Schuit, van Loon, Tijhuis, & Ocke, 2002), risks associated with poverty (e.g., McLeod & Almazan, 2003), environmental risks involving exposures to toxins and social indicators of poor living conditions (e.g., Evans, 2006; deFur et al., 2007), and psychosocial risks for psychopathology (e.g., Copeland, Shanahan, Costello, & Angold, 2009). In these and other instances, specific environmental candidates do not occur in isolation but rather tend to coalesce with other factors to comprise the individual's social context.

Such a view of environmental candidates may suggest a high level of conditioning among environmental candidates, meaning that, although environmental candidates may have significant additive effects, analyses should also consider the possibility that their full significance depends on how they interact. A systems view need not imply interaction, however;

environmental candidates could also act additively. Many studies suggest that risks tend to co-occur and to have a fairly straightforward dose-response relationship to indicators of health and well-being (e.g., Larson, Russ, Crall, & Halfon, 2008). A variation of this theme is that multiple co-occurring risks accumulate, and their effects tend to be nonlinear, reflecting ceiling and/or threshold mechanisms (see Kendler's discussion of ceiling and floor effects in dose-response relationships in Chapter 1). Although not reflecting high levels of interaction per se, these accumulation models still suggest a potential limit to reliance on one marker of the environment.

A second theme of life-course studies refines this systems view of environmental candidates by emphasizing the dynamic nature of social context. That is, environmental candidates are organized into systems that exhibit meaningful patterns of change and stability over time, and these patterns may help explain people's experiences with enhanced precision. Environmental candidates are thus not discrete occurrences but, in some instances, need to describe circumstances before, during, and after. With respect to the Great Depression, the father's pre-loss psychological profile moderated the experience of income loss on his behavior. The child's post-Depression experiences, particularly during World War II, greatly shaped his or her pathway into adulthood. The quality of experiences during the Depression also influenced the fathers and their children, but largely when viewed in the context of these pre- and post-Depression realities.

Similarly, the consequences of many psychosocial experiences depend on people's preexisting resources, reactions to the experiences (both individual and group), and subsequent pathways. This point was ably demonstrated with respect to antisocial behaviors by Rutter's studies of pathways of institutionalized children into adulthood (Rutter, 1986), Laub and Sampson's (2003) mapping of criminal careers from childhood into old age, and studies of prisoners released from institutions (Maruna, 2001). In all such instances, extensive data covering considerable periods of time are necessary, and rarely does a single environmental candidate emerge as decisive.

A very large and expansive literature also investigates the temporal properties of social roles—particularly work and family roles—and documents that such distinctions are often important to behavioral adjustment (e.g., Elder & Shanahan, 2006). By *temporal properties* we refer to descriptors of when roles or environmental markers occur in one's life, as well their duration, location in a pathway or sequence, and relationship to an initial condition. *Sensitive periods* have experienced a recent renaissance, as studies suggest mechanisms for the biological programming hypothesis (Tremblay & Hamet, 2008). For example, Miller et al. (2009) report that poverty in early childhood permanently alters stress reactivity (via regulation of glucocorticoid receptors) into adulthood for some people, controlling for adult socioeconomic status (see Cole, Chapter 9). Given the increasing number of epidemiological papers linking socioeconomic status at different stages of life to indicators of cardiovascular disease and other measures of health, the initial interest in biological programming in early childhood will likely expand into other phases of the life course.

Life-course sociology offers numerous concepts with which to analyze longitudinal properties of social context, with different concepts proving salient in different empirical contexts (Shanahan & Macmillan, 2007). The major point, however, is that such longitudinal properties of context are an extension of the view that environmental markers often need to be studied as part of a system of interrelated factors and, specifically, a system that exhibits patterns of change and stability that may often help in understanding people's experiences.

LIFE EVENTS AS AN ENVIRONMENTAL MARKER

Life events are an illustrative empirical example of these thematic issues and are particularly relevant because they represent a well-studied environmental candidate that is implicated in risk for depression. Indeed, a considerable body of research focuses on the role of life events in a gene–environment interaction involving the serotonin transporter gene and the prediction of various clinical manifestations of depression.

As with children who grew up during the Great Depression, people who are exposed to stressors such as life events exhibit a wide range of responses. Indeed, many scholars have speculated why the magnitude of association between life events and forms of distress tends to be modest, reflecting this variability in response. Such explanations have focused on differential vulnerability or resilience of individuals to stressors; the intrinsic stressfulness of an event, as indicated by its uncontrollability and other inherent properties; and the stress burden of the event, as determined by prior circumstances, how life events fit in the person's overall system of environmental factors, and subsequent resolution.

Gene–environment research typically is best characterized by the first category, based on the assumption that genetic risk creates a psychological diathesis that in turn leads to heightened responses to stressors, thus increasing the likelihood for behavior indicating distress. Nonhuman animal research—which allows for experimental designs—supports the validity of the serotonin candidate (and other genetic candidates, including, for example, polymorphisms associated with *BDNF* and *DRD4*) being linked to stress responsivity, although this body of work is somewhat limited by the fact that such animals do not experience life events per se.

Human research has relied largely on the strategy of summing life-events inventories. Even when the inherent properties of the life event (e.g., uncontrollability) are taken into consideration, this approach fails to consider the life-course mechanisms postulated by the third category—that the stress burden created by each life event depends upon prior circumstances, its meaning in the present, and its subsequent resolution. The implication of this perspective is noteworthy: A simple summation of life events that have been experienced may fail to capture these complexities, resulting in biased estimates of the relationship between life events and forms of distress. Growing evidence now suggests ways in which these considerations can be operationalized, creating opportunities for the study of gene–environment processes involving life events or any class of stressor.

Pre-event Circumstances

By examining the surrounding circumstances, Brown and Harris (1978) were among the first to gauge the stress burden created by life events. Their highly nuanced analyses document that, prior to events, the women whom they studied differed in their vulnerability. Particularly salient vulnerability factors were the death of the mother before age 11, low intimacy with the husband, three or more children at home, and/or lack of employment outside the home. Undoubtedly, other vulnerability factors may be salient beyond their samples, possibly including severe childhood maltreatment or exposure to extreme brutality and death. According to Brown and Harris, the crucial point is that these vulnerabilities lead to a diminished sense of self, which creates a diathesis for depression.

The significance of circumstances preceding life events was further studied by Wheaton (1990), who showed that typical, "negative" life events could actually dampen the distressful reactions. The findings are suggestive and intuitively plausible because they emphasize the importance of prior chronic stressors and contemporaneous role configurations in jointly determining the stress burden created by life events. With respect to prior circumstances, for example, a high level of chronic work stress attenuates the distressful reaction (as gauged by the General Health Questionnaire and the Center for Epidemiological Studies Depression Scale) to job loss.

With respect to role configurations, this chronic stress × job interaction tends to differ by age and marital status, with distress dampening for unmarried men and even improving for younger married women. In the cohorts studied by Wheaton, younger married women likely had a working spouse, so that losing a stressful job was unlikely to cause serious financial strains at home and might even reduce tensions spilling over from the workplace.

In the case of unmarried men, they did not occupy the bread-winner role for a family, so their job loss did not threaten the livelihood of other family members. Thus, pre-event circumstances and present circumstances and role sets greatly determined the meaning of events for the person and hence their distressful quality, if any.

The Wheaton findings suggest two possibilities that run contrary to the assumption that life events are stressors that invariably increase the likelihood of distress. First, the stress-attenuation hypothesis refers to situations in which the event disrupts chronic strain and distress is attenuated (i.e., its rate of increase lessens). Second, the stress-relief hypothesis suggests that the disruption in chronic stress may actually improve mental health, significantly decreasing distress when compared to its pre-event level. Wheaton's findings are somewhat complex—involving three-way interactions—but the overall conclusion is clear: job loss, by itself, is a poor marker of stress burden, which depends on preexisting chronic strains associated with the event and the meaning of the event with respect to career progression and obligations to others.

Although Wheaton's study examined several life events, its progeny focused largely on parental divorce and its effects on spouses and children. Subsequent findings suggest that stress attenuation is the rule and stress relief is not commonly observed. For example, with respect to spouses, Overbeeck and his colleagues' (2006) large-scale epidemiological study in the Netherlands using a *Diagnostic and Statistical Manual of Mental Disorders* (DMS-III-R)-based interview is revealing. The authors examined prospective associations between divorce and mood, anxiety, and substance disorders by drawing on three waves of data. Time 1 (1996) data were used to assess marital quality, Time 2 (1997) data assessed divorce, and incidence rates of the disorders were assessed from Time 2 to Time 3 (1999). Divorce was associated with increased risk for dysthymia and the onset of alcohol abuse and the broad category of substance disorders. Controlling marital quality at Time 1 fully attenuated the relationships involving dysthymia and alcohol use, however. A very similar pattern emerged for new case incidence between Times 2 and 3. Although the findings do not bear directly on the stress-attenuation versus stress-relief hypotheses, they do point to a central commonality with Wheaton: that the stress potential of this life event depends on its relational context.

With respect to the children of parental divorce, Strohschein (2005) drew on three waves of data from a nationally representative Canadian sample to study parental divorce and subscales of the Child Behavior Checklist. The study's design is unique in its ability to examine initial levels of anxiety/depression and antisocial behavior and then how these trajectories might be altered by subsequent parental separation. Children of parents who divorce have higher levels of anxiety/depression and antisocial behavior before the divorce, but only anxiety/depression increases further after the separation. However, all of the initial differences could be accounted for by "predivorce parental resources," which refer to parental social class, marital satisfaction, family dysfunction, and parental depression. Notably, there was some support for the stress-relief hypothesis for antisocial behavior, with children from highly dysfunctional families showing improvements in antisocial behavior after separation. Strohschein speculates that this improvement would be more dramatic as time from separation elapses. Like the Wheaton study, Strohschein (2005) suggests that the stress burden associated with this life event depends on levels of pre-event chronic stressors.

Collectively, these and related studies document significant interactions involving the experience of a life event and pre-event circumstances in determining their meaning and hence distressfulness. In some instances, negative life events actually improve mental health.

Meaning and Resolution

Other researchers have attempted to directly assess the subjective meaning of life events. As mentioned, Brown and Harris (1978) proposed, following the work of Aaron Beck, that self-conceptions are integral to creating pre-stressor vulnerabilities and moderating the effects of stressors on distress. Particularly salient were events associated with loss, disappointment, humiliation, and/or threat (Brown, Harris & Epworth, 1995); such events, when judged as severe and posing long-term threat, predicted depressive episodes (Brown, Adler, & Bifulco, 1988). A related approach to subjective meaning focuses on disruptions in how the person makes sense of self and the world. According to this perspective, stressors are agents that disrupt a homeostasis involving sense-making with respect to self, others, and life, and it is this anomic disruption that is distressful.

Thoits (1991) proposed that people maintain multiple identities with differing levels of salience (or importance) and that salient identity-threatening stressors are most likely to create distress. Subsequently, Thoits (1995) found little support for this expectation, but follow-up analyses were quite revealing. First, some "inconsistent cases" (involving salient identity-threatening events with little distress) involved situations of stress relief from preexisting chronic strains (consistent with Wheaton). Second, some inconsistent cases (involving positive identity-salient events with little improvement in distress) involved substantial, ongoing chronic stressors. And third, many cases involved complex combinations of positive and negative events that, because of their sequencing, resulted in high-salience events having little consequence for distress. Echoing Brown and Harris (1978), Thoits concluded that the importance of identity salience to the stress process cannot actually be examined without adequate consideration of the surrounding circumstances.

Reynolds and Turner (2008) take a different approach to sense-making, focusing on "crises," life events that fundamentally alter the person's "assumptive states" or basic beliefs about reality. Controlling a count measure of life events, they observed that the experience of a crisis had a substantial effect on depressive symptom scores. Reynolds and Turner also examined the effect of a crisis, depending on whether it was judged by the person to be successfully resolved. By itself, the successful resolution of an event has no effect on depression, but the successful resolution of a crisis was associated with a large decrement in symptom scores.

The importance of event resolution is also underscored by Kalmijn and Monden's (2006) study of the effects of divorce on spouses. The study examines the stress-relief hypothesis, in this case, that escape from a low-quality marriage would lead to improvements in depressive symptoms. Although the results did not support this hypothesis, they nevertheless revealed interactions between marital quality and event resolution. People who were highly satisfied with their marriage showed greatest increase in depressive symptoms if they did not repartner (i.e., remarried or started living together again), but no increase in symptoms if they repartnered. A similar pattern was observed for people in marriages in which the person felt treated fairly and people in marriages marked by physical aggression. That is, the effects of divorce depend on both preexisting strains (i.e., presence of unfair treatment, aggression, low satisfaction) and how the event resolves (repartnering or not).

The foregoing review, despite its brevity, suggests that the implications of life events for the person are highly contingent and depend on six categories of variables. First, prior to events, vulnerabilities/protective factors and chronic strains need to be considered. Vulnerabilities create diatheses, protective factors are sources of resilience, and chronic strains may actually render the event a source of stress relief. Second, during events, the person's role obligations and connections to others shape the meaning of the event. Events that threaten one's obligations to others (notably, involving the family and workplace) may be especially injurious.

Third, subjective meaning is also important and may reflect identity salience (a plausible possibility for which there is presently insufficient evidence); disruptions in self-concept, particularly stemming from a sense of loss and disappointment; and disruption in the fundamental way that people view themselves and life (i.e., a crisis). Brown and his colleagues (1995) likewise suggested the central role of humiliation, shame, and threat. Fourth, the stress burden created by life-events may depend on their resolution, especially how the events alter relationships with others, one's place in groups and organizations, and how one makes sense of the world. Fifth, the significance of any single event may well depend on its relationship to other events and chronic strains, as Thoits' (1995) qualitative study showed. Positive events may be of little consequence in the wake of ongoing chronic strains, whereas seemingly unimportant events may be consequential if accompanied by the same strains or other events. Sixth, reactions to life events may vary greatly with how much time has elapsed since the event. All other things being equal, the likelihood of stress relief or attenuation increases with the passage of time. A seventh set of relevant factors include social supports and coping mechanisms, which we have not reviewed given, the widespread acknowledgement of their substantive importance (see Perepletchikova and Kaufman, Chapter 12).

These categories all represent ways to describe the stress burden associated with life events and thus are ways of gauging their severity. These categories—pre-event vulnerabilities, level of threat to role obligations and relationships, subjective meaning for identity and cultural understanding, resolution, presence of chronic strains and sequencing of stressors, and coping mechanisms and social supports—were all identified based on well-done empirical research showing significant interactions. However, research that has examined LE*HTT has not been informed by these nuances (a noteworthy exception being Kendler et al., 2005), thus raising the possibility that the explanatory power of life events has been underestimated.

SIMULATION STUDY: THE SEVERITY OF LIFE EVENTS AND LE*HTT

The foregoing identifies a set of factors that likely moderate the effect of life events on indicators of distress. Put differently, a specific life event may create very different stress burdens, depending on these factors. But does failure to consider them make a difference in gene–environment research? We consider this question with specific reference to one of the most-studied gene–environment interactions, that involving life events and the serotonin transporter gene. Specifically, we conducted a simulation study to analyze change in the magnitude of the interaction between life events and the serotonin transporter gene in the prediction of depression, given different distributions of the severity of the life events. The design and implementation of the simulation reflects two over-arching goals: (1) reasonable parsimony and simplicity, and (2) reliance on the best estimates of parameters and distributions presently available in published data.

A relatively simple way to address the question is to consider the case of

$$\text{Logit}[y_i] = \alpha + \beta_1 X_1 + \beta_2 X_2 + \beta_3 X_1 X_2, \qquad \text{Eq. 3.1}$$

where y_i is a dichotomous measure of a clinical manifestation of depression (0 = does not meet diagnostic criteria; 1 = meets diagnostic criteria), X_1 is number of life events (0, 1, 2, 3, 4 or more) occurring in a discrete period of time, and X_2 is a dichotomous indicator of genetic risk for the serotonin transporter (0 = l/l, 1 = s/l or s/s).

Given our interest in the consequences of the severity of life events, we generate the data for our simulation study using a modified version of Equation 3.1:

$$\text{Logit}[y_i] = \alpha_{10} + \beta_{10} X_1 + \beta_{20} X_2 + \beta_{30} X_1 X_2 \text{ if } X_3=0, \qquad \text{Eq. 3.2}$$
$$\text{Logit}[y_i] = \alpha_{11} + \beta_{11} X_1 + \beta_{21} X_2 + \beta_{31} X_1 X_2 \text{ if } X_3=1, \qquad \text{Eq. 3.3}$$
$$\text{Logit}[y_i] = \alpha_{12} + \beta_{12} X_1 + \beta_{22} X_2 + \beta_{32} X_1 X_2 \text{ if } X_3=2, \qquad \text{Eq. 3.4}$$

where the parameters in the model are indexed by X_3, a grouping variable that reflects the average severity of life events (0 = mild, 1 = moderate, 2 = severe). That is, X_1 reflects number of events and X_3 indicates their average severity. β_1 and β_3 are assumed to vary by X_3 (i.e., the effect of life events and their interaction with HTT vary by the severity of events). Thus, although the foregoing section identifies many factors that contribute to how severe a life event is, we simplify the simulation to focus on three conditions: events of average mild, moderate, and severe severity. This simplification is consistent with Brown's (1989) suggestions about the measurement of life events, which involves the collection of extensive information about life events and then the use of resulting data to create an ordinal measure of severity for each event.

Distribution of X_1 and Parameter Estimate β_1 (Life Events)

Specifying the distribution of events is complicated by the time frame covered (e.g., life events experienced in the past 5 years, 1 year, 6 months, etc.) and the method used to collect the information. Longer time frames covered by the life events measure will presumably result in more observed events, although their salience likely dissipates over perhaps a year. Further, semistructured interviews may be superior to self-report (Uher & McGuffin, 2008).

To obtain a distribution for life events, we consulted all extant LE*HTT studies using semistructured interviews (Caspi et al., 2003, hereafter "Caspi 2003;" Kendler et al., 2005; Mandelli et al., 2007; Wilhelm et al., 2006; Zalsman et al., 2006). The studies differed in the time frames covered, and two studies reported no relevant descriptive information. Despite these inconsistencies, data from the Caspi 2003, Wilhelm, and Mandelli studies are consistent with Caspi 2003's reported frequency distribution, which we use in this simulation: 0 events = 30%, 1 event = 25%, 2 events = 22% 3 events = 11%, and 4 or more events = 13%.

The effect of life events is assumed to vary by their severity (i.e., X_3), an assumption supported by the empirical studies reviewed in the previous section of this chapter. Risch et al.'s (2009) meta-analysis notes an overall odds ratio (OR) of 1.41 with a 95% confidence interval (CI) of 1.27–1.57. Drawing on this information, we thus assign the mild life event parameter to b = .24 (i.e., OR = 1.27), moderate b = .34 (OR = 1.41), and severe b = .45 (OR = 1.57).

Distribution of X_2 and Parameter Estimate β_2 (Gene Risk)

Although many studies examine s/s, l/s, and l/l separately, we collapse s/s and s/l, consistent with biological evidence indicating that both genotypes represent risk. To obtain a distribution for this variable, we consulted nationally representative data, including two interaction studies (Caspi 2003 and Gillespie, Whitfield, Williams, Heath, & Martin, 2005) and data from the National Longitudinal Study of Adolescent Health. All three data sources report very similar distributions, converging on a distribution of s/s and s/l = 69% and l/l = 31%.

The effect of this variable is based on Risch's meta-analytic estimate of OR = 1.05 (b = .50), the only published attempt to obtain an averaged effect across multiple (14) published studies. This point estimate is likely to be low, however, as suggested by commentary accompanying their meta-analysis.

Parameter Estimate for β_3 (Interaction of Life Events and 5-HTT)

The Risch meta-analysis suggests an aggregate estimate of OR = 1.01, with a CI of 0.94–1.10. The present simulation assumes that the lower bounds of the CI represents a true lower bound to the estimate and indeed reflects a sample of people who are experiencing mild stressors. At the upper end, however, we draw directly on the estimate obtained by Wilhelm et al. (2006), one of the highest estimates obtained among the replications for LE*HTT. Wilhelm measured life events using a semistructured interview, and this fact, plus other considerations, suggests that the study is of sufficient quality to assume that the magnitude of interaction represents the highest possible estimate for the interaction. Given these considerations, the simulation assigns β_3 the following values: .00 (mild severity of life events); .10 (moderate severity); and .51 (severe severity).

Distribution of X_3

Of central importance, the distribution of the severity of life events, the grouping variable X_3, is varied in the simulation in order to examine how the interaction term and predicted value of the probability of a diagnosis would change with changes in the stressfulness of the events. Very little data have been published that can guide our choice of distributions, but we chose to simulate two conditions, circumstances of highly and normatively stressful events. This choice reflects three rather different considerations. First, prior research indicates that clinical samples report significantly more stressors than controls or community samples. Second, sociological work shows that life events are distributed quite differently by socioeconomic status, race, gender, and age (e.g., Turner & Avison, 2003; Turner & Turner, 2005). Thus, the high-stress circumstance could describe the circumstances of people in clinical samples, and/or people at the social margins of society. Third, macro research suggests that distribution of the severity of life events can change markedly with social, economic, or political turmoil or because of sociocultural differences. For example, Brown (2002) reports annual rates of "irregular or disruptive severe events" ranging from 87 per 100 for women in Zimbabwe, 31 in London, to 0 in rural Basque (with corresponding rates of depression being 30, 15.1, and 2.4 per 100). The low-stress group, in contrast, describes a nonclinical sample or may also be viewed as an essentially middle-class (or higher) group of people who are well-integrated and essentially successful in their relatively stable communities.

Drawing on Brown's (2002) London observations, we define the high-stress group's distribution as: mild, 34%; moderate, 33%; and severe 33%. The low-stress group's distribution, based on the work of Turner, was defined as: mild, 70%; moderate 20%; and severe 10%. To determine these distributions, we first randomly assign cases to one of the three categories. Then, in the clinical sample, we reassign 75% of the cases with 3+ life events to the severe category and, in the community sample, we reassign 25% of the cases with 3+ life events to the severe category. This reassignment process reflects the increased likelihood that having multiple stressful life events will result in a higher average severity, especially in the case of clinical samples (e.g., Flouri & Kallis, 2007; Lenze, Cyranowski, Thompson, Anderson, & Frank, 2008) (Fig. 3.1).

Simulation Design and Results

In addition to the distribution of X_3, we also explore the effect of different sample sizes. We consider samples of 1,000, 2,500, and 5,000 subjects each.

The smallest sample approximates Caspi et al.'s study (2003), and the largest is somewhat bigger than Surtees et al. (2006; $n = 4,175$), the largest sample used among the published replications. Most attempted replications to date have been based on samples smaller than Caspi et al. (2003), but preliminary analyses indicated that the 1,000–5,000 range illustrates the role of sample size.

With consideration of two distributions for X_3 and three sample sizes, we thus have $2 \times 3 = 6$ conditions. For each condition, we generate 1,000 datasets using Equations 3.2–3.4 and estimate the model in Equation 3.1. We collect β_3 from each run and examine the distribution of $\exp(\beta_3)$ across the 1,000 datasets for each of the six conditions. We also document the proportion of samples that return a significant estimate for β_3. This design allows us to explore the range of results that are possible for the estimate of the interaction across a range of sample sizes, given a distribution of severity of life events in the population that is not taken into account in the model.

Beginning with the clinical sample, we find a substantial amount of variability in the estimate of $\exp(\beta_3)$ at a sample size of 1,000 (see Fig. 3.1). The average estimate across the samples is 1.24, but this comes with a standard deviation of 0.16. In fact, in some samples, the estimate of $\exp(\beta_3)$ is below 1. Across all of these samples, only 33% generated a significant estimate for $\exp(\beta_3)$. Not surprisingly, as sample size increases, the variability in the estimate of $\exp(\beta_3)$ decreases. With 2,500 subjects, the average estimate changes little ($\exp(\beta_3) = 1.22$), but the standard deviation falls to 0.10. Even at this sample size, however, only 69% of the samples return a significant estimate for $\exp(\beta_3)$. Finally, at a sample size of 5,000, the average estimate for $\exp(\beta_3)$ is unchanged, and the standard deviation drops to 0.06. At this sample size, over 90% of the samples return a significant estimate for $\exp(\beta_3)$.

The community sample represents a population exposed to fewer severe life events. Consequently, as one would expect, $\exp(\beta_3)$ attenuates. With a sample size of 1,000, we find an average estimate for $\exp(\beta_3)$ of 1.06 with a standard deviation of 0.12 (see Fig. 3.1). This pattern is notable in two respects. First—and the primary point of this exercise—$\exp(\beta_3)$ decreases noticeably in magnitude, on average, about 15%. Second, the effect of sample size is pronounced. With 1,000 subjects, less than 1% of the samples return a significant result for $\exp(\beta_3)$. Even with 5,000 subjects, only 18% of the samples return a significant estimate for $\exp(\beta_3)$. These results suggest that detecting a significant LE*HTT, even with community sample sizes of 5,000, will be unusual, but not impossible, when studying a population with relatively normative levels of severity of life events.

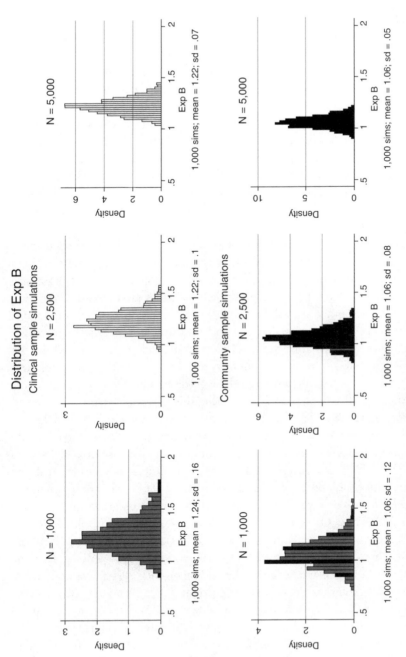

Figure 3.1. Simulated effect of Life-Events*5-*HTTLPR* on depression by levels of stress and sample size

CONCLUSION

In many respects, the major points of this chapter run parallel to arguments being advanced by geneticists who study behavior. First, just as the genome is subject to influence over time (e.g., by way of epigenetic processes), there is now abundant evidence that people's social circumstances are best described longitudinally. A person's status, interpersonal relationships, and roles all are subject to patterns of change and stability, patterns that are often significant in determining the implications of environment for behavior. Second, just as genes may often work in networks (e.g., that regulate neural circuits and modulate communication between different areas of the brain), specific features of the environment are often highly intercorrelated.

This complexity is obviously a serious methodological challenge for nonexperimental research designs that aspire to causal statements. But there is a measurement point as well: The study of context as sets of organized environmental candidates is a potentially valuable next step in improving gene–environment research. Just as these newly appreciated complexities will improve the conceptualization and measurement of genetic processes, so too they hold promise for environmental mechanisms. Indeed, in the specific case of life events, we identify six dynamic and systemic features that have demonstrably improved their prediction of depression.

What are the implications of ignoring the systemic, dynamic features of environmental candidates for gene–environment research? We consider the question by way of a simulation study that continues our example of life events and focuses specifically on the much-discussed case of Caspi et al.'s (2003) LE*HTT finding. Two substantive conclusions are warranted if one accepts the simulation's metrics and design as reasonable approximations. *First, the distribution of the severity of events is likely to influence the magnitude of the interaction.* Of course, the simulation was designed to illustrate this very point, but the magnitude of this difference was not obvious at the outset. The simulation is based on the reviewed research showing that the main effect of life events changes depending on their systemic and dynamic features. By extension, the simulation postulates that the interaction involving life events will also be sensitive to these features. The difference in $\exp(\beta_3)$ (1.24 vs. 1.06) is of notable magnitude, with an attenuation of 15%.

As noted, however, the distribution for the clinical sample likely understates the frequency of severe events. Further, the assigned parameter estimate for β_3 when $X_3 = 2$ (i.e., the effect size for LE*HTT when life events are severe) may actually be low. The estimate is based on Wilhelm et al.'s study (2006), but that sample does not suggest notably high levels of stressors. Thus, there are likely populations for which the distribution of life events is more severe than that found in our simulated clinical sample. In turn, the average decrement in LE*HTT between our exemplar clinical and community samples may approach 25% or more.

Second, sample size is potentially quite important, and apparently increases in salience as the distribution of life events becomes less severe. Even in the clinical setting with 2,500 subjects, only 69% of the obtained estimates of β_3 would be significant. However, Uher and McGuffin's (2008) review of past research identifies 15 (out of 17) replication studies as having a sample size of less than 1,000, and none of the studies reports a sample size exceeding that of Surtees et al. (2006), $n = 4,175$. The problem of the inability to detect potentially real effects is compounded because only five of the 17 reviewed replications might qualify as a clinical sample. Interestingly, four of these five report a statistically significant interaction. The fifth is Surtees et al. (2006), with the largest sample size, but the sample comprises neurotic extremes, which may contaminate the respondents' self-reports.

Given limited resources with which to collect data, further discussion of the relative benefits of increasing the quality of measures—by considering the dynamic and systemic properties of social experiences—versus increasing sample size may be productive. In a structural equation framework, Matsueda and Bielby (1986) showed that corrections for measurement error effectively increase statistical power. Wong et al. (2003; as cited in Uher & McGuffin, 2008) conclude that smaller samples with better measures are to be preferred over larger samples with poorer measures in the studies of gene–environment interaction. Consistent with the simulation, however, future research should also consider a third factor (in addition to better measurement and adequate sample size): the distribution of qualities of the environmental candidates in the population being studied. A study of stressors (and genes) will ideally consider, for example, the distribution of severity of stressors. The magnitude of the interaction in a population with relatively modest levels of stressors will be smaller than that found in a population with higher levels of stressors.

Our conceptual argument and simulation poses an intriguing implication, a challenge to the notion that there is one population estimate of LE*HTT that meaningfully reveals *the* effect size of the interaction. If a G × E effect reflects the distribution of these environmental features and their qualities (e.g., indicators of the stressfulness of purported stressors), then the discussion should be recast from the identification of *the* interaction effect to gene–environment interactions for specific populations. (This point is actually widely appreciated when interpreting heritability studies, but has not been acknowledged much, if at all, when interpreting molecular studies.) In summary, gene–environment research needs to (a) sharpen measures of environmental candidates to reflect the systemic and dynamic properties that many such candidates are known to have, and (b) interpret results with specific reference to populations whose environments are well-described.

ACKNOWLEDGMENTS

The authors wish to thank George Brown for helpful comments. This work is supported by 1R21HD050261-01A2 from NICHD.

REFERENCES

Alamian, A., & Paradis, G. (2009). Clustering of chronic disease behavioral risk factors in Canadian children and adolescents. *Preventive Medicine, 48*(5), 493–499.

Brown, G. W. (1989). Life events and measurement. In G. W. Brown & T. Harris (Eds.), *Life Events and Illness* (pp. 3–48). New York: Guilford.

Brown, G. W. (2002). Social roles, context, and evolution in the origins of depression. *Journal of Health & Social Behavior, 43*, 255–276.

Brown, G. W., Adler, Z., & Bifulco, A. (1988). Life-events, difficulties and recovery from chronic depression. *British Journal of Psychiatry, 152*, 487–498.

Brown, G. W., & Harris, T. O. (1978). *Social origins of depression: a study of psychiatric disorders in Women* (5th ed.). London: Routledge.

Brown, G. W., Harris, T. O., & Hepworth, C. (1995). Loss, humiliation and entrapment among women developing depression: A patient and non-patient comparison. *Psychological Medicine, 25*(1), 7–21.

Caspi, A., McClay, J., Moffitt, T. E., Mill, J., Martin, J., Craig, I. W., et al. (2002). Role of genotype in the cycle of violence in maltreated children. *Science (New York, N.Y.), 297*(5582), 851–854.

Caspi, A., Sugden, K., Moffitt, T. E., Taylor, A., Craig, I. W., Harrington, H., et al. (2003). Influence of life stress on depression: Moderation by a polymorphism in the 5-HTT gene. *Science, 301*(5631), 386–389.

Copeland, W., Shanahan, L., Costello, E. J., & Angold, A. (2009). Configurations of common childhood psychosocial risk factors. *Journal of Child Psychology & Psychiatry, 50*(4), 451–459.

deFur, P. L. Evans, G. W., Cohen Hubai, E. A., Kyle, A. D., Morello-Frosch, R. A., & Williams, D. R. (2007). Vulnerability as a function of individual and group resources in cumulative risk assessment. *Environmental Health Perspectives, 115*(5), 817–824.

Elder, G. H., Jr. (1974/1999). *Children of the Great Depression: Social change in life experience, 25th anniversary edition*. Boulder, CO: Westview.

Elder, G. H., Jr., & Shanahan, M. J. (2006). The life course and human development. In R. Lerner (Ed.), *Handbook of child psychology, volume 1, theory* (pp. 665–715). Hoboken, NJ: John Wiley & Sons.

Evans, G. W. (2006). Child development and the physical environment. *Annual Review of Psychology, 57*, 423–451.

Flouri, E., & Kallis, C. (2007). Adverse life events and psychopathology and prosocial behavior in late adolescence: testing the timing, specificity, accumulation, gradient, and moderation of contextual risk. *Journal of the American Academy of Child and Adolescent Psychiatry, 46*(12), 1651–1659.

Gillespie, N. A., Whitfield, J. B., Williams, B., Heath, A. C., & Martin, N. G. (2005). The relationship between stressful life events, the serotonin transporter (5-HTTLPR) genotype and major depression. *Psychological Medicine, 35*(1), 101–111.

Kalmijn, M., & Monden, C. W. S. (2006). Are the effects of divorce on well-being dependent on marital quality? *Journal of Marriage and Family, 68*, 1197–1213.

Kaufman, J., Gelernter, J., Kaufman, A., Caspi, A., & Moffitt, T. (in press). Arguable assumptions, datable conclusions. *Biological Psychiatry*. Epub ahead of print.

Kendler, K. S., Kuhn, J. W., Vittum, J., Prescott, C. A., & Riley, B. (2005). The interaction of stressful life events and a serotonin transporter polymorphism in the prediction of episodes of major depression: a replication. *Archives of General Psychiatry, 62*(5), 529–535.

Kim-Cohen, J., Caspi, A., Taylor, A., Williams, B., Newcombe, R., Craig, I. W., et al. (2006). MAOA, maltreatment, and gene-environment interaction predicting children's mental health: New evidence and a meta-analysis. *Molecular Psychiatry, 11*(10), 903–913.

Larson, K., Russ, S. A., Crall, J. J., & Halfon, N. (2008). Influence of multiple social risks on children's health. *Pediatrics, 121*(2), 337–344.

Laub, J. H., & Sampson, R. J. (2003). *Shared beginnings, divergent lives: delinquent boys to age 70.* Cambridge: Harvard.

Lenze, S. N., Cyranowski, J. M., Thompson, W. K., Anderson, B., & Frank, E. (2008). The cumulative impact of nonsevere life events predicts depression recurrence during maintenance treatment with interpersonal psychotherapy. *Journal of Consulting and Clinical Psychology, 76*(6), 979–987.

Mandelli, L., Serretti, A., Marino, E., Pirovano, A., Calati, R., & Colombo, C. (2007). Interaction between serotonin transporter gene, catechol-O-methyltransferase gene and stressful life events in mood disorders. *The International Journal of Neuropsychopharmacology/Official Scientific Journal of the Collegium Internationale Neuropsychopharmacologicum (CINP), 10*(4), 437–447.

Maruna, S. (2001). *Making good: how ex-convicts reform and rebuild their lives.* Washington, DC: American Psychological Association Books.

Matsueda, R. L., & Bielby, W. T. (1986). Statistical power in covariance structure models. *Sociological Methodology, 16*, 120–258.

McLeod J. D., & Almazan, E. P. (2003). Connections between childhood and adulthood. In J. T. Mortimer & M. J. Shanahan (Eds.), *Handbook of the Life Course* (pp. 391–411). New York: Kluwer-Plenum.

Miller, G. E., Chen, E., Fok, A. K., Walker, H., Lim, A., Nicholls, E. F., et al. (2009). Low early-life social class leaves a biological residue manifested by decreased glucocorticoid and increased proinflammatory signaling. *Proceedings of the National Academy of Sciences of the U.S.A., 106*(34), 14716–14721.

Mortimer, J. T., & Shanahan, M. J. (Eds.). (2003). *Handbook of the life course.* New York: Plenum Press.

Overbeek, G., Vollebergh, W., de Graaf, R., Scholte, R., de Kemp, R., & Engels, R. (2006). Longitudinal associations of marital quality and marital dissolution with the incidence of DSM-III-R disorders. *Journal of Family Psychology : JFP : Journal of the Division of Family Psychology of the American Psychological Association (Division 43), 20*(2), 284–291.

Reynolds, J. R., & Turner, R. J. (2008). Major life events: Their personal meaning, resolution, and mental health significance. *Journal of Health and Social Behavior, 49*(2), 223–237.

Risch, N., Herrell, R., Lehner, T., Liang, K. Y., Eaves, L., Hoh, J., et al. (2009). Interaction between the serotonin transporter gene (5-HTTLPR), stressful life events, and risk of depression: A meta-analysis. *JAMA : The Journal of the American Medical Association, 301*(23), 2462–2471.

Rutter, M. L. (1986). Child psychiatry: The interface between clinical and developmental research. *Psychological Medicine, 16*(1), 151–169.

Schuit, A. J., van Loon, A. J., Tijhuis, M., & Ocke, M. (2002). Clustering of lifestyle risk factors in a general adult population. *Preventive Medicine, 35*(3), 219–224.

Shanahan, M. J., & Macmillan, R. (2007). *Biography & the sociological imagination: Contexts and contingencies.* New York: W. W. Norton & Co.

Strohschein, L.(2005). Parental divorce and child mental health trajectories. *Journal of Marriage and Family, 67*(5), 1286–1300.

Tremblay, J., & Hamet, P. (2008). Impact of genetic and epigenetic factors from early life to later disease. *Metabolism: Clinical and Experimental, 57*(Suppl 2). S27–31.

Turner, H. A., & Turner, R. J. (2005). Understanding variations in exposure to social stress. *Health, 9*(2), 209–240.

Turner, R. J., & Avison, W. R. (2003). Status variations in stress exposure: Implications for the interpretation of research on race, socioeconomic status, and gender. *Journal of Health and Social Behavior, 44*(4), 488–505.

Uher, R., & McGuffin, P. (2008). The moderation by the serotonin transporter gene of environmental adversity in the aetiology of mental illness: review and methodological analysis. *Molecular Psychiatry, 13*(2), 131–146.

Wilhelm, K., Mitchell, P. B., Niven, H., Finch, A., Wedgwood, L., Scimone, A., et al. (2006). Life events, first depression onset and the serotonin transporter gene. *The British Journal of Psychiatry : The Journal of Mental Science, 188*, 210–215.

Wong, M. Y., Day, N. E., Luan, J. E., Chan. K. P., & Wareham, N. J. (2003). The detection of gene-environment interactions for continuous traits: should we deal with measurement error by bigger studies or better measurement? *International Journal of Epidemiology 32*, 51–57.

Zalsman, G., Huang, Y. Y., Oquendo, M. A., Burke, A. K., Hu, X. Z., Brent, D. A., et al. (2006). Association of a triallelic serotonin transporter gene promoter region (5-HTTLPR) polymorphism with stressful life events and severity of depression. *The American Journal of Psychiatry, 163*(9), 1588–1593.

4

Genotype–Environment Correlations

Definitions, Methods of Measurement, and Implications for Research on Adolescent Psychopathology

SARA R. JAFFEE

The literature on adolescent psychopathology abounds with papers that identify individual, family, peer, neighborhood, and school risk factors for common mental health problems that emerge in adolescence. However, because relatively little of this research is experimental or quasi-experimental, the causal status of these risk factors is largely unknown. In most cases, correlated risk factors could confound the association with adolescent psychopathology (Fig. 4.1). That is, as seen in the upper part of Figure 4.1, the environment might directly cause mental illness or, as seen in the lower part of the figure, the two might not be causally related, but result from a set of correlated risk factors. This chapter focuses on an individual's genetic makeup as one such factor that could confound the relationship between candidate risk factors and mental health outcomes in adolescence—a phenomenon known as *genotype–environment correlation*. Genotype–environment correlations refer to genetic differences in exposure to particular environments (Kendler & Eaves, 1986; Plomin, DeFries, & Loehlin, 1977). Such environments may increase risk for adolescent psychopathology.

The goal of this chapter is to define genotype–environment correlations, discuss the implications of genotype–environment correlations for research on psychosocial risk factors for adolescent psychopathology, and

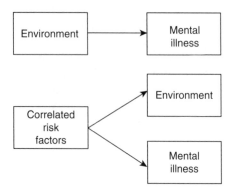

Figure 4.1. **Causal and non-causal models.** In the causal model, specific aspects of the environment directly cause mental illness. In the noncausal model, risk factors that are correlated with these specific aspects of the environment confound the observed relationship between the environment and mental illness. Correlated risk factors could include individual characteristics like genotype or other aspects of the environment.

describe some research designs that either estimate genotype–environment correlations directly or eliminate genotype–environment correlations as a potential confounder of the relationship between a candidate environment and some aspect of youth behavior. In this chapter, the terms "candidate environment" and "putative environment" are used interchangeably to denote exposures within the family or the broader social context that are assumed to have a direct, causal effect on youth outcomes, but that may be confounded by genotype.

GENOTYPE–ENVIRONMENT CORRELATIONS: DEFINITIONS

Genotype–environment correlations refer to genetic differences in exposure to particular environments (Kendler & Eaves, 1986; Plomin et al., 1977). Psychologists and psychiatrists commonly refer to three types of genotype–environment correlation. *Passive genotype–environment correlation* refers to the association between the genotype a child inherits from his or her parents and the environment in which the child is raised. For example, because parents who have histories of antisocial behavior (which is moderately heritable) are at elevated risk of abusing their children, maltreatment may be a marker for a genetic risk that parents transmit to children, rather than a causal risk factor for children's conduct problems (DiLalla & Gottesman, 1991). *Evocative (or reactive) genotype–environment correlation* refers to the association between an individual's genetically influenced behavior and others' reactions to that behavior. For example, although persistent conflict with a parent or romantic partner may cause an adolescent to become depressed, it is equally plausible that individuals who are prone to depression tend to provoke arguments with significant others, thus calling into question the direction of the effect. *Active (or selective) genotype–environment correlation* refers to the association between an individual's genetic propensities and the environmental niches that individual selects. For example, individuals who are characteristically extroverted may seek out social environments very different from those who are shy and withdrawn. These forms of genotype–environment correlation are depicted in Figure 4.2. Although evocative and active genotype–environment correlations are not distinguished in the figure, they reflect different processes whereby individuals create or modify their environments. For example, because children choose their friends, but not their parents, active genotype–environment correlations are more likely to involve the peer environment than the family environment.

To say that the relationship between a candidate environment and some aspect of adolescent psychopathology is "genetically mediated" implies that common genetic factors account for the relationship between the

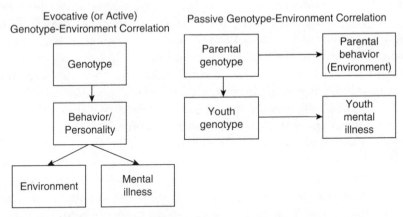

Figure 4.2. **Types of genotype-environment correlation.** In the case of active or evocative genotype–environment correlations, an individual's genotype confounds the relationship between some aspect of the individual's environment and his or her behavior or abilities. For example, genes that predispose an individual to sensation-seeking could lead the individual to seek out deviant peers and to engage in substance use. In the case of passive genotype–environment correlation, parental genotype confounds the relationship between some aspect of the family environment and the child's behavior or abilities. For example, genes that predispose a parent to use harsh, physical discipline with a child could also predispose the child to aggression.

environment and the mental health outcome. That is, genotype–environment correlation is present. To say the relationship is "environmentally mediated" implies that genotype–environment correlation is not present and that the candidate environment either directly influences symptoms of psychopathology or that some other aspect of the environment explains the effect of the candidate environment on psychopathology symptoms.

Genotype–environment correlation differs from gene × environment interaction (G × E), which refers to genetic differences in sensitivity to particular environmental effects (Kendler & Eaves, 1986) (see also Chapter 2, by Rudolf Uher). Genotype–environment correlations explain why individuals who have a genetic propensity to engage in sensation-seeking behaviors affiliate with drug-abusing peers (Yanovitzky, 2006). G × E explains why heavy drug use is most likely to lead to psychosis only among individuals with a particular genotype (Caspi et al., 2005).

THE IMPORTANCE OF ACCOUNTING FOR GENOTYPE–ENVIRONMENT CORRELATIONS IN RESEARCH ON ADOLESCENT PSYCHOPATHOLOGY

An underlying assumption in much of the research on psychosocial risk factors for psychopathology is that the social environment can directly bring about symptoms of mental health problems. For example, it is presumed that youth are more likely to engage in antisocial behavior when their parents get divorced, as a response to conflict and tension in the home and lower levels of parental monitoring (DeGarmo & Forgatch, 2005). If this is true, then modifying exposure to the risk environment—that is, preventing marital separations or mitigating the conflict that results from marital separations—should reduce levels of psychopathology. However, if genotype–environment correlation exists, then the relationship between psychosocial risk factors like divorce and psychiatric outcomes will be confounded by genotype. In this case, modifying the putative risk environment will have little effect on psychopathology. For example, a history of antisocial behavior increases the likelihood that a person will become divorced (Emery, Waldron, Kitzmann, & Aaron, 1999; Smith & Farrington, 2004). Thus, divorce could simply reflect a genetic propensity for antisocial behavior (Rhee & Waldman, 2002) that parents transmit to their children. If this is true, then reducing divorce rates or minimizing family conflict in the wake of a divorce may have little effect on youth behavior.

Research designs are needed that can test whether various social experiences are likely to be true causes of adolescent psychopathology or whether the relationship is spurious and confounded by genotype or other aspects of the environment. Experimental designs accomplish this goal by randomly assigning individuals to treatment versus control conditions, thus eliminating any systematic association between genotype (or any other characteristic of the individual) and treatment exposure. However, experiments that randomly expose individuals to psychosocial risk factors for psychopathology (e.g., divorce, family poverty) are generally neither practical nor ethical. Preventive interventions in the form of randomized control trials (RCTs) are an ethical alternative and allow for relatively strong causal inference about factors that produce change in outcome (e.g., comparing pharmacological versus psychotherapeutic treatments for symptoms of depression), but these also have their limitations (Academy of Medical Sciences Working Group, 2007). Most salient among these is that those factors that produce change in some behavior (e.g., effective parenting leading to reductions in adolescent delinquency) may not be the same factors that produced the behavior in the first place (Academy of Medical Sciences Working Group, 2007). For example, the combination of biological diatheses and harsh, inconsistent parenting practices tends to produce conduct problems that emerge in early childhood. These conduct problems are often exacerbated when children reach school age and are rejected by prosocial peers (as well as other adults) and form alliances with other deviant peers. At school age, interventions aimed at promoting positive peer relations and reducing deviant peer affiliations may lead to reductions in conduct problems, but these are not the factors that produced the antisocial behavior in the first place. From a purely clinical perspective, a substantial reduction in conduct problems is significant in its own right, but for researchers interested in understanding the etiology of problem behaviors, RCTs might not identify the right variables. There are also questions about the external validity of RCTs if the treatment and control groups are not representative of the population. Given these limitations of experimental designs, quasi-experimental designs (as described below) can be useful for testing whether candidate environments are causally related to adolescent outcomes or whether the association is confounded by genotype–environment correlation or other aspects of the environment.

Designs are also required that distinguish among different forms of genotype–environment correlation. Studies that identify evocative genotype–environment correlations and can distinguish evocative from passive genotype–environment correlations may help researchers resolve the direction of effects between exposures and outcomes and thus provide insight into effective versus ineffective targets for intervention. For example, Narusyte et al. (2008) showed that, although adolescents whose parents were emotionally overinvolved (i.e., overprotective and self-sacrificing) had elevated levels of internalizing problems, this was due to depressed and anxious youth eliciting overprotective behavior in parents, rather than the reverse.

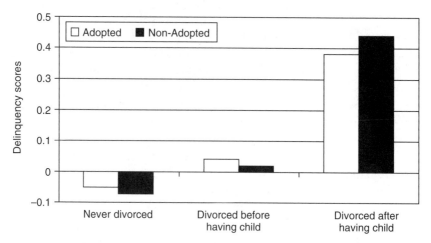

Figure 4.3. **Delinquency and parental divorce.** Levels of delinquency are highest among the adopted *and* nonadopted offspring of parents who divorced during their child's lifetime. Figure modified from Burt, S. A., Barnes, A. R., McGue, M., & Iacono, W. G. (2008). Parental divorce and adolescent delinquency. *Developmental Psychology, 44*, 1668–1677, with permission.

RESEARCH DESIGNS TO TEST GENOTYPE–ENVIRONMENT CORRELATION

Adoption, child- and parent-based twin, sibling, molecular genetic, and other novel research designs have been used to test for genotype–environment correlation. Examples of each design are provided, as they relate to the literature on psychosocial risk factors for adolescent psychopathology, as well as a description of the strengths and limitations of each design.

Adoption Design

The adoption design capitalizes on the fact that adoptive parents and their adopted children are not genetically related. Thus, any effects of the family environment on adopted offspring behavior must be environmentally rather than genetically mediated. For example, Burt et al. (2008) used an adoption design to test whether parental divorce was likely to be a cause of adolescent delinquency or whether it is a marker for antisocial behavior that parents transmit to children. Their sample included 406 biologically unrelated and 204 biologically related families. In some of these families, the parents had never been divorced; in others, the parents divorced during the child's lifetime; and in still others, parents had been divorced in previous relationships before the child was born. Burt et al. (2008) hypothesized that if divorce caused adolescent delinquency, then levels of delinquency would be highest among adolescents who experienced their parents' divorce, regardless of youth adoption status. In contrast, if genes common to parents and children mediated the association between divorce and adolescent delinquency then (a) divorce would only be correlated with youth delinquency in biologically related families, and (b) divorce would be correlated with youth delinquency regardless of whether the divorce happened before or after the child was born.

The data were consistent with the hypothesis that divorce was a cause of adolescent delinquency. As shown in Figure 4.3, regardless of whether adolescents and parents were biologically related, adolescents whose rearing parents divorced after the adolescent was born had significantly higher levels of delinquency than did adolescents whose parents never divorced or than did adolescents whose rearing parents were divorced in a prior relationship. In contrast, adolescents whose parents were divorced in a prior relationship before the adolescent was born did not differ significantly from adolescents whose parents never divorced.

The adoption design can also be used to test for evocative genotype–environment correlations. In this case, information about the adoptees' biological parents must be available to derive an estimate of the adoptee's genetic risk for some outcome. For example, Ge et al. (1996) used information about antisocial personality disorder and alcohol and drug abuse/dependence among biological parents to identify adoptees at high versus low genetic risk for psychopathology. High-risk adoptees experienced significantly higher levels of harsh, inconsistent discipline and lower levels of warmth and nurturance from their adoptive parents than did low-risk adoptees, and the relationship between adoptee risk status and adoptive parent behavior was mediated by adoptee antisocial behavior. Thus, a genetic propensity to engage in antisocial behavior made adolescents difficult to manage, resulting in less warmth and more negativity in the parent–adolescent relationship—evidence of evocative genotype–environment correlation.

A major strength of the adoption design is that it severs the genetic relationship between parent and child, eliminating the possibility that passive genotype–environment correlations account for any effect of the family environment on youth behavior. When information about biological parents is available, the design can also be used to test for the presence of evocative genotype–environmental correlations, although the resulting index of genetic risk is only a rough proxy measure and the design cannot test for reciprocal parent–child influences in the absence of longitudinal data. Additionally, the test of evocative genotype–environment correlation rests on the assumption that the genes that influence the (biological) parent phenotype are the same as those that influence the child phenotype. This assumption may be plausible for certain phenotypes, but not others. For example, Dick et al. (2006b) showed that a variant of the γ-aminobutyric acid A α2 receptor (*GABRA2)* gene (rs279871) was associated with alcohol dependence in adults, but not in children and adolescents, where it was instead associated with conduct disorder symptoms.

One of the strongest criticisms of the adoption design is that it reflects a restricted range of environments (Stoolmiller, 1999). Because adoptive families are heavily screened by social service agencies, very high-risk environments may not be included in samples of adoptive families. However, a recent study showed that restriction of range—although present—did not influence the degree of sibling phenotypic similarity in adoptive families, nor did it affect the coefficient of the regression of adolescent behavioral and cognitive outcomes on measures of family environment (McGue et al., 2007). Thus, although adoptive families may indeed be less "risky" than nonadoptive families, this has relatively few implications for studies concerned with identifying the effect of family environments on adolescent adjustment.

The adoption design rests on additional assumptions about the relationship between birth and adoptive parents, and birth parents and their children. First, the design assumes that birth and adoptive parents have not been matched by social service agencies for characteristics that might influence child outcomes. If they have, correlations between adopted children and the adoptive family environment could reflect genetic (or environmental) similarity rather than a purely environmental effect. Second, the design assumes that contact between adopted children and their biological parents is virtually nonexistent. If it is not, correlations between birth parents and their children could be inflated by their ongoing relationship. A study of adoptees that was recently launched failed to identify significant correlations between birth and adoptive parent demographic characteristics, demonstrating a lack of selective placement. Although they did not test whether correlations between adopted children and biological parents were greater among those who had more knowledge of or contact with each other, they did show that the degree of "openness" in the adoption was not correlated with birth parent or adoptive family measures (Leve et al., 2007).

Child- and Parent-based Twin Designs

Child-based twin designs refer to studies of twin children, whereas parent-based twin designs refer to studies of twin adults who are parents. These designs are useful because they can determine whether putative environmental risks for adolescent psychopathology are influenced by child or parent genotype (Neiderhiser et al., 2004). That is, putative environments, such as conflict and negativity in the home, low levels of parental monitoring, and low levels of parental warmth are treated like any other trait or characteristic that varies in the population, and differences in the genetic relatedness of monozygotic (MZ) and dizygotic (DZ) pairs are used to identify genetic and environmental sources of variation. If genetic factors account for variation in these phenotypes, then genotype–environment correlations are present. In the parent-based twin design, genetic influences on the putative environment reflect the effect of parental genotype, whereas in the child-based design, genetic influences reflect the effect of child genotype (Fig. 4.4). The literature suggests that genetic influences on candidate environments that increase risk for psychopathology are common (Kendler & Baker, 2007; Plomin & Bergeman, 1991), with a recent review showing that genetic factors accounted for a little over a quarter of the variance, on average, in such risk environments (Kendler & Baker, 2007).

The univariate twin model (in which genetic and environmental influences on a single phenotype are estimated) can be expanded to test whether the relationship between the putative environment and youth behavior is genetically or environmentally mediated (Fig. 4.5). For example, Narusyte et al. (2007) used a child-based twin design to test whether the relationship between parental criticism of adolescents and adolescent delinquency was genetically or environmentally mediated. As illustrated in Figure 4.5, the association between parental criticism and delinquency could be mediated by the common genetic factor, the common shared environment, and/

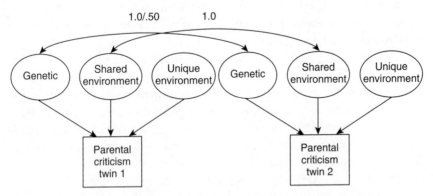

Figure 4.4. **Parent-based versus child-based designs.** In the parent-based design, parental criticism reflects each twin parents' use of criticism with their own children. In the child-based design, parental criticism reflects each twin child's experience of criticism from their parent. Genetic influences on parental criticism are correlated 1.0 for monozygotic and .50 for dizygotic twins, and shared environmental influences are correlated 1.0 for all twins.

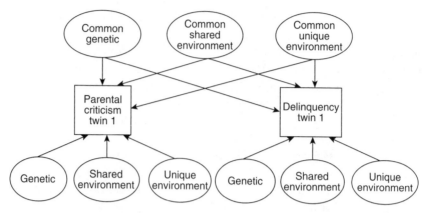

Figure 4.5. **Bivariate model of genotype and environmental influence.** The latent genetic and environmental factors in the upper portion of the figure refer to influences that are common to parental criticism and delinquency. The latent genetic and environmental factors in the lower portion of the figure refer to influences that are either specific to parental criticism or to delinquency. As shown in the figure, the relationship between parental criticism and delinquency can be accounted for by genetic or environmental factors that are common influences on each phenotype.

or the common unique environment. They showed that the relationship was genetically mediated and that approximately half of this genetic association was accounted for by early adolescent aggression. The findings suggest that youth with a genetic propensity for aggressive behavior both provoke parental criticism (consistent with an evocative genotype–environment correlation) and are at elevated risk of engaging in delinquency in later adolescence.

One limitation of the twin design is that child- and parent-based twin designs are each informative about only certain kinds of genotype–environment correlation—evocative genotype–environment correlation, in the case of the child-based twin design, and passive genotype–environment correlation, in the case of the parent-based design. However, when data from parent *and* child twins are available, these forms of genotype–environment correlation can be estimated simultaneously (Narusyte et al., 2008; Neiderhiser et al., 2004; Neiderhiser, Reiss, Lichtenstein, Spotts, & Ganiban, 2007). Another limitation is that genetic influences on the candidate environment may simply reflect genetically based individual differences in *perceptions* of the environment, rather than genetically based differences in the *actual experience* of the environment. For example, highly anxious adolescents may be especially likely to perceive rejection from peers even when peers do not actually engage in rejecting behavior. In this case, anxious adolescents may perceive nonexistent slights, but their anxiety is not actually causing them to be rejected. Therapists working with anxious youth might target slightly different processes, depending on whether anxiety causes behaviors that lead to peer rejection or whether anxiety generates perceptions of rejection (i.e., behavior around peers vs. cognitive distortions). Finally, the design does not test whether putative environments have direct effects on behaviors. Rather, it tests whether "black box" genetic or environmental factors account for the covariation between putative environments and youth outcomes. This is conceptually equivalent to testing for spurious association in a linear regression, when a researcher determines whether variable z accounts for the relationship between variable x and variable y. Again, however, when data from child and parent twins are available, these direct effects can be estimated (see discussion below of the extended children-of-twins model).

Discordant Monozygotic Twin Designs

Another use of child-based twin designs involves identifying MZ child twins who are discordant for some experience that is thought to increase risk for psychopathology. To the extent that within-pair differences in experience are correlated with within-pair differences in behavior, one can be confident that the effect of the experience is environmentally rather than genetically mediated (i.e., does not arise from evocative, active, or passive genotype–environment correlation). This is because MZ twins are genetically identical, so that differences in their experiences cannot logically arise from genetically based differences in their behavior.

When combined with longitudinal data involving repeated measures of experiences and youth behavior, the design offers a strong test of the direction of effects. For example, Burt et al. (2006) measured parent–adolescent conflict and adolescent antisocial behavior in a study involving 486 MZ twin pairs assessed at ages 11, 14, and 17 years. They found that among twins highly discordant for the experience of parent–child conflict (as well as for twins highly discordant for antisocial behavior), the twin who experienced higher levels of conflict at age 11 had significantly higher levels of antisocial behavior at age 14 than did his co-twin, controlling for both the stability of within-pair differences in antisocial behavior from early to middle adolescence, as well as for the stability of within-pair differences in parent–adolescent conflict. The results were consistent with the interpretation that parent–adolescent conflict was causally associated with youth antisocial behavior in middle adolescence, although the effect was only present for highly discordant pairs.

The discordant MZ twin approach can also be used to test whether differences in twins' behavior lead to differences in their development. For example, Lynskey et al. (2003) reported that MZ twins who used cannabis by age 17 years had elevated rates of drug use, alcohol dependence, and other drug abuse/dependence in young adulthood relative to their co-twin who did not use cannabis by age 17 years. These associations were not accounted for by other known risk factors for young adult substance use or abuse, and the findings imply that the effect of early cannabis use on later substance use is not accounted for by a common genetic or environmental liability to use substances.

As discussed by Vitaro, Brengden, and Arseneault (2009), the discordant MZ twins design—especially when it is embedded within a longitudinal design—can allow researchers to make powerful causal inferences about the relationship between experiences and behavior. This is because it eliminates genetic differences within pairs, as well as other potential family-wide confounds that might otherwise explain the relationship. A limitation of the design is that it requires there to be sufficient within-pair variability in experiences or outcome and that this reflects "true" variability rather than measurement error. This requirement is especially challenging for research on adolescents because adolescents have greater autonomy than younger children to select and create their own experiences. To the extent that such selection is driven by genetically influenced characteristics of the adolescent, MZ twins will become increasingly similar for their experiences as they transition from childhood to adolescence. In addition, because the design eliminates so many potential confounds, the variance in outcome accounted for by within-pair differences in experience is often modest (less than 5%).

Finally, the design estimates those within-pair differences in exposure (e.g., to cannabis or to parental criticism) that it presumes to be the cause of within-pair differences in outcome (e.g., later drug use, delinquency), but the design cannot rule out the possibility that some other event not shared by twins is the real culprit. For example, twins may differ with respect to their peer relationships, with one twin associating with peers who promote both cannabis and later drug use and the other twin associating with a more salubrious social group. Thus, although the design rules out the possibility that genetic differences within pairs account for the relationship between within-pair exposures and within-pair outcomes, it cannot definitively identify the exposure as the causal agent of within-pair differences in outcome.

Children-of-Twins Design

The children-of-twins (CoT) design involves parents who are twins and their offspring. The design is premised on the observation that, in MZ pairs, the magnitude of the genetic relationship between aunts or uncles and nieces or nephews is the same as that between parents and offspring. In DZ pairs, aunts or uncles are more closely related to their own children than to their nieces and nephews. The design also capitalizes on the fact that twin pairs may be discordant for aspects of the environment in which they raise their children. For example, one twin might be divorced, whereas her twin sister's marriage is intact. If the offspring of the twin who has not divorced have fewer problem behaviors than their cousins whose parents have divorced, then there is a strong basis for presuming that divorce causes off-spring maladjustment, because the design automatically controls for genetic differences in pairs discordant for divorce (completely, in the case of MZ pairs, wherein the nieces and nephews of the divorced twin will have inher-ited the same genetic liability as the sons and daughters of the divorced twin) and for environmental confounds shared by twins (e.g., ethnicity, socioeconomic background). Figure 4.6 illustrates the pattern of group dif-ferences indicative of an environmentally mediated effect of divorce. Comparing the white bars, we see that cousins who are discordant for the experience of divorce and whose mothers are MZ twin sisters, differ signifi-cantly in their level of behavior problems. Here, behavior problems are stan-dardized with a mean of zero. Children whose mother's marriage is intact (Mother Married/Aunt Divorced) have significantly fewer behavior prob-lems than do their cousins whose mother is divorced (Mother Divorced/Aunt Married), despite sharing the same genetic liability. The same pattern

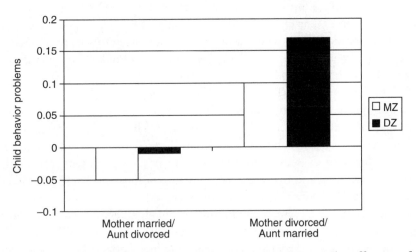

Figure 4.6. **Example of the Children of Twins Design.** Among the offspring of monozygotic (MZ) and dizygotic (DZ) pairs, those whose own parents are married but whose aunts are divorced have fewer behavioral problems than their cousins, pointing to environmentally mediated effects of divorce on offspring adjustment. Data from Table 3 in D'Onofrio et al., 2005 (p. 578).

of effects is observed for the offspring of DZ twins as well (black bars in Figure 4.6).

In a series of papers and two different samples, D'Onofrio and colleagues (D'Onofrio et al., 2005, 2006, 2007a,b) compared the offspring of twins discordant for divorce (see also Chapter 15 by D'Onofrio et al.). The results were consistent with a causal effect of divorce—particularly before the age of 16—on young adult psychopathology, early sexual intercourse, and educational problems. However, the effects of divorce on early cohabitation and earlier initiation of drug use were genetically mediated (D'Onofrio et al., 2006) and evidence for both causation and genetic mediation was found in the case of offspring tendency to divorce (D'Onofrio et al., 2007a) and to experience internalizing symptoms (D'Onofrio et al., 2005, 2007b).

The CoT design has been used to study the relationship between other candidate environments and offspring outcomes, although the vast majority of these studies have followed up offspring in young adulthood rather than adolescence. These include studies of teen motherhood (Harden et al., 2007a), marital conflict (Harden et al., 2007b), harsh punishment (Lynch, Turkheimer, D'Onofrio, Mendle, & Emery, 2006), parental alcohol use disorder (Knopik et al., 2006; Slutske et al., 2008), family structure (Lynch et al., 2006), and smoking during pregnancy (D'Onofrio et al., 2003).

The CoT design is a special case of a broader category known as the *children of siblings* (CoS) design. The CoS design, like the CoT design, uses information from sibling dyads who vary in their genetic relatedness (e.g., full siblings vs. half siblings) to estimate the magnitude of genetic and environmental confounds on observed associations between exposures and offspring outcomes. For example, Mendle et al. (2009) used a CoS design to test the relationship between father absence and age at first intercourse in the offspring of women raised in the same household which comprised twins, full siblings, half siblings, and cousins.

Strengths of the CoT design include strenuous controls for genetic and environmental factors correlated with the candidate environment, automatically ruling out most unmeasured confounders. However, the design is also characterized by a number of limitations. First, it does not automatically control for unmeasured environments that are correlated with the candidate environment and that are *not* common to both twins. These nonshared factors must be measured explicitly and included as covariates in the design. Second, when the candidate environment is likely to be influenced by both parents (i.e., the twin and his or her spouse), the design may not reliably distinguish between direct and genetically mediated effects of the candidate environment on the offspring phenotype (Eaves, Silberg, & Maes, 2005). This problem might be mitigated by including measured characteristics of spouses as covariates (D'Onofrio et al., 2005; Rutter, Pickles, Murray, &

Eaves, 2001). Third, the design does not control for reciprocal effects of the child's behavior on the family environment, although the direction of effects can be addressed in the extended children-of-twins design (see below).

Fourth, the design—like all twin designs—assumes the absence of assortative mating, even though assortative mating is moderately high for most psychiatric disorders and antisocial behavior (Galbaud du Fort, Bland, Newman, & Boothroyd, 1998; Krueger, Moffitt, Caspi, Bleske, & Silva, 1998; Maes et al., 1998). Assortative mating could confound correlations between the twin parent and his or her offspring if the spouse's behavior is strongly associated with the twin parent's behavior. Assortative mating in the grandparental and twin parental generations could also lead to violations of the assumption that DZ twins share half of their segregating genes, on average, and that the offspring of DZ twins will share an eighth of their segregating genes, on average. With assortative mating, DZ twins and or cousins who are offspring of DZ twins will have a greater coefficient of genetic relatedness, with corresponding implications for estimates of genetic and environmental influences on phenotypic variance.

Fifth, in the absence of extremely large samples and large parent–offspring correlations, the CoT design cannot reliably estimate which confound—genetic or environmental—accounts for the largest portion of the association between the environment and offspring behavior (Slutske et al., 2008).

Extended Children of Twins Design

The extended CoT design combines data from samples of adult twins and their offspring and child twins and their parents (Narusyte et al., 2008). The design estimates direct effects of parenting on children, direct effects of child characteristics on parenting, and passive and evocative genotype–environment correlation. The ability to distinguish passive from evocative genotype–environment correlations is a unique characteristic of the model. A simulation study established that the extended CoT model was stable and recovered the true parameter values with high precision under the assumption that individual differences in parenting were entirely accounted for by additive genetic and nonshared environmental factors (Narusyte et al., 2008). The model is able to detect relatively small genotype–environment correlations.

As reviewed by Narusyte et al. (2008), the requirement that the combined samples be comparable in terms of size, participant characteristics, and measured constructs is a major limitation of the extended CoT design. Samples of twin children and their parents on the one hand, and twin parents and their children (who are cousins) on the other hand, may produce correlations among youth behavior or parenting constructs that differ in magnitude simply because cousins are likely to be less similar to each other than siblings and because cousins will vary in age whereas twin siblings will not. Such differences could result in larger residual error terms and poorer model fit. Many researchers will not have access to data that meet these requirements for comparability. The design also assumes that those genes that influence the parent phenotype are the same as those that influence the child phenotype. As discussed in the section on adoption designs, this assumption may be plausible for certain phenotypes, but not others. To the extent that different genes influence the parental versus the child phenotype, passive genotype–environment correlation may be underestimated.

Sibling Comparison Designs

Sibling comparison designs involve studies of siblings within the same nuclear family who vary in their exposure to some aspect of the environment (e.g., exposure to alcohol in utero, family income in early childhood). Because the design compares children within the same family, it automatically rules out the kinds of between-family differences that typically confound observed relationships between candidate environments and youth outcomes. Moreover, when candidate environments are characteristics of parents (e.g., the tendency to smoke while pregnant), and candidate environments temporally precede the measure of adolescent behavior, the design can rule out all three types of genotype–environment correlation. This is because the process of meiosis randomly distributes genetic variants among siblings (so there should be no systematic association between the environment experienced by the child and the genotype inherited by the child), and genetically influenced child characteristics cannot logically affect environments that temporally preceded them (e.g., adolescent conduct problems cannot have caused their mothers to smoke while pregnant). In Chapter 15, by D'Onofrio, Rathouz, and Lahey, there is an extensive discussion of sibling comparison studies of adolescent antisocial behavior.

As reviewed by Lahey, D'Onofrio, and Waldman (2009), one strength of the sibling comparison design is that it does not rely on relatively rare events (e.g., twin births, adoptions). A limitation of the design is that the conditions under which it allows researchers to make the strongest causal inferences are somewhat constrained. If, for instance, researchers are interested in the relationship between parenting in early childhood and adolescent adjustment, efforts must be made to control for sibling differences in early childhood temperament that could not only explain variations in how they were parented, but also variations in adolescent behavior. In addition, the design requires that there be multiple children per family and that there be variation within families in exposure to the candidate environment. Thus, the findings may not generalize to single-child families or to families in which exposure to the candidate environment is constant across children.

Other Novel Research Designs

Recently, a novel research design used data from families undergoing assisted reproductive technologies (ART) to test the effects of the prenatal environment on offspring development (Rice et al., 2009). This study used data from 779 children born through in vitro fertilization whose prenatal environment was either provided by a biologically related mother or a biologically unrelated mother (e.g., a surrogate or the recipient of an egg donor). This design was used to test whether smoking during pregnancy was likely to be a causal factor in low birth weight or offspring antisocial behavior and hyperactivity in middle childhood. Data consistent with this causal hypothesis would show that observed associations were significant in biologically related and unrelated mother–offspring pairs. The alternative hypothesis was that genes common to mothers and their offspring would account for observed associations between smoking during pregnancy and offspring outcomes.

The results showed that infants whose mothers smoked during pregnancy had significantly lower birth weights than did infants whose mothers did not smoke during pregnancy, regardless of whether the mother and fetus were biologically related (Rice et al., 2009). In contrast, children whose mothers smoked during pregnancy had significantly higher levels of antisocial behavior and hyperactivity than did children whose mothers did not smoke, only when the mother–offspring pair was biologically related (Rice et al., 2009; Thapar et al., 2009). Thus, smoking during pregnancy is likely to be one cause of low birth weight, but not offspring antisocial behavior or hyperactivity.

Although highly innovative, one limitation of the design is that families who undergo ARTs may not be representative of families in the population. Women who use ARTs tend to be older and better-educated (Källén, Finnström, Nygren, & Olausson, 2005; Schieve et al., 2007), and this might not only influence their behavior while pregnant, but also the relationship between their behavior while pregnant and offspring behavior. In addition, although data are limited, women who have taken on the physical, financial, and emotional challenges of ARTs and become pregnant are less likely than other women to engage in behaviors that would put the fetus at risk (Källén et al., 2005). Thus, rates of risky behaviors, such as smoking or drinking alcohol during pregnancy, are likely to be very low in these samples, thus compromising statistical power, and women who do engage in these behaviors are likely to be heavily dependent on substances (Thapar et al., 2009).

Molecular Genetic Designs

A final method of estimating genotype–environment correlations is to genotype individuals and then correlate genetic variants with measured environments. As reviewed by Jaffee and Price (2007) efforts to identify genotype–environment correlations (mainly within studies of genotype × environment interaction) have largely produced null results. However, significant genotype–environment associations have been reported between *GABRA2* and marital status (Dick et al., 2006a), the serotonin transporter receptor 2_A (*5HT2A*) gene and popularity (Burt, 2008), the catechol-O-methyltransferase gene and daily hassles (van IJzendoorn, Bakermans-Kranenburg, & Mesman, 2008), brain-derived neurotrophic factor and childhood adversity (Wichers et al., 2008), and the dopamine D_2 receptor gene and both parenting behavior (Lucht et al., 2006) and marital status (Waldman, 2007). Although these associations require replication, they suggest that it may be possible to identify specific genotypes that correlate with measured environments.

One limitation to studying genotype–environment correlations in molecular genetic studies is that any effect of genotype on the environment must be behaviorally mediated (Jaffee & Price, 2007). Genes affect behavior, and behavior, in turn, is what creates experiences. However, correlations between single genes and behavior are typically small in magnitude and, thus, difficult to detect (Kendler, 2005). Correspondingly, genotype–environment correlations will be difficult to detect. Of the studies just cited, only Burt (2008) demonstrated behavioral mediation of the genotype–environment correlation; men who were heterozygous or homozygous for the G allele of the G1438A *5HT2A* polymorphism were better-liked by their peers than were men who were homozygous for the A allele because they engaged in higher levels of rule-breaking behavior (Burt, 2008).

In addition, the pathways by which genetic variants eventuate in specific experiences are likely to be dauntingly complex. The value of conducting such studies may depend greatly on the candidate environment in question. For example, "environments" that are likely to reflect dyadic processes (e.g., marital conflict, rejection by peers) or, even more directly, individual behavior (drinking or smoking during pregnancy, creating an in utero environment) are potentially better candidates for molecular genetic studies than are more distal environments (e.g., neighborhood poverty). In addition, molecular genetic studies of genotype–environment correlation may be informative for studies of genotype × environment interaction.

STATISTICAL IMPLICATIONS OF GENOTYPE– ENVIRONMENT CORRELATIONS

Genotype–environment correlations have several implications for studies of G × E (G × E models are described in greater detail in Chapters 1 and 2). First, as discussed throughout this chapter, measured environments may reflect genetic influence. If so, genotype × environment interactions would more accurately reflect genotype × genotype interactions (Moffitt, Caspi, & Rutter, 2005). Designs that estimate genotype–environment correlations are informative about which candidate environments are truly environmentally mediated.

Second, genotype–environment correlations and genotype × environment interactions can be impossible to distinguish in certain tests of G × E. In the case-only design, for example, power to detect G × E is amplified by focusing only on the group with the disorder and testing whether group members vary in terms of environmental or genetic risk exposure. Thus, in a sample of depressed adolescents, the majority may have been exposed to both genetic risk (e.g., the short form of the serotonin transporter polymorphism) and environmental risk (life stressors). One interpretation of the data could be that depression is produced by the combination of genetic and environmental risk factors (i.e., G × E), but a second interpretation is that genetic and environmental risk factors are correlated in individuals with depression.

Third, other methods of estimating G ×E, such as those that model the interaction between a measured environment and a latent genetic factor, produce biased estimates of heritability and mask G × E if genotype–environment correlation is present and is not modeled explicitly (Purcell, 2002; Rathouz, van Hulle, Rodgers, Waldman, & Lahey, 2008).

CONCLUSION

Genotype–environment correlations matter because they potentially confound observed associations between putative environmental risk factors and adolescent psychopathology. Without a clear and accurate understanding of the causal origins of mental health problems, intervention and prevention scientists will be poorly placed to develop practices aimed at reducing the burden of psychiatric illness in adolescence. Given the stability of mental health problems from adolescence into adulthood (Kim-Cohen et al., 2003), research is needed that has the potential to identify modifiable targets for intervention that will result in lower rates of disorder.

A number of designs exists that deal with genotype–environment correlations, either by measuring them directly or by ruling out the possibility that they confound the relationship between a candidate environment and youth outcomes. Such designs have already contributed to research and clinical practice by identifying family risk factors that are likely to be causally implicated in offspring adjustment (e.g., parental divorce) and others that are unlikely to be causally implicated (e.g., smoking during pregnancy and offspring conduct problems; see Chapter 15). These designs each have their own strengths and limitations, and replication of findings across different samples, using different methods, will increase confidence in causal inference. Such research could also instigate new investigations of the pathways that mediate causal risks for adolescent psychopathology.

Finally, as discussed in Chapter 1, genotype–environment correlations demonstrate that the effects of gene expression do not necessarily "stop at the skin." Our genes affect our behavior and humans—like other animals—behave in ways that shape the contours of our environments. Although this ability of humans to transform the environment is highly adaptive from an evolutionary perspective, it greatly complicates efforts to determine the relationship between persons and environments in the development of psychopathology.

ACKNOWLEDGMENTS

Preparation of this chapter was supported by grant RES-062-23-1583 from the Economic and Social Research Council and grant CPF/35191 from the Nuffield Foundation.

REFERENCES

Academy of Medical Sciences Working Group (2007). *Identifying the environmental causes of disease: how should we decide what to believe and when to take action?* London: Academy of Medical Sciences.

Burt, S. A. (2009). A mechanistic explanation of popularity: Genes, rule-breaking, and evocative gene-environment correlations. *Journal of Personality and Social Psychology, 96,* 783–794.

Burt, S. A., Barnes, A. R., McGue, M., & Iacono, W. G. (2008). Parental divorce and adolescent delinquency. *Developmental Psychology, 44,* 1668–1677.

Burt, S. A., McGue, M., Iacono, W. G., & Krueger, R. F. (2006). Differential parent-child relationships and adolescent externalizing symptoms: cross-lagged analyses within a monozygotic twin differences design. *Developmental Psychology*, *42*, 1289–1298.

Caspi, A., Moffitt, T. E., Cannon, M., McClay, J., Murray, R., Harrington, H., et al. (2005). Moderation of the effect of adolescent-onset cannabis use on adult psychosis by a functional polymorphism in the catechol-O-methyltransferase gene: longitudinal evidence of a gene x environment interaction. *Biological Psychiatry*, *57*, 1117–1127.

D'Onofrio, B. M., Turkheimer, E., Emery, R. E., Harden, K. P., Slutske, W. S., Heath, A. C., et al. (2007a). A genetically informed study of the intergenerational transmission of marital instability. *Journal of Marriage and Family*, *69*, 793–809.

D'Onofrio, B. M., Turkheimer, E., Emery, R. E., Maes, H. H., Silberg, J., & Eaves, L. J. (2007b). A children of twins study of parental divorce and offspring psychopathology. *Journal of Child Psychology and Psychiatry*, *48*, 667–675.

D'Onofrio, B. M., Turkheimer, E., Emery, R. E., Slutske, W. S., Heath, A. C., Madden, P. A., et al. (2006). A genetically informed study of the processes underlying the association between parental marital instability and offspring adjustment. *Developmental Psychology*, *42*, 486–499.

D'Onofrio, B. M., Turkheimer, E., Emery, R. E., Slutske, W. S., Heath, A. C., Madden, P. A., et al. (2005). A genetically informed study of marital instability and its association with offspring psychopathology. *Journal of Abnormal Psychology*, *114*, 570–586.

D'Onofrio, B. M., Turkheimer, E. N., Eaves, L. J., Corey, L. A., Berg, K., Solaas, M. H., et al. (2003). The role of the Children of Twins design in elucidating causal relations between parent characteristics and child outcomes. *Journal of Child Psychology and Psychiatry*, *44*, 1130–1144.

DeGarmo, D. S., & Forgatch, M. S. (2005). Early development of delinquency within divorced families: Evaluating a randomized preventive intervention trial. *Developmental Science*, *8*, 229–239.

Dick, D. M., Agrawal, A., Shuckit, M. A., Bierut, L., Hinrichs, A., Fox, L., et al. (2006a). Marital status, alcohol dependence, and *GABRA2:* Evidence for gene-environment correlation and interaction. *Journal of Studies on Alcohol*, *67*, 185–194.

Dick, D. M., Bierut, L., Hinrichs, A., Fox, L., Bucholz, K. K., Kramer, J., et al. (2006b). The role of GABRA2 in risk for conduct disorder and alcohol and drug dependence across developmental stages. *Behavior Genetics*, *36*, 577–590.

DiLalla, L. F., & Gottesman, I. I. (1991). Biological and genetic contributions to violence: Widom's untold tale. *Psychological Bulletin*, *109*, 125–129.

Eaves, L. J., Silberg, J. L., & Maes, H. H. Revisiting the children of twins: Can they be used to resolve the environmental effects of dyadic parental treatment on child behavior? *Twin Research and Human Genetics*, *8* (in press).

Emery, R. E., Waldron, M., Kitzmann, K. M., & Aaron, J. (1999). Delinquent behavior, future divorce or nonmarital childbearing, and externalizing behavior among offspring: a 14-year prospective study. *Journal of Family Psychology, 13,* 568–579.

Galbaud du Fort, G., Bland, R. C., Newman, S. C., & Boothroyd, L. J. (1998). Spouse similarity for lifetime psychiatric history in the general population. *Psychological Medicine, 28,* 789–803.

Ge, X., Conger, R. D., Cadoret, R. J., Neiderhiser, J. M., Yates, W., Troughton, E., et al. (1996). The developmental interface between nature and nurture: a mutual influence model of child antisocial behavior and parent behavior. *Developmental Psychology, 32,* 574–589.

Harden, K. P., Lynch, S. K., Turkheimer, E., Emery, R. E., D'Onofrio, B. M., Slutske, W. S., et al. (2007a). A behavior genetic investigation of adolescent motherhood and offspring mental health problems. *Journal of Abnormal Psychology, 116,* 667–683.

Harden, K. P., Turkheimer, E., Emery, R. E., D'Onofrio, B. M., Slutske, W. S., Heath, A. C., et al. (2007b). Marital conflict and conduct problems in children of twins. *Child Development, 78,* 1–18.

Jaffee, S. R. & Price, T. S. (2007). Gene-environment correlations: A review of the evidence and implications for prevention of mental illness. *Molecular Psychiatry, 12,* 432–442.

Källén, B., Finnström, O., Nygren, K. G., & Olausson, P. O. (2005). In vitro fertilization in Sweden: maternal characteristics. *Acta Obstetricia et Gynecologica Scandinavica, 84,* 1185–1191.

Kendler, K. S. (2005). "A gene for...": the nature of gene action in psychiatric disorders. *American Journal of Psychiatry, 162,* 1243–1252.

Kendler, K. S., & Baker, J. H. (2007). Genetic influences on measures of the environment: a systematic review. *Psychological Medicine, 37,* 615–626.

Kendler, K. S., & Eaves, L. J. (1986). Models for the joint effect of genotype and environment on liability to psychiatric illness. *American Journal of Psychiatry, 143,* 279–289.

Kim-Cohen, J., Caspi, A., Moffitt, T. E., Harrington, H., Milne, B. J., & Poulton, R. (2003). Prior juvenile diagnoses in adults with mental disorder. *Archives of General Psychiatry, 60,* 709–717.

Knopik, V. S., Heath, A. C., Jacob, T., Slutske, W. S., Bucholz, K. K., Madden, P. A. F., et al. (2006). Maternal alcohol use disorder and offspring ADHD: disentangling genetic and environmental effects using a children-of-twins design. *Psychological Medicine, 36,* 1461–1471.

Krueger, R. F., Moffitt, T. E., Caspi, A., Bleske, A., & Silva, P. A. (1998). Assortative mating for antisocial behavior: developmental and methodological implications. *Behavior Genetics, 28,* 173–186.

Lahey, B. B., D'Onofrio, B. M., & Waldman, I. D. (2009). Using epidemiologic methods to test hypotheses regarding causal influences on child and adolescent mental disorders. *Journal of Child Psychology and Psychiatry, 50,* 53–62.

Leve, L. D., Neiderhiser, J. M., Ge, X., Scaramella, L. V., Conger, R. D., Reid, J. B., et al. (2007). The early growth and development study: a prospective adoption design. *Twin Research and Human Genetics, 10,* 84–95.

Lucht, M., Barnow, S., Schroeder, W., Grabe, H., Finckh, U., John, U., et al. (2006). Negative perceived paternal parenting is associated with dopamine D2 receptor exon 8 and GABA(A) Alpha 6 receptor variants. *American Journal of Medical Genetics Part B: Neuropsychiatric Genetics, 141B,* 167–172.

Lynch, S. K., Turkheimer, E., D'Onofrio, B. M., Mendle, J., & Emery, R. E. (2006). A genetically informed study of the association between harsh punishment and offspring behavioral problems. *Journal of Family Psychology, 20,* 190–198.

Lynskey, M. T., Heath, A. C., Bucholz, K. K., Slutske, W. S., Madden, P. A. F., Nelson, E. C., et al. (2003). Escalation of drug use in early-onset cannabis users vs. co-twin controls. *Journal of the American Medical Association, 289,* 427–433.

Maes, H. H. M., Neale, M. C., Kendler, K. S., Hewitt, J. K., Silberg, J. L., Foley, D. L., et al. (1998). Assortative mating for major psychiatric diagnoses in two population-based samples. *Psychological Medicine, 28,* 1389–1401.

McGue, M., Keyes, M., Sharma, A., Elkins, I., Legrand, L., Johnson, W., et al. (2007). The environments of adopted and non-adopted youth: evidence on range restriction from the Sibling Interaction and Behavior Study. *Behavior Genetics, 37,* 449–462.

Mendle, J., Harden, K. P., Turkheimer, E., Van Hulle, C. A., D'Onofrio, B. M., Brooks-Gunn, J., et al. (2009). Associations between father absence and age of first sexual intercourse. *Child Development, 80,* 1463–1480.

Moffitt, T. E., Caspi, A., & Rutter, M. (2005). Strategy for investigating interactions between measured genes and measured environments. *Archives of General Psychiatry, 62,* 473–481.

Narusyte, J., Andershed, A. K., Neiderheiser, J. M., & Lichtenstein, P. (2007). Aggression as a mediator of genetic contributions to the association between negative parent-child relationships and adolescent antisocial behavior. *European Child & Adolescent Psychiatry, 16,* 128–137.

Narusyte, J., Neiderhiser, J. M., D'Onofrio, B. M., Reiss, D., Spotts, E. L., Ganiban, J., et al. (2008). Testing different types of genotype-environment correlation: An extended Children of Twins Model. *Developmental Psychology, 44,* 1591–1603.

Neiderhiser, J. M., Reiss, D., Lichtenstein, P., Spotts, E. L., & Ganiban, J. (2007). Father-adolescent relationships and the role of genotype-environment correlation. *Journal of Family Psychology, 21,* 560–571.

Neiderhiser, J M., Reiss, D., Pedersen, N. L., Lichtenstein, P., Spotts, E. L., Hansson, K., et al. (2004). Genetic and environmental influences on mothering of adolescents: a comparison of two samples. *Developmental Psychology, 40,* 335–351.

Plomin, R., & Bergeman, C. S. (1991). The nature of nurture: Genetic influence on "environmental" measures. *Behavioral and Brain Sciences, 14,* 373–427.

Plomin, R., DeFries, J. C., & Loehlin, J. C. (1977). Genotype-environment interaction and correlation in the analysis of human behavior. *Psychological Bulletin, 84,* 309–322.

Purcell, S. (2002). Variance components models for gene-environment interaction in twin analysis. *Twin Research, 5,* 554–571.

Rathouz, P. J., van Hulle, C. A., Rodgers, J. L., Waldman, I.D., & Lahey, B. B. (2008). Specifying, testing, and interpretation of gene-by-measured-environment interaction models in the presence of gene-environment correlation. *Behavior Genetics, 38,* 301–315.

Rhee, S. H., & Waldman, I. D. (2002). Genetic and environmental influences on antisocial behavior: a meta-analysis of twin and adoption studies. *Psychological Bulletin, 29,* 490–529.

Rice, F., Harold, G. T., Boivin, J., Hay, D. F., Van den Bree, M., & Thapar, A. (2009). Disentangling prenatal and inherited influences in humans with an experimental design. *Proc. Natl. Acad. Sci., 106,* 2464–2467.

Rutter, M., Pickles, A., Murray, R., & Eaves, L. (2001). Testing hypotheses on specific environmental causal effects on behavior. *Psychological Bulletin, 127,* 291–324.

Schieve, L. A., Cohen, B., Nannini, A., Ferre, C., Reynolds, M. A., Zhang, Z., et al. (2007). A population-based study of maternal and perinatal outcomes associated with assisted reproductive technology in Massachusetts. *Maternal and Child Heath Journal, 11,* 517–525.

Slutske, W. S., D'Onofrio, B. M., Turkheimer, E., Emery, R. E., Harden, K. P., Heath, A. C., et al. (2008). Searching for an environmental effect of parental alcoholism on offspring alcohol use disorder: a genetically informed study of children of alcoholics. *Journal of Abnormal Psychology, 117,* 534–551.

Smith, C. A., & Farrington, D. P. (2004). Continuities in antisocial behavior and parenting across three generations. *Journal of Child Psychology and Psychiatry, 45,* 230–247.

Stoolmiller, M. (1999). Implications of the restricted range of family environments for estimates of heritability and nonshared environment in behavior-genetic adoption studies. *Psychological Bulletin, 125,* 392–409.

Thapar, A., Rice, F., Hay, D., Boivin, J., Langley, K., Van den Bree, M., et al. (2009). Prenatal smoking might not cause attention-deficit/hyperactivity disorder: evidence from a novel design. *Biological Psychiatry, 66,* 722–727.

van IJzendoorn, M. H., Bakermans-Kranenburg, M. J., & Mesman, J. (2008). Dopamine system genes associated with parenting in the context of daily hassles. *Genes, Brain, and Behavior, 7,* 403–10.

Vitaro, F., Brendgen, M., & Arsenault, L. (2009). The discordant MZ-twin method: one step closer to the holy grail of causality. *International Journal of Behavioral Development, 33,* 376–382.

Waldman, I. D. (2007). Gene-environment interactions reexamined: does mother's marital instability interact with the dopamine receptor D2 gene in the etiology of childhood attention deficit/hyperactivity disorder? *Development and Psychopathology, 19,* 1117–1128.

Wichers, M., Kenis, G., Jacobs, N., Mengelers, R., Derom, C., Vlietinck, R., et al. (2008). The BDNF Val(66)Met x 5-HTTLPR x child adversity interaction and depressive symptoms: an attempt at replication. *American Journal of Medical Genetics Part B: Neuropsychiatric Genetics, 147,* 120–123.

Yanovitzky, I. (2006). Sensation seeking and alcohol use by college students: examining multiple pathways of effects. *Journal of Health Communication, 11,* 269–280.

5

The Social Dynamics of the Expression of Genes for Cognitive Ability

WILLIAM T. DICKENS, ERIC TURKHEIMER
AND CHRISTOPHER BEAM

Human gene expression takes place in the context of an extraordinarily complex system—the human body. But for many traits, and particularly behavioral traits, the context is a much larger and more complex system for gene expression—society and our social interactions. How we behave shapes our environment, and our environment, in turn, shapes how we behave. Therefore, if we wish to understand how genes affect behavior, we have to do so in the context of reciprocal effects between individuals and their social milieu. These reciprocal effects produce both active and evocative genotype–environment correlation (see Jaffee, Chapter 4). Cognitive ability is one of the most intensely studied behavioral traits. It may be the best example of the social dynamics of gene expression, because of the importance of environment in shaping cognitive ability and the importance of ability in determining the nature of our environment.

In this chapter, we first describe the *reciprocal effects model* of cognitive ability. We then show how it can explain a number of otherwise puzzling facts about how genes affect cognitive ability. One important implication of the model is that environmental effects decay over time. Using data from two different twin studies, we show evidence of the decay of environmental influence in older adults and children. However, effects decay extremely quickly in children, whereas they decay only very slowly in older adults. The rate of decay in older adults is sufficiently slow that it calls into question some of the important implications of the reciprocal effects model. We thus propose an amendment to the reciprocal effects model that preserves its ability to explain the full range of facts it addresses, as well as the very slow rate of decay of environmental influences in adults.

THE RECIPROCAL EFFECTS MODEL

Dickens and Flynn (2001) proposed the reciprocal effects model of gene expression for cognitive ability as an explanation for what they called "the IQ paradox." There is very strong evidence from twin and adoption studies that a preponderance of the variation in cognitive ability in adults is explained by differences in genetic endowment (Plomin, DeFries, McClearn, & McGuffin, 2001). Nonetheless, there is also powerful evidence from early childhood interventions, from data on changes in average IQ test scores over generations, and from other sources as well, that measures of cognitive ability are malleable to a degree that seems impossible in light of the dominant role of genes in explaining variance in cross-sections of adults. Dickens and Flynn illustrated how both facts could be accommodated in a reciprocal effects model, by way of a sports analogy.

Consider a young man with a small genetic predisposition toward greater height and faster reflexes. When he is young, he is likely to be slightly better than his playmates at basketball. His reflexes make him generally better at sports, and his height provides a particular advantage when it comes to shooting, passing, catching, and rebounding. These advantages by themselves confer only a very modest edge, but they may be enough to make the game more rewarding for him than for the average person, thus encouraging him to play more than do his friends. After a while, he will be significantly better than the average player of his age, making it likely that he will be picked first for teams and perhaps receive more attention from gym teachers. Eventually, he might be put on a school team where he gets exhaustive practice and professional coaching. His basketball ability is now far superior to that of his old playmates. Through a series of feedback loops, his initial minor physical advantage has been multiplied into a huge overall advantage. This process produces both active and evocative gene–environment correlation. When an individual with an advantageous genetic endowment is more likely to *actively* seek out environments that will further increase his ability over time, the result is an *active gene–environment correlation*. When the environment selects that person for special attention, it results in *evocative gene–environment correlation*. In contrast, a child who started life with even a slight predisposition to be pudgy, slow, and small would be very unlikely to enjoy playing basketball, get much practice, or receive coaching. He would therefore be unlikely to improve his skills. Assuming children with a range of experience between these two extremes, a twin or adoption study would attribute a great deal of variation in basketball skills to genetic predisposition. In a sense, it would be correct to do so. However, this most certainly would not mean that the basketball skills of short kids without lightning reflexes could not be enormously improved with practice and coaching. To reiterate, the genes that predispose someone to be good at basketball get a lot of help from correlated environmental causes.

Figure 5.1 illustrates the process that was just described, as it would apply to cognitive ability. Genetic endowment and environment both affect cognitive ability. We split environmental influences into two types: *exogenous environmental influences*, which are those aspects of environment that are not affected by individual ability, and *endogenous environmental influences*, which are those that are affected by individual ability. Note that if improved cognitive ability mainly leads one into environments that demand and support higher levels of ability, and that if the presence in such environments leads to the development and maintenance of higher ability, then the influences of both genetic endowment and exogenous environment may become magnified by the feedback process between cognitive ability and endogenous environment.

Thus, the basketball analogy illustrates how the IQ paradox can be resolved. First, genes tend to get *matched* to complementary environments,[1] giving rise to gene–environment correlation. As a result, the full power of the environment is not reflected in estimates from twin or adoption studies. For example, genes are given credit for any similarity between identical twins raised apart, whereas environment only gets credit for their differences. However, environment, responding to genetically determined differences in ability, may be the immediate cause of much of the similarity of twin pairs. It understates the potency of environment to give it credit only

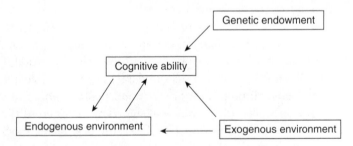

Figure 5.1. **Reciprocal effects model of cognitive ability.**

---

[1] This is not to say that the opposite is never true—that in some places, at some times, effort is made to compensate for less good genes with better environment. All that is necessary for reciprocal effects to enhance genetic predisposition is if, on balance, environment is more complementary than compensatory.

for twin differences. Second, positive feedback loops multiply the effects of genetic differences. Small initial differences are expanded by processes whereby people's initial variations in ability are matched to complementary environments, which in turn cause further divergence in their abilities.

In theory, this same multiplier process could be driven by small environmental differences as well. However, to drive the multiplier to its maximum, the environmental advantage would have to be as constant over time as the genetic difference, because in the absence of the initial advantage a tendency will exist for the whole process to unwind. For example, suppose that midway through high school the basketball enthusiast suffers a leg injury that makes him less steady and offsets his original advantage in height and reflexes. Due to all his practice and learning, he will still be a superior player. But his small decrement in performance could mean discouragement, more bench time, or not making the cut for the varsity team. This could lead to a further deterioration of his skills and further discouragement until he gives up on playing on the team entirely. Although each individual's experience will differ, the theory Dickens and Flynn (2001) laid out would have people with average physical potential reverting to average ability over time on average, unless they had a sustained environmental advantage or disadvantage. The effect is like that of exercise on muscles. Exercise can build muscles up, but we would not expect a program in calisthenics undertaken 15 years ago to have detectible effects today.

Lack of physical exercise leads to muscle atrophy, but does lack of cognitive exercise lead to a loss of cognitive skills? Several pieces of evidence suggest that it does. First, most studies find that the effect of shared family environment on cognitive ability completely disappears in the late teens or early twenties after children leave home.[2] Second, the near universal fade-out of the effects of preschool programs on cognitive ability fits this pattern (Lazar & Darlington 1982). Again, statistically significant persistent effects on behavior and achievement are sometimes found (Barnett 2004), but considering all the evidence together, if there are long-lasting effects on cognitive ability, they must be small to moderate at best. Third, Cardon, Fulker, DeFries, and Plomin (1992) fit a model to data on the cognitive ability of young twins followed through time. They found that the effect of shared environment (those aspects of environment shared by members of the same family) is nearly the same from ages 1 to 7, but the best fitting model had no observation-to-observation correlation for nonshared environment (those

---

[2] See Plomin et. al. (2001) survey pp. 173–177. This is not true of some other traits for which statistically significant effects of family environment can be found in adulthood, including years of school attended (Behrman and Taubman, 1989) and some aspects of personality (Loehlin 1992).

aspects of environment unique to each family member).[3] If environmental effects were cumulative or long-lived, nonshared environmental effects would be correlated over time.

If environmental differences are transient, relative to genetic influences, then they will not get the full benefit of the multiplier effect of the positive feedback loop and may thus explain less variance. But that does not mean that a *sustained* exogenous environmental effect, such as the difference in technology and culture between generations, could not have huge effects. Thus, the IQ paradox is resolved.

Dickens and Flynn (2001) were not the first to propose that active or evocative gene–environment correlations might be important. Both Jensen (1973a, p. 235; 1973b, p. 417) and Scarr (1985) warned against interpreting correlations between environmental factors and IQ as proof of environmental potency, and have emphasized the possibility of reverse causation (Jensen, 1998, pp. 179–181; Scarr, 1992; Scarr & McCartney, 1983). Bell (1968) and Bell and Harper (1977) examined the role that even very young children may play in shaping their environments. Jencks (1972, pp. 66–67; 1980), Jensen (1975), and Goldberger's (1976) response to Jensen (1975) show that all three authors had an understanding of how correlation between genetic endowment and environment could mask environmental effects. The notion of reciprocal causation of IQ and environment is at the core of Bronfenbrenner's bioecological model of development (1989) and kindred work by Ceci (1990), and Bronfenbrenner and Ceci (1994). Kohn and Schooler (1983) and Schooler, Mulatu, and Oates (1999) estimated reciprocal effects for the complexity of work and a measure of individual intellectual flexibility. The latter article also showed that intellectual flexibility is highly correlated with several more standard measures of cognitive ability. Harris (1995, 1999, p. 247) and Harris and Liebert (1991, p. 58) also discuss reciprocal effects and describe feedback loops and vicious cycles in the development of behavioral traits. Turkheimer and Gottesman (1996) and Turkheimer (2004) developed and simulated an even more complex model of reciprocal effects and gene expression showing that high heritability could coexist with large environmental effects. Winship and Korenman (1999) work out an example of environmental multiplier effects due to reciprocal causation between IQ and schooling. What makes Dickens and Flynn's approach unique is that they specified and analyzed a mathematical model of the process that allowed them to see implications of reciprocal effects that others had missed—in particular, the importance of the relative

---

[3] Typically, any variance due to error in measuring the outcome is also included in the share attributed to nonshared environment, as it is typically estimated as the residual variance after all other sources are identified.

persistence of genetic effects and the transience of environmental effects in explaining the IQ paradox.

Dickens and Flynn's model has not gone without criticism. Rowe and Rogers (2002) and Loehlin (2002) present separate criticisms of the work. Dickens and Flynn (2002) responded to them in the same issue of *Psychological Review*. More recently, Mingroni (2007) and Rushton and Jensen (2006) have also written critiques of the work. Dickens (2009) has replied to both.

OTHER APPLICATIONS OF THE RECIPROCAL EFFECTS MODEL OF COGNITIVE ABILITY

In addition to explaining the IQ paradox, the reciprocal effects model sheds light on a number of phenomena and changes the way we understand others. These include changes in the heritability of cognitive ability with age, the decline of the shared environmental component in adulthood, the structure of correlations across different measures of intellectual functioning that give rise to the phenomena of general intelligence (or *g*), and even the black–white IQ gap and Jensen's "Spearman's hypothesis."

UNDERSTANDING HERITABILITY AND ITS CHANGES WITH AGE

In the simple linear decomposition of variance that is the workhorse model of behavior genetics, three causal factors are at work shaping the phenotype: additive genetic effects, shared environmental effects, and nonshared environmental effects. The fraction of variance explained by the first category of effects is called the trait's *heritability* and is denoted as h^2.[4] The fractions of variance explained by shared and nonshared environments are denoted as C^2 and E^2, respectively. Estimates of the heritability of cognitive ability— the fraction of variance explained by differences in genetic endowment— vary from about 40% in children to 60%–80% in adults. The fraction of variance explained by both shared and nonshared environment decline to accommodate the increase in heritability with age, but in most studies it is the decline in the role of shared environment to statistical insignificance that allows most of the expansion (Neisser et al. 1996; Plomin et al. 2001).

The reciprocal effects model changes our interpretation of these numbers and can help us understand how heritability might change with age. In the

---

[4] A lowercase h^2 used to denote narrow heritability (the variance explained by linear genetic effects). The uppercase H^2 is reserved for broad-form heritability that includes nonlinear genetic effects such as dominance and epistasis.

simple linear decomposition of variance model, the impact of a 1 standard deviation increase in genetic endowment on phenotype is equal to the square root of the fraction of variance explained by genetic endowment. For cognitive ability in adults heritability (h^2s) of .6 or more would give a value of around .77 or higher. With both types of environment combined explaining only about 20%–40% of the variance in cognitive ability in adults, a standard deviation increase in the quality of environment would increase cognitive ability by only about .6 standard deviations or less. These numbers are the basis for what has been termed *genetic pessimism*: the view that the large role of genes in shaping cognitive ability makes it unlikely that environmental change could improve cognitive ability much. Moreover, Jensen (1973a,b) pointed out that many aspects of the environment may not be subject to manipulation, so that changing what can be changed would produce still smaller effects. However, viewed through the lens of the reciprocal effects model, these values take on a different meaning.

There is a sense in which the effect size for genetic endowment (h) is still quite meaningful in the context of the reciprocal effects model. It is the reduced form effect of genetic endowment—that is, the sum of all direct *and* indirect effects. As such, it tells us what the expected impact of a 1 standard deviation change in genetic endowment for cognitive ability, in the environment in which it was measured, would be. However, the environmental impacts are highly misleading.

The main problem with the measured effect sizes of shared and non-shared environment is how much of the environment is being considered in computing them. In particular, to the extent that environment responds to genetic endowment, it will not be counted. Return for a moment to the basketball example and imagine what would happen if the only determinant of who played basketball a lot was height—a trait that is over 90% heritable. Assume that no environmental factors, including the family in which one was brought up, had any impact on how much basketball one played. Imagine also that height itself made only a small difference in basketball playing ability, so that no matter how tall one was, if one played basketball a lot one would be very good at it, but not quite as good as someone a bit taller. In that case, the estimated h^2 for basketball ability would be near 1 and that would be, in a sense, appropriate because anyone born with the genes for height would likely be a very good basketball player. However, the estimates for C^2 and E^2 would be near 0 and that obviously paints a very misleading picture of the impact of environment. Were society to decide to have everyone play varsity basketball, everyone would become better at the game. If families differ in their propensity to play basketball, the picture becomes more complicated, but there would still be an important sense in which the full *potential* impact of the environment would not be picked up by estimates of C^2

and E^2 as long as there is a gene–environment correlation induced by reciprocal effects between the phenotype and the environment.

Now, suppose that families do differ in how much they want their children to play basketball. Some parents may go to great lengths to get their children into after-school leagues and practice with them in the evenings and weekends, whereas others view the effort as a bother and want nothing to do with it. Others may want their children to play other sports and could even actively discourage their children from playing basketball. We would then see a substantial C^2 component—the family you came from would help to explain variance across children in how good they were at basketball. However, as children grow into adolescence and then young adulthood, the influence of their family would likely dwindle. More and more, the children may pursue their own interests, which are determined, to some extent, by their individual talents. Children with a natural ability to play basketball will gravitate to it no matter what their family background, and in the end C^2 declines and h^2 grows. But this is not necessarily because genes somehow assert themselves more in the body. In the reciprocal effects model, the decline in C^2 happens because the constraints the family imposes on children weaken, allowing their individual characters to take over in shaping their environments. Dickens and Flynn (2001) describe how this would work in the context of their model, but Jensen (1998, pp. 179–181) and Neisser et al. (1996, p. 86) preceded Dickens and Flynn in suggesting that reciprocal effects could explain the rise in heritability and the decline in C^2. But neither recognized the implications for other phenomena, including the structure of cognitive ability.

CORRELATIONS ACROSS TESTS AND g

The correlation that exists across all measures of cognitive functioning gave rise to the notion that there exists a single powerful cause underlying individual differences in cognitive ability (Spearman, 1927). Differences between people are measured by their scores on a factor derived either as the first principal component of the correlation matrix or by extracting higher-order factors from a hierarchical factor analysis (Carroll, 1993). Scores on this factor are thought to measure general intelligence (g). Many have taken seriously the notion that the pattern of correlations across abilities reflects something about the structure of the brain, but the reciprocal effects model suggests something entirely different. In this view, the pattern of correlation across different abilities may have more to do with what abilities are needed in specific types of circumstances in society rather than with the underlying brain structures (Dickens, 2004, 2009).

Consider again the basketball analogy. Imagine that, for very young children who have never played basketball, all of the many skills that contribute to basketball proficiency are uncorrelated. That would mean that those who can shoot from a distance are no better at lay-ups, passing, or dribbling, on average, than other children. No "basketball g" would exist for these young children. But there certainly would be a "basketball g" by the time these children were teens. Those who played basketball more would get practice in all of the skills and become better at all of them. Those with an advantage in any area of basketball might find the game more enjoyable and play more and get more coaching. Those who had no talent in any area would not. After several years, those with the initial advantage in any skill would have developed their abilities in all areas. Skills would now be correlated, and there would be a "basketball g."

The above example naturally applies to cognitive ability. Those with any advantage in any intellectual ability are likely to find themselves associating with smarter friends, getting assigned to more advanced tracks in school, and choosing leisure activities that challenge many of their cognitive faculties. Although they may get more practice in the activities for which they are innately inclined, activities that draw on *any* cognitive facility tend to put demands on all of them and will therefore develop a broader set of skills than the one that first got them into the activity. Thus, a g factor will tend to emerge even if mental skills are completely independent. Dickens (2010) presents a mathematical model and shows what conditions are necessary to prove this point rigorously.

Reciprocal effects can explain how correlations across tests might arise, but they probably are not the only source of these correlations. It seems very likely that many mental tests are related because they draw on the same mental resources—such as working memory, the language centers in the dominant frontal and temporal lobes, or the spatial capabilities of the parietal lobes. But if reciprocal effects between phenotype and environment play an important role in shaping people's cognitive ability, it is likely that these functional overlaps are not the only source of the correlation. Thus, the g factor would represent an amalgam of causes. The more important are reciprocal effects, the greater the degree to which the pattern would reflect the social reinforcement system rather than the structure of ability in the brain.

RACE, "SPEARMAN'S HYPOTHESIS," AND RECIPROCAL EFFECTS

Jensen (1973a) and Herrnstein and Murray (1994, pp. 298–299) both argued that high heritability estimates make purely environmental explanations for the approximately 1-standard deviation gap between black and white cognitive ability difficult at best. But just as the reciprocal effects model explains how environment can account for the large IQ gains between generations, it also allows for large environmentally induced differences between races. To do so, it hypothesizes that environmental differences within races tend to be transient, whereas average differences between races are longer lasting. As a result, the between-race differences receive the full benefit of individual and social multipliers, whereas the relatively transient within-race differences do not. As a result, large environmentally induced differences between the races are possible.

The reciprocal effects model can also defuse another potent argument often made against an environmental explanation for black–white IQ differences. Jensen pointed out that black–white differences tend to be largest on the most g-loaded tests and therefore argued that this indicates that black–white differences are differences in g. Further, he and Rushton (2005) argue that those subtests with the highest g loadings are also the most heritable. However, Dickens (2010) showed that the reciprocal effects model can explain both findings in a framework in which g simply reflects the pattern of the degree to which different skills are demanded by more cognitively complex environments. If blacks are discriminated against in access to more cognitively demanding environments, then they will have less opportunity to practice the full set of cognitive skills and will not develop them to the same degree as whites, on average. In particular, those skills that are most essential in cognitively complex environments will receive the least practice relative to whites, so that the gap on those skills will be largest. These skills will also tend to be those that are most highly g loaded. In addition, Dickens (2010) demonstrated that, under certain conditions, these skills will tend to be the most highly heritable.

Testing the Transience of Environment

From the previous discussion, it should be clear that the assumption that most within-group environmental differences tend to be transient is very important for the ability of the reciprocal effects model to explain a wide range of phenomena. Testing that assumption is therefore crucial to validating the model. We can go further than just saying that environmental effects should be transient. In the model, the removal of an exogenous environmental advantage or disadvantage causes its influence to unwind over time through a reversing of the process by which any multiplier effects were felt. If the mathematical form of the model in Dickens and Flynn (2001) is taken literally, it implies exponential decay in the effects of exogenous environmental influences over time. Dickens, Turkheimer, and Beam (in preparation) provide a rigorous demonstration of this.

As discussed above, there is considerable evidence that shared environmental influences diminish considerably, and possibly disappear, once children are no longer under the influence of their families. Evidence also suggests that nonshared environment is not persistent in children (Hoekstra, Bartels, & Boomsma, 2007). But there is no evidence on this point in adult samples, and no attempts have been made to test the hypothesis of exponential decay. The notion that nonshared environment should be transient is also antithetical to much psychological theory on the origins of these effects.

Plomin and Daniels' (1987) seminal article on nonshared environment laid out a puzzle. In adults, quantitative genetic analysis of the sources of variance in a wide range of individual characteristics shows a large role for environment, but a minor role for aspects of environment shared by people raised in the same family. Plomin and Daniels suggested that the source of nonshared environmental influences could be differences between the environments of siblings raised in the same family.

This speculation has generated a great deal of research. Turkheimer and Waldron (2000) reviewed this literature and concluded that the effects of environmental differences are small or nonexistent once genetic controls are included in the research designs. This leaves the vast majority of nonshared variance unexplained. An alternative to sibling contrasts is peer group influences. Harris (1995, 1999) proposes that siblings are exposed to different peer groups and that the roles they adopt in these peer groups are the source of nonshared environmental variance.

Turkheimer (1991) identifies a problem with Plomin and Daniels' approach, which is also a problem for Harris's view. If contrasts between siblings are important in explaining traits later in life, similarities should be important as well. Extreme assumptions about the nature of shared versus nonshared environments in childhood are required for one to be important in adulthood but not the other. Turkheimer has suggested that the problem may be that the same objective environment can cause different behavior in people with different genotypes (G × E interactions) or different histories of development (phenotype × environment interaction), and has developed simulation models of such processes (Turkheimer & Gottesman 1996; Turkheimer 2004).

Yet another explanation for the puzzle is proposed by Molenaar et al. (1993), and has been expanded on by Jensen (1997). These authors suggest that nonshared environment may be a misnomer in that the source of variance that has been labeled "nonshared environment" could be random differences in physical development.

Loehlin, Neiderhiser, and Reiss (2003) argue that the nonshared environment is nothing more than random measurement error, or the unreliability of repeated measurements. They suggest that idiosyncratic experiences between siblings and developmental processes altered by chance factors, as well as G × E interactions do not have a strong impact on within-sibling variation.

As noted above, the reciprocal effects model suggests one more solution to this puzzle. If environmental effects are relatively transient, then when the exogenous source of an environmental influence is removed, its effect is expected to wither and eventually disappear completely. The nonshared environment mainly reflects the influence of relatively recent environmental shocks. In addition, to some extent, nonshared environment reflects the unreliability of the measure of the characteristic being studied. Standard measures of reliability do not take into account sources of variation such as significant variation in testing conditions, experimenter error, or coding error. Thus, even if estimates are corrected for reliability using standard procedures, nonshared environmental variance probably still reflects some measurement error.

As demonstrated by Dickens, Turkheimer, and Beam (in preparation) all three qualitatively different hypotheses about the nature of nonshared environments can be nested in the same model. That model implies the following equation explaining the observed covariances for nonshared environment:

$$\text{cov}(N_{it}, N_{it-k}) \cong m_1^k m_u + m_n + m_e i(k = 0) \qquad \text{Eq. 5.1}$$

where N_{it} is the nonshared environmental component of cognitive ability of person i at time t, k represents the time gap between the two periods being considered (for example, if we are examining the covariance of nonshared environment between children when they are 6 and 9, k would equal 3), $i(k = 0)$ is an indicator function equal to 1 when k equals 0 (that is, when we are examining the variance of N) and 0 otherwise, and m_1, m_u, m_n, and m_e are parameters to be estimated.

The last term in Equation 5.1 captures measurement error and any other source of variance in nonshared environment that is completely uncorrelated across time periods. Thus, m_e is equal to that variance and, if variances are standardized, is equal to the variance share attributable to such sources. The first term allows for the exponentially decaying component predicted by the reciprocal effects model. If m_1 is between 0 and 1, and m_u is positive, then the further apart two observations on nonshared environment are in time (the larger is k), the smaller the covariance between them will be. One minus m_1 is interpreted as the rate of exponential decay in the effects of nonshared environmental innovations that are neither due to measurement error nor are permanent. The parameter m_u represents the variance from such sources and is equal to the fraction of variance explained by such innovations if the variance of nonshared environment is standardized. Finally, the second term (m_n) represents unchanging effects on the nonshared environment, such as those that would be predicted in adults by the sibling or peer-group contrast theories or the theory that nonshared environment represents random differences in physical development of the brain. The parameter m_n would represent the share of variance in nonshared environments that could be attributed to such sources if the variance of nonshared environments was standardized.

We took two approaches to exploring the implications of the model. First, we simply estimated the covariances between the nonshared environmental components of the measures of cognitive ability in our data at different points in time. We then fit the more restrictive model that required the covariances to be explained by Equation 5.1. We used standard twin methods to decompose the observed twin covariances at each measurement occasion into genetic (A), shared environmental (C), and nonshared environmental (E) components. All statistical models were estimated with Mplus 5.2 (Muthén & Muthén, 2008) using full information maximum likelihood.

Data

We used two datasets to study the longitudinal structure of the nonshared environment: one consisting of child and adolescent data, and another consisting of data from middle-aged and older adults. The younger dataset comes from twin-pair and cross-twin/cross-age correlation matrices of the Netherlands Twin Register (NTR) data, published in Hoekstra, Bartels, and Boomsma (2007). The adult sample is from the Swedish Adoption/Twin Study of Aging (SATSA; Finkel & Pedersen, 2004). Neither of these datasets is ideal for the task at hand. Rather, they were chosen because they were the only longitudinal datasets easily available when we were preparing this preliminary analysis. For this chapter, we did not have access to the individual data for the NTR and had to estimate our models from published figures. Although access to the individual data would have allowed us to estimate other models, the published data were adequate to construct full information maximum likelihood estimates of the model above. The SATSA is also not ideal as it is limited to a relatively elderly and very homogenous population, which limits potential sources of environmental variation. We are attempting to obtain other more appropriate and complete data, but they were not ready in time for inclusion in this chapter.

THE NETHERLANDS TWIN REGISTER

The NTR is an ongoing longitudinal twin study that focuses on the genetic and environmental contributions to cognitive development and problem behavior across the lifespan. In the current study, we used the published twin-pair and cross-twin/cross-age nonverbal IQ correlation matrices published in Hoekstra et al. (2007) to study the longitudinal structure of the nonshared environment from early childhood to young adulthood. Beginning in 1992, measurements were collected at age 5 (mean age 5.3 years, SD = 0.2; Hoekstra et al., 2007), and subsequently at ages 7, 10, 12, and 18. The study included 89 monozygotic (MZ) twin pairs and 120 dizygotic (DZ) twin pairs. Further details of the sample and measures of intelligence may be found in the original published study (Hoekstra et al., 2007).

SWEDISH ADOPTION TWIN STUDY OF AGING

The SATSA is an ongoing longitudinal twin study devoted to gerontological genetics, which began in 1984 (Finkel & Pedersen, 2004). Across all measurement occasions, the ages of the participants ranged from 42.02 to 96.40 years. Data collection consisted of two components, mail-out questionnaires and in-person testing. The in-person testing component began between 1986 and 1988 and included 162 MZ twins and 276 DZ twins, focusing on measures of cognitive ability and health. Originally, five measurement occasions were scheduled to be collected every 3 years, but because of funding considerations, the fourth measurement was conducted via telephone and only included a brief battery of cognitive measures. The fourth measurement occasion is omitted from the present analyses. Therefore, the data we used consists of four measurement occasions over 13 years on ten cognitive ability subtests, including measures of verbal ability, processing speed, and memory. Data were collected every 3 years for the first 6 years. The final measurement occasion was in 1999, 7 years later.

For scaling and interpretative purposes, we used the first principal component score of the ten subtests as a general measure of cognitive ability. The principal components analysis (PCA) was conducted in R (R-Project, 2009); the first PC accounted for approximately 50% (range 45.55%–51.32%) of the total variance in the ten tests, pooling across time and twins.

Results

THE NETHERLANDS TWIN REGISTER

For our first set of estimates, we decomposed each of the measurements at the five different points in time into genetic, shared environmental, and nonshared environmental components, and allowed the off-diagonal elements of the longitudinal covariance matrices for the A, C, and E terms to be unconstrained.[5] In Table 5.1, the diagonal elements show the fraction of variance explained by each component, whereas the off-diagonal elements show the estimated correlation for each component between each two ages.[6] Genetic sources accounted for the most variance in the data at all ages. The correlations among the A terms were substantial, and did not vary as a function of the interval between measurements. Shared environmental variation accounted for the smallest proportion of the total variation at each measurement occasion. However, the longitudinal covariances among the shared environmental terms are also relatively stable over time, until they drastically decrease around the age of 18. The nonshared environment accounted for a substantial proportion of the variation in the individual measurements. Its covariances across measurement occasions were small and not significantly different from 0, but as the scatterplot in Figure 5.2 shows an inverse relationship exists between the correlations and the interval between the measurements. The reciprocal effects model predicts this relationship, whereas models that imply either an accumulating effect of environment or that equate nonshared environment with either measurement error or very short-lived effects do not. On the other hand, no model predicts the negative correlations seen in Figure 5.2.

Next, we evaluated the fit of the reciprocal effects model to the longitudinal covariance matrix of the nonshared environment component. The reciprocal effects model does a notably worse job of fitting the covariances than does the unrestricted model.[7] The major source of the lack of fit in the structured model appears to be the negative covariances with the nonshared environment term at age 18. As can be seen in Figure 5.2, the exponential decay term in the reciprocal effects model predicts covariances that approach 0 at long intervals, but they cannot become negative.

The left three columns of Table 5.2 present the variance estimates for the constant, measurement error, and occasion-specific components, as

---

[5] The phenotypic variances were constant across time, as the cognitive measures were all standardized. The reciprocal effects model constrained the E variances to be constant over time. We also constrained all the correlations to be less than 1, as without this some were estimated to be slightly greater than 1. Sampling error can lead to this outcome in models such as this if constraints are not imposed.

[6] Although the model was being fit to a correlation matrix, the lack of fit arising from the constraints of the reciprocal effects model resulted in diagonal elements that were slightly less than 1.0.

[7] The unstructured covariance matrix had 11 parameters (the ten covariances among five measurement occasions, plus one variance constrained to be the same at each occasion). The reciprocal effects model structured the longitudinal nonshared covariance matrix with four parameters, leaving 7 df in the comparison. The chi-square change indicated a significant decrement in fit ($\Delta\chi^2=19.32$, df=8, p < .02).

Table 5.1. Estimated Netherlands Twin Registry Intertemporal Correlations
and Variance Shares

Additive Genetic (A)

	Age 5	Age 7	Age 10	Age 12	Age 18
Age 5	**46.86%**				
Age 7	0.78	**62.18%**			
Age 10	0.88	0.98	**57.04%**		
Age 12	0.80	0.87	0.88	**58.18%**	
Age 18	0.94	0.93	0.97	0.93	**65.50%**

Shared Environmental (C)

	Age 5	Age 7	Age 10	Age 12	Age 18
Age 5	**25.23%**				
Age 7	0.98	**10.19%**			
Age 10	0.61	0.56	**15.69%**		
Age 12	0.79	0.66	0.75	**14.97%**	
Age 18	0.15	-0.03	0.53	0.73	**7.48%**

Nonshared Environmental (E)

	Age 5	Age 7	Age 10	Age 12	Age 18
Age 5	**27.91%**				
Age 7	0.03	**27.62%**			
Age 10	0.01	0.20	**27.26%**		
Age 12	-0.10	0.06	0.14	**26.86%**	
Age 18	-0.23	-0.08	-0.09	-0.21	**27.02%**

Variance shares are given on main diagonal, correlations are given on off-diagonals.

well as for m_l, the time-decay parameter. The reciprocal effects model
would be called into question if nonshared environmental effects (NSE)
were either too persistent or nothing but measurement error—that is, if
either m_n or m_e was too large relative to m_u or if the rate of decay $(1 - m_l)$
was too fast or too slow. In fact, the results are quite hospitable to the
reciprocal effects model. The constant component (m_n) accounted for none
of the total nonshared environmental variance. When the model was estimated, the variance of the constant component was always driven to 0, so
it was removed from the model. The measurement error/very-short-term

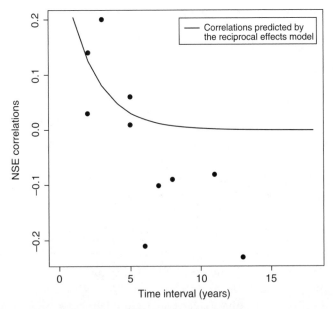

Figure 5.2. **Netherlands Twin Registry.** Scatterplot of the unstructured nonshared environment intercorrelations against the intervals between measurements, overlaid with the model-predicted decay of transient exogenous effects over time. The line labeled "predicted" was generated by multiplying the m_u component by m_l^k, whereby $k > 0$ represents the annual interval between measurements, for increasing values of k, and then dividing the product by the total NSE variance. The steep decline observed in the correlations is inconsistent with the prediction from the reciprocal effects model.

effects term (m_e) accounted for approximately 67%[8] of the nonshared environmental variance at each measurement, suggesting that a major portion of transient environmental effects is unique to the moment. The variance component subject to exponential decay (m_u) accounted for about 33% of the nonshared variation, but estimate was not very precise as the 90% confidence interval (CI) includes 0. The decay parameter (m_l) was estimated to be .63 per year, with a 95% CI ranging from .28 to .97, thus providing evidence of positive correlations between the nonshared components at different points in time. As this parameter can be interpreted as 1 minus the rate of decay of the time-limited component of NSE, our

---

[8] The values presented in the table are the variances of the terms. Since the variance of the non-shared environmental influences is less than one, the variances do not sum to one and cannot be interpreted as variance shares. To get the variance share for each component we divide the variance of the component by the sum of the variance accounted for by all components of non-shared environment.

Table 5.2. Parameter Estimates for the Reciprocal Effects Model

| | Netherlands Twin Registry | | | Swedish Adoption/Twin Study of Aging | | |
	Estimate	SE	p-value	Estimate	SE	p-value
m_n	NA	NA	NA	0.00	0.00	0.001
m_e	0.19	0.07	0.010	0.23	0.04	0.001
m_u	0.09	0.07	0.215	0.49	0.08	0.001
m_l	0.63	0.18	0.001	0.97	0.02	0.001

m_n, constant NSE component; m_e, unreliability of the NSE (or measurement error) component; m_u, importance of exponential decay terms contribution to explaining NSE variance ($k = 0$); m_l, 1 minus the rate of exponential decay in the covariance ($k > 0$).

estimates imply that, in children, the effects of nonshared environmental effects that are not measurement error have a half-life of about 1.5 years. Thus, although our estimates are not precise enough to rule out other models, they fit best with the predictions of the reciprocal effects model. The largest fraction of the variance is attributable to measurement error or factors whose influence disappears completely between measurements. There is no evidence of any nondecaying error component. The component that is subject to decay accounts for a substantial fraction of the variance and has a very plausible half-life of about 1.5 years.

SWEDISH ADOPTION TWIN STUDY OF AGING

We applied the same analysis to the elderly sample. First, we decomposed each of the measurements at each of the 4 years into genetic and nonshared environmental components, and allowed the longitudinal covariance matrices for the A and E terms to be unconstrained. No evidence for a significant role for shared environment was found in preliminary work, so it was dropped from the model. The model without shared environment fit the data well, with a comparative fit index of .98 and a root mean square error of approximation of .05 (90% CI: .03–.07). As shown in Table 5.3, genetic sources accounted for approximately 86% of the variance. The correlations across time among the A terms were substantial, and did not vary as a function of the interval between measurements. The nonshared environment accounted for a small proportion (approximately 14%) of the variation at the individual measurements. In contrast to the NTR data, the nonshared environmental longitudinal covariances were positive, sizeable, and significantly different from 0. The scatterplot in Figure 5.3 shows that the longitudinal correlations have an inverse relationship with the interval between test occasions. Although the rate of decay of the effects of nonshared

Table 5.3. Estimated Swedish Adoption/Twin Study of Aging Intertemporal Correlations and Variance Shares

Additive Genetic (A)

	1986	1989	1992	1999
1986	85.27%			
1989	1.00	86.08%		
1992	0.97	1.00	87.49%	
1999	0.97	1.05	1.01	86.26%

Nonshared Environmental (E)

	1986	1989	1992	1999
1986	14.73%			
1989	0.61	13.92%		
1992	0.67	0.68	12.51%	
1999	0.44	0.68	0.46	13.74%

Variance shares are given on main diagonal, correlations are given on off-diagonals. No evidence of shared environmental effects were found in any specification, so they were excluded from the modeling.

environmental influences is much slower in older adults than in children, correlations between measurements of longer intervals are smaller. Once again, the systematic pattern of interrelations conforms to the expectations of the reciprocal effects model.

We evaluated the fit of the reciprocal effects model to the covariance matrix for the nonshared environment component. The restricted model fit nearly as well as the unrestricted one ($\Delta\chi^2 = 5.31$, df = 3, $p = .15$), so we cannot reject the hypothesis that the reciprocal effects model reproduces the interrelations of the observed data as well as the unstructured model.[9]

The right three columns of Table 5.2 present the variance estimates for the constant and measurement error (or occasion-specific) components, as well as the time-limited environmental decay parameter. Again, the constant component variance (m_n) was driven to 0 consistently and was removed from the model. The measurement error/very-short-lived effects term (m_e) accounted for approximately 32% of the nonshared environmental variance

---

[9] The unstructured covariance matrix for nonshared environment had seven parameters (the six covariances among four measurement occasions, plus one variance constrained to be the same at each occasion). The reciprocal effects model structured the longitudinal nonshared covariance matrix with four parameters, leaving 3 df in the comparison.

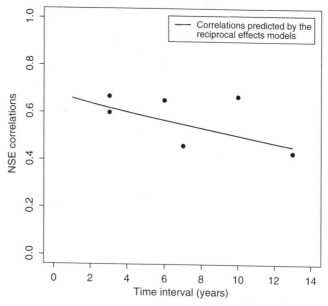

Figure 5.3. **Swedish Adoption/Twin Study of Aging.** Scatterplot of the unstructured nonshared environment intercorrelations against the intervals between measurements, overlaid with the model-predicted decay of transient exogenous effects over time. The predicted line was generated as per Figure 5.2.

at each point in time and has a 99% CI that is bounded away from 0. The component that is subject to exponential decay (m_u) accounted for 48% of the nonshared variation and also has a 99% CI bounded away from 0. The time-limited decay parameter was equal to .97 per year, with a 90% CI running from .94 to slightly less than 1. This coefficient is interpreted as 1 minus the annual rate of decay of nonshared environmental effects. Thus, in late adulthood, the rate at which environmental influences decay is slower than in children and adolescents, a difference that is addressed below.

Implications for the Reciprocal Effects Model

With this very preliminary data analysis, we have attempted to provide evidence for one of the primary assumptions of the reciprocal effects model— that the effects of environment are not only not permanent, but sufficiently transient that they would not last long enough to get full benefit from multiplier effects due to the reciprocal effects between ability and environment. The results are mixed. On the one hand, in both datasets, we found no evidence that a significant fraction of the variance in nonshared environment is explained by a permanent component, as would be predicted by a number of theories.[10] Also, in both datasets, we find that point estimates suggest that a transient exogenous environmental component that is subject to exponential decay plays an important role, as predicted by the reciprocal effects theory. At least for children, nonshared environment is nothing if not transient. According to the point estimates, about half the variance is due to a component that is entirely gone in a year, whereas the other half is due to a component whose influence decays at the rate of about 37% a year. Thus, for children, transient environmental effects act quickly, persist for a short while, and then disappear completely. Although these point estimates are consistent with the reciprocal effects model, we could not rule out the possibility that all of the effects of innovations in the nonshared environment were gone within a year, and that would be inconsistent with the reciprocal effects model.

For the adults in the SATSA, environmental effects are very long lasting. Even though our estimates suggest that they decay over time, the half-life of an environmental influence on cognitive ability estimated in the SATSA is 23 years. The slow decay among older adults suggests that there ought to be enough time for the reciprocal effects multiplier to amplify exogenous environmental effects. But if that were the case, we would expect that both shared and nonshared environment would play a much bigger role in explaining variance in adults in cognitive ability. Yet the nonshared environment explains only 14% of the variance in cognitive ability in these data, and shared environment apparently explains none. If a quarter of the impact of an environmental innovation from 46 years ago is still present, we should be able to see some impact from shared family environment even in these data. However, we observed none.

CONCLUSION

Do these results breathe new life into the "IQ Paradox"? Possibly. As noted above, these are not the ideal data for examining this question. In particular, the SATSA is a sample with particularly homogenous environments. Sweden is a very egalitarian society, and volunteers for intensive scientific studies tend to be middle class. It is not surprising then that the fraction of variance attributed to environment is very small in this study—far less than typical estimates for adult populations in the United States, which are typically at

---

[10] It is also worth noting that the NTR results provide no support for a view that physiological changes are accumulating over time. If that were the case, we would expect that the correlations in nonshared environment between measurements would grow as the children age. So, for example, we would expect that the correlations between observations at ages 18 and 12 would be notably larger than those between 12 and 5. Nothing in the data supports this.

least twice as large (Plomin et al., 2001). Therefore, it is possible that SATSA understates the persistence of nonshared environmental effects.

But what if the persistence of environmental effects for Swedish adults is typical? Does that mean that the reciprocal effects model could not explain the phenomena that it was designed to explain? Not necessarily. Whether a half-life of 23 years for the effects of an exogenous environmental effect is sufficiently long for the full multiplier effect to develop depends on how quickly the feedback loop between phenotype ability and environment proceeds. If environment adapts quickly to ability, and ability adapts quickly to environment, then multipliers should work themselves out quickly. But what if the multiplier process slows down as people age? In particular, suppose that changes in environment happen less frequently with age. Then, an environmental shock could persist for a long time without producing the multiple rounds of feedback necessary to produce a full multiplier effect. Although variation with age in the rate at which feedback takes place was not part of the original Dickens and Flynn model, it would be relatively easy to incorporate.

It makes sense that this might happen. When children are growing up, their cognitive ability is constantly improving. As a result, the games that suited them just months ago now seem too simple. The TV programs they enjoy turn over quickly. Once they enter school, they are faced with a constantly changing curriculum. Friends come and go, interests come and go, and changes happen at a breathtaking pace. Typically, this is not the case for adults. Sometime between the end of school and the late 30s, individuals settle into a groove that does not allow much change. People tend to stay with the same job, keep the same hobbies, and settle with a spouse. Once the developmental process runs its course, the need for environmental change, and consequently the opportunities to match environment with current ability, diminishes. If this change does take place between the time people leave home and their early 30s, it could explain why there are few, if any, vestiges of shared family environment in adulthood. There may be adequate distance between the time when family exerts a major influence on one's environment, and when the process finally slows to that of a typical adult, to allow the home influences to fade.

If this is correct, it suggests an interesting possibility. For older adults, it may be very difficult to change their cognitive ability because they are locked within a web of environments that do not respond to small changes in ability. But, for young adults who have not yet reached the point in which their lives have stabilized, there may be a "critical period" in late adolescence and early adulthood in which environmental impacts on cognitive ability get "locked in."

It remains to be shown that a formal model could be developed along these lines, and if it could be, if further analysis would validate it. Obviously, data that cover only childhood/adolescence and middle/late adulthood are inadequate. The ideal dataset would have rich cognitive measures on a genetically informed representative longitudinal sample that covered at least from the teen years through the 30s. Such data might allow us to gain considerably more insight into how genes are expressed in the interplay of environment and phenotype ability.

REFERENCES

Barnett, W. S. (2004). Maximizing returns from prekindergarten education. In *Federal Reserve Bank of Cleveland research conference: education and economic development* (pp. 5–18). Cleveland, OH: Federal Reserve Bank of Cleveland.

Behrman, J. R., & Taubman, P. (1989). Is schooling "mostly in the genes"? Nature-nurture decomposition using data on relatives. *The Journal of Political Economy*, 97, 1425–1446.

Bell, R. Q. (1968). A reinterpretation of the direction of effects in studies of socialization. *Psychological Review*, 75, 81–95.

Bell, R. Q., & Harper, L. (1977). *Child effects on adults*. Lincoln: University of Nebraska Press.

Bronfenbrenner, U. (1989). Ecological system theory. In R. Vasta (Ed.), *Six theories of child development: revised formulations and current issues* (pp. 185–246). Greenwich, CT: JAI Press.

Bronfenbrenner, U., & Ceci, S. J. (1994). Nature-nurture reconceptualized in developmental perspective: a bioecological model. *Psychological Review, 101,* 568–586.

Cardon, L. R., Fulker, D. W., DeFries, J. C., & Plomin, R. (1992). Continuity and change in general cognitive ability from 1 to 7 years of age. *Developmental Psychology, 28,* 64–73.

Carroll, J. B. (1993). *Human cognitive abilities: a survey of factor-analytical studies*. New York: Cambridge University Press.

Ceci, S. J. (1990). *On intelligence–more or less: a bio-ecological treatise on intellectual development*. Englewood Cliffs, NJ: Prentice Hall.

Dickens, W. T. (2004). The malleability of cognitive ability in adults. A proposal to the National Institute of Ageing, Brookings. Available from author on request.

Dickens, W. T. (2009). A response to recent critics of Dickens and Flynn (2001), http://www.brookings.edu/ /media/Files/rc/articles/2001/0401IQ_critics_dickens/0401IQ_critics_dickens.pdf

Dickens, W. T. (2010). What is g? Brookings mimeo. Available at: http://www.brookings.edu/~/media/Files/rc/papers/2007/0503education_dickens/20070503.pdf

Dickens, W. T., & Flynn, J. R. (2001). Heritability estimates versus large environmental effects: the IQ paradox revisited. *Psychological Review, 108,* 346–369.

Dickens, W.T., & Flynn, J.R. (2002). The IQ paradox is still resolved: Reply to Loehlin (2002) and Rowe and Rodgers (2002). *Psychological Review, 109,* 764–771 .

Finkel, D., & Pedersen, N. L. (2004). Processing speed and longitudinal trajectories of change for cognitive abilities: the Swedish adoption/twin study of aging. *Aging, Neuropsychology, and Cognition, 11,* 325–345.

Goldberger, A. S. (1976). On Jensen's method for twins. *Educational Psychologist, 12,* 79–82.

Harris, J. R. (1995). Where is the child's environment? A group socialization theory of development. *Psychological Review, 102,* 458–489.

Harris, J. R. (1999). *The nurture assumption.* New York: Simon and Schuster.

Harris, J. R., & Liebert, R. M. (1991). *The child: a contemporary view of development* (3rd ed.). Englewood Cliffs, NJ: Prentice Hall.

Herrnstein, R. J., & Murray, C. (1994). *The bell curve: intelligence and class structure in American life.* New York: Simon and Schuster.

Hoekstra, R. A., Bartels, D., & Boomsma, D. I. (2007). Longitudinal genetic study of verbal and nonverbal IQ from early childhood to young adulthood. *Learning and Individual Differences, 17,* 97–114.

Jencks, C. (1972). *Inequality: a reassessment of the effects of family and schooling in America.* New York: Basic Books.

Jencks, C. (1980). Heredity, environment, and public policy reconsidered. *American Sociological Review, 45,* 723–736.

Jensen, A. R. (1973a). *Educability and group differences.* New York: Harper and Row.

Jensen, A. R. (1973b). *Educational differences.* London: Methuen.

Jensen, A. R. (1975). The meaning of heritability in the behavioral sciences. *Educational Psychologist, 11,* 171–183.

Jensen, A. R. (1997). Adoption data and two g-related hypotheses. *Intelligence, 25,* 1–6.

Jensen, A. R. (1998). *The g factor: the science of mental ability.* Westport, CT: Praeger.

Kohn, M. L., & Schooler, C. (1983). Work and personality: an inquiry into the impact of social stratification. Norwood, NJ: Ablex.

Lazar, I., & Darlington, R. (1982). Lasting effects of early education: a report from the consortium for longitudinal studies. *Monographs of the Society for Research in Child Development, 47,* 2–3.

Loehlin, J. C. (1992). *Genes and environment in personality development.* Newbury Park, CA: Sage.

Loehlin, J. C. (2002). The IQ paradox: Resolved? Still an open question. *Psychological Review, 109,* 754–758.

Loehlin, J. C., Neiderhiser, J. M., & Reiss, D. (2003). Genetic and environmental components of adolescent adjustment and parental behavior: a multivariate analysis. *Child Development, 76,* 1104–1115.

Mingroni, M. A. (2007). Resolving the IQ Paradox: Heterosis as a Cause of the Flynn Effect and Other Trends. *Psychological Review, 114* (3), 806–829.

Molenaar, P. C. M., Boomsma, D. I., Dolan, C. V. (1993). A third source of developmental differences. *Behavior Genetics, 23,* 519–524.

Muthén, L. K., & Muthén, B. O. (2008). MPlus (5.2) [Computer software]. Los Angeles, CA: Muthén & Muthén.

Neisser, U., Boodoo, G., Bouchard, T. J., Jr., Boykin, A. W., Brody, N., Ceci, S. J., et al. (1996). Intelligence: knowns and unknowns. *American Psychologist, 51,* 77–101.

Plomin, R., DeFries, J. C., McClearn, G. E., & McGuffin, P. (2001). *Behavioral genetics* (4th ed.). New York: Worth Publishers.

Plomin, R., & Daniels, D. (1987). Why are children in the same family so different from one another? *Behavioral and Brain Sciences, 14,* 373–427.

R Development Core Team (2009). R: A language and environment for statistical computing. R Foundation for Statistical Computing, Vienna, Austria. ISBN 3-900051-07-0, URL http://www.R-project.org.

Rowe, D. C., & Rodgers, J. L. (2002). Expanding variance and the case of historical changes in IQ means: a critique of Dickens and Flynn (2001). *Psychological Review, 109,* 759–763.

Rushton, J. P., & Jensen, A. R. (2005). Thirty years of research on race differences in cognitive ability. *Psychology, Public Policy, and Law, 11,* 235–294.

Rushton, J.P., & Jensen, A.R. (2006). The totality of available evidence shows the race IQ gap still remains. *Psychological Science, 17,* 921–922.

Scarr, S. (1985). Constructing psychology: making facts and fables for our times. *American Psychologist, 40,* 499–512.

Scarr, S. (1992). Developmental theories for the 1990s: development and individual differences. *Child Development, 63,* 1–19.

Scarr, S., & McCartney, K. (1983). How people make their own environments: a theory of genotype → environment effects. *Child Development, 54,* 424–435.

Schooler, C., Mulatu, M. S., & Dates, G. (1999). The continuing effects of substantively complex work on the intellectual functioning of older workers. *Psychology and Aging, 14,* 483–506.

Spearman, C. (1927). *The abilities of man: their nature and measurement.* New York: Macmillan.

Turkheimer, E. (1991). Individual and group differences in adoption studies of IQ. *Psychological Bulletin, 110,* 392–405.

Turkheimer, E. (2004). Spinach and ice cream: why social science is so difficult. In L. DiLalla (Ed.), *Behavior genetics principles: perspectives in development, personality, and psychopathology* (pp. 161–89). Washington, DC: American Psychological Association.

Turkheimer, E., & Gottesman, I. I. (1996). Simulating the dynamics of genes and environment in development. *Development and Psychopathology, 8,* 667–677.

Turkheimer, E., & Waldron, M. (2000). Nonshared environment: a theoretical, methodological, and quantitative review. *Psychological Bulletin, 126,* 78–108.

Winship, C., & Korenman, S. (1999). Economic success and the evolution of schooling and mental ability. In S. Mayer & P. E. Peterson (Eds.), *Earning & Learning: how Schools Matter* (pp. 49–78). Washington, DC: Brookings.

6

Family Relationship Influences on Development: What Can We Learn from Genetic Research?

BRIANA HORWITZ, KRISTINE MARCEAU,
AND JENAE M. NEIDERHISER

There is a well-established literature documenting that the emotional quality of family relationships, including the parent–child relationship (e.g., parents' warmth and negativity toward their children) and the marital relationship (e.g., marital satisfaction and conflict) are linked to the adjustment of individual family members, including children and spouses. For example, one of the most studied family relationships in relation to child development is that between parents and children. Numerous studies have indicated that the quality of the parent–child relationship is associated with child and adolescent adjustment, with more warmth and support linked to better child outcomes and harsh and negative parenting with the development of behavioral and emotional problems (e.g., Demo & Cox, 2004; Fletcher, Steinberg, & Williams-Wheeler, 2004). Marital conflict has also been tied to an increase in the likelihood of the development of adjustment problems in children (e.g., Cummings & Davies, 2002), and the quality of the marital relationship has also been strongly tied to spouses' mental health (e.g., Proulx, Helms, & Buehler, 2007). What is less clear from this work is the extent to which family relationship quality is linked to family members' adjustment through both genetic and environmental pathways, although a number of studies have attempted to specify the mechanisms involved.

Numerous studies using behavior genetic samples, in which family members vary in their genetic relatedness (e.g., twin and sibling pairs), have examined the degree to which variability in family relationship quality, including the quality of parenting behaviors and marital quality arises from genetic, shared environmental (nongenetic factors contributing to twin or sibling similarity) and nonshared environmental influences (nongenetic factors contributing to twin or sibling dissimilarity). Within this research, studies have demonstrated that genetic influences contribute to the quality of family relationships, including parent–child and marital relationships (e.g., Kendler & Baker, 2007; Ulbricht & Neiderhiser, 2009). It makes sense to interpret these genetic influences as genotype–environment correlation (rGE), a process by which individuals' heritable characteristics are associated with their exposure to certain environments (e.g., Moffitt, 2005; Plomin, DeFries, & Loehlin, 1977; Scarr & McCartney, 1983). For example, a child's heritable characteristics may influence a parent's responses to the child, thereby affecting the quality of the parent's behaviors toward the child. A similar example can be applied to the marital relationship, such that spouses' heritable characteristics may impact the behaviors of their partners, and, therefore, the spouses' heritable characteristics are influencing the quality of the marriage. As such, behavior genetic research provides an important tool for illustrating how children's and spouses' heritable characteristics can influence the emotional quality of their family relationships.

In this chapter, we review behavior genetic studies that have examined how family relationship quality, including parenting behaviors and marital quality, is shaped and how the quality of these family relationships is linked to the adjustment of children and spouses. Specifically, we will focus on clarifying how forms of gene–environment interplay, including rGE and genotype × environment interaction (G × E), are involved in shaping these associations. Accumulating evidence suggests that rGE contributes to family relationship quality and, importantly, associations among family relationship quality and children's and spouses' adjustment; thus, we review this work first. Next, we review studies focused on G × E, in which family relationship quality is the moderating variable. This is followed by a review of studies that have considered the role of both rGE and G × E. In closing, we discuss the potential intervention implications of this work and propose a strategy for continuing to delineate how genetic and environmental influences operate synergistically to shape links between family relationship quality and the adjustment of children and spouses.

GENOTYPE–ENVIRONMENT CORRELATION (rGE)

This purpose of this section is to review studies examining how genetic and environmental factors contribute to family relationships and their associations to individual adjustment. We first review an emerging literature that has employed an array of creative and advanced methodological approaches to understand the nature of the genetic influences on parenting and on the association between parenting and child adjustment by disentangling passive from evocative *r*GE. Briefly, passive *r*GE is the result of parents and children sharing genes, and these genes are correlated with their environment. For example, negative emotionality may be passed on to the child both by genetic transmission of a parent's heritable negative emotionality and because the parent who is high in this characteristic may also tend to behave more negatively toward his or her child. This correlation between the parent's genes (and possibly also the child's) and the environment that the parent provides is *passive* *r*GE. On the other hand, *evocative* *r*GE arises when an individual's heritable characteristics evoke a particular response from his or her environment. For example, a child high on emotional reactivity may evoke more negative responses from a parent than would a child with a more easy-going temperament. Next, we review work that has assessed the marital relationship and its association with children's or spouses' adjustment. Studies examining how the marital relationship is tied to child outcomes have focused on delineating the direct effects of marital difficulties on child adjustment due to passive *r*GE, as in genetic influences that are common to marital difficulties and child adjustment. Work investigating how the marital relationship is connected to spouse outcomes has distinguished the environmental effects of the marital relationship on spouses' adjustment from *r*GE. In the case of the marital relationship, *r*GE may encompass evocative *r*GE, in which spouses' heritable characteristics evoke responses from their partners that, in turn, influence the quality of the marriage and/or assortative mating, in which individuals' selection of mates with similar characteristics affect the quality of the marriage and their adjustment.

Parenting

Numerous studies have reported genetic and environmental influences on parenting using child-based designs, in which the children are twins or siblings of varying genetic relatedness who report on the quality of their parents' behaviors (e.g., McGue, Elkins, Walden, & Iacono, 2005), and using parent-based designs, in which parents are twins or siblings of varying genetic relatedness who report on the quality of their own behaviors toward their children (e.g., Kendler, 1996). The presence of genetic influences on parenting behaviors in child- and parent-based designs are reflective of the contribution of evocative and passive *r*GE, respectively. Figure 6.1, based loosely on the determinants of parenting described by Belsky (1984),

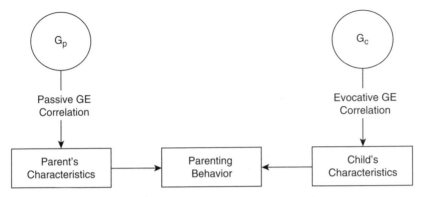

Figure 6.1. **Passive *r*GE and evocative *r*GE mechanisms of parenting.** G_p represents the parents' genotype, and the degree to which genotype influences their parenting is indicated by the path from G_p to parenting characteristics. Any influence of parental genotype to parenting behavior indicates passive *r*GE. G_c represents the child's genotype, and the influence of the child's genotype on parenting behavior is indicated by the path from G_c to parenting. Evocative *r*GE occurs when the path from G_c to parenting is significantly different from 0.

illustrates how evocative and passive *r*GE can operate to influence parenting behaviors.

A recent set of studies, comparing results from compatible child- and parent-based samples, attempted to specify the roles of both passive and evocative *r*GE on parental positivity, monitoring, negativity, and control (Neiderhiser, Reiss, Lichtenstein, Spotts, & Ganiban, 2007; Neiderhiser et al., 2004). Findings suggested that mothers' positive parenting and monitoring was, in part, due to their own heritable characteristics, but control and negative parenting behaviors were better explained as a response to their adolescents' heritable characteristics than by mothers' genes (Neiderhiser et al., 2004). There were some notable differences in mothers' and fathers' positivity and negativity (Neiderhiser et al., 2007). In particular, whereas mothers' positivity was primarily ascribed to passive *r*GE, fathers perceived themselves as parenting more positively in response to their adolescents' heritable characteristics, as in evocative *r*GE. The converse was true for negativity, with evidence of both passive and evocative *r*GE for a father's reports of his own negativity and primarily evocative *r*GE for a mother's negativity. Finally, a notable similarity is that passive *r*GE was detected for both maternal and paternal monitoring. That is, parents who monitored their children were also likely to have children who were responsive to their monitoring. This is of interest because parents' reports of parental monitoring have been critiqued for being less useful than child reports due to social desirability (e.g., Dishion & McMahon, 1998). Because both parent and adolescent reports of parental monitoring suggest passive rGE, the likelihood that social desirability is driving these findings is small. In addition, the finding of passive rGE in this behavior suggests that some of the benefits of parental monitoring may derive from genetic similarities between parents and children, rather than from the direct effects of monitoring. In sum, the work by Neiderhiser and colleagues (2004, 2007) has illustrated the role of passive and evocative rGE for different dimensions of parenting and for different reporters.

Parenting and Child Adjustment

Other work has used multivariate quantitative genetic models to investigate the degree to which the association between parenting and child adjustment can be explained by common genetic and environmental sources (see Chapter 4 by Jaffee, Fig. 4.5). If genetic influences are found to explain the correlation between parenting and child adjustment in a child-twin design, this indicates that the heritable characteristics influencing how a parent behaves toward his or her child also influences the child's adjustment. This is best explained by evocative rGE. Specifically, the child's behavior evokes a response from his or her parents and this is due to genetic influences on the child. On the other hand, if shared environmental influences account for the association between parenting and child adjustment, this illustrates that parenting behaviors that are common to each child in the same family also influence similarities in the adjustment of each child. Finding evidence for shared environmental influences on these associations is consistent with the assumptions of much research on the impact of parenting on child functioning. These shared environmental influences could be due to things like similarity in parenting and socioeconomic status. Finally, evidence of nonshared environmental influences on the association between parenting and child adjustment indicates that parenting behaviors unique to each child contribute to differences in the adjustment of each child. Nonshared environmental influences that are correlated for parenting and child adjustment may be due to things like differential parenting, child age or sex differences, and/or idiosyncratic factors in the child (e.g., an illness or injury). Studies that have utilized multivariate genetic models have demonstrated that genetic influences often account for a large portion of the overlap between parental warmth and positive child adjustment (Reiss, Neiderhiser, Cederblad, & Lichtenstein, 2000), as well as for parental negativity and corporal punishment and child adjustment problems (e.g., Burt, Krueger, McGue, & Iacono, 2003; Jaffee et al., 2004; Reiss et al., 2000). Therefore, these findings suggest that children's heritable characteristics play an important role in influencing the quality of the parenting they receive and their own adjustment, via evocative rGE.

To better clarify how both parents and children contribute to parenting behaviors and child adjustment, researchers have taken advantage of longitudinal biometric models using child-based designs. Longitudinal biometric models provide a creative approach to examining the bidirectional effects of parenting and child development, in which parenting and child adjustment both influence, and are influenced by, each other. If parenting predicts child adjustment for genetic reasons, this suggests that unmeasured children's heritable characteristics are influencing parenting, via evocative rGE and their own adjustment, in turn. If child adjustment predicts parenting for genetic reasons, this implicates the measured dimension of child adjustment as a heritable characteristic involved in evoking parenting over time. Furthermore, if parenting predicts child adjustment for environmental reasons, this suggests that parenting has a direct environmental influence on child development. Finally, if child adjustment predicts parenting for environmental reasons, this suggests that unmeasured environmental influences on child adjustment are shaping parenting behaviors over time. One such study indicated a bidirectional association between parent–child conflict and child externalizing behaviors over time, such that these variables independently predicted one another 3 years later (Burt, McGue, Krueger, & Iacono, 2005). In addition, parent–child conflict exerted a direct environmental influence on child externalizing behaviors, whereas child externalizing behaviors influenced parent–child conflict for genetic reasons. Likewise, another study indicated that parental negativity exerted an environmental influence on children's antisocial behavior, and genetic influences on children's antisocial behavior evoked parental negativity (Larsson, Viding, Rijsdijk, & Plomin, 2008). Considered together, findings from these longitudinal models illustrate that children's heritable characteristics influence parents' behaviors over time, via evocative rGE. These studies further implicate parental negativity as a direct environmental influence on child development. Of note, however, neither the bivariate biometric models nor the longitudinal biometric models that have been employed can test for passive rGE, as they have relied on child-based designs.

Another powerful approach for examining how parenting behaviors are linked to child adjustment is within the framework of the children-of-twins (CoT) design (Rutter, Pickles, Murray, & Eaves, 2001; Silberg & Eaves, 2004) (see also Chapters 4 by Jaffee and 16 by Kendler). The CoT design compares differentially exposed offspring of monozygotic (MZ) and dizygotic (DZ) twins. The offspring of MZ pairs share 50% of their genes with their parents and 50% with their parents' co-twin (the offsprings' aunt/uncle).

On the other hand, the offspring of DZ pairs share 50% of their genes with their parents, but 25% with their parents' co-twin (like any niece/nephew and aunt/uncle pair). This design can thus distinguish direct environmental influences of parenting from shared genotypic factors (indicating passive rGE), and those unmeasured shared environmental factors making the parent co-twins similar (D'Onofrio et al., 2003; Rutter et al., 2001). Within one study, results revealed that offspring of twins who were exposed to harsher parenting had greater antisocial problems compared to their cousins (genetic half-twins) who were exposed to less harsh parenting (Lynch et al., 2006). Thus, findings provide strong evidence for a direct causal link between harsh parenting and offsprings' problem behaviors. Findings also underscore the value of the CoT model in disentangling mechanisms underlying the association between parenting and offspring adjustment. However, in the CoT design, the power to detect evocative rGE is relatively weak because the offspring cousins share 25% and 12.5% of their genes, respectively. Given such a modest variation in the cousins' degree of genetic relatedness, the effects of children's genotype influencing parenting are difficult to detect with reasonable sample sizes.

More recently, a novel extension of CoT, the extended CoT (ECoT) model, combines multiple samples to facilitate the estimation of both evocative and passive rGE (Narusyte et al., 2008). The ECoT extends the CoT design to incorporate children who are twins within a single nested model. In this way, the ECoT design is able to capitalize on the full range of genetic variance for estimating the effects of parents' genes (using the twin parents), as well as for estimating the effects of the children's genes (using the child-based twin sample, as well as the cousin pairs from the parent-based twin sample). A strength of the ECoT design is its ability to distinguish direct environmental effects from evocative and passive rGE. The ECoT model includes two phenotypes: one describing parenting and one reflecting child adjustment. Both parenting and child adjustment are influenced by genetic, shared environmental and nonshared environmental influences. Because samples of twins who are parents, along with their children, and of children who are twins, along with their parent(s) are used in this design, reciprocal paths between parenting and child adjustment can be examined, and the genetic and environmental contributions to these reciprocal effects can also be estimated.

Narusyte and colleagues (2008) applied this model to detect the direction of influence between mothers' emotional overinvolvement and child internalizing behaviors. Findings indicated that child internalizing problems were influenced both by genetic effects contributed by the mother, and by child-specific genetic effects, with little nonshared environment influence. This pattern of findings indicated evocative rGE. In contrast, passive rGE did not contribute to the association between maternal emotional overinvolvement and child internalizing behaviors, as the direct environmental path from parenting to child problems was nonsignificant. Therefore, maternal emotional overinvolvement may be understood as a response that is evoked from the child's heritable internalizing behaviors. A noteworthy limitation to this study is that the ECoT model requires large sample sizes for detailed analyses of family relationships. Although these data were corrected for child's age and sex, separate analyses on the relationship between mothers and their sons or daughters would illustrate potentially different processes involved, depending on a child's sex. At the same time, this model is important for advancing our understanding of the development of parent–child relationships, and more specifically, for directly differentiating evocative from passive rGE. Considered together, genetically informed studies examining parenting as well as associations between parenting and child adjustment help to clarify the role of genetic and environmental factors.

Marital Relationship

Although parenting has been one of the most thoroughly studied family relationships in quantitative genetic research, other family relationships have been examined, including the marital relationship. These studies have focused on general aspects of marriage, including the propensity to be married, risk of divorce, remarriage, and marital instability, as well as on more specific marital relationships qualities, like satisfaction and conflict about specific children (e.g., D'Onofrio et al., 2005; Johnson, McGue, Krueger, & Bouchard, 2004; Reiss et al., 2000; Spotts, Neiderhiser, Towers et al., 2004b). Results have typically indicated both genetic and nonshared environmental influences on both the general and specific marital constructs, but little evidence of shared environmental influences. A few studies have attempted to specify the genetic and nonshared environmental influences on marital constructs by examining personality characteristics of the twin-spouse as potential mediators. These studies found that genetic influences explain a bulk of the association between spouses' personality traits and the risk of divorce (Jockin, McGue, & Lykken, 1996) and the propensity to be married (Johnson et al., 2004). Another study examining wives who are twins and their husbands found that wives' personality characteristics explain genetic and nonshared environmental influences in their own marital satisfaction, their husbands' marital satisfaction, and the agreement between the spouses on the quality of their marriage (Spotts, Lichtenstein et al., 2005a). The common genetic influences suggest that wives' heritable personality traits are influencing the emotional tenor of the marriage, via rGE pathways that could include evocative rGE and/or assortative mating. Together, these results suggest that aspects of the marital relationship arise from partners' heritable characteristics and from experiences unique to each twin (in particular, the spouse).

Marital Relationship and Child Adjustment

Using the CoT design, researchers have addressed how the marital relationship shapes offspring adjustment (D'Onofrio et al., 2005, 2006; Harden et al., 2007) (see chapter 15 by D'Onofrio, Rathouz, & Lahey). For example, a recent study attempted to disambiguate the direct environmental impact of marital conflict on offspring adjustment problems from passive rGE (Harden et al., 2007). That is, is the observed correlation a result of marital conflict directly causing offspring adjustment problems, or a shared genetic liability that is common to parents' marital conflict and offsprings' adjustment problems? Results showed that offspring of MZ twins differentially exposed to varying levels of marital conflict had similar rates of antisocial problems. However, the antisocial behavior of the offspring of DZ twins was associated with differences in marital conflict within the twin family. This indicates that the association between marital conflict and offsprings' antisocial behavior is accounted for by children's inheritance of genetic liabilities common to their psychopathology and their parents' conflict. Thus, the CoT model is an important tool for examining the interplay of genetic factors and the marital relationship for shaping offspring adjustment.

Marital Relationship and Spouse Adjustment

Within a sample of MZ and DZ pairs of wives who are twins and their husbands (who are not twins), researchers assessed pathways linking marital satisfaction, adequacy of social support, and twin wives' mental health, including their depressive symptoms and positive mental health (Spotts, Neiderhiser, Ganiban et al., 2004a; Spotts, Pederson et al., 2005b). Because twin wives are the source of genetic influences in this sample, detecting genetic influences on assessments of marital satisfaction reflects the wives' heritable characteristics shaping their marital relationship via rGE. Findings revealed that associations among wives' reported marital satisfaction, social support, and mental health were influenced by common genetic effects. That is, the same genetic influences contributing to a wife's perceived marital satisfaction also influenced her social support and mental health. Replacing the wives' reported marital satisfaction with their husbands' reported marital satisfaction revealed a different pattern of findings. Specifically, although genetic influences contributed to husbands' reported marital satisfaction, nonshared environmental influences explained the variance common to husbands' reported marital quality and wives' reported social support and mental health. On the other hand, wives' reported social support and mental health retained common genetic influences, independent of the husbands' reported marital quality. Findings indicate that wives' genetically influenced characteristics play an important role in shaping general perceptions of their interpersonal relationships and mental health. In addition, husbands also appear to represent an important source of nonshared environmental influence on their wives' mental health. Considered together, genetically informed studies examining associations between family relationships and the outcomes of individual family members help to clarify the mechanisms through which relationships may have an impact on adjustment by specifying the role of genetic and environmental factors. One clear message from this work is that the characteristics of the individuals involved in these relationships play an important role in their adjustment and the adjustment of their family members. It is also the case that different mechanisms appear to be involved in shaping associations among different outcomes and that these mechanisms also differ for different family members.

In sum, there is now a sizable literature that has examined genetic and environmental influences on the quality of family relationships, including parenting behaviors and marital quality, and on the association between family relationship quality and the adjustment of children and spouses. The clear conclusion of this work is that rGE is important for parenting and the marital relationship. Furthermore, rGE plays an important role in shaping the association between family relationship quality and family members' adjustment. However, there is also evidence that certain aspects of family relationships (i.e., harsh parenting and marital satisfaction) have direct environmental influences on children's and spouses' adjustment, respectively. Together, this research highlights the importance of investigating how distinct aspects of different family relationships are linked to the adjustment of individual family members. Although some of this work has relied on "traditional" twin or sibling studies, the use of novel and combination designs (i.e., CoT and ECoT) allows us to better understand the processes involved in rGE and permits us to have more confidence in our findings.

GENOTYPE X ENVIRONMENT INTERACTION (G X E)

This section includes a review of behavior genetic research that has used twin or adoption designs to examine how the quality of family relationships (i.e., parenting behaviors) can influence child development through G × E. A number of studies have focused on the extent to which differences in genetic, shared environmental, and nonshared environmental influences on child adjustment are moderated by varying levels of parenting behaviors, typically described as G × E (Purcell, 2002). Researchers have also taken advantage of the adoption design, in which children are adopted at birth and placed with unrelated adoptive parents, thus affording the effects of parenting (indexed by the adoptive parents' behaviors) and the effects of genetic factors (indexed by the biological parents' characteristics) and their interaction to be disentangled. Next, we discuss the importance of taking rGE into account in these analyses, as rGE can be confounded with G × E, allowing the former to mask the latter.

Parenting As a Moderator

Studies that have considered family relationships as moderators of the genetic and environmental influences on child adjustment have often focused on the moderating role of parenting behavior (e.g., Dick et al., 2007; Feinberg, Button, Neiderhiser, Reiss, & Hetherington, 2007) (see Chapter 14 by Dick). For example, one such investigation assessed whether parenting behaviors (negativity and warmth) moderated genetic and environmental influences on adolescent antisocial behavior and depressive symptoms (Feinberg et al., 2007). Both parental negativity and warmth were found to moderate genetic influences on adolescent antisocial behavior, including aggressive and nonaggressive forms of antisocial behavior. Specifically, genetic influences assumed a greater role for adolescent antisocial behavior when parenting behaviors were more negative or less warm. This indicates that genetic effects have a stronger influence on antisocial behavior in the presence of parental negativity and lack of warmth. In contrast, nonshared environmental effects on adolescents' depressive symptoms increased in the context of greater parental negativity. As such, experiences of parental negativity unique to each sibling take on an increasingly important role with increasing levels of parental negativity.

Of note, a focus on standardized components, as is typical in most behavior genetic reports, can lead to misinterpretation (Purcell, 2002). Say, for example, there are very low levels of variance in a given phenotype at one level of a moderator, but large levels of phenotypic variance at another level of the moderator. A genetic component of variance may account for all of the small amount of variance at the first point but for only half the variance at the second point. A standardized estimate does not yield a sense of the absolute magnitude of the variance accounted for, but only the proportion of variance. Thus, a standardized heritability estimate in this example would be 1.0 at the first point and 0.5 at the second point. Without understanding the magnitude of the variance at each level of the moderator, it is possible to infer that genetic factors are twice as strong at the first point, when, in fact, genetic factors account for more actual variance at the second point.

Two more recent studies have taken advantage of a prospective adoption design, in which infants are adopted at birth and placed with unrelated adoptive parents (Leve et al., 2009; Natsuaki et al., 2010). One such study assessed the interaction between adoptive parents' structured parenting and genetic risk for psychopathology (as indexed by birth parents' psychopathology) on toddlers' behavior problems (Leve et al., 2009). Findings showed that the effects of structured parenting on toddler behavior problems varied as a function of genetic risk for psychopathology. Specifically, structured parenting was beneficial for toddlers at high genetic risk for psychopathology but was positively related to behavior problems for those toddlers at low genetic risk. Thus, children at risk for psychopathology benefit from more structured environments, whereas children at low genetic risk benefit from less structured environments. This is an example of a crossover type 2 interaction (see Chapter 1 by Kendler). The second study investigated the effects of the parenting environment (indexed by adoptive parents' depressive symptoms and responsiveness) and genetic risk of depression (reflected by biological parents' major depressive disorder) on the development of fussiness between 9 and 18 months of age in adopted children (Natsuaki et al., 2010). Results revealed that children at genetic risk showed higher levels of fussiness at 18 months of age when adoptive mothers had been less responsive at 9 months of age. In contrast, children at genetic risk did not show increased levels of fussiness at 18 months in the context of high levels of adoptive mothers' responsiveness. This is illustrative of a fan-shaped interaction (see Chapter 1 by Kendler), and these findings suggest that genetic liability for depression is accentuated by unresponsive caregiving. Considered together, these studies suggest that the etiologies of child or adolescent adjustment are better understood when taking the context of the family environment or family relationships into consideration.

Considering Both rGE and G × E

To date, most studies of G × E that include a measure of family environment or family relationships use a strategy that first accounts for the main effects of the moderator on the behavior under study (Purcell & Koenen, 2005), although, as noted by Price and Jaffee (2008) this does not control for the effects of *r*GE. Indeed, the presence of *r*GE has important consequences for understanding the contribution of G × E, as *r*GE can confound the interpretation of G × E (see Chapters 15 by D'Onofrio, Rathouz, & Lahey, 4 by Jaffee, and 1 by Kendler). Specifically, if *r*GE is present but is not explicitly modeled, the interaction between the latent genetic factor and the measured environmental moderator will produce biased estimates of heritability and mask the effects of G × E (Purcell, 2002; Rathouz, Van Hulle, Rodgers, Waldman, & Lahey, 2008). Similarly, if *r*GE is not controlled for in analysis, findings may suggest G × E interaction where none exists. Thus, as argued by Jaffee and Price (2007), a more nuanced approach is needed for considering both *r*GE and G × E.

CONCLUSION

The purpose of this chapter was to provide an overview of work focused on understanding how genes and environments work synergistically to influence family relationship quality and associations between family relationship quality and child adjustment. The role of family relationship quality has been a guiding theme throughout, both because of its importance in helping to shape the adjustment of children and spouses and because there is now evidence that family relationships are linked to family members' adjustment, via *r*GE and G × E. Within studies that have focused on the role of *r*GE, a clear pattern of results indicates that *r*GE is operating to shape parenting behaviors and the marital relationship. Furthermore, *r*GE plays an important role in shaping the association between family relationship quality and family members' adjustment. However, there is also evidence that certain aspects of family relationships (i.e., harsh parenting and marital satisfaction) have direct environmental influences on children's and spouses' adjustment, respectively. Within research that has addressed the role of G × E, findings indicate that parenting behaviors also play an important role in moderating the genetic and environmental influences on child adjustment outcomes. Finally, an emerging literature has investigated how *r*GE and G × E operate together to shape child development. An important direction for future research will be to examine how a variety of interpersonal relationships (e.g., peer groups, sibling relationships) are linked to the adjustment of individual family members, gene environment interplay, and the joint effects of *r*GE and G × E.

Until fairly recently, research in behavioral genetics and in prevention science has been conducted independently, with little effort to use findings across areas to inform interpretation or design. There have been a number of more recent reviews and calls to action (e.g., Leckman & Yanki, 2010; Leve et al., 2010; Reiss & Leve, 2007), as well as at least one empirical report (Beach et al., 2009), that attempt to use information from these areas, as well as more general developmental research, to advance research and our understanding of mechanisms. Research focused on understanding the interplay of genes and environments has the potential to inform prevention and intervention by helping to better specify where to target interventions. Although there has been a great deal of focus on G × E, studies of rGE also can provide directions for intervention. For example, if passive rGE is found to best explain the impact of a mother's positive parenting on her child's internalizing behavior, a strategy focused more on changing the child's behavior than on changing the mother's parenting may be more effective. On the other hand, when we find evidence of evocative rGE, parent training—in which parents are taught to respond differently to their children—could be a viable approach to changing the child's behavioral outcomes. At this point, using the findings of genetic research to direct intervention on an individual level is not feasible. We suggest, however, that using findings from genetically informed research designs can be used to help guide intervention strategies. Most importantly, genetic research, especially behavioral genetic research, underscores the need to consider individual differences and highlights how approaches that allow for differences among individuals are likely to be most effective.

ACKNOWLEDGMENTS

Data collection for the TOSS was supported by R01MH54601 (NIMH); for the NEAD project by R01MH43373, R01MH48825, and R01MH59014 (NIMH) and the William T. Grant Foundation; and for the EGDS by R01HD042608 (NICHD, NIDA, and Office of the Director) and R01DA020585 (NIDA, NIMH, and Office of the Director).

REFERENCES

Belsky, J. (1984). The determinants of parenting: a process model. *Child Development*, 55(1), 83–96.

Beach, S. R. H., Brody, G. H., Kogan, S. M., Philibert, R. A., Chen, Y.-f., & Lei, M. K. (2009). Change in caregiver depression in response to parent training: genetic moderation of intervention effects. *Journal of Family Psychology*, 23, 112–117.

Burt, S. A., Krueger, R. F., McGue, M., & Iacono, W. G. (2003). Parent-child conflict and the comorbidity among children externalizing disorders. *Archives of General Psychiatry*, 60, 505–513.

Burt, S. A., McGue, M., Krueger, R. F., & Iacono, W. G. (2005). Sources of covariation among the child-externalizing disorders: informant effects and the shared environment. *Psychological Medicine*, 35(8), 1133–1144.

Cummings, E. M., & Davies, P. T. (2002). Effects of marital conflict on children: recent advances and emerging themes in process-oriented research. *Journal of Child Psychology and Psychiatry, 43*(1), 31–63.

D'Onofrio, B. M., Turkheimer, E., Emery, R. E., Slutske, W. S., Heath, A. C., Madden, P. A., et al. (2005). A genetically informed study of marital instability and its association with offspring psychopathology. *Journal of Abnormal Psychology, 114*(4), 570–586.

D'Onofrio, B. M., Turkheimer, E., Emery, R. E., Slutske, W. S., Heath, A. C., Madden, P. A., et al. (2006). A genetically informed study of the processes underlying the association between parental marital instability and offspring adjustment. *Developmental Psychology, 42*(3), 486–499.

D'Onofrio, B. M., Turkheimer, E. N., Eaves, L. J., Corey, L. A., Berg, K., Solaas, M. H., et al. (2003). The role of the children of twins design in elucidating causal relations between parent characteristics and child outcomes. *Journal of Child Psychology and Psychiatry, 44*(8), 1130–1144.

Demo, D. H., & Cox, M. J. (2004). Families with young children: a review of research in the 1990's. *Journal of Marriage and Family, 62,* 876–895.

Dick, D. M., Pagan, J. L., Viken, R., Purcell, S., Kaprio, J., Pulkkinen, L., et al. (2007). Changing environmental influences on substance use across development. *Twin Research and Human Genetics, 10*(2), 315–326.

Dishion, T. J., & McMahon, R. J. (1998). Parental monitoring and the prevention of child and adolescent problem behavior: a conceptual and empirical formulation. *Clinical Child and Family Psychology Review, 1*(1), 61–75.

Feinberg, M. E., Button, T. M., Neiderhiser, J. M., Reiss, D., & Hetherington, E. M. (2007). Parenting and adolescent antisocial behavior and depression: evidence of genotype x parenting environment interaction. *Archives of General Psychiatry, 64*(4), 457–465.

Fletcher, A. C., Steinberg, L., & Williams-Wheeler, M. (2004). Parental influences on adolescent problem behavior: revisiting Stattin and Kerr. *Child Development, 75,* 781–796.

Harden, K. P., Turkheimer, E., Emery, R. E., D'Onofrio, B. M., Slutske, W. S., Heath, A. C., et al. (2007). Marital conflict and conduct problems in children of twins. *Child Development, 78*(1), 1–18.

Jaffee, S. R., Caspi, A., Moffitt, T. E., Polo-Tomas, M., Price, T. S., & Taylor, A. (2004). The limits of child effects: evidence for genetically mediated child effects on corporal punishment but not on physical maltreatment. *Developmental Psychology, 40*(6), 1047–1058.

Jaffee, S. R., & Price, T. S. (2007). Gene-environment correlations: a review of the evidence and implications for prevention of mental illness. *Molecular Psychiatry, 12*(5), 432–442.

Jockin, V., McGue, M., & Lykken, D. T. (1996). Personality and divorce: a genetic analysis. *Journal of Personality and Social Psychology, 71*(2), 288–299.

Johnson, W., McGue, M., Krueger, R. F., & Bouchard, T. J., Jr. (2004). Marriage and personality: a genetic analysis, *Journal of Personality and Social Psychology, 86*(2), 285–294.

Kendler, K. S. (1996). Parenting: a genetic-epidemiologic perspective. *American Journal of Psychiatry, 153*(1), 11–20.

Kendler, K. S., & Baker, J. H. (2007). Genetic influences on measures of the environment: a systematic review. *Psychological Medicine, 37*(5), 615–626.

Larsson, H., Viding, E., Rijsdijk, F. V., & Plomin, R. (2008). Relationships between parental negativity and childhood antisocial behavior over time: a bidirectional effects model in a longitudinal genetically informative design. *Journal of Abnormal Child Psychology, 36*(5), 633–645.

Leckman, J. F., & Yanki, Y. M. (2010). Editorial: Developmental transitions to psychopathology: from genomics and epigenomics to social policy. *Journal of Child Psychology and Psychiatry, 51*(4), 333–340.

Leve, L. D., Harold, G. T., Ge, X., Neiderhiser, J. M., & Patterson, G. (2010). Refining intervention targets in family-based research: lessons from quantitative behavioral genetics. *Perspectives in Psychological Science, 5*(5), 516–526.

Leve, L. D., Harold, G. T., Ge, X., Neiderhiser, J. M., Shaw, D., Scaramella, L. V., et al. (2009). Structured parenting of toddlers at high versus low genetic risk: two pathways to child problems. *Journal of the American Academy of Child and Adolescent Psychiatry, 48*(11), 1102–1109.

Lynch, S. K., Turkheimer, E., D'Onofrio, B. M., Mendle, J., Emery, R. E., Slutske, W. S., et al. (2006). A genetically informed study of the association between harsh punishment and offspring behavioral problems. *Journal of Family Psychology, 20*(2), 190–198.

McGue, M., Elkins, I., Walden, B., & Iacono, W. G. (2005). Perceptions of the parent-adolescent relationship: a longitudinal investigation. *Developmental Psychology, 41*(6), 971–984.

Moffitt, T. E. (2005). The new look of behavioral genetics in developmental psychopathology: gene-environment interplay in antisocial behaviors. *Psychological Bulletin, 131*(4), 533–554.

Narusyte, J., Neiderhiser, J. M., D'Onofrio, B. M., Reiss, D., Spotts, E. L., Ganiban, J., et al. (2008). Testing different types of genotype-environment correlation: an extended children-of-twins model. *Developmental Psychology, 44*(6), 1591–1603.

Natsuaki, M. N., Ge, X., Leve, L. D., Neiderhiser, J. M., Shaw, D. S., Conger, R. D., et al. (2010) press). Genetic liability, environment, and the development of fussiness in toddlers: the roles of maternal depression and parental responsiveness. *Developmental Psychology, 46*(5), 1147–1158.

Neiderhiser, J. M., Reiss, D., Lichtenstein, P., Spotts, E. L., & Ganiban, J. (2007). Father-adolescent relationships and the role of genotype-environment correlation. *Journal of Family Psychology, 21*(4), 560–571.

Neiderhiser, J. M., Reiss, D., Pedersen, N. L., Lichtenstein, P., Spotts, E. L., Hansson, K., et al. (2004). Genetic and environmental influences on mothering of adolescents: a comparison of two samples. *Developmental Psychology, 40*(3), 335–351.

Plomin, R., DeFries, J. C., & Loehlin, J. C. (1977). Genotype-environment interaction and correlation in the analysis of human behavior. *Psychological Bulletin, 84*(2), 309–322.

Price, T. S., & Jaffee, S. R. (2008). Effects of the family environment: gene-environment interaction and passive gene-environment correlation. *Developmental Psychology*, *44*(2), 305–315.

Proulx, C. M., Helms, H. M., & Buehler, C. (2007). Marital quality and personal well-being: a meta-analysis. *Journal of Marriage and Family*, *69*(3), 576–593.

Purcell, S. (2002). Variance components models for gene-environment interaction in twin analysis. *Twin Research*, *5*(6), 554–571.

Purcell, S., & Koenen, K. C. (2005). Environmental mediation and the twin design. *Behavior Genetics*, *35*(4), 491–498.

Rathouz, P. J., Van Hulle, C. A., Rodgers, J. L., Waldman, I. D., & Lahey, B. B. (2008). Specification, testing, and interpretation of gene-by-measured-environment interaction models in the presence of gene-environment correlation. *Behavior Genetics*, *38*(3), 301–315.

Reiss, D. & Leve, L. D. (2007). Genetic expression outside the skin: clues to mechanisms of genotype x environment interaction. *Development and Psychopathology*, *19*(4), 1005–1027.

Reiss, D., Neiderhiser, J., Cederblad, M., & Lichtenstein, P. (2000). *The Relationship Code: deciphering genetic and social influences on adolescent development.* Cambridge, MA: Harvard University Press.

Rutter, M., Pickles, A., Murray, R., & Eaves, L. (2001). Testing hypotheses on specific environmental causal effects on behavior. *Psychological Bulletin*, *127*(3), 291–324.

Scarr, S., & McCartney, K. (1983). How people make their own environments: a theory of genotype → environment effects. *Child Development*, *54*(2), 424–435.

Silberg, J. L., & Eaves, L. J. (2004). Analysing the contributions of genes and parent-child interaction to childhood behavioural and emotional problems: a model for the children of twins. *Psychological Medicine*, *34*(2), 347–356.

Spotts, E. L., Lichtenstein, P., Pedersen, N. L., Neiderhiser, J. M., Hansson, K., Cederblad, M., et al. (2005a). Personality and marital satisfaction: a behavioural genetic analysis. *European Journal of Personality*, *19*, 205–227.

Spotts, E. L., Neiderhiser, J. M., Ganiban, J., Reiss, D., Lichtenstein, P., Hansson, K., et al. (2004a). Accounting for depressive symptoms in women: a twin study of associations with interpersonal relationships. *Journal of Affective Disorders*, *82*(1), 101–111.

Spotts, E. L., Neiderhiser, J. M., Towers, H., Hansson, K., Lichtenstein, P., Cederblad, M., et al. (2004b). Genetic and environmental influences on marital relationships. *Journal of Family Psychology*, *18*(1), 107–119.

Spotts, E. L., Pederson, N. L., Neiderhiser, J. M., Reiss, D., Lichtenstein, P., Hansson, K., et al. (2005b). Genetic effects on women's positive mental health: do marital relationships and social support matter? *Journal of Family Psychology*, *19*(3), 339–349.

Ulbricht, J. A., & Neiderhiser, J. M. (2009). *Genotype-environment correlation and family relationships.* In Y.K. Kim (Ed.), *Handbook of Behavior Genetics* (pp. 455–571). New York: Springer.

7

Epigenetic Effects on Gene Function and their Role in Mediating Gene–Environment Interactions

JONATHAN MILL

Recent years have seen tremendous progress in our ability to interrogate genetic variation; since the completion of the first-draft human genome sequence in 2001, huge advances have been made in genotyping and sequencing technology. It is now economically feasible to perform genome-wide association (GWA) studies using high-resolution microarrays to tag nearly all common variants across the genome. The development of next-generation sequencing technology means that we are at the dawn of an era in which it will be possible to sequence entire genomes on an industrial scale. The advent of GWA studies has allowed a systematic, hypothesis-free exploration of the genes associated with mental illness. Given the high heritability of disorders such as schizophrenia, bipolar disorder, and major depression, it is somewhat surprising, therefore, that few novel susceptibility loci have been uncovered using genetic research paradigms. Although molecular genetic studies have uncovered polymorphisms in a number of loci that appear to mediate risk for disorders such as schizophrenia (Stefansson et al., 2009) and bipolar disorder (Ferreira et al., 2008), these findings are characterized by small effect sizes and considerable heterogeneity, and await replication (Kato, 2007; Purcell et al., 2009; Riley & Kendler, 2006). Structural genomic alterations, such as copy number variations (CNVs), have also been implicated, but these de novo events are extremely rare and only found in a very small number of patients (Merikangas, Corvin, & Gallagher, 2009). Despite considerable research effort, we are thus still a long way from realizing the postgenomic promises of novel diagnostic and therapeutic strategies for psychiatric illness.

Until relatively recently, research into the factors influencing susceptibility to mental illness was directed by a perceived dichotomy between "nature" and "nurture" that we now know to be false. It is clear that neither the genome nor the environment acts in isolation. With the exception of a few rare Mendelian disorders that are caused by nonsynonymous changes to the function of specific genes, it is likely that most "susceptibility" genes act to moderate an individuals' response to the environment, rather than playing a direct pathological role (Caspi & Moffitt, 2006). As the chapters in this book exemplify, research investigating such *interactions* between genetic variation and the environment has been fully embraced by both quantitative and molecular geneticists. We are now at an exciting stage in the development of gene–environment interaction (G × E) research, in which a number of replicated findings have been uncovered. One major caveat to current G × E approaches is that the interactions being identified are purely *statistical* in nature, providing few clues about the actual molecular *mechanisms* operating to mediate susceptibility. There must be, for example, some tangible effect of early-life stress occurring specifically in individuals carrying the "risk" genotype of the serotonin transporter gene (*SERT*), which can explain the interactive effect of these two risk factors, as reported by Caspi and colleagues (Caspi et al., 2003). It has yet to be elucidated whether the environment impacts directly upon the function of *SERT* or operates on pathways further downstream, but it appears that early-life stress can induce long-term neurobiological alterations that increase the risk of developing affective disorders in genetically vulnerable individuals. It is apparent that G × E interactions are highly complex, often being sex-specific or restricted to environmental exposure at certain key developmental stages (Uher & McGuffin, 2010). Elucidating the pathway(s) through which G × E interactions operate is clearly vital if any diagnostic, therapeutic, or preventative benefits are to result from this line of research.

One problem with traditional genetic research paradigms is the assumption that stable alterations to the DNA sequence are the primary cause of interindividual transcriptomic (and thus phenotypic) variation. In this chapter, I describe how DNA should instead be seen as a nonstatic entity, and I stress the importance of looking beyond the series of nucleotides that comprise our genomic sequence. I will introduce the role that epigenetic processes play in dynamically regulating gene expression and demonstrate how these mechanisms are highly labile, responsive to the environment, and provide a potential mechanistic substrate for G × E interactions.

BEYOND DNA SEQUENCE: THE EPIGENETIC
REGULATION OF GENE FUNCTION

Traditional approaches to studying the genome, especially within the context of understanding disease etiology, have tended to consider DNA solely as a series of A's, C's, T's, and G's. With the goal of identifying alterations in the sequence of these bases that are statistically associated with disease, either directly or via an interaction with the environment, these studies assume that the genome is a static entity, that functionally relevant information is only stored in the genomic sequence, and that transgenerational inheritance is solely mediated by the DNA sequence. Evidence is mounting to support the notion that all these assumptions are incorrect.

With the exception of a few rare somatic mutation events, it is true that the sequence of nucleotides making up an individuals' genome is identical across all cells in the body and remains unchanged from the moment of conception onward. But DNA is structurally much more complex than a simple string of nucleotides, and at a functional level, the genome is anything but static. Although every cell in our bodies contains the same DNA sequence, each has its own unique phenotype, characterized by a specific pattern of gene expression that is in a constant state of flux. It's not only the gene-encoding DNA sequence that is important in determining the phenotype of a cell, but the degree to which specific genes are functionally active at any particular time in development. In this regard, sequencing the genome was only the first step in our quest to understand how genes are expressed and regulated. Sitting above the DNA sequence is a second layer of information (the so-called *epigenome*) that regulates a range of genomic functions, including when and where genes are turned on or off.

British biologist Conrad Waddington first coined the term "epigenetics" (literally meaning "above genetics") in the mid-20th century. Waddington developed the concept of an epigenetic landscape to describe the ways in which cell fates are established during development, thus enabling the tissues and organs of complex organisms to develop from an initially undifferentiated mass of cells. A contemporary definition regards epigenetics as the reversible regulation of gene expression, occurring independently of DNA sequence, mediated principally through changes in DNA methylation and chromatin structure (Jaenisch & Bird, 2003). Epigenetic processes are essential for normal cellular development and differentiation, and allow the long-term regulation of gene function through nonmutagenic mechanisms (Henikoff & Matzke, 1997). For a glossary of epigenetic mechanisms referred to in this chapter, and a basic description of the genomic functions they perform, see Table 7.1.

Table 7.1. Glossary of Epigenetic Terms

Term	Definition	Key Reference(s)
Chromatin	The complex of DNA, histones, and other proteins that make up chromosomes. Chemical modifications to both DNA and histone proteins are important in regulating the structure of chromatin. Condensed chromatin (heterochromatin), in which the DNA and histone proteins are tightly packed, acts to block the access of transcription factors and other instigators of gene expression. Open chromatin (euchromatin) allows the cells' transcriptional machinery to access DNA and drive transcription.	Berger, 2007
DNA methylation	The addition of a methyl group at position 5 of the cytosine pyrimidine ring in CpG dinucleotides in a reaction catalyzed by DNA methyltransferases. DNA methylation disrupts the binding of transcription factors and attracts methyl-binding proteins that are associated with gene silencing and chromatin compaction.	Jaenisch & Bird, 2003
Epigenetic inheritance	Epigenetic signals are transmitted mitotically through cell lineages, but are generally assumed to be reset during gametogenesis and thus not transmitted meiotically. In other words, these signals are thought to be maintained through normal cell reproduction (mitosis), but not transmitted to the next generation. Evidence is mounting, however, that the epigenetic marks of at least some mammalian genes are not fully erased during meiosis, thus suggesting the possibility of transgenerational epigenetic inheritance.	Richards, 2006
Epigenetics	The heritable, but reversible, regulation of various genomic functions that occur independently of the DNA sequence. Epigenetic regulation is primarily mediated by DNA methylation, physical changes to chromatin structure, and the action of non-coding RNA molecules.	Henikoff & Matzke, 1997
Genomic imprinting	An epigenetic process that alters the expression of genes in a parent-of-origin specific manner. Genomic imprinting is fundamental to normal mammalian development and growth.	Davies et al., 2005

Term	Description	Reference
Histone modifications	A number of covalent post-translational histone modifications, occurring at specific residues, have been described (e.g., acetylation, methylation, phosphorylation, SUMOylation, and ubiquitylation). These modulate gene expression via alterations in chromatin structure. Like DNA methylation, histone modifications are highly dynamic and actively regulated by a host of catalytic enzymes.	Berger, 2007
Nucleosome	Nucleosomes are the fundamental packaging units of DNA in the nucleus. DNA is wrapped around a histone protein core, compacting DNA and enabling the regulation of gene expression via epigenetic processes.	Kornberg & Lorch, 1999
Transcription factor	A protein that binds to specific DNA sequence motifs to induce the expression of downstream genes.	Latchman, 1997
X-inactivation	X-chromosome inactivation silences genes on one X-chromosome in females to ensure dosage compensation with males via a process involving hypermethylation of CpG islands. X-inactivation in any given cell is typically random, and is maintained once established, so that the inactivated allele is transcriptionally silenced for the lifetime of that cell.	Avner & Heard, 2001

Epigenetic Modifications to DNA and Histones

Structurally, we know that DNA is much more than a sequence of DNA bases; in its mitotic state, for example, each DNA molecule is packaged into a chromosome that is 10,000 times shorter than its extended length. DNA is coiled tightly around histone proteins to form nucleosomes in a formation that has been likened to a series of beads on a string (see Figure 7.1), which in turn are condensed together to form *chromatin*, the complex combination of DNA, RNA, and protein that makes up chromosomes. It is this ornate packaging that allows each mammalian cell to contain approximately 2 meters of DNA in the nucleus in a form that is easily accessible and able to regulate cellular phenotype. Chemical modifications to both DNA and histone proteins are important in regulating how accessible the genome is to the cells' transcriptional machinery (see Figure 7.1). Condensed chromatin (*heterochromatin*), in which the DNA and histone proteins are tightly packed, acts to block the access of transcription factors and other instigators of gene expression to DNA, and is thus associated with repressed transcription. Conversely, an open chromatin conformation

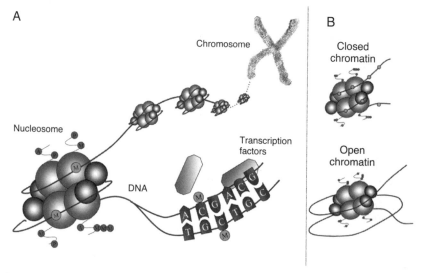

Figure 7.1. **Epigenetic modifications and chromatin structure.** (A) DNA is wrapped around a histone core, forming a nucleosome. DNA methylation (M) at CpG dinucleotides and covalent modifications to histone tails (e.g., acetylation [A], methylation [M], phosphorylation [P]) alter the physical structure of the chromosome (B), and mediate the access of transcription factors that drive gene expression. DNA methylation also disrupts transcription directly by blocking transcription factor binding sites. Reproduced from Pidsley & Mill (2010), with permission.

(*euchromatin*) allows the cells' transcriptional machinery to access DNA and drive transcription.

The best-understood and most stable epigenetic process that modulates the transcriptional plasticity of mammalian genomes is cytosine methylation. This process involves the transfer of a methyl group to position 5 of the cytosine pyrimidine (C) ring of a cytosine-guanine dinucleotide (CpG) catalyzed by a group of enzymes called *DNA methyltransferases* (Dnmts) (Klose & Bird, 2006). The methylation of CpG sites can act to disrupt the activity of the cells' transcriptional machinery, for example by blocking the binding of transcription factors, those proteins that bind upstream of specific genes to instigate gene expression. DNA methylation also attracts methyl-binding proteins that initiate chromatin compaction and bring about gene silencing (Klose & Bird, 2006). Because methylated cytosines are liable to spontaneous mutation at a much higher frequency than other nucleotides, CpG dinucleotides are less common in the genome than would be predicted by chance, and primarily occur in largely unmethylated clusters called *CpG islands*, which are often found around the promoters of constitutively expressed housekeeping genes (Bird, 1986). DNA methylation at these CpG islands has been the predominant focus of epigenetic research and is associated with chromatin remodelling and reduced gene expression at proximal genes (Jaenisch & Bird, 2003). It is increasingly apparent, however, that DNA methylation at CpGs occurring outside these CpG islands (~70% of total CpGs), also has an important (perhaps even more important) role in regulating genomic function, particularly during development (Shen, Chow, Wang, & Fan, 2006). A recent study utilizing novel methods for profiling DNA methylation across the genome, for example, highlighted how epigenetic variation at so-called CpG island "shores" may be important regulators of tissue-specific cellular phenotype (Irizarry et al., 2009).

The chemical modification of histones, the basic proteins around which DNA is wrapped to form nucleosomes, comprises the other major type of epigenetic mechanism related to gene expression. A number of covalent histone modifications, occurring at specific residues, have been described (e.g., acetylation, methylation, phosphorylation, SUMOylation, and ubiquitylation), which together constitute a complex "histone code" modulating gene expression via alterations in chromatin structure (Berger, 2007). Like DNA methylation, histone modifications are highly dynamic and actively regulated by a host of catalytic enzymes, such as histone acetyltransferases (HATs) and histone deacetylases (HDACs), which add and remove acetyl groups respectively (Saha & Pahan, 2006).

Although often investigated independently, epigenetic modifications to DNA and histones are not mutually exclusive and clearly interact in a number of ways. It is apparent that the classification of epigenetic mechanisms in terms of either gene activation or suppression is too simplistic (Berger, 2007). Recently, a third epigenetic system involving small noncoding RNA molecules has been described (Hamilton, Voinnet, Chappell, & Baulcombe, 2002). It has been shown, for example, that these can suppress the activity of specific genes via targeted RNA interference, a mechanism important in the developmental regulation of gene expression, mediated by both DNA methylation and histone modifications.

The Dynamic Epigenome

Like the DNA sequence, the epigenetic profile of somatic cells is transmitted from maternal to daughter chromatids during mitosis (i.e., the normal process of cell duplication occurring within an individual throughout the lifespan). Unlike the DNA sequence, which is stable and strongly conserved, epigenetic processes are tissue-specific, developmentally regulated, and highly dynamic. This notion is supported by a growing body of evidence suggesting that epigenetic variation is strongly associated with age (Bjornsson et al., 2008; Christensen et al., 2009; Flanagan et al., 2006; Fraga et al., 2005; Siegmund et al., 2007). As will be discussed below, and in other chapters in this book, increasing evidence suggests that influences external to the organism can directly interact with the epigenome, altering epigenetic processes and thus gene expression. In addition, random stochastic and developmental epigenetic changes are also likely to be important. Experiments tracking the inheritance of epigenetic marks through generations of genetically identical cells in tissue-culture, for example, have indicated that considerable infidelity exists in the maintenance of methylation patterns in mammalian cells, and that de novo methylation events are fairly common during mitosis (Riggs, Xiong, Wang, & LeBon, 1998; Ushijima et al., 2003). Because epigenetic processes are integral in determining when and where specific genes are expressed, such epigenetic metastability, environmentally or stochastically induced, may have profound phenotypic effects on gene expression in the cell. The dynamic nature of the epigenome calls into question many of our basic assumptions about the origins of phenotypic variance. Although largely unstudied in nonmalignant phenotypes, epigenetic lability can offer new insights about the non-Mendelian patterns of inheritance often observed for a wide range of complex traits and diseases (Mill & Petronis, 2007; Petronis, 2003; Petronis, 2004).

Dynamic Epigenetic Changes May Mediate Phenotypic Discordance Between Genetically Identical Individuals

Numerous examples of differences between genetically identical inbred animals (Gartner, 1990), genetically cloned animals (Cibelli et al., 2002; Tamashiro et al., 2003), and monozygotic (MZ) twins (Wong, Gottesman, & Petronis, 2005) have been reported, supporting the notion that nonge-netic factors can strongly influence phenotype. The dynamic nature of the epigenome and its significant role in regulating gene expression suggest that epigenetic variation could mediate a proportion of this phenotypic discordance. Indeed, emerging evidence suggests that significant epigenetic differences exist between genetically identical twins, leading to speculation that such variation could account for MZ twin discordance in disease susceptibility. A recent study has demonstrated that fairly profound epigenetic differences across the genome arise during the lifetime of MZ twins, thus highlighting the dynamic nature of epigenetic processes (Fraga et al., 2005). Kaminsky and colleagues investigated genome-wide epigenetic differences in MZ and dizygotic (DZ) twin pairs using microarrays that can simultaneously assess DNA methylation across thousands of promoter regions (Kaminsky et al., 2009) and found a large degree of MZ co-twin DNA methylation variation in all three tissue types investigated. Moreover, there was significantly higher epigenetic difference between DZ co-twins compared with MZ co-twins, suggesting that molecular mechanisms of heritability may not be limited to DNA sequence differences, but to some extent by epigenetic differences. Interestingly, MZ twin methylation differences have been reported for CpG sites in a number of specific genes previously implicated in psychiatric disorders, including the dopamine D_2 receptor (DRD2) gene (Petronis et al., 2003), the catechol-O-methyltransferase (COMT) gene (Mill et al., 2006) and the serotonin transporter (SERT) gene (Wong et al., 2010). To date, few empirical epigenetic studies have been performed on discordant MZ twin-pairs, although this is an area of considerable research interest. Kuratomi and colleagues investigated DNA methylation differences between MZ twins discordant for bipolar disorder (Kuratomi et al., 2007) and reported evidence for disease-associated epigenetic changes in several genes.

Epigenetics: Linking the Environment to Phenotype

Mounting evidence suggests that epigenetic processes can be induced following exposure to a range of environmental insults. DNA methylation, for example, has been shown to vary as a function of numerous nutritional, chemical, physical, and psychosocial factors. Because such epigenetic changes are inherited as cells reproduce mitotically during development, and can have important effects on normal gene function, they provide a mechanism by which the environment can lead to long-term alterations in phenotype. Fraga and colleagues examined DNA methylation and histone acetylation in 80 pairs of MZ twins, ranging from 3 to 74 years of age, using a combination of global and locus-specific methods (Fraga et al., 2005). They found that one-third of MZ twins had a significantly dissimilar epigenetic profile, with older twins and those with a history of nonshared environments being the most disparate, thus suggesting that environmental factors may shape the epigenome over the life course. Although this study highlighted mounting epigenetic discordance with age, a recent longitudinal twin study by Wong et al (2010) found that DNA methylation differences are apparent already in early childhood, even between genetically identical individuals, and that individual differences in DNA methylation are not stable over time. These data suggest that environmental influences are important factors accounting for interindividual DNA methylation differences, and that these influences differ across the genome. Given the role that prenatal and early-life environmental factors appear to play in the etiology of mental illness, it is pertinent that the epigenome appears to be particularly labile during a number of key developmental periods (Waterland & Michels, 2007). This is particularly the case during embryogenesis, when the rate of DNA synthesis is high and the epigenetic marks needed for normal tissue differentiation and development are being established (Dolinoy, Weidman, & Jirtle, 2007). It is thus plausible that epigenetic changes in response to an adverse in utero environment may affect normal patterns of neurobehavioral development and increase susceptibility to disease after birth (see Figure 7.2). A growing number of environmental factors have been shown to dynamically mediate the epigenome, and these are reviewed in detail elsewhere (e.g., Dolinoy & Jirtle, 2008; Jirtle & Skinner, 2007). Here, I will briefly discuss a number of examples that are particularly pertinent for common complex psychiatric disorders.

Environmental Effects on the Epigenome Implicated in Mental Disorders

A major external influence on DNA methylation involves the availability of external methyl donors and co-factors, usually contained in the diet, that are required for the correct pattern of gene expression during embryogenesis and development. One such effect involves the formation of S-adenosyl

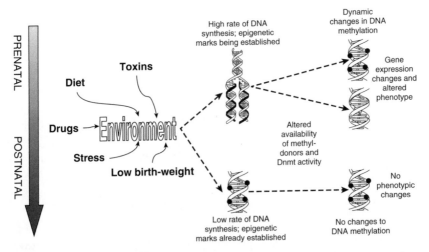

Figure 7.2. **Environmental insults during certain key developmental periods may have important effects on the epigenome.** One critical window for such effects is likely to be during early embryonic development in utero, when the rate of DNA synthesis is high and the epigenetic marks needed for normal tissue development are being established (reproduced from Mill & Petronis [2008], with permission).

methionine (SAM), which acts as a methyl donor for the methylation of cytosine DNA residues, a reaction catalyzed by a family of enzymes called Dnmts. De novo DNA methylation is initiated by Dnmt3a and Dnmt3b, and maintained in mitotic lineages by Dnmt1 (Bestor, 2000). Examples of dietary factors required for the formation of SAM include folate, methionine, choline, vitamin B_{12}, vitamin B_6, and vitamin B_2. It has been postulated that exposure to a diet lacking such components at specific developmental time points could have profound phenotypic effects (see also Chapter 8, by Glymour, Velling, and Susser). Extreme folate deficiency, for example, is known to cause depletion of SAM, resulting in genome-wide DNA hypomethylation, the activation of oncogenes, and cancer (Duthie et al., 2004). Interestingly, it has been shown that a dietary deficiency of vitamin B_{12} and folate is linked to impaired central nervous system development and a number of psychiatric conditions (Reynolds, 2006). Also of interest is the epidemiological evidence from two large independent population samples in the Netherlands and China suggesting that in utero nutritional deficiency, as indicated by maternal exposure to severe famine during pregnancy, is associated with an increase of schizophrenia in adult life (St. Clair et al., 2005;

Susser et al., 1996; Susser & Lin, 1992). Exposure to famine at this time has also been associated with loss of imprinting at insulin growth factor-2 (IGF2) (Heijmans et al., 2008), an epigenetic change correlated with brain size (Pidsley et al., 2009).

Various other external environmental factors have been shown to epigenetically alter gene expression in animals, including a range of toxins and chemicals. Of interest given the recent link between prenatal hexachlorobenzene exposure and behavioral disorders (Ribas-Fito et al., 2007) is the observation that the offspring of pregnant rats exposed to the endocrine disruptor vinclozolin, another agrochemical used as a fungicide in crops, have altered DNA methylation profiles that correlate with adverse phenotypic changes (Chang, Anway, Rekow, & Skinner, 2006).

It is not just the chemical environment that can cause long-lasting epigenetic changes. Evidence suggests that even the psychosocial environment during key developmental periods early in life can epigenetically mediate gene expression (Szyf et al., 2007). This is potentially interesting, given studies linking early psychosocial adversity to an increased risk of developing a range of neuropsychiatric conditions. Research by Weaver and colleagues (Weaver et al., 2004), for example, has shown that immediate postnatal maternal care in rats, as measured by increased pup licking, grooming, and arch-backed nursing, leads to epigenetic modification of a NGF1-A transcription factor binding site in the promoter region of the glucocorticoid receptor gene (Nr3c1), directly affecting gene expression and stress-related phenotypes in offspring. Interestingly, like the other examples discussed above, these epigenetic changes only occurred during a specific critical period—in this case, immediately after birth. Subsequent transcriptomic studies by the same group identified over 900 genes in the hippocampus that are stably regulated by maternal care (Weaver, Meaney, & Szyf, 2006), suggesting an even more profound effect of the early social environment on gene expression through the life course. A recent study in humans investigated epigenetic differences in a homologous Nr3c1 promoter region, comparing DNA methylation in postmortem hippocampus samples obtained from suicide victims with a history of childhood abuse to that seen in samples from either suicide victims with no childhood abuse or controls. In line with the animal findings, abused suicide victims had increased CpG methylation of the Nr3c1 promoter, with concomitant changes in mRNA (McGowan et al., 2009). Of course, the limitations of using human clinical samples mean that it is often hard to conclusively prove whether the impact of the environment is mediated through the epigenetic changes. The use of animal models, as demonstrated by the studies of Meaney and colleagues, allows a more controlled investigation of these effects, and it is likely that the coming years will see exciting advances in this area of research.

Transgenerational Environmental Effects?

It is conventionally believed that epigenetic profiles are fully erased during gametogenesis, preventing the meiotic transmission of epigenetic information between generations. Evidence is mounting, however, that the epigenetic marks of at least some mammalian genes are not fully erased during meiosis and can thus be transmitted from generation to generation (Klar, 1998; Rakyan & Whitelaw, 2003; Richards, 2006; Sandovici et al., 2005). The meiotic transmission of epigenetic marks is not actually necessary for the transgenerational inheritance of environmentally induced phenotypic effects. In the example of maternal care in rats given above, for instance, infants cared for by high-licking and -grooming mothers themselves went on to provide high levels of care to their offspring, an effect apparently mediated by epigenetic changes upstream of the glucocorticoid receptor gene (Weaver et al., 2004). In other words, epigenetic effects can be transmitted across generations through a purely experiential effect on gene expression. This blurs the demarcation between epigenetic- and DNA sequence–based inheritance and, although still a controversial notion, challenges the assumption that the "heritable" component to complex psychiatric disorders is entirely genetic. In fact, given the growing evidence that environmental factors can influence epigenetic modifications, this so-called "soft inheritance" supports the Lamarckian concept that environmental exposures in one generation could mediate phenotype in subsequent generations (Richards, 2006). Interestingly, a growing body of epidemiological evidence supports the notion that environmental exposures in ancestral generations can have phenotypic effects several generations later (Pembrey et al., 2006).

Epigenetic Dysfunction and Mental Illness

It is well established that the regulation of gene activity is critically important for normal functioning of the genome. It follows that even genes that carry no mutations or disease-predisposing polymorphisms can be rendered harmful if they are not expressed in the appropriate amount in the correct type of cell at the right time of the cell cycle. Because epigenetic processes are integral for cellular development and function, aberrant DNA methylation signatures and histone modifications are hypothesized to be involved in a diverse range of complex human pathologies (Hatchwell & Greally, 2007), including cancer (Jones & Baylin, 2007) and several rare imprinting disorders (Feinberg, 2007).

Numerous epidemiological, clinical, and molecular peculiarities associated with psychiatric disorders are also suggestive of an epigenetic contribution to etiology. These include the incomplete concordance between MZ twins, a fluctuating disease course with periods of remission and relapse, sexual dimorphism, peaks of susceptibility to disease coinciding with major hormonal rearrangements, and parent-of-origin effects (Petronis, 2004). These observations have led to speculation about the importance of epigenetic factors in mediating susceptibility to a number of psychiatric conditions including schizophrenia (Petronis, 2004), bipolar disorder (Petronis, 2003), major depression (Mill & Petronis, 2007), drug addiction (Renthal & Nestler, 2008), Alzheimer disease (Wu, Basha, & Zawia, 2008), attention-deficit hyperactivity disorder (ADHD) (Mill & Petronis, 2008), and autism (Schanen, 2006).

Despite sharing the same DNA sequence, the concordance rate between MZ twins for most psychiatric disorders is far from 100%: for example, between 41% and 65% in schizophrenia (Cardno & Gottesman, 2000), approximately 60% in bipolar disorder (Craddock & Jones, 1999), and 45%–61% in Alzheimer disease (Gatz et al., 2006). Furthermore, studies of MZ twin-pairs concordant for these disorders suggest there is often considerable variation in how symptoms become manifest. The age of onset of Alzheimer disease can vary by as much as 20 years between concordant MZ twins (Cook, Schneck, & Clark, 1981; Nee & Lippa, 1999). Even for autism, in which heritability estimates are extremely high, MZ-concordance rates are still below 100%, and the phenotypic expression of the disorder varies widely within MZ twin pairs (Le Couteur et al., 1996). Such discordance between MZ twins is generally attributed to nonshared environmental factors that, as discussed below, may exert their influence upon phenotype via epigenetic processes. Epigenetic factors may also cause phenotypic changes independently of the environment, however, and random stochastic and developmental epigenetic changes are known to be a cause of phenotypic variability. This could be one explanation for why the empirical evidence for such a large environmental contribution to any psychiatric disorder is lacking. A review of numerous behavioral studies measuring the environments of twins and nontwin siblings, and relating them to differences in their developmental outcomes, has shown that although over 50% of phenotypic variance is accounted for by factors attributed to be nonshared environment, actual measured environmental variables account for only a very small proportion of this variability, with no specific environmental risk factors being conclusively linked to etiology (Turkheimer & Waldron, 2000). As described below, the partial stability of epigenetic signals provides an alternative explanation for phenotypic discordance between MZ twins.

Psychiatric disorders are also often characterized by dramatic sexual dimorphism in disease prevalence (Kaminsky, Wang, & Petronis, 2006). Autism and ADHD, for example, show a much higher prevalence in males, whereas depression is more prevalent in females. Sex differences are also a feature of disease progression: in schizophrenia, for example, men generally develop symptoms between 15 and 25 years of age, but for women the period of maximum onset is around age 30, with a smaller peak around 50 (i.e., menopausal age). Interestingly, it has been shown that sex hormones such as estrogen often act by altering the molecular epigenetic signatures of specific chromosomal regions, modulating the access of transcription factors and producing long-lasting sex-specific epigenetic effects on gene transcription (Kaminsky et al., 2006).

Another observation indicative of an epigenetic contribution to the etiology of psychiatric phenotypes is the abnormal levels of folate and homocysteine, a marker indicative of dysregulated DNA methylation, in the plasma of affected individuals for several disorders including Alzheimer disease (Tchantchou, 2006), autism (Rogers, 2008), and schizophrenia (Haidemenos et al., 2007). Finally, epigenetic mechanisms may also be behind the parental-origin effects often observed in molecular genetic association studies of psychiatric illnesses (e.g., Hawi et al., 2005). One explanation for such effects is genomic imprinting—i.e., differential gene expression at either a chromosomal or allelic level, depending on whether the genetic material has been inherited from the paternal or maternal side. Genomic imprinting is fundamental to normal mammalian development and growth, and it plays an important role in brain function and behavior (Davies, Isles, & Wilkinson, 2005). It has been demonstrated that loss of imprinting at *IGF2* in the cerebellum is correlated with brain weight in males (Pidsley et al., 2009), an interesting observation given the association between brain size and disorders such as schizophrenia (Harrison, Freemantle, & Geddes, 2003). Of note, several studies report that interindividual differences in IGF2 methylation may be environmentally mediated, particularly by prenatal exposure to famine (Heijmans et al., 2008), which is of particular relevance given the known link between prenatal factors (including nutrition) and the risk of developing psychiatric illness as an adult.

Compared to other biomedical disorders, particularly cancer, relatively few studies have empirically investigated the role of epigenetic factors in the etiology of psychiatric illness. Most focus has been on the major psychotic disorders of schizophrenia and bipolar disorder, with the majority of studies focusing on the role of DNA methylation changes in the vicinity of specific candidate genes. Early studies reported DNA methylation differences associated with schizophrenia in the vicinity of both *COMT* (Abdolmaleky et al., 2006) and reelin (*RELN*) (Grayson et al., 2005), although these findings were not confirmed by other groups using fully quantitative methylation profiling methods (Dempster, Mill, Craig, & Collier, 2006; Tochigi et al., 2008). Furthermore, cortical γ-aminobutyric acid (GABA)-ergic neurons in schizophrenia have been shown to express increased levels of DNA-methyltransferase-1 (DNMT1), which is associated with altered expression of both reelin and GAD67 (Grayson et al., 2005; Veldic, Guidotti, Maloku, Davis, & Costa, 2005).

Two recent studies have employed genome-wide approaches to identify DNA methylation changes associated with major psychosis. The first investigated DNA methylation differences between MZ twins discordant for bipolar disorder (Kuratomi et al., 2007). They found evidence for increased DNA methylation in affected twins in the region upstream of the transcription start-site of the spermine synthase gene (SMS) and lower DNA methylation upstream of the peptidyl prolyl isomerase E-like gene (PPIEL). Although DNA methylation upstream of SMS was not correlated with expression of the gene, a strong inverse correlation between PPIEL expression and DNA methylation was observed. More recently, Mill et al. (2008) utilized frontal cortex brain tissue from patients with schizophrenia and bipolar disorder to assess DNA methylation across approximately 12,000 regulatory regions of the genome using CpG island microarrays (Mill et al., 2008). Consistent with increasing evidence for altered glutamatergic and GABAergic neurotransmission in the pathogenesis of major psychosis (Benes & Berretta, 2001; Coyle, 2004), this study identified epigenetic changes in loci associated with both these neurotransmitter pathways. Glutamate is the most abundant fast excitatory neurotransmitter in the mammalian nervous system, with a critical role in synaptic plasticity. Several lines of evidence link the glutamate system to psychosis, in particular the observation that glutamate receptor agonists can cause psychotic symptoms in unaffected individuals.

X-chromosome inactivation silences genes on one X-chromosome to ensure dosage compensation with males via a process involving hypermethylation of CpG islands. X-inactivation in any given cell is typically random, and is maintained once established, so that the inactivated allele is transcriptionally silenced for the lifetime of that cell (Avner & Heard, 2001). Skewed X-chromosome inactivation, in which either the maternally inherited or paternally inherited X is preferentially silenced, is an epigenetic process that could partially explain female MZ twin discordance for complex psychiatric disorders (Craig et al., 2004). The potential role of skewed X-inactivation in major psychotic disorders was highlighted by a study examining X-chromosome inactivation in a series of 63 female MZ twin pairs concordant or discordant for bipolar disorder or schizophrenia and healthy MZ controls (Rosa et al., 2007). They found that discordant female bipolar disorder twins showed greater differences in the methylation of the maternal and paternal chromosome X alleles than did concordant twin pairs, thus suggesting a potential contribution from X-linked loci to discordance within twin pairs for bipolar disorder.

The Role of Epigenetics in Mediating G × E

It is often suggested that environmental mediation of the epigenome, such as in the examples described above, provides a mechanism behind genotype–environment (G × E) interactions. Strictly, this is not the case, as G × E interactions, by definition, also require the involvement of a specific genotype (Caspi & Moffitt, 2006). The examples given above are all illustrations of how the environment can directly alter gene function (and phenotype) via epigenetic processes, but none is an instance of true G × E interaction as there is no evidence that these effects are mediated by genotype. There are, however, several pathways through which epigenetic changes could mediate G × E interactions:

1. Genotype influences gene function via epigenetic effects (e.g., if a genetic variant directly alters DNA methylation), which in turn are susceptible to environmental mediation (Fig. 7.3A).
2. The environment can mediate the expression of functional polymorphisms via epigenetic changes (Fig. 7.3B and 7.3C).
3. Genotype alters the sensitivity of promoter regulatory regions to epigenetic changes in response to the environment.

At the most basic level, it is intuitive that the pathogenic effect of a polymorphism associated with disrupted gene function is likely to be dependent upon the degree to which that particular variant is actually expressed. It is thus plausible that risk could be exaggerated or suppressed if expression is directly influenced by environmental factors via processes such as DNA methylation.

Of particular interest are so-called "metastable epialleles"; loci that can be epigenetically modified to produce a range of phenotypes from genetically identical cells (Rakyan, Blewitt, Druker, Preis, & Whitelaw, 2002). Many of these loci have been shown to be environmentally sensitive, and particularly affected by the prenatal environment of the developing fetus. A classic example of how such a mechanism could explain gene–environment interactions is provided by the agouti viable yellow allele (A^{vy}) inbred mouse strain, which demonstrates a range of coat color and metabolic phenotypes, depending upon the epigenetic state of a large transposable element inserted upstream of the agouti gene. The transposon contains a cryptic promoter, which expresses a phenotype characterized by yellow fur and various detrimental metabolic features, such as diabetes and obesity. When the transposon is methylated, this phenotype is not expressed, and the mice have brown fur and are metabolically healthy. Interestingly, DNA methylation across this region (and thus phenotype) can be manipulated in offspring by altering the diet of pregnant mothers (Cooney, Dave, & Wolff, 2002; Dolinoy,

A

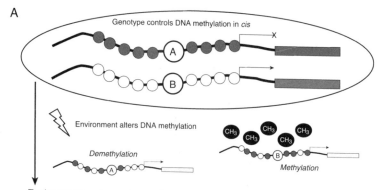

Environment and genotype act on a common pathway to control gene expression

B

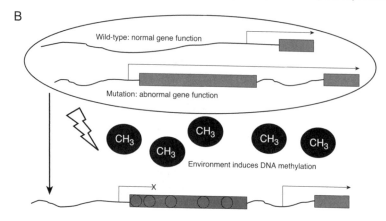

Mutation silenced by DNA methylation: normal gene function

C

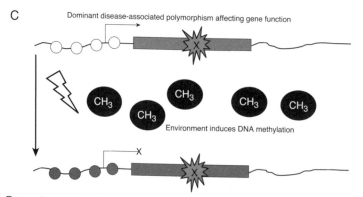

Expression of risk polymorphism silenced by environmentally-induced DNA methylation

Figure 7.3. **Epigenetic alterations may mediate gene–environment interactions in a number of ways.** (A) Genotype may exert an effect on the expression of nearby genes (i.e., in cis) via epigenetic effects in an allele-specific manner,

Figure 7.3. (*continued*)
which in turn may be susceptible to environmental mediation. In this figure, the red bar represents repressed gene expression from one allele that is associated with high levels of promoter DNA methylation (*filled circles*). The other allele, on the other hand, is expressed at high levels (*green bar*) and is characterized by low levels of promoter DNA methylation (*empty circles*). Such cis effects on the epigenome may be subsequently influenced by the environment—e.g., via the addition of methyl groups (CH_3) from the diet. In this way, genotype and the environment act on a common pathway to control gene expression. (**B**) There are many instances in which genetic polymorphisms create abnormal gene products. Epigenetic processes may mediate gene–environment interactions if, for example, methyl groups (CH_3) in the environment act to silence abnormal gene products associated with genomic mutations. (**C**) Similarly, the effect of dominant disease-associated polymorphisms may be silenced by epigenetic processes occurring as a result of environmental factors. (See Plate 1 for a color version of this figure.)

Weidman, Waterland, & Jirtle, 2006). Enriching the maternal diet with methyl-donor supplements increases offspring DNA methylation, leading to gene expression changes associated with brown fur and metabolic health. Gene–environment interactions may also result when genetic polymorphisms alter the ability of a specific region of the genome to be epigenetically altered in *response* to an environmental pathogen. The interplay between the genome, the environment, and epigenetic processes is further complicated by the fact that some DNA alleles and haplotypes are themselves associated with a specific epigenetic profile (Schalkwyk et al., 2010). For example, allele-specific epigenetic modifications have been associated with "risk" polymorphisms in psychiatric candidate genes including the serotonin receptor gene (*5HTR2A*) (Polesskaya, Aston, & Sokolov, 2006) and the encoding brain-derived neurotrophic factor gene (*BDNF*) (Mill et al., 2008). Given the known influence of environmental factors on epigenetic regulation, the direct regulation of DNA methylation by genetic variation would suggest a common pathway behind both genetic and environmental effects and a potential mechanism for G × E interaction (Meaburn, Schalkwyk & Mill, 2010).

CONCLUSION

Implications for Etiological Studies in Psychiatry

Despite extensive research efforts to determine the genetic and environmental factors that underlie psychiatric illness, the etiology of these disorders remains poorly understood. Although interactions between the genome and the environment are key mediators of disease risk, there is a growing consensus that epigenetic variation should be included in etiological models, as epigenetic factors have the potential to integrate and mediate the effect of genetic and environmental factors upon the phenotype. The study of epigenetics in complex disease is still a relatively new field, yet there are already empirical studies that have identified associations between epigenetic modifications and mental illness. Recent advances in technology mean that we are now at a stage where it is feasible to start investigating the ways in which environmental factors act upon the genome to bring about epigenetic changes in gene expression and behavior. Understanding the epigenetic processes involved in linking specific environmental pathogens to an increased risk for psychiatric disorders may offer new possibilities for preventative and therapeutic intervention. In addition to highlighting novel neurobiological pathways involved in mental illness that could inform the development of new drugs, the discovery of epigenomic dysfunction would be particularly exciting in this regard, given the dynamic regulation of epigenetic phenomena. Importantly, the dynamic nature of the epigenome means that, unlike pathogenic DNA sequence mutations, epigenetic disruption is potentially reversible, and thus a realistic target for pharmacological intervention. In fact, a number of currently prescribed treatments for several neuropsychiatric disorders, including psychosis, have been shown to have lasting effects on the epigenome. Valproic acid, which is widely used to treat symptoms of psychosis, for example, is a potent inhibitor of histone deacetylating enzymes and induces widespread epigenetic reprogramming (Milutinovic, D'Alessio, Detich, & Szyf, 2007). The recent growth in epigenetics research, combined with the emerging new high-throughput technologies for epigenomic profiling currently being developed, promise to bring us closer to understanding the molecular etiology of psychiatric disorders and the development of novel prevention and treatment measures.

REFERENCES

Abdolmaleky, H. M., Cheng, K. H., Faraone, S. V., Wilcox, M., Glatt, S. J., Gao, F., et al. 2006. Hypomethylation of MB-COMT promoter is a major risk factor for schizophrenia and bipolar disorder. *Human Molecular Genetics, 15*(21), 3132–3145.

Avner, P., Heard, E. (2001). X-chromosome inactivation: counting, choice and initiation. *Nature Reviews. Genetics, 2*(1), 59–67.

Benes, F. M., Berretta, S. (2001). GABAergic interneurons: implications for understanding schizophrenia and bipolar disorder. *Neuropsychopharmacology, 25*(1), 1–27.

Berger, S. L. (2007). The complex language of chromatin regulation during transcription. *Nature, 447*(7143), 407–412.

Bestor, T. H. (2000). The DNA methyltransferases of mammals. *Human Molecular Genetics, 9*(16), 2395–2402.

Bird, A. P. (1986). CpG-rich islands and the function of DNA methylation. *Nature, 321*(6067), 209–213.

Bjornsson, H. T., Sigurdsson, M. I., Fallin, M. D., Irizarry, R. A., Aspelund T., Cui, H., et al. (2008). Intra-individual change over time in DNA methylation with familial clustering. *Journal of the American Medical Association, 299*(24), 2877–2883.

Cardno, A. G., & Gottesman, I. I. (2000). Twin studies of schizophrenia: from bow-and-arrow concordances to star wars Mx and functional genomics. *American Journal of Medical Genetics, 97*(1), 12–17.

Caspi, A., & Moffitt, T. E. (2006). Gene-environment interactions in psychiatry: joining forces with neuroscience. *Nature Reviews. Neuroscience, 7*(7),583–590.

Caspi, A., Sugden, K., Moffitt, T. E., Taylor, A., Craig, I. W., Harrington, H., et al. (2003). Influence of life stress on depression: moderation by a polymorphism in the 5-HTT gene. *Science, 301*(5631), 386–389.

Chang, H. S., Anway, M. D., Rekow, S. S., & Skinner, M. K. (2006). Transgenerational epigenetic imprinting of the male germline by endocrine disruptor exposure during gonadal sex determination. *Endocrinology, 147*(12), 5524–5541.

Christensen, B. C., Houseman, E. A., Marsit, C. J., Zheng, S., Wrensch, M. R., Wiemels, J. L., et al. (2009). Aging and environmental exposures alter tissue-specific DNA methylation dependent upon CpG island context. *PLoS Genetics, 5*(8), e1000602.

Cibelli, J. B., Campbell, K. H., Seidel, G. E., West, M. D., & Lanza, R. P. (2002). The health profile of cloned animals. *Nature Biotechnology, 20*(1),13–14.

Cook, R. H., Schneck, S. A., & Clark, D. B. (1981). Twins with Alzheimer's disease. *Archives of Neurology, 38*(5), 300–301.

Cooney, C. A., Dave, A. A., & Wolff, G. L. (2002). Maternal methyl supplements in mice affect epigenetic variation and DNA methylation of offspring. *Journal of Nutrition, 132*(8 Suppl), 2393S–2400S.

Coyle, J. T. (2004). The GABA-glutamate connection in schizophrenia: which is the proximate cause? *Biochemical Pharmacology, 68*(8), 1507–1514.

Craddock, N., Jones, I. (1999). Genetics of bipolar disorder. *Journal of Medical Genetics, 36*(8), 585–594.

Craig, I. W., Harper, E., & Loat, C. S. (2004). The genetic basis for sex differences in human behaviour: role of the sex chromosomes. *Annals of Human Genetics, 68*(Pt 3), 269–284.

Davies, W., Isles, A. R., & Wilkinson, L. S. (2005). Imprinted gene expression in the brain. *Neuroscience and Biobehavioral Reviews, 29*(3), 421–430.

Dempster, E. L., Mill, J., Craig, I. W., & Collier, D. A. (2006). The quantification of COMT mRNA in post mortem cerebellum tissue: diagnosis, genotype, methylation and expression. *BMC Medical Genetics, 7*, 10.

Dolinoy, D. C., & Jirtle, R. L. (2008). Environmental epigenomics in human health and disease. *Environmental and Molecular Mutagenesis, 49*(1), 4–8.

Dolinoy, D. C., Weidman, J. R., & Jirtle, R. L. (2007). Epigenetic gene regulation: linking early developmental environment to adult disease. *Reproductive Toxicology, 23*(3), 297–307.

Dolinoy, D. C., Weidman, J. R., Waterland, R. A., & Jirtle, R. L. (2006). Maternal genistein alters coat color and protects Avy mouse offspring from obesity by modifying the fetal epigenome. *Environmental Health Perspectives, 114*(4), 567–572.

Duthie, S. J., Narayanan, S., Sharp, L., Little, J., Basten, G., & Powers, H. (2004). Folate, DNA stability and colo-rectal neoplasia. *Proceedings of the Nutrition Society, 63*(4), 571–578.

Feinberg, A. P. (2007). Phenotypic plasticity and the epigenetics of human disease. *Nature, 447*(7143), 433–440.

Ferreira, M. A., O'Donovan, M. C., Meng, Y. A., Jones, I. R., Ruderfer, D. M., Jones, L., et al. (2008). Collaborative genome-wide association analysis supports a role for ANK3 and CACNA1C in bipolar disorder. *Nature. Genetics, 40*(9), 1056–1058.

Flanagan, J. M., Popendikyte, V., Pozdniakovaite, N., Sobolev, M., Assadzadeh, A., Schumacher, A., et al. (2006). Intra- and interindividual epigenetic variation in human germ cells. *American Journal of Human Genetics, 79*(1), 67–84.

Fraga, M. F., Ballestar, E., Paz, M. F., Ropero, S., Setien, F., Ballestar, M. L., et al. (2005). Epigenetic differences arise during the lifetime of monozygotic twins. *Proceedings of the National Academy of Sciences of the U.S.A., 102*(30), 10604–10609.

Gartner, K. (1990). A third component causing random variability beside environment and genotype. A reason for the limited success of a 30 year long effort to standardize laboratory animals? *Laboratory Animals, 24*(1), 71–77.

Gatz, M., Reynolds, C. A., Fratiglioni, L., Johansson, B., Mortimer, J. A., Berg, S., et al. (2006). Role of genes and environments for explaining Alzheimer disease. *Archives of General Psychiatry, 63*(2), 168–174.

Grayson, D. R., Jia, X., Chen, Y., Sharma, R. P., Mitchell, C. P., Guidotti, A., et al. (2005). Reelin promoter hypermethylation in schizophrenia. *Proceedings of the National Academy of Sciences of the U.S.A., 102*(26), 9341–9346.

Haidemenos, A., Kontis, D., Gazi, A., Kallai, E., Allin, M., & Lucia, B. (2007). Plasma homocysteine, folate and B12 in chronic schizophrenia. *Progress in neuro-psychopharmacology & biological psychiatry, 31*(6), 1289–1296.

Hamilton, A., Voinnet, O., Chappell, L., & Baulcombe, D. (2002). Two classes of short interfering RNA in RNA silencing. *EMBO Journal, 21*(17), 4671–4679.

Harrison, P. J., Freemantle, N., & Geddes, J. R. (2003). Meta-analysis of brain weight in schizophrenia. *Schizophrenia Research, 64*(1), 25–34.

Hatchwell, E., & Greally, J. M. (2007). The potential role of epigenomic dysregulation in complex human disease. *Trends in Genetics, 23*(11), 588–595.

Hawi, Z., Segurado, R., Conroy, J., Sheehan, K., Lowe, N., Kirley, A., et al. (2005). Preferential transmission of paternal alleles at risk genes in attention-deficit/hyperactivity disorder. *American Journal of Human Genetics,* *77*(6), 958–965.

Heijmans, B. T., Tobi, E. W., Stein, A. D., Putter, H., Blauw, G. J., Susser, E. S., et al. (2008). Persistent epigenetic differences associated with prenatal exposure to famine in humans. *Proceedings of the National Academy of Sciences of the U.S.A.,* *105*(44), 17046–17049.

Henikoff, S., & Matzke, M. A. (1997). Exploring and explaining epigenetic effects. *Trends in Genetics,* 13(8), 293–295.

Irizarry, R. A., Ladd-Acosta, C., Wen, B., Wu, Z., Montano, C., Onyango, P., et al. (2009). The human colon cancer methylome shows similar hypo- and hyperm-ethylation at conserved tissue-specific CpG island shores. *Nature. Genetics,* *41*(2), 178–186.

Jaenisch, R., & Bird, A. (2003). Epigenetic regulation of gene expression: how the genome integrates intrinsic and environmental signals. *Nature. Genetics,* 33, 245–254.

Jirtle, R. L., & Skinner, M. K. (2007). Environmental epigenomics and disease sus-ceptibility. *Nature Reviews. Genetics,* *8*(4), 253–262.

Jones, P. A., & Baylin, S. B. (2007). The epigenomics of cancer. *Cell,* *128*(4), 683–692.

Kaminsky, Z., Wang, S. C., & Petronis, A. (2006). Complex disease, gender and epigenetics. *Annuls of Medicine,* *38*(8), 530–544.

Kaminsky, Z. A., Tang, T., Wang, S. C., Ptak, C., Oh, G. H., Wong, A. H., et al. (2009). DNA methylation profiles in monozygotic and dizygotic twins. *Nature. Genetics,* *41*(2), 240–245.

Kato, T. (2007). Molecular genetics of bipolar disorder and depression. *Psychiatry and Clinical Neurosciences,* *61*(1), 3–19.

Klar, A. J. (1998). Propagating epigenetic states through meiosis: where Mendel's gene is more than a DNA moiety. *Trends in Genetics,* *14*(8), 299–301.

Klose, R. J., & Bird, A. P. (2006). Genomic DNA methylation: the mark and its mediators. *Trends in Biochemical Sciences,* *31*(2), 89–97.

Kornberg, R. D., & Lorch, Y. (1999). Twenty-Five Years of the Nucleosome, Fundamental Particle of the Eukaryote Chromosome. *Cell,* *98,* 285–294.

Kuratomi, G., Iwamoto, K., Bundo, M., Kusumi, I., Kato, N., Iwata, N., et al. (2007). Aberrant DNA methylation associated with bipolar disorder identified from discordant monozygotic twins. *Molecular Psychiatry,* *13,* 429–441.

Latchman, D. S. (1997). Transcription factors: an overview. *International journal of biochemistry & cell biology,* *29*(12),1305–1312.

Le Couteur, A., Bailey, A., Goode, S., Pickles, A., Robertson, S., Gottesman, I., et al. (1996). A broader phenotype of autism: the clinical spectrum in twins. *Journal of Child Psychology and Psychiatry, and Allied Disciplines,* *37*(7), 785–801.

McGowan, P. O., Sasaki, A., D'Alessio, A. C., Dymov, S., Labonte, B., Szyf, M., et al. (2009). Epigenetic regulation of the glucocorticoid receptor in human brain associates with childhood abuse. *Nature Neuroscience,* *12*(3), 342–348.

Merikangas, A. K., Corvin, A. P., & Gallagher, L. (2009). Copy-number variants in neurodevelopmental disorders: promises and challenges. *Trends in Genetics, 25* (12), 536–544.

Meaburn, E. L., Schalkwyk, L. C., & Mill, J. (2010). Allele-specific methylation in the human genome Implications for genetic studies of complex disease. *Epigenetics, 5*(7).

Mill, J., & Petronis, A. (2007). Molecular studies of major depressive disorder: the epigenetic perspective. *Molecular Psychiatry, 12*(9), 799–814.

Mill, J., & Petronis, A. (2008). Pre- and peri-natal environmental risks for attention-deficit hyperactivity disorder (ADHD): the potential role of epigenetic processes in mediating susceptibility. *Journal of Child Psychology and Psychiatry, and Allied Disciplines, 49*(10), 1020–1030.

Mill, J., Tang, T., Kaminsky, Z., Khare, T., Yazdanpanah, S., Bouchard, L., et al. (2008). Epigenomic profiling reveals DNA methylation changes associated with major psychosis. *American Journal of Human Genetics, 82*(3), 696–711.

Milutinovic, S., D'Alessio, A.C., Detich, N., & Szyf, M. (2007). Valproate induces widespread epigenetic reprogramming which involves demethylation of specific genes. *Carcinogenesis, 28*(3), 560–571.

Nee, L. E., & Lippa, C. F. (1999). Alzheimer's disease in 22 twin pairs—13-year follow-up: hormonal, infectious and traumatic factors. *Dementia and Geriatric Cognitive Disorders, 10*(2), 148–151.

Ooi, S. K., Qiu, C., Bernstein, E., Li, K., Jia, D., Yang, Z., et al. (2007). DNMT3L connects unmethylated lysine 4 of histone H3 to de novo methylation of DNA. *Nature, 448*(7154), 714–717.

Pembrey, M. E., Bygren, L. O., Kaati, G., Edvinsson, S., Northstone, K., Sjostrom, M., et al. (2006). Sex-specific, male-line transgenerational responses in humans. *European Journal of Human Genetics, 14*(2), 159–166.

Petronis, A. (2003). Epigenetics and bipolar disorder: new opportunities and challenges. *American Journal of Medical Genetics. Part C, Seminars in Medical Genetics, 123*C(1), 65–75.

Petronis, A. (2004). The origin of schizophrenia: genetic thesis, epigenetic antithesis, and resolving synthesis. *Biological Psychiatry, 55*(10), 965–970.

Petronis, A., Gottesman, I., Kan, P., Kennedy, J. L., Basile, V. S., Paterson, A. D., & Popendikyte, V. (2003). Monozygotic twins exhibit numerous epigenetic differences: Clues to twin discordance? *Schizophrenia Bulletin, 29*(1), 169–178.

Pidsley, R., Dempster, E. L., & Mill, J. (2009). Brain weight in males is correlated with DNA methylation at IGF2. *Molecular Psychiatry, 15,* 880–881.

Pidsley, R., & Mill, J. (2010) Epigenetic Studies of Psychosis: Current Findings, Methodological Approaches, and Implications for Postmortem Research. *Biological Psychiatry.* May 24. [Epub ahead of print]

Polesskaya, O. O., Aston, C., & Sokolov, B. P. (2006). Allele C-specific methylation of the 5-HT2A receptor gene: evidence for correlation with its expression and expression of DNA methylase DNMT1. *Journal of Neuroscience Research, 83*(3), 362–373.

Purcell, S. M., Wray, N. R., Stone, J. L., Visscher, P. M., O'Donovan, M. C., Sullivan, P. F., et al. (2009). Common polygenic variation contributes to risk of schizophrenia and bipolar disorder. *Nature, 460*(7256), 748–752.

Rakyan, V., & Whitelaw, E. (2003). Transgenerational epigenetic inheritance. *Current Biology, 13*(1), R6.

Rakyan, V. K., Blewitt, M. E., Druker, R., Preis, J. I., & Whitelaw, E. (2002). Metastable epialleles in mammals. *Trends in Genetics, 18*(7), 348–351.

Renthal, W., & Nestler, E. J. (2008). Epigenetic mechanisms in drug addiction. *Trends in Molecular Medicine, 14*(8), 341–350.

Reynolds, E. (2006). Vitamin B12, folic acid, and the nervous system. *Lancet Neurology, 5*(11), 949–960.

Ribas-Fito, N., Torrent, M., Carrizo, D., Julvez, J., Grimalt, J. O., & Sunyer, J. (2007). Exposure to hexachlorobenzene during pregnancy and children's social behavior at 4 years of age. *Environmental Health Perspectives, 115*(3), 447–450.

Richards, E. J. (2006). Inherited epigenetic variation—revisiting soft inheritance. *Nature Reviews. Genetics, 7*(5), 395–401.

Riggs, A. D., Xiong, Z., Wang, L., & LeBon, J. M. (1998). Methylation dynamics, epigenetic fidelity and X chromosome structure. *Novartis Foundation Symposium, 214,* 214–225; discussion 225–232.

Riley, B., Kendler, K. S. (2006). Molecular genetic studies of schizophrenia. *European Journal of Human Genetics, 14*(6), 669–680.

Robertson, K. D., Wolffe, A. P. (2000). DNA methylation in health and disease. *Nature Reviews. Genetics, 1*(1), 11–19.

Rogers, E. J. (2008). Has enhanced folate status during pregnancy altered natural selection and possibly autism prevalence? A closer look at a possible link. *Medical Hypotheses, 71*(3), 406–410.

Rosa, A., Picchioni, M. M., Kalidindi, S., Loat, C. S., Knight, J., Toulopoulou, T., et al. (2007). Differential methylation of the X-chromosome is a possible source of discordance for bipolar disorder female monozygotic twins. *American Journal of Medical Genetics. Part B, Neuropsychiatric Genetics, 147B*(4), 459–462.

Saha, R. N., & Pahan, K. (2006). HATs and HDACs in neurodegeneration: a tale of disconcerted acetylation homeostasis. *Cell Death and Differentiation, 13*(4), 539–550.

Sandovici, I., Kassovska-Bratinova, S., Loredo-Osti, J. C., Leppert, M., Suarez, A., Stewart, R., et al. (2005). Interindividual variability and parent of origin DNA methylation differences at specific human Alu elements. *Human Molecular Genetics, 14*(15), 2135–2143.

Schalkwyk, L. C., Meaburn, E. L., Smith, R., Dempster, E. L., Jeffries, A. R., Davies, M.N., Plomin, R., & Mill, J. (2010). Allelic skewing of DNA methylation is widespread across the genome. *American Journal of Human Genetics, 86*(2): 196–212.

Schanen, N. C. (2006). Epigenetics of autism spectrum disorders. *Human Molecular Genetics, 15*(Suppl 2), R138–150.

Shen, Y., Chow, J., Wang, Z., & Fan, G. (2006). Abnormal CpG island methylation occurs during in vitro differentiation of human embryonic stem cells. *Human Molecular Genetics, 15*(17), 2623–2635.

Siegmund, K. D., Connor, C. M., Campan, M., Long, T. I., Weisenberger, D. J., Biniszkiewicz, D., Jaenisch, R., et al. (2007). DNA methylation in the human cerebral cortex is dynamically regulated throughout the life span and involves differentiated neurons. *PLoS One, 2*(9), e895.

St. Clair, D., Xu, M., Wang, P., Yu, Y., Fang, Y., Zhang, F., et al. (2005). Rates of adult schizophrenia following prenatal exposure to the Chinese famine of 1959–1961. *Journal of the American Medical Association, 294*(5), 557–562.

Stefansson, H., Ophoff, R. A., Steinberg, S., Andreassen, O. A., Cichon, S., Rujescu, D., et al. (2009). Common variants conferring risk of schizophrenia. *Nature, 460*(7256), 744–747.

Susser, E., Neugebauer, R., Hoek, H. W., Brown, A. S., Lin, S., Labovitz, D., et al. (1996). Schizophrenia after prenatal famine. Further evidence. *Archives of General Psychiatry, 53*(1), 25–31.

Susser, E. S., & Lin, S. P. (1992). Schizophrenia after prenatal exposure to the Dutch Hunger Winter of 1944–1945. *Archives of General Psychiatry, 49*(12), 983–988.

Szyf, M., Weaver, I., & Meaney, M. (2007). Maternal care, the epigenome and phenotypic differences in behavior, *Reproductive Toxicology*, 24(1), 9–19.

Tamashiro, K. L., Wakayama, T., Yamazaki, Y., Akutsu, H., Woods, S. C., Kondo, S., Yanagimachi, R., et al. (2003). Phenotype of cloned mice: development, behavior, and physiology. *Experimental Biology and Medicine (Maywood, N.J)*, 228(10), 1193–1200.

Tchantchou, F. (2006). Homocysteine metabolism and various consequences of folate deficiency. *Journal of Alzheimers Disease, 9*(4),421–427.

Tochigi, M., Iwamoto, K., Bundo, M., Komori, A., Sasaki, T., Kato, N., et al. (2008). Methylation status of the reelin promoter region in the brain of schizophrenic patients. *Biological Psychiatry, 63*(5), 530–533.

Turkheimer, E., & Waldron, M. (2000). Nonshared environment: a theoretical, methodological, and quantitative review. *Psychological Bulletin, 126*(1), 78–108.

Uher, R., & McGuffin, P. (2010). The moderation by the serotonin transporter gene of environmental adversity in the etiology of depression: 2009 update. *Molecular Psychiatry, 15*(1), 18–22.

Ushijima, T., Watanabe, N., Okochi, E., Kaneda, A., Sugimura, T., & Miyamoto, K. (2003). Fidelity of the methylation pattern and its variation in the genome. *Genome Research,13*(5), 868–874.

Veldic, M., Guidotti, A., Maloku, E., Davis, J. M., & Costa E. (2005). In psychosis, cortical interneurons overexpress DNA-methyltransferase 1. *Proceedings of the National Academy of Sciences of the U.S.A.,102*(6), 2152–2157.

Waterland, R. A., & Michels, K. B. (2007). Epigenetic Epidemiology of the Developmental Origins Hypothesis. *Annual Review of Nutrition, 27*, 363–388.

Weaver, I. C., Cervoni, N., Champagne, F. A., D'Alessio, A. C., Sharma, S., Seckl, J. R., Dymov, S., et al. (2004). Epigenetic programming by maternal behavior. *Nature Neuroscience, 7*(8), 847–854.

Weaver, I. C., Meaney, M. J., & Szyf, M. (2006). Maternal care effects on the hippocampal transcriptome and anxiety-mediated behaviors in the offspring that are reversible in adulthood. *Proceedings of the National Academy of Sciences of the U.S.A., 103*(9), 3480–3485.

Wong, A. H., Gottesman, I. I., & Petronis, A. 2005. Phenotypic differences in genetically identical organisms: the epigenetic perspective. *Human Molecular Genetics,14*(Suppl 1), R11–R18.

Wong, C. C., Caspi, A., Williams, B., Craig, I. W., Houts, R., Ambler, A., Moffitt, T. E., & Mill, J. (2010). A longitudinal study of epigenetic variation in twins. *Epigenetics, 5*(6).

Wu, J., Basha, M. R., & Zawia, N. H. (2008). The environment, epigenetics and amyloidogenesis. *Journal of Molecular Neuroscience, 34*(1), 1–7.

8

Integrating Knowledge of Genetic and Environmental Pathways to Complete the Developmental Map

M. MARIA GLYMOUR, WIM VELING, AND EZRA SUSSER

The goal of clinical and applied research is to identify intervention opportunities to prevent or treat disease and promote health. To identify such opportunities, we must elaborate the causal mechanisms that lead to poor health. Causal inference is notoriously difficult, but we argue in this chapter that integrating research on genetic and environmental pathways in development can help illuminate the intersecting causal steps in both types of pathways. Although we may have only very incomplete understanding of mechanisms in either genetic or environmental domains, we can use the information we *do* have on environmental determinants to help test hypotheses about genetic determinants, and vice versa.

We begin this chapter by briefly describing an example in which an established link between an environmental exposure (famine) and a mental disorder (schizophrenia) can be used to illuminate genetic influences on that same mental disorder (Gottesman, 1991). We first summarize the evidence for the pathways linking prenatal famine to schizophrenia and explain how this can inform research on the genetic foundations of schizophrenia. We then turn to an example in which research moves in the opposite direction: leveraging genetic information to discover environmental pathways. For this, we focus on childhood neurodevelopmental outcomes instead of schizophrenia. We explain how the hypothesized effects of folate on neurodevelopmental outcomes can be examined using an approach called *Mendelian randomization* (MR). We will spend much of the chapter describing the central idea in MR and explaining how capitalizing on emerging information on gene–environment interactions could potentially provide stronger MR designs in the next generation of studies. Finally, we describe what we consider an important step in MR applications: tests for violations of the assumptions. We discuss approaches to such evaluations, in particular applications based on understanding of gene–environment interactions.

USING ENVIRONMENTAL EFFECTS TO DISCOVER GENETIC PATHWAYS

Starvation and Schizophrenia: Evidence for a Folate Pathway

Three studies have reported a link between periconceptional or early gestational exposure to starvation and risk of schizophrenia in offspring (Brown & Susser, 2008; St Clair et al., 2005; Susser et al., 1996; Xu et al., 2009). The first of these studies was based on the Dutch Hunger Winter of 1944–1945. The results indicated that persons who were conceived at the height of the famine had an increased risk of neural tube defects (NTDs) at birth and of schizophrenia in adulthood (Susser et al., 1996). For NTDs, it is now known that the etiologically relevant exposure period was periconceptional or early gestational, roughly meaning from 2 weeks prior to 4 weeks after conception. Randomized controlled trials (RCTs) have proven that maternal folate supplementation in the periconceptional period reduces the risk of NTDs (Group, 1991). The relevant exposure period for schizophrenia turned out to be very similar, in contrast to other outcomes such as depression, for which famine exposure later in pregnancy was most relevant (Brown, Susser, Lin, Neugebauer, & Gorman, 1995).

A subsequent study by an independent group in Anhui Province, China, tested the Dutch results in a cohort exposed to the drastic Chinese Famine of 1959–1961 (exact timing varied across regions) that followed the initiation of the Great Leap Forward (St Clair et al., 2005). The Chinese investigators found a two-fold increased risk of schizophrenia in the birth cohorts conceived during the height of the famine. A third study in a different region of China, in which one of us (ES) also participated, found the same result, and added some further refinements (e.g., greater effect in rural areas where famine was much more marked) that supported the validity of the relationship (Xu et al., 2009). Although the Chinese data were less specific than the Dutch with respect to timing of exposure, they were based on far larger numbers. Hence, the three studies had complementary features, and the remarkable concordance of their results provides strong evidence that early prenatal starvation was linked with the risk of schizophrenia. Their results are summarized in Figure 8.1 and Table 8.1.

These results do not, however, reveal the mechanism by which exposure to prenatal starvation is linked to schizophrenia. There are numerous possibilities, including micronutrient deficiencies, maternal stress effects on the neuroendocrine system, toxic effects of maternal ingestion of food substitutes, genetic selection (e.g., genetic differences between women who can and cannot ovulate and conceive during famine conditions), and preconceptional damage

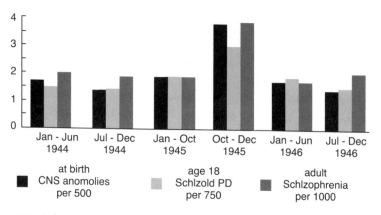

Figure 8.1. **Schizophrenia in adulthood in Dutch birth cohorts exposed and unexposed to prenatal famine.** The months of most extreme famine were February, March, and April 1945. Schizophrenia risk is elevated exclusively among birth cohorts exposed early in gestation, who also showed increased risk of neural tube defects.

Table 8.1. Studies on Chinese Famine and Risk for Schizophrenia

Publication	Year [1]	Cumulative incidence per 1,000 persons (No. cases)		Relative Risk (95% CI)
		Exposed	Unexposed	
St Clair et al., 2005	1960	14.0 (192)	8.3 (3,481)	2.3 (2.0–2.7)
	1961	11.8 (191)	8.3 (3,481)	1.9 (1.7–2.2)
Xu et al., 2009	1960	5.3 (355)	3.9 (3,223)	1.5 (1.4–1.7)
	1961	7.6 (451)	3.9 (3,223)	2.1 (1.9–2.3)

[1] Year of birth. A massive famine occurred in China between 1959 and 1961. Unexposed was defined as being born between 1956 and 1958 or between 1963 through 1965, well outside the famine period. Exposed were individuals conceived or in early gestation during the famine period.

to paternal spermatogonia, among others. Among these various plausible mechanisms linking starvation and schizophrenia, folate deficiency is an appealing explanation. For example, folate deficiency is indirectly implicated by the evidence that folate mediates the effects of famine exposure on NTDs, because of the close correspondence between the etiologic period for schizophrenia and that for NTDs in the Dutch study.

Evidence on Starvation and Schizophrenia Informs Hypotheses on the Genetics of Schizophrenia

Genetic association studies suggest an association between a polymorphism in the folate metabolic pathway (*MTHFR677TT*) and the risk of NTDs (Botto & Yang, 2000). Large-scale meta-analyses of genetic association studies have also linked schizophrenia with the same *MTHFR677TT* polymorphism, although these results cannot be considered definitive due to the possibility of publication bias (Allen et al., 2008). The concordance between the effects of famine on NTDs and schizophrenia provides additional support for the potential links between *MTHFR* polymorphisms and schizophrenia.

To test the hypothesis that folate influences schizophrenia, we can examine what we know about the folate metabolic pathway to illuminate the etiology of schizophrenia. It turns out that many of the physiologic processes influenced by folate are consistent with hypothesized causes of schizophrenia (Brown & Susser, 2008). The folate metabolic pathway is complex, and we refer readers elsewhere for a full discussion (Lucock, 2000). Here, we wish to emphasize three points that are directly relevant to our topic, as diagrammed in Figure 8.2. First, a pathway triggered by folate, and also regulated by *MTHFR*, is thought to convert homocysteine to methionine, via the *one-carbon pathway*. Methionine is required for DNA methylation, which plays a central role in the epigenetic processes shaping gene expression (see below). Second, if the transformation of homocysteine into methionine is slowed down, which can occur due to either folate restriction or mutations of *MTHFR*, homocysteine concentration is likely to increase. Homocysteine is thought to be toxic and has been linked to disturbances of glutamate receptor function (Lipton et al., 1997). As shown in Figure 8.2, this path is also regulated by *MTHFR*. Carriers of the *MTHFR677TT* allele are known to have slower conversion of 5,10 MTHFR into 5MTHF. This leads to slow homocysteine conversion, just as folate restrictions does. Very high concentrations of folate appear to shift the equilibrium of conversion of 5,10 MTHFR into 5MTHF and offset this effect of *MTHFR677TT*. Thus, among high-folate consumers, the effect of *MTHFR677TT* on slowing conversion of homocysteine is effectively silenced. Third, as shown in Figure 8.2, folate

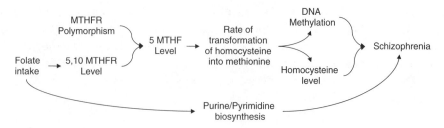

Figure 8.2. **Three causal pathways potentially linking folate intake and schizophrenia.** This figure shows causal pathways via which folate intake may affect schizophrenia. Two of these pathways are mediated by the transformation of homocysteine into methionine. The pathways mediated by homocysteine transformation are also influenced by *MTHFR* alleles, so there is potential for gene–environment interaction.

is key to synthesis of nucleotide bases and thus DNA synthesis and repair. Folate deficiency is thus associated with mutagenesis. This is important for studies of schizophrenia because recent work suggests that de novo mutations, including copy number variations, may play an important role in schizophrenia (McClellan, Susser, & King, 2006; Xu et al., 2008). Figure 8.2 illustrates these three pathways by exhibiting elements of the folate pathway that are relevant to this chapter. We will present variations of this figure later in the chapter, to clarify our logic, but we do not attempt to convey a full understanding of the one-carbon pathway here (readers may refer again to Lucock, 2000).

Of the three pathways potentially linking folate and schizophrenia, we focus on mediation by homocysteine transformation and epigenetic modifications. For more on mutagenesis, see our previous paper (McClellan et al., 2006). The indirect evidence from the famine studies implicating epigenetic pathways in schizophrenia is especially interesting because it seems both biologically plausible and could help explain some puzzles in schizophrenia research. However, epigenetic effects are only recently being recognized as a potentially important mechanism by which environmental conditions modify genetic influences on health.

How Environmental Effects May Be Mediated by Epigenetic Modifications

Epigenetics has been defined as modifications of the DNA or associated proteins, other than DNA sequence variations, that carry information content during cell division (Feinberg, 2007; see also Chapter 7, by Mill). We briefly describe key points relevant to our discussion. The main epigenetic mechanisms involve methylation of the DNA nucleotide cytosine and modification of histones that make up the nucleosome around which the DNA is coiled. Epigenetic marks are often tissue-specific and potentially reversible. The methylation and histone modifications modify transcriptional access to the DNA, thus they can effectively silence or activate genes. Several monogenetic epigenetic diseases have been described, such as Beckwith-Wiedemann syndrome, characterized by loss of normal imprinted gene regulation (DeBaun et al., 2002). Epigenetic mechanisms are important in cancer as well. Tumor development is regulated by activation of growth-promoting genes through both global and gene-specific hypomethylation, and by silencing of tumor suppression genes through hypermethylation (Feinberg, 2007). Relevant to the example we are using in this chapter, there is preliminary evidence that DNA-methylation changes may be associated with schizophrenia (Mill et al., 2008).

The epigenetic status of the genome is genetically programmed (Reik, Dean, & Walter, 2001), but is also influenced by exposure to environmental factors, including nutrition. Experiments with agouti mice showed that periconceptual/early gestational dietary methyl supplementation can produce persistent changes in DNA methylation with life-long phenotypic consequences. In the offspring of agouti mice that were assigned to the supplemented diet, the distribution of coat color shifted from yellow toward brown as a result of the increased methylation status and lower expression of the agouti gene (Waterland & Jirtle, 2003; Wolff, Kodell, Moore, & Cooney, 1998). The latter also has metabolic implications, as yellower agouti mice with overexpression of this gene had a higher prevalence of obesity and diabetes, and an increased susceptibility to tumors than did their browner siblings (Morgan, Sutherland, Martin, & Whitelaw, 1999).

As previously discussed, folate is a central nutrient in epigenetic gene regulation, because it plays a key role in the one-carbon pathway (Fig. 8.2). In this pathway, low folate intake leads to a decreased transformation of homocysteine into methionine, which would, in theory, result in lower DNA methylation and modify gene expression. The subsequent methylation/ expression changes could affect any gene in the genome: It is not yet understood if and how organism-level changes in methyl group availability are targeted to modify epigenetic marks on a specific gene. Indeed, individuals who were prenatally exposed to famine (and thus to folate deficiency) during the Dutch Hunger Winter had, six decades later, less DNA methylation of several genetic loci implicated in growth and metabolic disease (Tobi et al., 2009).

The epigenome can be modified in response to the environment throughout the life course, but it is most dynamic early in development. The epigenome appears to be particularly vulnerable to environmental influences in the first months of fetal development, after (most) parental epigenetic marks are erased (Reik et al., 2001). Since the Hunger Winter study showed that the risk for schizophrenia was increased for those individuals who had been exposed to famine periconceptually or in early pregnancy, but not later in pregnancy (Susser et al., 1996), epigenetic effects as a result of folate deficiency may very well be a mechanism through which the increased risk is shaped. This hypothesis was supported by the first epigenetic studies using data from the Hunger Winter cohort, which found that DNA methylation differences in the imprinted IGF2 gene were confined to periconceptual exposure to famine (Abel, 2004; Heijmans et al., 2008; Pidsley, Dempster, & Mill, 2009; Tobi et al., 2009).

Thus, by spelling out the mediating pathways that could link the environmental exposure (famine) to the outcome (schizophrenia), we provide clues to genetic polymorphisms such as *MTHFR677TT*, which may influence epigenetic effects on schizophrenia. At the same time, we add credence to the search for epigenetic effects that represent a response to the environment and do not require modifications of the nucleotide code. Moreover, this exercise points to powerful designs that could be used to test these hypotheses.

To confirm (or refute) genetic and epigenetic hypotheses implicated by the famine–schizophrenia link, we can focus our investigations on a population of cases enriched for the environmental exposure to prenatal famine. Among an ordinary population of cases, the environmental influences on disease risk will be heterogeneous. By contrast, among cases that were all exposed to prenatal famine, this is the dominant environmental factor. Consequently, we have much greater power to test hypotheses that pertain to the pathways linking prenatal famine and schizophrenia.

We can, for example, compare famine-exposed individuals with schizophrenia to their unexposed siblings with respect to epigenetic marks such as loss of imprinting on *IGF2* (Heijmans et al., 2008). In a pilot study led by one of us (ES) in China, we have collected data on 50 individuals who developed schizophrenia after prenatal famine, and on their family members. The analyses of these pilot data include a comparison of epigenetic marks in "exposed cases" versus their "unexposed siblings" (results are not yet available). The success of this approach will depend, of course, on the presence of the relevant epigenetic marks in peripheral blood cells (see Chapter 7).

USING GENETIC EFFECTS TO DISCOVER ENVIRONMENTAL PATHWAYS

In the previous section, we argued that information about environmental determinants of disease can inform genetic research. We now turn to how genetic information can help us understand environmental pathways to disease. Here, we continue to focus on the folate pathway, but we use childhood neurodevelopmental disorders rather than schizophrenia as our outcome. The reason for this switch to childhood outcomes is that studies have directly examined the relation of periconceptional folate to fetal and childhood neurodevelopment.

However, neither observational studies nor RCTs have resolved the question of whether periconceptional folate supplementation is related to childhood neurodevelopmental outcomes. We argue that MR can be a useful tool to study the interplay among folate, genes, and childhood neurodevelopmental outcomes.

In recent years, a near-revolution has occurred in thinking about causal inference, fomented by advances in cognitive science, philosophy of science, and epidemiology (Hernán, 2004; Pearl, 2000; Robins, 1987; Spirtes, Glymour, & Scheines, 1993). One resounding message from that research is the principle that one cannot draw causal inferences unless one is willing to make at least some causal assumptions; in other words, "no causal assumptions in, no causal inferences out" (Hernán, Hernandez-Diaz, Werler, & Mitchell, 2002; Robins & Wasserman, 1999). Causal assumptions can be pieced together from many types of research and observation, from bench science to RCTs. The good news about recent advances in genetics and genomics is that even modest causal information gained in these fields, such as the protein products of specific genes, may be useful as input assumptions to draw causal inferences about environmental influences on health (Khoury, Davis, Gwinn, Lindegren, & Yoon, 2005). In other words, burgeoning genetic information could be leveraged to identify modifiable determinants of disease, and we argue that these approaches may be most valuable when integrated with an understanding of how genetic and environmental pathways to disease intersect. There is a longstanding epidemiologic tradition of incorporating evidence on molecular- or cellular-level pathways to evaluate or justify the plausibility of arguments about causation in populations. The methods we discuss extend beyond this type of application to use the causal information drawn from bench science to evaluate causal claims about determinants of population health.

If properly used, this approach could provide a clever route around some nearly intractable problems in observational epidemiology. The many limitations of this sort of search for causal structures should be balanced against recognition of the limitations of conventional epidemiologic approaches. In observational epidemiology, the dominant analysis is to examine the association between a hypothesized exposure and the hypothesized disease outcome, often conditioning on a set of measured confounders. This approach has led to many spectacular successes, but also to many dismaying failures, in which strong observational evidence laid the foundation for unsuccessful trials of nutritional, pharmacologic, and psychosocial interventions (Albert et al., 2008; Berkman, 2009; Rossouw et al., 2002). For this reason, it is important to aggressively seek complementary approaches to evaluating causal claims. Among other advantages, this may enable us to make the best possible uses of observational data to justify RCTs and inform the design of those RCTs. Ethical concerns preclude using RCTs to evaluate the effects of many important exposures; when this is the case, observational evidence is especially important.

To illustrate how genetic information could be used as a valuable complement to conventional observational studies, we focus on the hypothesis that prenatal folate consumption affects neurodevelopment of the offspring via pathways mediated by homocysteine level. We begin by describing our background understanding of the substantive relationships, then discuss an MR approach, summarizing the necessary criteria for an MR study to work, and finally review some options to use additional information on genetic and environmental risk factors to evaluate bias in the MR study.

FOLATE SUPPLEMENTATION AND NEURODEVELOPMENT

By the 1990s, RCTs of periconceptional folate supplements had demonstrated beyond any doubt that these supplements reduce the risk of NTDs in offspring (MRC Vitamin Study Group, 1991). Neural tube defects were the most common congenital anomalies in developed countries, and supplements removed about 80% of the risk with no apparent adverse effect. This justified public health action, and from then on, all women considering pregnancy were advised to take folate supplements. Supplements prior to conception might be beneficial, and supplements after 28 days' gestation—which is often before pregnancy is recognized—have no preventive effect on neural tube closure. Despite the advisory, only a minority of women started folate supplements before knowledge of pregnancy. To enhance prevention of NTDs, the United States introduced mandatory fortification of grains and cereals with folate, which substantially increased folate levels in the general population. However, the recommended prepregnancy doses of folate dietary supplements are much higher than ordinarily achieved by diet alone, even with folate supplementation of foods, such as in the United States. Therefore, the prevention of NTDs is still only partially successful, and a public health campaign is devoted to increasing uptake of the preventive supplements.

Questions are now being raised, however, about other potential impacts of folate supplements on child health. There has been some investigation of potential adverse effects, such as increased vulnerability to atopy and asthma, as well as positive effects on neurodevelopment of children (Haberg, London, Stigum, Nafstad, & Nystad, 2009; Roza et al., 2010; Schlotz et al., 2009; Tamura et al., 2005; Whitrow, Moore, Rumbold, & Davies, 2009). None of these effects on child health has been proven to occur. Nonetheless, the biological plausibility of effects on neurodevelopment has been greatly enhanced by evidence that folate plays an important role in DNA methylation; that prenatal folate supplementation in animals influences DNA methylation in offspring; and most recently, by emerging evidence from human studies that prenatal folate supplements influence DNA methylation in offspring (Kaminsky et al., 2009; Morgan et al., 1999). High folate is known to reduce homocysteine levels and concurrently increase methionine availability, as discussed above, and this is likely to be the pathway by which folate impacts DNA methylation. The investigation of these potential effects has become vitally important to public health, in light of the public health campaign to encourage women of childbearing age to take folate supplements.

How can these questions be investigated? At present, RCTs of periconceptional folate supplements are not an ethical option, because of their proven efficacy in preventing NTDs. The individuals whose mothers participated in the definitive trials of folate supplementation to prevent NTDs have not been traced, so these trials do not provide information about outcomes after birth. This means that we may have to rely upon observational studies that identify women who do and do not take periconceptional folate supplements, and then follow their children for detection of both positive and adverse consequences.

The paramount difficulty in observational studies of folate supplements is that supplement use is a health-conscious behavior associated with numerous healthy behaviors, socioeconomic status, and other factors that may be intangible and/or difficult to measure (e.g., genetic influences, or the nature of mother–child interactions). Many of these factors may also promote improved neurodevelopmental outcomes in children (Fergusson, Horwood, & Boden, 2008). The observational studies suggesting that periconceptional folate supplements prevented NTDs did turn out to be correct (Milunsky et al., 1989), but there are many recent examples of observational findings on adult vitamin supplements and other lifestyle factors that have been refuted by subsequent RCTs. Indeed, a prime example is that numerous observational studies suggested that the low homocysteine levels induced by folate supplements in adulthood reduced the risk of cardiovascular disease (Ford et al., 2002), a result refuted by large randomized trials (Albert et al., 2008; Ebbing et al., 2008). Therefore, in a matter of such enormous public health import, it is hazardous to rely solely upon such observational findings, and we must seek every possible means to validate (or refute) a causal inference.

This is a situation in which we can potentially apply an MR approach to estimate the effect of homocysteine buildup on neurodevelopmental outcomes, avoiding bias due to the unobserved common causes of neurodevelopmental outcomes and either homocysteine levels or folate intake.[1] There are several groups applying MR to these outcomes, but the results are not yet published. Therefore, we use this example as a hypothetical illustration.

THE MENDELIAN RANDOMIZATION DESIGN

Mendelian randomization has recently been popularized as an approach to identifying the health consequences of exposures when the exposures are at least partially under genetic control (Brennan, 2004; Casas, Bautista, Hingorani, & Sharma, 2004; Katan, 2004; Keavney, 2004; Little & Khoury, 2003; Smith, 2004; Smith & Ebrahim, 2003, 2004; Smith, Harbord, & Ebrahim, 2004; Thomas & Conti, 2004; Tobin, Minelli, Burton, & Thompson, 2004; Wheatley & Gray, 2004).

Mendelian randomization is a special case of instrumental variables (IV) analysis, an econometric technique known since at least the 1920s. Instrumental variables analyses are based on finding an "exogenous factor" that influences an exposure or a biomarker of interest (e.g., homocysteine). Informally, this exogenous factor—the instrumental variable—must be uncorrelated with other unmeasured causes of the outcome of interest (e.g., neurodevelopment) and must not influence that outcome via any pathway

---

[1] This example differs from a typical MR design in a subtle way. In order to focus on the key points, we do not discuss these, but they may be of interest to readers who wish to explore this topic further. In normal physiologic systems, high levels of homocysteine indicate inadequate transformation of homocysteine to methionine, because homocysteine and methionine are mechanistically related. As noted earlier, it is hypothesized that the change in methionine availability resulting from limited homocysteine transformation will impact DNA methylation. For clarity here, we treat homocysteine level as the "exposure" whose effect on neurodevelopment we are seeking to estimate. We have very little direct evidence supporting the net effect of homocysteine on neurodevelopmental outcomes. Any possible observational evidence would be potentially biased to unmeasured common causes of folate consumption and neurodevelopmental outcomes. Homocysteine can also be metabolized by a second, alternative pathway involving the enzyme cystathionine B-synthase, which transforms homocysteine into cystathionine. It is possible to argue that homocysteine per se is not in the causal pathway to DNA methylation, but rather serves as a marker for methionine. Since methionine is generally derived from homocysteine, however, the MR study can be based on homocysteine levels, regardless of whether homocysteine is designated as a causal intermediate, as a marker for a causal intermediate (methionine), or both. Unlike methionine, homocysteine levels are routinely and reliably measured, and the relation of homocysteine levels to enzymatic activity of MTHFR has been extensively studied. Hence homocysteine level is both an appropriate proxy for the unobserved level of methionine available and could feasibly be implemented as the exposure in an MR study in many existing datasets.

except that mediated by the exposure. Randomized controlled trials are a special case of IV models familiar to epidemiologists: randomization influences a treatment of interest, and we use the association between randomized assignment and the outcome (i.e., the intent-to-treat estimate) to evaluate whether the treatment received affects the outcome (Rothman, Greenland, & Lash, 2008). In the case of RCTs, random assignment to treatment is termed the instrumental variable. Econometric applications of IV include using draft lottery numbers to estimate the effects of veteran status on earnings; date of birth to estimate the effects of age at school entry on academic performance; or changes in policies regulating years of compulsory schooling to estimate the effects of educational attainment on health (Angrist & Krueger, 2001).

The term *Mendelian randomization* was coined for the situation in which a genotype is the source of exogenous variation in an IV analysis. In economics, the exposure of interest—that is, the factor whose effect one wishes to evaluate—is often called the *endogenous variable*. In MR terminology, this variable is often referred to as the *phenotype*. For clarity, and to emphasize the parallelism with RCTs, we will refer to this variable as the *exposure*. In our example, the exposure is the biomarker homocysteine.

The essential elements of a causal structure in which MR is appropriate are shown in Figure 8.3: There is a hypothesized effect of biomarker X (homocysteine level) on Y (neurodevelopmental outcomes) that cannot be directly estimated from the association between homocysteine and neurodevelopment because of presumed unmeasured confounders. This confounding occurs because unmeasured factors (U) influence both homocysteine

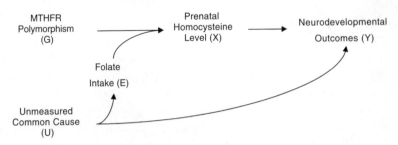

Figure 8.3. **Causal structure under which a Mendelian randomization study could be useful and valid.** We are interested in the effect of folate (and the mediating pathway homocysteine metabolism) on neurodevelopmental outcomes. The observed association between folate intake and neurodevelopment is potentially biased by an unmeasured common cause (U). *MTHFR* polymorphism affects the same mediating homocysteine metabolism pathway, and under the assumptions represented in this diagram, there are no confounders of the *MTHFR* genotype and neurodevelopmental outcomes.

and neurodevelopmental outcomes, so that any observed association between homocysteine and neurodevelopment could reflect the influence of U rather than homocysteine's direct effect on neurodevelopment (Hernán, Hernandez-Diaz, & Robins, 2004; Rothman et al., 2008). To apply an MR design to this problem, we must identify a gene, such as *MTHFR*, that influences homocysteine level, but is not associated with neurodevelopmental outcomes for any other reason. For example, it must not be associated due to coincidental population differences in the prevalence of *MTHFR* alleles and neurodevelopmental outcomes (a path from U to G), or because *MTHFR* has other consequences for neurodevelopment, via mechanisms unrelated to homocysteine (a direct path from G to Y).

Suppose our research goal is to test the existence of causal paths, from the exposure (X, in our case homocysteine) to the health outcome (Y, neurodevelopmental outcomes) and estimate the average magnitude of this path within some population.

Some key assumptions represented in Figure 8.3 are necessary for a valid use of MR to estimate the relation between homocysteine and neurodevelopment:

1. The genotype (*MTHFR*) predicts homocysteine level.
2. The genotype shares no unmeasured common causes with neurodevelopmental outcomes (in other words, there are no unmeasured confounders of the genotype and the outcomes).
3. All directed (causal) paths connecting the genotype to neurodevelopmental outcomes pass through homocysteine (in other words, the only mechanisms via which the genotype could influence neurodevelopment are mediated by homocysteine).

If these three assumptions are not met, we cannot use an MR design to estimate the effect of homocysteine on neurodevelopmental outcomes. The first assumption—that *MTHFR* genotype predicts homocysteine level—is well established. The other two assumptions may be controversial, calling into question the validity of an MR study in this case. We have drawn Figure 8.3 so that these assumptions are fulfilled, but we do not know if the figure represents the way the world actually works. If these assumptions are not met, the MR approach will provide a biased estimate of the effect of homocysteine on neurodevelopment, so it is important to incorporate assessments of these assumptions into MR analyses. We discuss approaches to evaluating these assumptions in later sections.

In an MR analysis, if homocysteine does not influence neurodevelopment (i.e., if, contrary to Figure 8.3, there were no arrow between homocysteine and neurodevelopment), we would anticipate that the *MTHFR* allele will not predict neurodevelopmental outcomes. This is analogous to an RCT: If the treatment received does not affect the outcome, then we expect that the random assignment to treatment will be statistically independent of the outcome. The unmeasured confounder of the homocysteine–outcome relationship does not confound the gene–outcome relationship any more than unmeasured confounding of treatment uptake and outcomes confounds the association between random assignment and the outcome in an RCT.

Mendelian randomization and IV methods in general are conceptually distinct from approaches based on assessing the association between the exposure and the outcome directly, including methods based on multiple regression or propensity scores. Such methods rely on the assumption that all common causes (i.e., confounders) of the exposure and the outcome have been identified. To identify the causal effect of the exposure on the outcome, either the common causes or some other factor on the confounding path must be included in the regression or propensity score model. This is a strong and essentially untestable assumption. Mendelian randomization methods depend on the three assumptions above, which do not require that confounders of the exposure and outcome be controlled. Mendelian randomization can identify a causal effect of the exposure on the outcome even if there are important unmeasured confounders of the exposure–outcome relationship. The MR methods also depend on strong and generally untestable assumptions, including no confounding of the instrument–outcome relationship, but they are *not* the same assumptions as those required for conventional observational models. Therefore these approaches are complementary.

Mendelian randomization applications have proliferated in recent years, including studies of the health consequences of alcohol use (Chen, Smith, Harbord, & Lewis, 2008; Heidrich, Wellmann, Doring, Illig, & Keil, 2007; Lewis & Smith, 2005), low-density lipoprotein (Kamstrup, Tybjaerg-Hansen, Steffensen, & Nordestgaard, 2009; Thanassoulis & O'Donnell, 2009), adiposity (Brennan et al., 2009; Timpson et al., 2009), inflammatory cytokines (Keavney, 2010), and sex-hormone binding globulin (Ding et al., 2009). Identifying opportunities to use MR designs depends on a detailed understanding of the determinants of the exposure under consideration.

Mendelian randomization analyses are frequently used to provide hypothesis tests of whether the exposure or biomarker of interest affects the outcome Y, without attempting to estimate the effect size. Under the assumptions for a valid instrument, this requires only assessing whether the *MTHFR* allele is statistically associated with neurodevelopmental outcomes. This is comparable to an intent-to-treat analysis in an RCT. To test the null hypothesis that received treatment does not affect the outcome, we need only examine whether randomly assigned treatment is statistically associated with the outcome, not whether treatment actually received is associated with the outcome. In the MR analogy, the genotype represents random assignment, and the exposure represents treatment received. Frequently, however, we are interested in more than simply testing the null hypothesis of "no effect": IV analyses can also be used to quantify the magnitude of the effect of the exposure on the outcome (Angrist, Imbens, & Rubin, 1996). Such estimates require additional assumptions and may, depending on the assumptions invoked, refer to either the effect size on average for the entire population or to the effect within a specific subgroup of the population (Greenland, 2000).

VIOLATIONS OF ASSUMPTIONS FOR MENDELIAN RANDOMIZATION

Mendelian randomization appears to offer a clever solution to a perennial problem in observational epidemiology: confounding by unmeasured factors. However, the applications of MR have been limited because the assumptions for MR turn out to be strong. Indeed, as discussed above, MR exchanges one set of strong assumptions (regarding the causal structures linking the exposure to the outcome) for another set of strong assumptions (regarding the causal structures linking the genotype and the outcome). Violations of the MR assumptions are usually classed into two categories. The first category, *population stratification*, occurs if the populations in which the genotype is most common differ from populations with a lower frequency of the genotype in ways that influence the outcome but are unrelated to the exposure, which would violate assumption 2 above. The 2nd category is *pleiotropy*, in which there are direct causal pathways from the genotype to the outcome, not mediated by the exposure of interest, which would violate assumption 3 above.

Figure 8.4 shows a causal diagram with examples of such violations. For clarity, we have labeled the pathways. Assumption 2 (no unmeasured common causes of the instrument and the outcome) is violated by the existence of a population group that is associated with *MTHFR* prevalence (pathway γ_1) and has a separate causal effect on neurodevelopment outcomes (pathway γ_2). In this diagram, we show population group as "influencing" genotype prevalence, but this is a matter of convention: The key point is that population group and genotype are statistically associated. Assumption 3 (no pathways linking the instrument and the outcome not mediated by the exposure of interest) is violated by the existence of another (unmeasured) intermediate that is affected by *MTHFR*, and in turn, affects neurodevelopmental

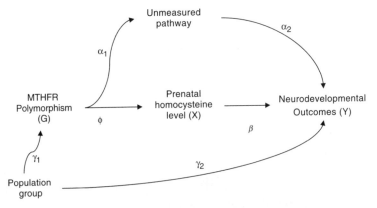

Figure 8.4. Violations of assumptions for a Mendelian randomization study. In this diagram, folate consumption is omitted for simplicity of presentation. There are two pathways linking *MTHFR* polymorphism to neurodevelopmental outcomes that would bias the Mendelian randomization effect estimate: one mediated by an unmeasured pathway (i.e., pleiotropy) and the other due to the common prior cause of population group. If either of these pathways is shown to exist, the MR effect estimate is probably biased.

outcome (pathways α_1 and α_2). If the biasing pathways shown in Figure 8.4 do exist (i.e., if MR assumptions 2 or 3 are violated), then the MR effect estimate will be biased by the other pathways between G and Y.

Mendelian randomization effect estimates can be quite sensitive to violations of the assumptions, so it is important to incorporate critical evaluations of the assumptions into any MR analysis. We desire tests that will help us determine whether the arrow sequences α_1/α_2 and γ_1/γ_2 shown in Figure 8.4 actually exist in the world. Two potential ways to evaluate the assumptions are discussed here. One depends on identifying additional IVs for overidentification tests, and the other depends on gene–environment interactions that can be used to identify violations of the assumptions.

Overidentification tests are a common econometric tool for testing IV assumptions, but have rarely been used in MR studies to date. The method is based on comparing IV effect estimates derived from alternative instruments (Wooldridge, 2003). Effect estimates from alternative instruments (e.g., two alternative genes or two polymorphisms in the same gene) should identify the same parameter, even if they have very different, or even qualitatively opposite, effects on the exposure. For example, if we could identify not just one but two or more genetic polymorphisms that affect homocysteine level, we could use these in combination in an overidentification test. Thus, the multiple MR effect estimates derived from the different instruments can be tested against each other (for an example of this intuition, see the application of Ding et al., [2009] using two polymorphisms).

This approach may not have been used extensively in MR because, in the past, we have often only known of a single genotype that influences the exposure of interest. As genetic information accumulates, there may be more opportunities to apply overidentification tests.

Gene–environment interactions could also be leveraged to help evaluate the validity of an MR study (Chen, Smith, Harbord, & Lewis, 2008). A good opportunity to apply this approach would be in the use of MR in the study of the effects of alcohol on cardiovascular disease (CVD). In most societies, the *ALDH* genotype predicts alcohol consumption, but in a society with strict social mores against alcohol use, *ALDH* would have no or very little effect on alcohol use. *ALDH* has a context-dependent effect on alcohol consumption. If we examined a society of teetotalers and found that, even in this population, *ALDH* genotype predicted CVD, we would conclude that *ALDH* influenced CVD via mechanisms unrelated to alcohol consumption. In this case, the MR effect estimates are biased.

Now consider the folate example. As discussed above, high levels of folate supplementation may overcome the *MTHFR* slow-acting allele and expedite the metabolism of 5,10 MTHFR into 5 MTHF. In other words, the enzyme catalysis rate is influenced by both features of the enzyme coded in the *MTHFR* gene and the concentration of the substrate (5,10 MTHFR), which is directly increased by folate supplementation. As a result, the *MTHFR* gene has a context-dependent effect on homocysteine metabolism, with dietary folate a key environmental modifier of the effects of *MTHFR* allele on homocysteine levels.

We describe folate supplementation as an environmental factor in which the effect of the *MTHFR* genotype on homocysteine is "silenced." Among women who take folate supplements in pregnancy, *MTHFR* allele is independent of homocysteine. In Figure 8.3, the arrow from G to X is eliminated among women who use folate supplementation but present for fetuses whose mothers did not use supplementation. In a silencing environment, the genotype should be independent of the exposure and therefore independent of the outcome if the assumptions of MR are fulfilled (Greenland, Pearl, & Robins, 1999).

Under the biased diagram in Figure 8.4, however, other (biasing) pathways connect G and Y, so that even in the silencing environment in which the genotype does not affect the exposure, the genotype is expected to predict the outcome. Thus, if the association between the genotype and the outcome deviates from 0 in an environmental stratum in which the genotype does not affect the exposure, the MR design is biased. If the association between the genotype and the neurodevelopmental outcome is close to 0 in the folate-supplemented condition, this provides support for the MR assumptions and suggests the MR effect estimate in the folate-restricted condition can be interpreted causally (there are some caveats to this argument, but detailed discussion is beyond the scope of this chapter). An analogous approach can be adopted if an environmental modifier does not silence the genotype–exposure association, but actually reverses the association.

CONCLUSION

We have shown how elucidating a rich causal story about the genetic and environmental pathways using what *is* known affords opportunities to test new hypotheses and extend our knowledge into new areas regarding both genetic and environmental determinants of disease. The link between exposure to famine and schizophrenia—an "environmental" pathway—can be used to broaden our understanding of the genetic factors that influence schizophrenia. Conversely, knowledge of genetic factors that influence the same homocysteine metabolism pathway as folate consumption can help evaluate the effects of folate consumption on neurodevelopmental outcomes. And understanding how environmental and genetic pathways intersect is key to evaluating the reliability of these methods.

As we apply these approaches, we need to be very cautious about each step and attempt to adopt critical evaluations of the assumptions. For example, although large meta-analyses of MR studies of homocysteine and CVD found no evidence to support a causal relationship (Lewis, Ebrahim, & Davey Smith, 2005), several early MR studies indicated that the relationship *was* causal (Wald, Law, & Morris, 2002), contrary to the findings of subsequent RCTs (Albert et al., 2008; Ebbing et al., 2008). It is therefore critical that the assumptions for MR be subject to "brittle" tests: We must elaborate the hypothesized causal structure to identify opportunities to demonstrate violations of the MR assumptions. Such brittle tests have rarely been possible in prior MR studies, but in this case, we propose two approaches, including capitalizing on information on multiple genetic risk factors and gene–environment interactions. If such tests are routinely adopted, the MR design will be a far more valuable tool for scientific progress.

The promise of the genomic age includes the possibility of leveraging genetic information to improve understanding of the effects of environmental conditions on complex diseases (Khoury et al., 2005; Khoury, Millikan, Little, & Gwinn, 2004). We have invested tremendous scientific and financial resources in collecting genetic information in large cohort studies. Integrating this information with current knowledge of environmental pathways and clear causal thinking could help overcome many nearly intractable challenges in observational epidemiology.

REFERENCES

Abel, K. M. (2004). Foetal origins of schizophrenia: testable hypotheses of genetic and environmental influences. *British Journal of Psychiatry, 184,* 383–385.

Albert, C. M., Cook, N. R., Gaziano, J. M., Zaharris, E., MacFadyen, J., Danielson, E., et al. (2008). Effect of folic acid and B vitamins on risk of cardiovascular events and total mortality among women at high risk for cardiovascular disease: a randomized trial. *Journal of the American Medical Association, 299*(17), 2027.

Allen, N. C., Bagade, S., McQueen, M. B., Ioannidis, J. P., Kawoura, F. K., Khoury, M. J., et al. (2008). Systematic meta-analyses and field synopsis of the genetic association studies in schizophrenia: the SzGene database. *Nature Genetics, 40,* 827–834.

Angrist, J. D., Imbens, G. W., & Rubin, D. B. (1996). Identification of causal effects using instrumental variables. *Journal of the American Statistical Association, 91*(434), 444–455.

Angrist, J. D., & Krueger, A. B. (2001). Instrumental variables and the search for identification: from supply and demand to natural experiments. *Journal of Economic Perspectives, 15*(4), 69–85.

Berkman, L. F. (2009). Social epidemiology: social determinants of health in the United States: are we losing ground? *Annual Review of Public Health, 30*(1), 27–41.

Botto, L. D., & Yang, Q. (2000). 5, 10-Methylenetetrahydrofolate reductase gene variants and congenital anomalies: a HuGE review. *American Journal of Epidemiology, 151,* 862–877.

Brennan, P. (2004). Commentary: Mendelian randomization and gene-environment interaction. *International Journal of Epidemiology, 33*(1), 17–21.

Brennan, P., McKay, J., Moore, L., Zaridze, D., Mukeria, A., Szeszenia-Dabrowska, N., et al. (2009). Obesity and cancer: Mendelian randomization approach utilizing the FTO genotype. *International Journal of Epidemiology, 38*(4), 971-975.

Brown, A. S., & Susser, E. S. (2008). Prenatal nutritional deficiency and risk of adult schizophrenia. *Schizophrenia Bulletin, 34,* 1054–1063.

Brown, A. S., Susser, E. S., Lin, S. P., Neugebauer, R., & Gorman, J. M. (1995). Increased risk of affective disorders in males after second trimester prenatal exposure to the Dutch hunger winter of 1944–45. *British Journal of Psychiatry, 166*(5), 601–606.

Casas, J. P., Bautista, L. E., Hingorani, A. D., & Sharma, P. (2004). Plasma homocysteine, and ischaemic stroke: "Mendelian randomization" provides further evidence of causal link. *Journal of Neurology Neurosurgery and Psychiatry, 75*(8), 1218–1218.

Chen, L., Smith, G. D., Harbord, R. M., & Lewis, S. J. (2008). Alcohol intake and blood pressure: A systematic review implementing a Mendelian Randomization approach. *PLoS Medicine, 5*(3), 461–471.

DeBaun, M. R., Niemitz, E. L., McNeil, D. E., Brandenburg, S. A., Lee, M. P., & Feinberg, A. P. (2002). Epigenetic alterations of H19 and LIT1 distinguish patients with Beckwith-Wiedemann syndrome with cancer and birth defects. *American Journal of Human Genetics, 70*, 604–611.

Ding, E. L., Song, Y., Manson, J. A. E., Hunter, D. J., Lee, C. C., Rifai, N., et al. (2009). Sex Hormone-Binding Globulin and Risk of Type 2 Diabetes in Women and Men. *The New England Journal of Medicine, 361*(12), 1152.

Ebbing, M., Bleie, O., Ueland, P. M., Nordrehaug, J. E., Nilsen, D. W., Vollset, S. E., et al. (2008). Mortality and cardiovascular events in patients treated with homocysteine-lowering B vitamins after coronary angiography: a randomized controlled trial. *Journal of the American Medical Association, 300* (7), 795.

Feinberg, A. P. (2007). Phenotypic plasticity and the epigenetics of human disease. *Nature, 447*, 433–440.

Fergusson, D. M., Horwood, L. J., & Boden, J. M. (2008). The transmission of social inequality: examination of the linkages between family socioeconomic status in childhood and educational achievement in young adulthood. *Research in Social Stratification and Mobility, 26*(3), 277–295.

Ford, E. S., Smith, S. J., Stroup, D. F., Steinberg, K. K., Mueller, P. W., & Thacker, S. B. (2002). Homocyst (e) ine and cardiovascular disease: a systematic review of the evidence with special emphasis on case-control studies and nested case-control studies. *International Journal of Epidemiology, 31*(1), 59.

Gottesman, I. (1991). *Schizophrenia genesis*. New York: W. H. Freeman.

Greenland, S. (2000). An introduction to instrumental variables for epidemiologists. *International Journal of Epidemiology, 29*(4), 722–729.

Greenland, S., Pearl, J., & Robins, J. M. (1999). Causal diagrams for epidemiologic research. *Epidemiology, 10*(1), 37–48.

MRC Vitamin Study Group. (1991). Prevention of neural tube defects: results of the Medical Research Council Vitamin Study. *Lancet, 338*, 131–137.

Haberg, S., London, S. J., Stigum, H., Nafstad, P., & Nystad, W. (2009). Folic acid supplements in pregnancy and early childhood respiratory health. *Archives of Disease in Childhood, 94*, 180–184.

Heidrich, J., Wellmann, J., Doring, A., Illig, T., & Keil, U. (2007). Alcohol consumption, alcohol dehydrogenase and risk of coronary heart disease in the MONICA/ KORA-Augsburg cohort 1994/1995-2002. *European Journal of Cardiovascular Prevention & Rehabilitation, 14*(6), 769–774.

Heijmans, B. T., Tobi, E. W., Stein, A. D., Putter, H., Blauw, G. J., Susser, E. S., et al. (2008). Persistent epigenetic differences associated with prenatal exposure to famine in humans. *Proceedings of the National Academy of Sciences, 105*, 17046–17049.

Hernán, M. A. (2004). A definition of causal effect for epidemiological research. *Journal of Epidemiology and Community Health, 58*(4), 265–271.

Hernán, M. A., Hernandez-Diaz, S., & Robins, J. M. (2004). A structural approach to selection bias. *Epidemiology, 15*(5), 615–625.

Hernán, M. A., Hernandez-Diaz, S., Werler, M. M., & Mitchell, A. A. (2002). Causal knowledge as a prerequisite for confounding evaluation: an application to birth defects epidemiology. *American Journal of Epidemiology, 155*(2), 176–184.

Kaminsky, Z. A., Tang, T., Wang, S. C., Ptak, C., Oh, G. H. T., Wong, A. H. C., et al. (2009). DNA methylation profiles in monozygotic and dizygotic twins. *Nature Genetics, 41*, 240–245.

Kamstrup, P. R., Tybjaerg-Hansen, A., Steffensen, R., & Nordestgaard, B. G. (2009). Genetically elevated lipoprotein (a) and increased risk of myocardial infarction. *Journal of the American Medical Association, 301*(22), 2331.

Katan, M. B. (2004). Commentary: Mendelian randomization, 18 years on. *International Journal of Epidemiology, 33*(1), 10–11.

Keavney, B. (2004). Commentary: Katan's remarkable foresight: genes and causality 18 years on. *International Journal of Epidemiology, 33*(1), 11–14.

Keavney, B. (2010). The interleukin-1 cluster, dyslipidaemia and risk of myocardial infarction. *BMC Medicine, 8*(1), 6.

Khoury, M. J., Davis, R., Gwinn, M., Lindegren, M. L., & Yoon, P. (2005). Do we need genomic research for the prevention of common diseases with environmental causes? *American Journal of Epidemiology, 161*(9), 799–805.

Khoury, M. J., Millikan, R., Little, J., & Gwinn, M. (2004). The emergence of epidemiology in the genomics age. *International Journal of Epidemiology, 33*(5), 936–944.

Lewis, S. J., Ebrahim, S., & Davey Smith, G. (2005). Meta-analysis of MTHFR 677C-> T polymorphism and coronary heart disease: does totality of evidence support causal role for homocysteine and preventive potential of folate? *British Medical Journal, 331*(7524), 1053.

Lewis, S. J., & Smith, G. D. (2005). Alcohol, ALDH2, and esophageal cancer: A meta-analysis which illustrates the potentials and limitations of a Mendelian randomization approach. *Cancer Epidemiology Biomarkers & Prevention, 14*(8), 1967–1971.

Lipton, S. A., Kim, W. K., Choi, Y. B., Kumar, S., D'Emilia, D. M., Rayudu, P. V., et al. (1997). Neurotoxicity associated with dual actions of homocysteine at the N-methyl-D-aspartate receptor. *Proceedings of the National Academy of Sciences, 94*, 5923–5928.

Little, J., & Khoury, M. J. (2003). Mendelian randomisation: a new spin or real progress? *The Lancet, 362*(9388), 930–931.

Lucock, M. (2000). Folic acid: nutritional biochemistry, molecular biology, and role in disease processes. *Molecular Genetics and Metabolism, 71*, 121–138.

McClellan, J. M., Susser, E., & King, M. C. (2006). Maternal famine, de novo mutations, and schizophrenia. *Journal of the American Medical Association, 296*, 582–584.

Mill, J., Tang, T., Kaminsky, Z., Khare, T., Yazdanpanah, S., Bouchard, L., et al. (2008). Epigenomic profiling reveals DNA-methylation changes associated with major psychosis. *American Journal of Human Genetics, 82*, 696–711.

Milunsky, A., Jick, H., Jick, S. S., Bruell, C. L., MacLaughlin, D. S., Rothman, K. J., et al. (1989). Multivitamin/folic acid supplementation in early pregnancy reduces the prevalence of neural tube defects. *Journal of the American Medical Association, 262*(20), 2847–2852.

Morgan, H. D., Sutherland, H. G., Martin, D. I., & Whitelaw, E. (1999). Epigenetic inheritance at the agouti locus in the mouse. *Nature Genetics, 23*, 314–318.

Pearl, J. (2000). *Causality*. Cambridge, UK: Cambridge University Press.

Pidsley, R., Dempster, E. L., & Mill, J. (2009). Brain weight in males is correlated with DNA methylation at IGF2. *Molecular Psychiatry, 18, 4046-4053.*

Reik, W., Dean, W., & Walter, J. (2001). Epigenetic reprogramming in mammalian development. *Science, 293,* 1089–1092.

Robins, J. (1987). A graphical approach to the identification and estimation of causal parameters in mortality studies with sustained exposure periods. *Journal of Chronic Diseases, 40*(Suppl 2), 139S–161S.

Robins, J., & Wasserman, L. (1999). On the impossibility of inferring causation from association without background knowledge. In C. Glymour & G. Cooper (Eds.), *Computation, Causation, and Discovery* (pp. 305–321). Menlo Park, CA and Cambridge, MA: AAAI Press/The MIT Press.

Rossouw, J. E., Anderson, G. L., Prentice, R. L., LaCroix, A. Z., Kooperberg, C., Stefanick, M. L., et al. (2002). Risks and benefits of estrogen plus progestin in healthy postmenopausal women: principal results From the Women's Health Initiative randomized controlled trial. *Journal of the American Medical Association, 288*(3), 321–333.

Rothman, K. J., Greenland, S., & Lash, T. L. (2008). *Modern Epidemiology* (3rd ed.). Philadelphia, PA: Lippincott Williams & Wilkins.

Roza, S. J., Van Batenburg-Eddes, T., Steegers, E. A. P., Jaddoe, V. W. V., Mackenbach, J. P., Hofman, A., et al. (2010). Maternal folic acid supplement use in early pregnancy and child behavioural problems: the generation R study. *British Journal of Nutrition, 103,* 445–452.

Schlotz, W., Jones, A., Phillips, D. I., Gale, C. R., Robinson, S. M., & Godfrey, K. M. (2009). Lower maternal folate status in early pregnancy is associated with childhood hyperactivity and peer problems in offspring. *Journal of Child Psychology and Psychiatry, published online,* 28 October 2009.

Smith, G. D. (2004). Genetic epidemiology: an "enlightened narrative"? *International Journal of Epidemiology, 33*(5), 923–924.

Smith, G. D., & Ebrahim, S. (2003). "Mendelian randomization": can genetic epidemiology contribute to understanding environmental determinants of disease? *International Journal of Epidemiology, 32*(1), 1–22.

Smith, G. D., & Ebrahim, S. (2004). Mendelian randomization: prospects, potentials, and limitations. *International Journal of Epidemiology, 33*(1), 30–42.

Smith, G. D., Harbord, R., & Ebrahim, S. (2004). Fibrinogen, C-reactive protein and coronary heart disease: does Mendelian randomization suggest the associations are non-causal? *QJM: An International Journal of Medicine, 97*(3), 163–166.

Spirtes, P., Glymour, C., & Scheines, R. (1993). *Causation, Prediction, and Search* (Vol. *81*). New York: Springer-Verlag.

St Clair, D., Xu, M., Wang, P., Yu, Y., Fang, Y., Zhang, F., et al. (2005). Rates of adult schizophrenia following prenatal exposure to the Chinese famine of 1959-1961. *Journal of the American Medical Association, 294*(5), 557–562.

Susser, E., Neugebauer, R., Hoek, H. W., Brown, A. S., Lin, S., Labovitz, D., et al. (1996). Schizophrenia after prenatal famine. *Further evidence. Archives of General Psychiatry, 53,* 25–31.

Tamura, T., Goldenberg, R. L., Chapman, V. R., Johnston, K. E., Ramey, S. L., & Nelson, K. G. (2005). Folate status of mothers during pregnancy and mental and psychomotor development of their children at five years of age. *Pediatrics, 116,* 703–708.

Thanassoulis, G., & O'Donnell, C. J. (2009). Mendelian Randomization: Nature's Randomized Trial in the Post-Genome Era. *Journal of the American Medical Association, 301*(22), 2386.

Thomas, D. C., & Conti, D. V. (2004). Commentary: The concept of "Mendelian randomization." *International Journal of Epidemiology, 33*(1), 21–25.

Timpson, N. J., Harbord, R., Davey Smith, G., Zacho, J., Tybjaerg-Hansen, A., & Nordestgaard, B. G. (2009). Does greater adiposity increase blood pressure and hypertension risk?: Mendelian randomization using the FTO/MC4R genotype. *Hypertension, 54(1),* 84.

Tobi, E. W., Lumey, L. H., Talens, R. P., Kremer, D., Putter, H., Stein, A. D., et al. (2009). DNA methylation differences after exposure to prenatal famine are common and timing- and sex-specific. *Human Molecular Genetics, 18,* 4046–4053.

Tobin, M. D., Minelli, C., Burton, P. R., & Thompson, J. R. (2004). Commentary: Development of Mendelian randomization: from hypothesis test to "Mendelian deconfounding." *International Journal of Epidemiology, 33*(1), 26–29.

Wald, D. S., Law, M., & Morris, J. K. (2002). Homocysteine and cardiovascular disease: evidence on causality from a meta-analysis. *British Medical Journal, 325*(7374), 1202.

Waterland, R. A., & Jirtle, R. L. (2003). Transposable elements: targets for early nutritional effects on epigenetic gene regulation. *Molecular and Cellular Biology, 23,* 5293–5300.

Wheatley, K., & Gray, R. (2004). Commentary: Mendelian randomization - an update on its use to evaluate allogeneic stem cell transplantation in leukaemia. *International Journal of Epidemiology, 33*(1), 15–17.

Whitrow, M. J., Moore, V. M., Rumbold, A. R., & Davies, M. J. (2009). Effect of supplemental folic acid in pregnancy on childhood asthma: a prospective birth cohort study. *American Journal of Epidemiology, 170,* 1486–1493.

Wolff, G. L., Kodell, R. L., Moore, S. R., & Cooney, C. A. (1998). Maternal epigenetics and methyl supplements affect *agouti* gene expression in A^{vy}/a mice. *FASEB Journal, 12,* 949–957.

Wooldridge, J. M. (2003). *Introductory econometrics: a modern approach* (2nd ed.). Mason, Ohio: Thomson Southwestern.

Xu, B., Roos, J. L., Levy, S., Van Rensburg, E. J., Gogos, J. A., & Karayiorgou, M. (2008). Strong association of *de novo* copy number mutations with sporadic schizophrenia. *Nature Genetics, 40,* 880–885.

Xu, M. Q., Sun, W. S., Liu, B. Q., Feng, B. Y., Yu, L., Yang, L., et al. (2009). Prenatal malnutrition and adult schizophrenia: further evidence from the 1959-1961 Chinese famine. *Schizophrenia Bulletin, 35,* 568–576.

9

Socioenvironmental Effects on Gene Expression

STEVE W. COLE

THE FLUID GENOME

We typically think of our bodies as stable biological entities that interact with the world around us, but are fundamentally separate from it. This chapter explores the implications of two biological facts at odds with that perspective—the fact that our physical bodies are involved in constant self-renewal, and the fact that the gene transcriptional dynamics mediating self-renewal are sensitive to our physical and social environments. The average half-life of a protein in the human body is approximately 80 days, implying that around 1% of our molecular being is actively reconstructed during each day of life. Self-renewal involves the expression of genes in our DNA, which are transcribed into RNA and subsequently translated into the proteins that ultimately confer the distinctive functional characteristics of our cells (Fig. 9.1). The ability of our brain cells to think, our muscle cells to move, and our immune cells to kill pathogens all depend on the coordinated expression of thousands of genes. However, the exact subset of our 22 thousand DNA genes that are actually expressed as RNA and protein differs markedly both across cell types (e.g., neuron, muscle, white blood cell) and over time (e.g., immune response genes can up-regulate by > 1,000-fold in response to a pathogen). Our DNA genome encodes many more potential proteins than are actually expressed in a given cell at a particular point in time, and our functional genome—the subset of total genomic potential that is actually realized—is far more fluid than we generally appreciate (Champagne & Mashoodh, 2009; Cole, 2009).

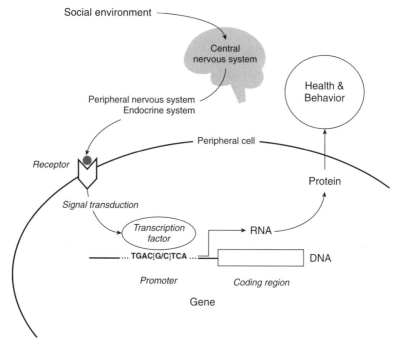

Figure 9.1. **Social signal transduction.** Socioenvironmental processes regulate human gene expression by activating central nervous system processes that subsequently influence hormone and neurotransmitter activity in the periphery of the body. Peripheral signaling molecules interact with cellular receptors to activate transcription factors, which bind to characteristic DNA motifs in gene promoters to initiate (or repress) gene expression. Only genes that are transcribed into RNA actually impact health and behavioral phenotypes. Individual differences in promoter DNA sequences (e.g., the [G/C] polymorphism shown here) can affect the binding of transcription factors, and thereby influence genomic sensitivity to socio-environmental conditions. Adapted from Cole, S. W. (2009). Social regulation of human gene expression. *Current Directions in Psychological Science, 18*(3), 132–137, with permission.

Some aspects of the functional genome are governed by a cell's individual transcriptional history. Liver cells continue to express liver cell characteristics and "biological identity" precisely because they tend to continue expressing liver cell-related gene transcripts and they do not express, for example, neuron-defining gene products. However, every liver cell carries the potential in its DNA to express all types of genes. All cells in the human body are also endowed with the capacity for *inducible gene expression*—to alter gene transcription in response to changes in the microenvironmental conditions surrounding the cell, such as pH, oxygenation, molecular interactions with neighboring cells, or signaling molecules such as hormones and neurotransmitters. Some of those microenvironmental conditions are, in turn, shaped by the broader physiologic macroenvironment of the body (e.g., the systemic circulation of hormones), and by extension, are subject to indirect regulation by the external conditions that surround us—including social signals. After all, social processes are well known to change internal physiologic parameters via central nervous system (CNS) control of peripheral neural activity and hormone levels (Sapolsky, 1994; Weiner, 1992). To the extent that our social environments regulate the general physiology of a body, and by extension, the local microenvironment of a cell, the social world acquires, in principle, the capacity to regulate the expression of our genes. Gene expression controls which of the many "potential bodies" encoded by our stable DNA genome is actually realized in our RNA transcriptome, raising the possibility that the historical distinction between "person" and "environment" may be more a product of our perceptual experience of ourselves than an intrinsic division within the biological world.

SOCIAL REGULATION OF GENE EXPRESSION

The possibility that social factors might regulate gene expression first emerged in the context of biobehavioral health research. Social stress and isolation have long been known to influence individual vulnerability to disease (Seeman, 1996). That effect is particularly strong for viral infections, in which adverse social conditions have been linked to increased replication of cold-causing rhinoviruses (Cohen, Doyle, Skoner, Rabin, & Gwaltney, 1997), the acquired immune deficiency syndrome (AIDS) virus HIV-1 (Cole, 2008a), and some cancer-related viruses (Antoni et al., 2006; Chang et al., 2005). Viruses are little more than small packages of 10–100 genes that hijack the protein production machinery of their host cells (us) to make more copies of themselves. As obligate parasites of our living cells, viruses evolved in a microenvironment structured by our own genome. If social factors can regulate the expression of viral genes, this suggests that social factors are likely to regulate the expression of our own genes as well.

One of the first studies to analyze relationships between social conditions and human gene expression analyzed transcriptional profiles in white blood cells (leukocytes) from healthy older adults who differed in the extent to which they felt socially connected to others (Cole et al., 2007). Long-term social isolation is both a well-established risk factor for physical illness (Seeman, 1996) and a personally-generated sociobehavioral phenotype that stems from individual differences in social cognition (Cacioppo & Hawkley,

2009), CNS neurobiological sensitivity to threat (Kagan, 1994; Schwartz, Wright, Shin, Kagan, & Rauch, 2003), and increased activity of threat-related peripheral signaling pathways such as the sympathetic nervous system (SNS) (Cole, Kemeny, Fahey, Zack, & Naliboff, 2003; Kagan, 1994; Kagan, Reznick, & Snidman, 1988). Motivated by the epidemiologic evidence in humans and experimental studies documenting adverse effects of social isolation on disease pathogenesis in animal models (summarized in Cole, 2008b; Cole et al., 2007), this study sought to identify alterations in immune cell gene expression that might mediate effects of social conditions on host resistance to infection and inflammation-related diseases such as cardiovascular disease, neurodegenerative disease, and certain types of cancer. Among the 22,283 genes assayed, 209 showed systematically different levels of expression[1] in people who consistently experienced themselves as lonely and distant from others over a span of 4 years (Fig. 9.2). These transcriptional alterations did not involve a random smattering of all human genes, but focally impacted three specific groups of genes. Genes supporting the early "accelerator" phase of the immune response—inflammation—showed selective up-regulation. However, two groups of genes involved in the subsequent "steering" of immune responses were markedly down-regulated: those involved in responses to viral infection (particularly Type I Interferons), and those involved in the production of antibodies. This initial portrait of a socially sensitive transcriptome provided a molecular framework for understanding why socially isolated individuals show heightened vulnerability to inflammation-driven cardiovascular diseases (i.e., excessive nonspecific immune activity) (Caspi, Harrington, Moffitt, Milne, & Poulton, 2006) and impaired responses to viral infections and vaccines (i.e., insufficient immune responses to specific pathogens) (Cohen et al., 1997; Cole, Kemeny et al., 2003; Cole, Kemeny, & Taylor, 1997; G. E. Miller, Cohen, Rabin, Skoner, & Doyle, 1999; Pressman et al., 2005). The key role of CNS

---

[1] If approximately 1% of the human genome is found to be differentially expressed in this study, how do we know those differences aren't simply due to the chance accumulation of spuriously "significant" results in 22,283 separate hypothesis tests? This study, like most other microarray analyses, employed false discovery rate (FDR) analyses (Benjamini & Hochberg, 1995) to hold the rate of false positive results at 5% or less. False discovery rate analyses replace the conventional statistical significance threshold of $p < .05$ (i.e., rate of false-positive findings as a fraction of truly null results) with a different "critical value" calculated to maintain the number of false-positive findings as a fraction of total declared positive findings at 5% (for further details on false-positive and false-negative errors in massively parallel microarray analyses, see Cole, Galic, & Zack, 2003). Based on that analysis, some of the more than 200 significant differences may have been due to chance, but at least 95% of them should be replicable. To empirically check that statistical estimate, this study also followed the standard practice of confirming several microarray-based indications of differential gene expression using an independent assay strategy (reverse transcriptase PCR). As would be expected based on FDR estimates, the vast majority of genes retested continued to show statistically significant differences.

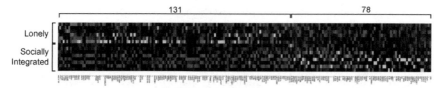

Figure 9.2. **Social regulation of gene expression in human immune cells.**
Expression of 22,283 human gene transcripts was assayed in 10 million blood
leukocytes sampled from each of 14 older adults who showed consistent
differences over 4 years in their level of subjective social isolation. Two hundred
nine gene transcripts showed 30% or greater difference in average expression
level in leukocytes from six people experiencing chronic social isolation versus
eight experiencing consistent social integration. In the heat-plot above, each row
represents data from one of the 14 study participants, each column contains
expression values for one of the 209 differentially active genes, and the coloring
of each cell represents the relative level of that gene's expression in a given
participant's leukocyte sample: Red = high expression, Black = intermediate
expression, Green = low expression. Adapted from Cole, S. W., Hawkley, L. C.,
Arevalo, J. M., Sung, C. Y., Rose, R. M., & Cacioppo, J. T. (2007). Social
regulation of gene expression in human leukocytes. *Genome Biology, 8*(9), R189,
with permission. (See Plate 2 for a color version of this figure.)

interpretation of social conditions was highlighted by the observation that
gene expression alterations were most strongly linked to a person's subjective
sense of isolation, rather than their objective number of social contacts.

Following the initial study of subjective social isolation, additional studies rapidly began to identify transcriptional correlates of other socioenvironmental conditions such as low socioeconomic status (SES) (Chen et al.,
2009; G. E. Miller et al., 2009) and the chronic threat of social loss (e.g.,
having a spouse with cancer) (G. E. Miller et al., 2008). These analyses also
found up-regulated expression of inflammation-related genes in leukocytes
and identified specific psychological processes that appeared to mediate
those dynamics. For example, among children with asthma, those from
lower SES backgrounds tended more often to interpret ambiguous social
stimuli as threatening, and that perception of threat was more strongly
linked to differential gene expression than was SES per se (Chen et al.,
2009). Subsequent studies have also begun to suggest that socioenvironmental conditions present early in life may induce a sort of "defensive programming" of the developing body that can persist for years after the
environmental risk factor itself has abated. For example, one analysis of leukocyte gene expression profiles in young adults found that those who had
lived in low SES conditions during early childhood continued to show
increased expression of inflammation-related genes, even if they had come

to enjoy favorable socioeconomic conditions in adulthood (G. E. Miller et al., 2009). Those findings are consistent with transcriptional studies of social mammals showing that social conditions in early life can persistently alter CNS gene expression in ways that govern behavioral and neuroendocrine responses to future challenges (Champagne & Mashoodh, 2009; Weaver, Meaney, & Szyf, 2006). Thus, the burgeoning literature on social regulation of gene expression may help illuminate the biological basis for developmentally plastic "critical periods" and their transition into long-term molecular "imprints" on physiology.

Coupled with the potential for long-lasting impact is a striking degree of short-term plasticity evident in some gene expression regimes. For example, the expressed transcriptome in circulating leukocytes can vary within 30 minutes after the onset of acute social stress (Morita et al., 2005; Nater et al., 2009), although that may reflect in part changes in cell trafficking into the circulating leukocyte pool, in addition to any per-cell changes in gene transcription. Other studies that control for cell trafficking dynamics have shown that transient events, such as one night of sleep loss, can alter gene expression profiles in the immune system (Irwin, Wang, Campomayor, Collado-Hidalgo, & Cole, 2006). Recent studies in rhesus macaques have also shown that changes in global social conditions can alter transcriptional profiles in the immune system within a matter of weeks (e.g., 4 weeks of unstable social hierarchy in adulthood (Cole, Sloan, Arevalo, & Capitanio, 2010; Sloan, Capitanio, Tarara, Mendoza et al., 2007) or unstable social conditions during the first 12 weeks of life (Cole, Arevalo, Ruggerio, Heckman, & Suomi, 2010). The capacity of relatively brief socioenvironmental shocks to induce persisting changes in gene expression stems from the recursive structure of gene regulatory networks, which involve highly interconnected feedback systems that can produce nonlinear "catastrophic" jumps from one regulatory equilibrium to another (Kauffman, 1993; Kim, Shay, O'Shea, & Regev, 2009).

Beyond their remarkable kinetics, social genomic relationships can also penetrate surprisingly deeply into our bodies. Experimental animal studies have shown that social conditions can regulate the expression of key neural genes such as *NGF* (encoding nerve growth factor [NGF]) (Sloan, Capitanio, Tarara, Mendoza et al., 2007), the glucocorticoid receptor gene (T. Y. Zhang et al., 2006), and the expression of hundreds of genes in CNS structures such as the hippocampus and prefrontal cortex (Karssen et al., 2007; Weaver et al., 2006). This is perhaps not too surprising, given the nervous system's key role in perceiving and responding to social stimuli. More surprising is the discovery that key genes involved in the immune system's response to pathogens are also sensitive to social conditions (Sloan, Capitanio, Tarara, Mendoza et al., 2007). Immune cells exert selective pressure on the evolution of viral genomes by killing human cells that actively express "foreign" viral genes. As a result of immune-mediated selective pressure against viral genomes, and immune cells' functional sensitivity to social conditions, many viruses also appear to have developed a genomic sensitivity to our social conditions as part of their immune-evasion strategy (i.e., activating viral replication during periods of stress-induced immunodeficiency, as reviewed above). However, even pathogens that escape the immune system's detection and are thus spared immune-mediated selective pressure may still modulate gene transcription in response to host stress and social conditions. Most human cancers are invisible to the immune system because they express no "foreign" genes for the immune system to attack. Nevertheless, some cancers still show significant changes in transcription in response to social stress (Antoni et al., 2006; Thaker et al., 2006). One recent study of women with ovarian cancer found that more than 220 genes were selectively up-regulated[1] in tumors from women with low levels of social support and high depressive symptoms (Lutgendorf et al., 2008). If our socially sensitive immune system is not conveying those effects via natural selection of tumor gene expression, how do social influences reach into the damaged genome of a cancer cell? New insights have come from bioinformatic analyses of *social signal transduction*.

SOCIAL SIGNAL TRANSDUCTION

Molecular biologists construe signal transduction as a micro-level process in which signaling molecules outside the cell interact with receptors to initiate a cascade of biochemical reactions inside the cell, ultimately activating a protein transcription factor to induce gene expression (Fig. 9.1). Transcription factors flag a particular stretch of DNA (the coding region of a gene) for transcription into RNA. Those genes that can be activated by a given transcription factor are determined by the nucleotide sequence of the gene's promoter—the stretch of DNA lying upstream of the coding region. For example, the transcription factor nuclear factor (NF)-κB binds to the nucleotide motif GGGACTTTCC, whereas the CREB transcription factor targets the motif TGACGTCA. These two transcription factors are activated by different receptor-mediated signal transduction pathways, providing distinct molecular channels by which extracellular events can regulate intracellular genomic response. The distribution of transcription factor–binding motifs across our ~ 22,000 gene promoters constitutes a type of "wiring diagram" that maps microenvironmental processes onto gene expression patterns. Additional layers of negative regulation can also superimpose on the positive activity of transcription factors to further trim the expressed transcriptome. For example, epigenetic modifications of DNA by methylation or histone-mediated sequestration can block transcription factor binding to promoters and thereby "veto" transcriptional stimulation that would otherwise occur (Champagne & Mashoodh, 2009; Weaver et al., 2006; T. Y. Zhang et al., 2006) (see also Chapter 7, by Mill). Transcription factors can also inhibit gene expression under certain circumstances, particularly when they bind to promoter sequences in a way that impedes access by other transcription factors that would otherwise act to up-regulate gene expression.

Given these cell-level microregulatory pathways, how can we account for the effects of macro-level influences from the social ecology? Key to that dynamic is the brain's role in translating environmental stimuli into changes in cellular function via the regulation of hormones, neurotransmitters, and other signaling molecules that disseminate throughout the body to activate cellular receptors and transcription factors. For example, the sympathetic nervous system (SNS) and the hypothalamic-pituitary-adrenal (HPA) axis represent two major pathways by which CNS-mediated perceptions of negative social conditions can alter gene transcription in a wide array of somatic cells (Sapolsky, 1994; Weiner, 1992). Positive psychological states may also regulate transcription (Dusek et al., 2008), although their molecular mediators are less well understood.

Links between social experiences and neural/endocrine responses have long been recognized (Sapolsky, 1994; Weiner, 1992), but the breadth of their impact on gene expression has only recently become apparent following the sequencing of the human genome. Early computational analyses of the human genome sequence suggested that promoter DNA sequences might provide for some degree of psychological specificity in transcriptional responses. For example, any gene bearing the motif GGTACAATCTGTTCT in its promoter might potentially be stimulated by severe, overwhelming stress experiences that release cortisol (e.g., defeat or bereavement) (Frankenhauser, 1975; Sapolsky, 1994; Weiner, 1992), because the cortisol-stimulated glucocorticoid receptor (GR) binds specifically to that DNA motif. In contrast, genes bearing the CREB promoter motif TGACGTCA would be predicted to activate in response to active-coping, fight-or-flight stress responses associated with catecholamine release and β-adrenergic receptor signaling (e.g., public speaking or physical exertion) (Dimsdale & Moss, 1980; Frankenhauser, 1975; Weiner, 1992). In the context of Figure 9.1, the production of catecholamines by the SNS and the production of cortisol by the HPA axis represent two biologically distinct "social signal transduction pathways," each of which activates a different transcription factor (i.e., CREB and the GR, the latter of which also functions as a transcription factor). Because the promoter motifs targeted by those transcription factors are distributed differentially across genes, the two distinct psychological stress experiences that activate those transcription factors may trigger very different transcriptional responses.[2] Genes predicted to be cortisol-responsive disproportionately encode receptors and other molecules involved in cellular "perception" of the physiologic microenvironment. Putative catecholamine-responsive genes include few receptors, but high concentrations of signal transduction molecules and transcription factors involved in cellular "decision-making" (i.e., converting receptor-mediated perception into changes in gene expression and cellular behavior). Thus, severe, overwhelming stress may trigger a cellular form of "denial" (altering perception via changes in receptor expression), while active-coping challenges induce something more akin to "sublimation" (altering cellular responses to perceptions via changes in transcription factor expression).

The sequencing of the human genome has also provided a new analytic infrastructure for mapping the molecular signaling pathways that convert socioenvironmental conditions into differential gene expression. One approach to this problem reverses the normal flow of biological information from the environment, through transcription factor activity, and into gene expression (Fig. 9.1). This analysis scans the promoters of differentially expressed genes (e.g., in Figure 9.2) to identify transcription factor–binding motifs that are over-represented in activated promoters, and thus reflect the specific transcription factors mediating the observed differences in gene expression (Cole, Yan, Galic, Arevalo, & Zack, 2005). Promoter-based bioinformatics have uncovered some surprising differences between the transcriptional signals "sent" by the brain and the transcriptional signals "heard" by the human genome in peripheral tissues. In studies of chronic loneliness, threat of social loss, and low SES (Cole et al., 2007; G. E. Miller et al., 2009; G. E. Miller et al., 2008), promoter analyses indicated that the inflammation-driving NF-κB transcription factor played a central role in orchestrating

---

[2] This does not imply that each unique facet of social experience will necessarily have a distinct transcriptional fingerprint. However, experimental studies show that there do exist some broad dimensions of variation in the nature of stressful experience (e.g., active vs. passive coping) that are associated with variations in neuroendocrine response (e.g., SNS vs. HPA axis) (Frankenhauser, 1975; Weiner, 1992). For example, stressing a mouse by physically restraining it has a very different impact on neuroendocrine activity and disease resistance than does rearing it in isolation or exposing it to a social competitor (Padgett et al., 1998; Sheridan, Stark, Avitsur, & Padgett, 2000; Stark et al., 2001). This implies that transcriptional responses to stress can vary as a function of the nature of the stressor and the individual's subjective capacity to surmount it (Sapolsky, 1994). However, many common social stressors may have similar (and cumulative) transcriptional effects because they tend to engage similar psychological and neuroendocrine responses (e.g., the similar transcriptional effects of low SES, anticipated bereavement, and subjective social isolation) (Cole et al., 2007; G. E. Miller et al., 2009; G. E. Miller et al., 2008). Severe, overwhelming threats are relatively uncommon in contemporary human societies, so their neuroendocrine and transcriptional effects may be empirically uncommon in human social genomics studies. Low-grade stressors are comparatively common, and thus more likely to empirically recur in studies.

the observed transcriptional alterations. Results from those studies also suggested that the GR was failing to inhibit the activity of NF-κB as it should (Pace, Hu, & Miller, 2007). Neither study found decreases in circulating cortisol levels that might explain the reduced activity of the GR. If the HPA axis were sending the proper anti-inflammatory cortisol signal, why would the leukocytes of stressed people not down-regulate NF-κB transcription of inflammatory genes? The answer appears to involve a reduction in the GR's sensitivity to cortisol, thus rendering the leukocyte transcriptome deaf to the brain's request to down-regulate proinflammatory

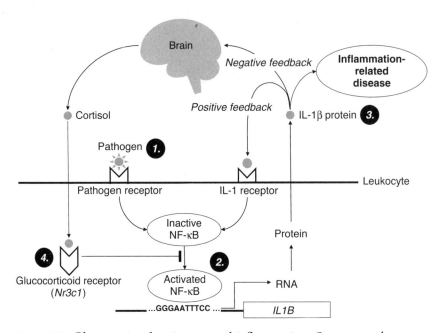

Figure 9.3. **Glucocorticoid resistance and inflammation.** Ongoing pathogen exposure (1.) activates the proinflammatory transcription factor nuclear factor (NF)-κB (2.), which induces expression of multiple genes involved in immune response (3.). Some of those genes encode proinflammatory cytokines (e.g., *IL1B*) that feed back to cellular receptors to positively propagate NF-κB activity. Deleterious hyperinflammatory responses are normally held in check by an additional negative feedback loop in which the brain detects high proinflammatory cytokine levels and produces cortisol, which stimulates the glucocorticoid receptor to inhibit NF-κB activity (4.). Extended periods of social adversity can desensitize the glucocorticoid receptor, disrupting the negative feedback circuit that would normally inhibit NF-κB and thereby facilitating unhealthy chronic inflammation that can promote the development of cardiovascular disease, certain types of cancer, and neurodegenerative diseases.

A

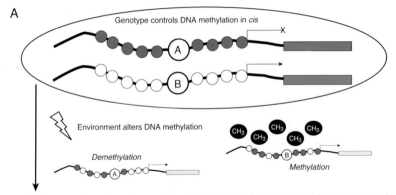

Genotype controls DNA methylation in *cis*

Environment alters DNA methylation

Demethylation

Methylation

Environment and genotype act on a common pathway to control gene expression

B

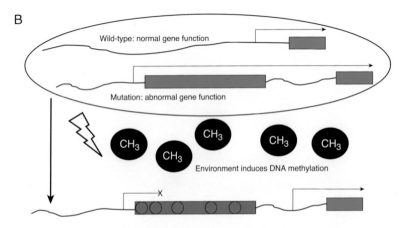

Wild-type: normal gene function

Mutation: abnormal gene function

Environment induces DNA methylation

Mutation silenced by DNA methylation: normal gene function

C

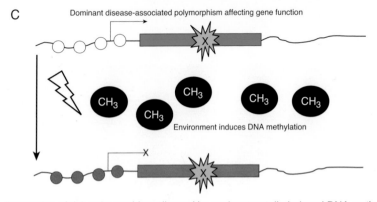

Dominant disease-associated polymorphism affecting gene function

Environment induces DNA methylation

Expression of risk polymorphism silenced by environmentally-induced DNA methylation

Plate 1. **Epigenetic alterations may mediate gene–environment interactions in a number of ways.** (A) Genotype may exert an effect on the expression of

Plate 1. (*continued*)

nearby genes (i.e., in cis) via epigenetic effects in an allele-specific manner, which in turn may be susceptible to environmental mediation. In this figure, the red bar represents repressed gene expression from one allele that is associated with high levels of promoter DNA methylation (*filled circles*). The other allele, on the other hand, is expressed at high levels (*green bar*) and is characterized by low levels of promoter DNA methylation (*empty circles*). Such cis effects on the epigenome may be subsequently influenced by the environment—e.g., via the addition of methyl groups (CH_3) from the diet. In this way, genotype and the environment act on a common pathway to control gene expression. (**B**) There are many instances in which genetic polymorphisms create abnormal gene products. Epigenetic processes may mediate gene–environment interactions if, for example, methyl groups (CH_3) in the environment act to silence abnormal gene products associated with genomic mutations. (**C**) Similarly, the effect of dominant disease-associated polymorphisms may be silenced by epigenetic processes occurring as a result of environmental factors.

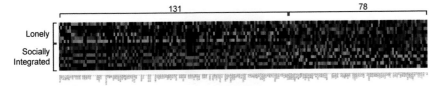

Plate 2. **Social regulation of gene expression in human immune cells.** Expression of 22,283 human gene transcripts was assayed in 10 million blood leukocytes sampled from each of 14 older adults who showed consistent differences over 4 years in their level of subjective social isolation. Two hundred nine gene transcripts showed 30% or greater difference in average expression level in leukocytes from six people experiencing chronic social isolation versus eight experiencing consistent social integration. In the heat-plot above, each row represents data from one of the 14 study participants, each column contains expression values for one of the 209 differentially active genes, and the coloring of each cell represents the relative level of that gene's expression in a given participant's leukocyte sample: Red = high expression, Black = intermediate expression, Green = low expression. Adapted from Cole, S. W., Hawkley, L. C., Arevalo, J. M., Sung, C. Y., Rose, R. M., & Cacioppo, J. T. (2007). Social regulation of gene expression in human leukocytes. *Genome Biology, 8*(9), R189, with permission.

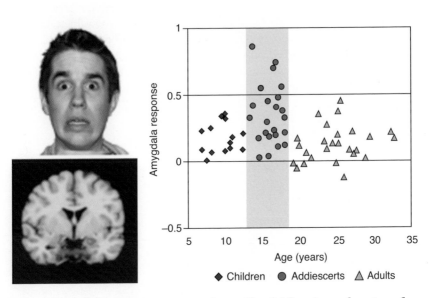

Plate 3. **Amygdala response to empty threat (fearful faces) as a function of age.** Adapted from Hare et al., 2008.

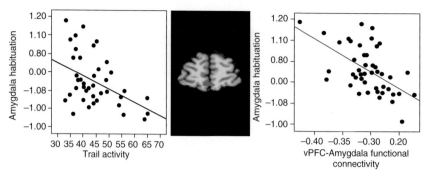

Plate 4. **Habituation of amygdala response is associated with anxiety and less ventral prefrontal cortex (vmPFC) activity.** Trait anxiety scores were negatively correlated with habituation (decrease from early to late trials) of amygdala activity ($r = -.447$; $p < .001$). **Left panel:** Scatter plot of the correlation between trait anxiety and amygdala habituation. The y-axis represents magnetic resonance (MR) signal in the left amygdala for early-late trials. The x-axis represents trait anxiety score. There was negative functional coupling between the amygdala and the ventral prefrontal cortex (vmPFC). The magnitude of activity in vmPFC (*middle panel*) and the strength of the connectivity between vmPFC and the amygdala were negatively correlated with amygdala habituation ($r = -.56$ $p < .001$). Right panel: Scatterplot of vmPFC–amygdala connectivity values versus amygdala habituation. The y-axis represents MR signal in the left amygdala for early-late trials. The x-axis represents Z-scored vmPFC–amygdala connectivity values. Adapted from Hare, T. A., Tottenham, N., Galvan, A., Voss, H. U., Glover, G. H., & Casey, B. J. (2008). Biological substrates of emotional reactivity and regulation in adolescence during an emotional go-nogo task. *Biological Psychiatry, 63*(10), 927–934, with permission.

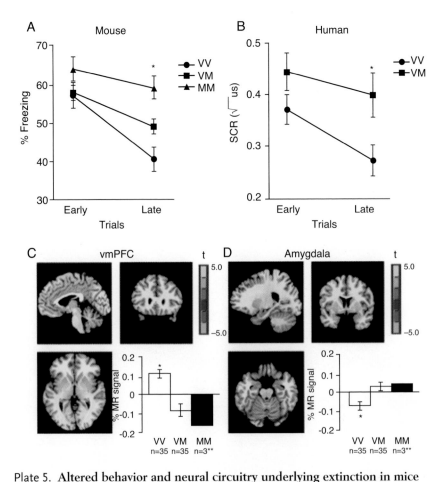

Plate 5. Altered behavior and neural circuitry underlying extinction in mice and humans with brain-derived neurotrophic factor (BDNF) Val66Met. Impaired extinction in Met allele carriers (Val/Met and Met/Met) as a function of time in 68 mice (**A**) and 72 humans (**B**) as indexed by percent time freezing in mice and skin conductance response (SCR) in humans to the conditioned stimulus when it was no longer paired with the aversive stimulus. (**C**) Brain activity as indexed by percent change in MR signal during extinction in the ventromedial prefrontal cortex (vmPFC) by genotype ($xyz = -4, 24, 3$), with Met allele carriers having significantly less activity than Val/Val homozygotes (VM < VV = blue), image threshold $p < 0.05$, corrected. (**D**) Genotypic differences in left amygdala activity during extinction ($xyz = -25, 2, -20$) in 70 humans, with Met allele carriers having significantly greater activity than Val/Val homozygotes (VM > VV = orange), image threshold $p < 0.05$, corrected. *$p < 0.05$. **MM were included in the analysis with VM, but plotted separately to see dose response. All results are presented as a mean ± SEM. VV = Val/Val; VM = Val/Met; MM = Met/Met. From Soliman, F., Glatt, C. E., Bath, K. G., Levita, L., Jones, R. M., Pattwell, S. S., Jing, D., et al. (2010). A genetic variant BDNF polymorphism alters extinction learning in both mouse & human. *Science, 327*(5967), 863–866, with permission.

genes (Fig. 9.3). Although the precise molecular mechanism underlying this effect is not fully defined, it appears that the activity of other physiologic signaling pathways can induce phosphorylation of the GR protein and thereby inhibit its activation by cortisol (Bamberger, Schulte, & Chrousos, 1996; Pace et al., 2007).

Animal models of immune regulation show that repeated social threat can induce GR desensitization (Cole, Mendoza, & Capitanio, 2009; Stark, Avitsur, Padgett, & Sheridan, 2001). Recent social genomics studies in humans have shown that chronic loneliness, threat of social loss, and low SES all appear to disconnect this key physiologic negative feedback system that would normally hold harmful inflammation in check (Cole et al., 2007; G. E. Miller et al., 2009; G. E. Miller et al., 2008). One implication of these results is that superficially different types of social adversity might induce a common pattern of regulatory alteration in immune cell gene expression and thereby increase the risk of inflammation-related disease (Seeman, 1996). Promoter bioinformatics have identified other alterations in transcription factor activity that may connect low SES to inflammatory gene expression in asthma (Chen et al., 2009), and connect low social support and depression to altered gene expression in ovarian cancer (Lutgendorf et al., 2008). By relating cell-level changes in gene transcription to neural and endocrine signaling pathways that ultimately convey socioenvironmental influences into the body, these early studies of social signal transduction have begun to provide a more comprehensive biological portrait of the pathways by which socioenvironmental conditions affect the molecular characteristics of the human body.

REMODELING THE BODY

Because RNA transcription shapes the protein complement of our cells, and those proteins mediate cellular function (Fig. 9.1), psychological regulation of gene expression implies that the social world can remodel the functional characteristics of the human body. One major goal of social genomics involves determining which specific aspects of human molecular function are most sensitive to socioenvironmental influences. In addition to the marked effects on immune system–related genes outlined above, neurobiological characteristics also appear to be particularly responsive. Consider *NGF*, the gene encoding the NGF protein—a key neurotrophic factor supporting the development and maintenance of SNS neural fibers. In rhesus macaques exposed to low-grade social stress for several months, transcription of *NGF* increased markedly in peripheral tissues such as lymph nodes (Sloan, Capitanio, & Cole, 2008; Sloan, Capitanio, Tarara, Mendoza et al., 2007).

As a result of that increased *NGF* expression, the number of SNS neural fibers present in the lymph node increases substantially, and the CNS thus acquires a greater capacity to regulate the cellular interactions that take place in this critical "convention center" for immune cell interaction. That change in neurobiological regulation in turn alters the nature of immune response genes expressed in the lymph node. One consequence is suppression of leukocyte *IFNB* expression, which plays a critical role in cellular defense against viral infections (Collado-Hidalgo, Sung, & Cole, 2006; Sloan, Capitanio, Tarara, Mendoza et al., 2007). As a result of those molecular interactions among *NGF*, SNS nerve fibers, and *IFNB*, socially stressed animals show increased replication of the primate version of HIV-1 (simian immunodeficiency virus) and increased damage to their immune systems as a result (Sloan et al., 2008; Sloan, Capitanio, Tarara, Mendoza et al., 2007). This series of studies not only provides a comprehensive map of the molecular mechanisms by which socioenvironmental conditions influence the pathogenesis of viral infection (Cole, 2008a; G. Miller, Chen, & Cole, 2009), but it also exemplifies the self-modifying recursive characteristics of transcription control pathways. To the extent that social signal transduction pathways such as the SNS increase transcription of the very genes that mediate their activity (e.g., up-regulating expression of the SNS-sustaining *NGF*), the functional alterations that result can become self-sustaining via positive feedback loops.

Recursive gene expression dynamics at the micro level (e.g., as in the *NGF/IFNB*/SIV example) can significantly influence the long-term temporal dynamics of macro-level development due to the persistence of their effects on protein expression and cell function. Figure 9.4 illustrates these dynamics in an abstract theoretical form. From a functional genomics perspective, we can think of our physical bodies as machines that convert environmental stimuli into changes in outputs, such as behavior and gene expression (RNA). Because the RNA "output" at one time point (e.g., RNA_1) influences the molecular characteristics of the body at future time points (e.g., $Body_2$)—due to the 80-day average half-life of transcriptionally induced proteins—that future body may respond to a subsequently encountered $Environment_2$ differently than it would if the body's current molecular characteristics were shaped by a different RNA history. The self-modifying characteristics of social signal transduction pathways can continue propagating into the future, with $Body_3$ responding to $Environment_3$ differently depending upon the earlier RNA_2, and thereby producing a different set of $Behavior_3$ and RNA_3 outputs.

In the *NGF/IFNB*/SIV system, for example, the socioenvironmental up-regulation of *NGF* at Time 1 induces a denser neural network within the lymph node at Time 2. That denser neural network provides a stronger

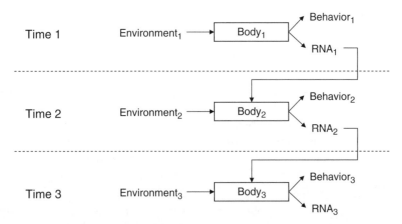

Figure 9.4. **RNA as a molecular medium of recursive development.** Social conditions at one point in time (Environment$_1$) are transduced into changes in behavior (Behavior$_1$) and gene expression (RNA$_1$) via central nervous system perceptual processes that trigger systemic neural and endocrine responses (mediated by Body$_1$). Those RNA transcriptional dynamics can alter the molecular characteristics of cells involved in environmental perception or response, resulting in a functionally altered Body$_2$. Body$_2$ may respond differently to a given environmental challenge than would the previous Body$_1$, resulting in different behavioral (Behavior$_2$) and transcriptional responses (RNA$_2$). The effect of RNA transcriptional dynamics on persisting cellular protein and functional characteristics provides a molecular framework for understanding how socioenvironmental conditions in the past may continue to affect current behavior and health, and how those historical conditions interact with current environments to shape our future trajectories (e.g., Body$_3$, Behavior$_3$, RNA$_3$). Because gene transcription serves as both a cause of social behavior (by shaping Body) and a consequence of social behavior (a product of Environment × Body), RNA serves as the physical medium for a recursive developmental trajectory that integrates genetic characteristics and historical-environmental regulators to shape individual biological and behavioral responses to current environmental conditions. Adapted from Cole, S. W. (2009). Social regulation of human gene expression. *Current Directions in Psychological Science*, 18(3), 132–137, with permission.

signaling pathway by which a stressful event at Time 2 can release the SNS neurotransmitter norepinephrine (NE) into the lymph node. As a consequence of increased NE signaling, the immune system mounts a poorer *IFNB* response to a viral infection at Time 2 solely because lymph node innervation was remodeled by differing social conditions at Time 1. In other words, social stress at Time 1 (Environment$_1$) is transmitted through the nervous system (Body$_1$) into behavioral stress responses (Behavior$_1$) and increased *NGF* expression (RNA$_1$). Up-regulated *NGF* at Time 1 increases SNS innervation of the lymph node, and thereby alters the functional relationship between the nervous and immune systems over the life of the newly expressed proteins/neural fibers (Body$_2$). When that functionally remodeled Body$_2$ encounters a new viral infection in Environment$_2$, the increased SNS neurotransmitter release inhibits transcription of anti-viral *IFNB* (RNA$_2$). The antiviral response is impaired, and the resulting increase in viral replication alters both gene expression and health/behavior in the future (RNA$_3$ and Behavior$_3$). Because gene expression alterations can change the functional characteristics of gene regulation pathways, the experience of Environment$_1$ not only "gets inside the body" but "stays there" in a concrete molecular sense that propagates through multiple gene transcriptional responses, physiologic systems, and time epochs. Recursive social signal transduction dynamics provide one major pathway by which environmental conditions present early in life can exert a persisting impact on biological function decades later in adulthood (G. E. Miller et al., 2009).

Environmental induction of stable epigenetic alterations provides another biological mechanism for long-term change in gene expression (Champagne & Mashoodh, 2009; T. Y. Zhang et al., 2006). Socioenvironmental conditions can also regulate the molecular composition of CNS cells, and thereby alter psychological, behavioral, and neuroendocrine-mediated transcriptional responses to future environments (T. Y. Zhang et al., 2006). Each of these biological mechanisms can propagate a transient change in gene expression into a self-perpetuating cycle that persists even after its initial enviornmental stimulus has abated. As a result, your body may respond very differently to a challenge today than it would have had you experienced a different history of environmental exposures and transcriptional responses.

Because the molecular composition of our cells constitutes the physical machinery by which we perceive and respond to the world around us ("Body" in Figure 9.4), and that molecular composition is itself subject to remodeling by socioenvironmental influences, gene expression constitutes both a cause and a consequence of behavior. RNA can be construed as the physical medium of a recursive developmental system in which social, behavioral, and health *outcomes* at one point in time also constitute *inputs* that shape our future responses to the environment (e.g., as in Heckman's model of human capability development, which analyzes how capacities developed at Time$_1$ impact our ability to capitalize on environmental opportunities at Time$_2$ (Heckman, 2007). Future research will push these models out of accessible immune cells and into the more sensitive CNS structures that shape social, cognitive, and affective processes. It will also be critical to define the particular features of social environments that trigger transcriptional remodeling of specific cells. Given the key role of neuroendocrine responses in mediating these effects, the most decisive influences likely involve our psychological reactions to social conditions rather than the properties of the external condition per se. After all, it is the subjective perception of conditions as threatening or uncertain that directly triggers SNS and HPA responses (Sapolsky, 1994; Weiner, 1992). Our genome's social sensitivity ultimately stems from the capacity of social conditions to affect CNS perceptions (e.g., of safety vs. threat) (Dickerson & Kemeny, 2004) and thereby trigger biological stress responses that alter gene transcription.

THE NEW GENETICS

With genes and environments now operating in parallel to shape our RNA-driven bodies, the integration of those two streams of influence has become a central challenge in biological analyses of human health and behavior. The regulatory paradigm outlined in Figure 9.1 provides a framework for analyzing the interplay of those factors in the context of gene × environment interactions. For example, variations in the DNA sequence of promoters can affect the binding of environmentally activated transcription factors and thereby buffer the effects of adverse social environments on behavioral responses (e.g., averting stress-induced depression via increased expression of the serotonin-regulating *SLC6A4/5HTT*) (Caspi et al., 2003; Champoux et al., 2002) or on immunologic responses (e.g., averting SNS induction of the proinflammatory cytokine *IL6*) (S. Cole et al., 2010). These "regulatory polymorphisms" in DNA sequence essentially add or remove a connection within the complex wiring diagram of a cell.

Figure 9.5 illustrates one example in which the most prevalent version of the human *IL6* gene shows substantial transcriptional activation in response to the SNS neurotransmitter NE. In the wiring diagram of the ancestral human genome, SNS activation is connected to *IL6* transcription because the GATA1 transcription factor stimulated by NE can bind with high affinity to the GATG motif present in the *IL6* promoter. However, people whose *IL6* promoters bear the variant CATG allele are much less susceptible to stress-induced up-regulation of *IL6* because NE-activated GATA1 does not bind efficiently to that alternative DNA motif. The G > C substitution essentially disconnects the regulatory circuit linking stressful experience to *IL6* transcription via SNS activation of GATA1. Presumably as a result of that differential transcriptional response to SNS activation, people bearing G versus C allele *IL6* promoters show differing social epidemiologic relationships between stressful life circumstances and the incidence of inflammation-related disease and mortality. Stressful conditions sufficient to induce symptoms of depression are associated with a significant

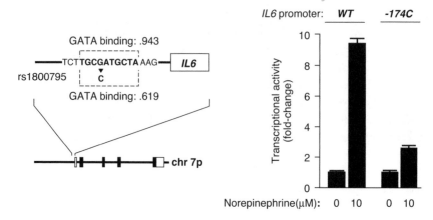

Figure 9.5. **Effect of the human *IL6* regulatory polymorphism on norepinephrine-induced transcription.** The left panel shows bioinformatic analyses comparing the predicted binding of the GATA1 transcription factor (0–1 metric) to either the ancestral G allele of the *IL6* promoter or the minor C allele. Affinity values of less than .75 are functionally inactive. The second panel shows the results of in vitro biochemical analyses modeling the effects of stress on *IL6* expression. *IL6* promoter sequences driving the expression of a reporter gene were transfected into human ovarian cancer cells, which were then stimulated with 0 or 10 μM concentrations of the sympathetic nervous system neurotransmitter norepinephrine (NE) to activate GATA1. Consistent with the bioinformatic predictions, NE efficiently activated transcription from the GATA1-sensitive G allele promoter, but had minimal effect on the GATA1-insensitive C allele promoter. The G > C substitution essentially disconnected *IL6* transcription from stress-induced regulation by the sympathetic nervous system.

increase in inflammation-related mortality risk in people bearing two copies of the GATA1-sensitive *IL6* G allele promoter (S. Cole et al., 2010). However, heterozygotes and those homozygous for the GATA1-insensitive C allele show no relationship between stress/depression and mortality. The *IL6* regulatory polymorphism blocks the capacity of adverse environmental conditions to activate the expression of this key disease-related gene, and thus renders carriers less vulnerable to socioenvironmentally mediated health risks.

Regulatory polymorphisms may have little effect on behavioral or bio-logical phenotypes if the affected physiologic function continues to be maintained through other redundant regulatory pathways. However, regula-tory polymorphisms that affect physiologically influential genes with no functional back-ups can significantly alter relationships between environ-mental risk factors and survival-related phenotypes, such as cardiovascular disease, inflammation-related cancers, or viral infections (Capitanio et al., 2008; S. Cole et al., 2010). Because disease phenotypes influence survival and reproduction (i.e., fitness), individual differences in transcriptional responses to environmental conditions have the potential to shape the evolutionary trajectory of our DNA genome at the population level. In integrating the molecular biology of gene structure (DNA), the environmental control of gene expression (RNA), and the social biology of individual behavior and survival, the *SLC6A4/5HTTLPR* serotonin transporter polymorphism and the *IL6* regulatory polymorphism exemplify a new "environmentally conscious" conception of genetics in which cellular and organismic behaviors constitute the fundamental units of evolutionary selection, and genes and environments depend mutually on one another to shape those behaviors by structuring our brains and bodies.

IMPLICATIONS OF A DYNAMIC GENOME FOR MENTAL HEALTH AND DEVELOPMENT

Research in social genomics has now shown that the interpersonal world exerts biologically significant effects on the molecular composition of the human body. These effects typically target a nonrandom 1%–5% of the human genome, although often a different 1%–5% depending upon the social circumstances and cell type studied. Moreover, the 1%–5% of genes that are socially sensitive may disproportionately influence health and development. Ongoing research will more fully map the specific genes that are subject to social regulation, clarify the types of socioenvironmental conditions that alter their transcription, define psychological and biological pathways mediating those effects, and identify DNA polymorphisms that moderate or intensify their impact. However, even at this early stage of discovery, it is worth considering how the burgeoning literature on socioenvironmental control of gene expression might influence our conceptions of mental health, social development, and human identity.

Social Therapeutics

One significant implication of social genomics involves the prospects for environmentally mediated change in the molecular biology of health and development. Our RNA functional genome is far more fluid and situationally malleable than is generally appreciated, and the experimental studies reviewed above show that some key aspects of the RNA "realized genome" can be altered via social signal transduction. Transient alterations can also persist over time due to the temporal characteristics of gene regulatory dynamics (e.g., recursive feedback, stable epigenetic modifications, molecular remodeling of CNS structures involved in environmental interpretation, etc.). To the extent that we can actively create socioenvironmental conditions favoring healthy development, either societally or individually, and either consciously or unintentionally (e.g., via gene–environment correlation), we can potentially redirect the intraindividual evolution of our RNA transcriptomes in ways that not only enhance current health but also create a molecular resilience to future challenges (Heckman, 2007). Our DNA genome itself can be construed as a certain kind of challenge, to the degree that our structural genetics are riddled with risk-conferring genes. As Suomi and colleagues have shown in experimental analyses of gene ×–environment interactions (Stevens, Leckman, Coplan, & Suomi, 2009), the imposition of "good environments" can potentially override the effects of "bad genes." Mapping the "active ingredients" of salutary environments—the specific social and psychological variables that trigger healthy, protective gene expression dynamics—provides a major opportunity for translating basic genomic research into advances in human health. Strategic shaping of social environments to mitigate the effects of adverse life histories and DNA risk factors represents one of the most significant opportunities to emerge from a functional genomics perspective.

Personalized Targeting

The impact of social genomic interventions may be magnified through selective targeting based on individual measures of genetic (DNA) and functional genomic (RNA) risk factors. As an example, consider the gene–environment interaction involving stress and the *IL6* promoter polymorphism. Risk of the adverse outcome—an inflammation-mediated disease, such cancer or cardiovascular disease—increases in the simultaneous presence of a genetic risk factor (presence of the G allele promoter) and a functional genomic risk factor (stress-induced SNS activation of the GATA1 transcription factor). To the extent that either risk factor is mitigated, no biological interaction would occur, and the risk of the adverse outcome would abate. It may be possible to reduce stress-induced sympathetic nervous system activation of GATA1 (e.g., via treatment with β-blockers or global lifestyle alterations), but such approaches are not likely to be feasible or cost effective at the population level. With regard to the genetic risk factor, we have no effective strategies for altering the risky *IL6* promoter G nucleotide at the level of somatic DNA. However, it may be feasible to deploy resource-intensive GATA1-inhibiting strategies selectively toward the approximately 25% of individuals who are homozygous for the *IL6* G allele and thus at genetic risk for the toxic interaction. If we further narrow our intervention target to those who show empirical evidence of GATA1 activation (e.g., measured via promoter-based bioinformatic analysis of leukocyte gene expression profiles (S. Cole et al., 2010)), the health return on our intervention investment may increase considerably.

Let's plausibly assume that approximately one-third of *IL6* G homozygotes also show evidence of GATA1 activation. If resource-intensive protective interventions are focused on the 33% (GATA1 activation) × 25% (GG heterozygote) = ~8% of the total population bearing both elements of that toxic gene–environment interaction, the per capita health yield of the intervention would increase by more than ten-fold. We need not rely solely on the random distribution of good environments to overcome the effects of bad genes, or on the random distribution of good genes to protect some of us from bad environments. Resource-intensive active interventions could become economically viable if genomics-based targeting can focus their impact on those who stand to benefit the most.

Transcriptional Therapeutics

Functional genomics also suggests new opportunities for targeting molecular therapeutics (e.g., antidepressant drugs) and new biomarkers of their impact. Current drug development efforts generally focus on a single gene product, such as the serotonin transporter molecule encoded by *SLC6A4*. However, environmental influences on gene expression are mediated by transcription factors that affect the expression of tens to thousands of genes simultaneously. Therapeutic agents that selectively target transcription factors may provide substantially greater leverage over biological, behavioral, or psychological phenotypes than do agents targeting a single gene product. For example, positive responses to pharmacologic and psychotherapies for depression have been linked to activation of the CREB transcription factor (Blendy, 2006; Koch et al., 2009; Koch, Kell, & Aldenhoff, 2003; Koch, Kell, Hinze-Selch, & Aldenhoff, 2002). CREB regulates the expression of more than 4,000 human genes, or about 20% of the total human genome (X. Zhang et al., 2005). Agents that act directly on CREB might provide a more effective strategy for ameliorating depression than do current antidepressants targeting a single neurotransmitter receptor or transporter. Direct assessment of CREB activity may also provide a useful predictive biomarker for clinical response to treatment (Koch et al., 2009; Koch et al., 2002) that could potentially help optimize treatment regimens for the neurobiological characteristics of different patients. After all, social genomics tells us that the activity of some transcription factors can provide an integrated biological assessment of the total "environment" surrounding a cell, including the effects of pharmacologic agents, individual psychological conditions, individual transcriptional "life histories," and purely symbolic therapeutic interventions (Koch et al., 2009).

Penetrance

Viewed from a DNA-centered perspective, functional genomics also provides new opportunities to unpack the mysteries of gene penetrance. Most genetic risk factors are only weakly penetrant—the probability of manifesting a given phenotype generally differs by less than 20% as a function of genotypic variation at any single locus (i.e., risk ratios < 1.2, or $r \le .10$). The relatively small effect sizes for most individual genetic risk factors implies that the vast majority of total genetic influence involves complex effects generated through large numbers of small genetic effects or interactions with other genes or environments. Efforts to understand the weak penetrance of DNA characteristics into behavioral phenotypes often focus on translating behavioral phenotypes into neurobiological "endophenotypes" that may show greater correlation with genetic characteristics (Meyer-Lindenberg & Weinberger, 2006) or discovering combinations of DNA polymorphisms that act synergistically to affect behavior (i.e., epistatic interactions) (Lykken, 2006; Sullivan et al., 2004; York et al., 2005). A functional genomic perspective would emphasize the effects of environmental variation in determining whether or not a DNA-endowed genetic potential is actually realized in RNA, protein, and phenotype. This approach grounds the statistical concept of gene–environment interaction in specific molecular hypotheses regarding social and biological signal transduction. By focusing on specific biological hypotheses, a functional genomics-based approach greatly reduces the statistical challenges that would otherwise plague a hypothesis-free blind search for penetrance modifiers over the vast search space defined by the combination of tens to hundreds of possibly relevant environmental parameters and the estimated 10 million DNA polymorphisms present in the human genome. Instead of decimating statistical power through a blind search of all 10^8–10^9 potential interactions, this approach explicitly tracks the flow of socioenvironmental information through the CNS and into the signal transduction machinery that shapes gene expression and cellular function, and thereby identifies specific features of the environment that enable or impede the expression of DNA genotypes as health or behavioral phenotypes. This approach has already begun to shed light on the environmental basis for the conditional penetrance of the

5HTTLPR and *IL6* promoter polymorphisms (Caspi et al., 2003; Champoux et al., 2002; S. Cole et al., 2010). Recent advances in computational modeling of physical interactions between environmentally induced transcription factors and DNA polymorphisms could considerably accelerate the genome-wide application of this approach (S. Cole et al., 2010).

Molecular Development

A functional genomic perspective also provides a concrete molecular basis for understanding individual continuity and change over time. Unlike individually stable DNA genomes, RNA transcriptomes evolve over the course of an individual lifespan in response to "selective pressure" from changing environmental conditions (Choi & Kim, 2007; Fraga et al., 2005). Figure 9.4 emphasizes the two general factors shaping transcriptome evolution: (1) regulatory inputs from the environment, and (2) previous transcriptional regimes that structure the present molecular characteristics of the cell. The second stream of influence implies an "autoregressive" temporal structure, in which the condition of the system at the present time t is shaped in part by the state of the system at t-1, and that state at t-1 is in turn influenced by the state of the system at t-2, etc. Autoregressive RNA dynamics stem from positive feedback loops in which transcription factors regulate (1) their own expression, or that of functionally redundant transcriptional mediators (gene $\rightarrow$ gene loops); and (2) the expression of environment-transducing receptors and environment-modifying effector molecules that induce stable patterns of interaction between individual behavior and environmental niches (i.e., gene $\rightarrow$ environment $\rightarrow$ gene loops, or gene–environment correlation). Both of these self-perpetuating dynamics embed a "transcriptional memory" of previous environmental conditions within the current transcriptome.

One intriguing implication of transcriptional memory is that we should be able to detect the historical experience of biologically significant environmental shocks within current gene expression profiles. For example, the persisting effects of early social deprivation on repression of the glucocorticoid-responsive transcription factor *Nr3c1* (Weaver et al., 2006; T. Y. Zhang et al., 2006) might allow us to infer that individuals who underexpress this gene later in life were exposed to a substantial period of environmental stress during early development. Some recent evidence suggests that this type of "transcriptional back-casting" may indeed be possible. In the study of early childhood SES noted earlier (G. E. Miller et al., 2009), reduced *Nr3c1*-mediated gene expression remained evident decades later in adulthood. The persisting effects of social adversity in rewiring SNS control of the immune response via *NGF* transcription (described above and in Sloan et al. [2008] and Sloan, Capitanio, Tarara, Mendoza et al. [2007]) provide another example in which adult transcriptional profiles are shaped by earlier socioenvironmental conditions (Sloan, Capitanio, Tarara, Mendoza et al., 2007).

Molecular Forecasting

Autoregressive transcriptional dynamics also imply the potential to predict future states of the system (e.g., at $t + 1$, $t + 2$, etc.) based on its current state at time t. That opens an opportunity for molecular forecasting of individual developmental trajectories over time. Because RNA is a more proximal determinant of phenotype (Fig. 9.1), it should provide more accurate predictions of health and behavioral traits than would more distal DNA characteristics. RNA expression also integrates the effects of structural genomic (DNA) and environmental influences, providing a more comprehensive platform for prediction. RNA-based prognostics are already being applied to somatic diseases such as cancer (e.g., forecasting the likelihood of future disease recurrence or treatment response based on current gene expression profiles in breast tumors) (Buyse et al., 2006; Mamounas et al., 2010; Paik et al., 2004). Accurate forecasting of future psychosocial characteristics such as depression, alcoholism, or antisocial behavior would require substantial advances in our understanding of the specific transcriptional dynamics underlying those phenotypes. However, one precedent for this approach is emerging in the use of CREB activity in leukocytes to predict the remission of major depression in response to antidepressant medications or psychotherapy (Koch et al., 2002, 2009). CREB mediates its effects by altering gene expression, and it is possible that prognostic approaches based on differential expression of CREB target genes might provide even better predictive power. To the extent that other conserved transcriptional alterations underlie normal human development and pathological alterations, forecasting future behavioral health and development trajectories based on current gene expression profiles could eventually become feasible.

One major challenge in molecular forecasting of behavioral outcomes involves the inaccessibility of key CNS tissues mediating behavioral responses. Structural genetic influences on CNS function can be analyzed using readily accessible proxy cells, such as circulating leukocytes, because all of an individual's cells share a common DNA genome. However, the RNA transcriptome differs greatly in leukocytes and neurons, and even across different subtypes of leukocyte such as B, T, and NK cells (Palmer, Diehn, Alizadeh, & Brown, 2006; Whitney et al., 2003). Tissue heterogeneity in RNA expression does not completely thwart molecular forecasting because some important behavioral phenotypes may show reliable transcriptional correlates in proxy cells (e.g., leukocytes) even if other cells, such as neurons, are more directly responsible for the phenotype. This can happen when systemic hormone dynamics impact both the behaviorally causal cell type and noncausal "reporter cells" (e.g., leukocytes transduce signals from a diverse array of hormones and neuromediators that also affect CNS cells). For example, several studies noted above have empirically linked social adversity to blunted *Nr3c1*-mediated gene transcription in circulating leukocytes (Cole et al., 2007; G. E. Miller et al., 2008, 2009), even though impaired leukocyte transcription of *Nr3c1* target genes is unlikely to directly mediate behavioral responses to adversity. Likewise, CREB activity in circulating leukocytes appears to predict behavioral responses to antidepressant therapy (Koch et al., 2002, 2009), even though increased leukocyte transcription of CREB target genes is unlikely to directly mediate those changes in behavior. Tissue transcriptional heterogeneity may not prevent statistical forecasting, but it does complicate efforts to understand the basic biology of causal cells through extrapolation from accessible proxy cells. Empirical calibration of proxy cell transcriptional responses (e.g., in leukocytes) against target cell responses (e.g., in CNS neurons) or direct analyses of target tissues in animal models will generally be required for true biological insight.

Molecular Dynamics in Personality

Because our molecular characteristics shape stable individual differences in behavior, functional genomics may also provide some new perspectives on the nature of human personality. The complex systems dynamics of transcriptional networks (Kauffman, 1993) provide a natural explanation for the paradoxical kinetics of human personality, which is marked by long periods of apparent continuity punctuated by occasional marked shifts (i.e., "catastrophic" dynamics). Extended periods of stability emerge because transcription-regulatory networks show attractor-like dynamics in which gene expression profiles tend to self-perpetuate in the absence of external shocks. Those stable equilibria are driven by positive feedback at the cellular level (gene → gene loops) and by self-reinforcing selection of environmental niches at the organismic level (gene → environment → gene circuits, or gene–environment correlation). For example, socially inhibited temperament is thought to stem from a neurobiologically rooted aversion to novelty that fails to habituate during development because inhibited individuals selectively avoid high-novelty environments (Kagan, 1994; Kagan et al., 1988; Schwartz et al., 2003) (i.e., a self-propagating gene → environment →

gene circuit). The impact of serotonin-targeted agents in ameliorating social inhibition (Kramer, 1992) implies a molecular basis for that ecological self-selection dynamic (and suggests candidate genes to analyze in the context of the gene → environment → gene circuit).

When personality change does occur, it is often experienced as a discontinuous "jump" in response to an environmental shock. However, rapid transitions from one attractor state to another are a hallmark of networks that bear the recursive regulatory characteristics of transcriptional systems (Kauffman, 1993). Even if the environmental inputs into the system grow linearly over time (e.g., as our bank of social experience gradually grows), the self-reinforcing dynamics of transcriptional networks would tend to avert any phenotypic change until the net force of environmental press crosses a critical "saddle point" in the topography of possible gene expression regimes. Once environmental pressure does force a transcriptional regime shift, that new regime may again self-propagate for an extended period before being pushed into a new equilibrium by gradually accumulating environmental pressure. The promise of functional genomics in understanding personality dynamics involves concretizing the abstract theoretical conceptions of dynamic systems theory in empirically measurable behaviors, environments, and molecular characteristics. Those parameters provide the data that would ultimately be required to validate a dynamic systems-based account of personality and individual differences.

Molecular Dimensions of Personality

An RNA-rooted conception of personality might also imply that additional dimensions of personality exist at the molecular level and may shape individual differences in response to biological challenges. For example, consider the implications of the inhibited temperament theory discussed above for the socially mediated alterations in SNS regulation of immune responses to viral infection described earlier. To the extent that recurrent experiences of social threat are both a hallmark of social inhibition (Kagan, 1994) and drive hyperinnervation of immune system organs by the SNS (Sloan et al., 2008; Sloan, Nguyen et al., 2007), socially inhibited individuals might be expected to show increased SNS-mediated alterations in the expression of immune-response genes and disease resistance. That is precisely what has emerged in studies relating individual differences in social behavior to SNS activity and gene expression in rhesus macaques (Capitanio, Mendoza, & Baroncelli, 1999; Sloan, Capitanio, Tarara, & Cole, 2007) and viral pathogenesis in humans (Cohen et al., 1997; Cole, Kemeny et al., 2003). Personality likely arises in part from individual variations in gene expression, so it is to be expected that personality has molecular manifestations deep within our bodies and well outside our awareness.

The Self

Social genomic relationships belie two fundamental tenants of our individualistic conception of the self: continuity and independence. We generally experience ourselves as stable individual beings who live in the world but are fundamentally separate from it. Nevertheless, all the data reviewed above show that the world around us can alter the activity of some of our most fundamentally internal and self-defining features: the activity of our genes. Far from being "outside" of us, the social world blows through our permeable bodies like a breeze, quietly reshaping who we are at the molecular level. We are who we are biologically *in part* because of who we have spent time with, the trail of physical and social ecologies we have inhabited, and our subjective psychological reactions to our social-environmental life histories.[3] These notions may seem preposterous and perhaps even a bit frightening in the context of the Western individualist conception of identity. After all, this implies that some molecular characteristics of our bodies exist at the whim of external conditions over which we may have little control. But, from another perspective, social regulation of gene expression also provides a serendipitous opportunity to shape the molecular composition of our bodies, and the specific realization of our DNA-endowed potential, by strategically altering the characteristics of our physical and social environments. As noted at the outset of this chapter, the events of today will influence the molecular characteristics of your body for the next 2–3 months. Plan your day accordingly.

ACKNOWLEDGMENTS

Supported by grants CA116778 and AG028748 from the National Institutes of Health.

---

[3] Of course, our DNA genotypes do matter too!

REFERENCES

Antoni, M. H., Lutgendorf, S. K., Cole, S. W., Dhabhar, F. S., Sephton, S. E., McDonald, P. G., et al. (2006). The influence of bio-behavioural factors on tumour biology: pathways and mechanisms. *Nature Reviews. Cancer, 6*(3), 240–248.

Bamberger, C. M., Schulte, H. M., & Chrousos, G. P. (1996). Molecular determinants of glucocorticoid receptor function and tissue sensitivity to glucocorticoids. *Endocrine Reviews, 17*(3), 245–261.

Benjamini, Y., & Hochberg, Y. (1995). Controlling the false discovery rate: A practical and powerful approach to multiple testing. *Journal of the Royal Statistical Society, Series B, 57,* 289–300.

Blendy, J. A. (2006). The role of CREB in depression and antidepressant treatment. *Biological Psychiatry, 59*(12), 1144–1150. Epub 2006 Feb 2.

Buyse, M., Loi, S., van't Veer, L., Viale, G., Delorenzi, M., Glas, A. M., et al. (2006). Validation and clinical utility of a 70-gene prognostic signature for women with node-negative breast cancer. *Journal of the National Cancer Institute, 98*(17), 1183–1192.

Cacioppo, J. T., & Hawkley, L. C. (2009). Perceived social isolation and cognition. *Trends in Cognitive Sciences, 13*(10), 447–54. Epub 2009 Aug 31.

Capitanio, J. P., Abel, K., Mendoza, S. P., Blozis, S. A., McChesney, M. B., Cole, S. W., et al. (2008). Personality and serotonin transporter genotype interact with social context to affect immunity and viral set-point in simian immunodeficiency virus disease. *Brain, Behavior, and Immunity, 22*(5), 676–89. Epub 2007 Aug 23.

Capitanio, J. P., Mendoza, S. P., & Baroncelli, S. (1999). The relationship of personality dimensions in adult male rhesus macaques to progression of simian immunodeficiency virus disease. *Brain, Behavior, and Immunity, 13*(2), 138–154.

Caspi, A., Harrington, H., Moffitt, T. E., Milne, B. J., & Poulton, R. (2006). Socially isolated children 20 years later: risk of cardiovascular disease. *Archives of Pediatrics & Adolescent Medicine, 160*(8), 805–811.

Caspi, A., Sugden, K., Moffitt, T. E., Taylor, A., Craig, I. W., Harrington, H., et al. (2003). Influence of life stress on depression: moderation by a polymorphism in the 5-HTT gene. *Science, 301*(5631), 386–389.

Champagne, F. A., & Mashoodh, R. (2009). Genes in context: gene-environment interplay and the origins of individual differences in behavior. *Current Directions in Psychological Science, 18*(3), 127–131.

Champoux, M., Bennett, A., Shannon, C., Higley, J. D., Lesch, K. P., & Suomi, S. J. (2002). Serotonin transporter gene polymorphism, differential early rearing, and behavior in rhesus monkey neonates. *Molecular Psychiatry, 7*(10), 1058–1063.

Chang, M., Brown, H., Collado-Hidalgo, A., Arevalo, J., Galic, Z., Symensma, T., et al. (2005). Beta-adrenoreceptors reactivate KSHV lytic replication via PKA-dependent control of viral RTA. *Journal of Virology,79*(21), 13538–13547.

Chen, E., Miller, G. E., Walker, H. A., Arevalo, J. M., Sung, C. Y., & Cole, S. W. (2009). Genome-wide transcriptional profiling linked to social class in asthma. *Thorax, 64*(1), 38–43. Epub 2008 Nov 10.

Choi, J. K., & Kim, S. C. (2007). Environmental effects on gene expression phenotype have regional biases in the human genome. *Genetics, 175*(4), 1607–1613. Epub 2007 Jan 21.

Cohen, S., Doyle, W. J., Skoner, D. P., Rabin, B. S., & Gwaltney, J. M. (1997). Social ties and susceptibility to the common cold. *Journal of the American Medical Association, 227*, 1940–1944.

Cole, S., Arevalo, J., Takahashi, R., Sloan, E. K., Lutgendorf, S., Sood, A. K., et al. (2010). Computational identification of Gene-Social Environment interaction at the human IL6 locus. *Proceedings of the National Academy of Sciences, U.S.A., in press.*

Cole, S. W. (2008a). Psychosocial influences on HIV-1 disease progression: neural, endocrine, and virologic mechanisms. *Psychosomatic Medicine, 70*(5), 562–568.

Cole, S. W. (2008b). Social regulation of leukocyte homeostasis: The role of gluco-corticoid sensitivity. *Brain, Behavior, and Immunity, 22*(7), 1049–1055.

Cole, S. W. (2009). Social regulation of human gene expression. *Current Directions in Psychological Science, 18*(3), 132–137.

Cole, S. W., Arevalo, J. M., Ruggerio, A. M., Heckman, J. J., & Suomi, S. (2010). Transcriptional modulation of the developing immune system by early life social adversity. Under review.

Cole, S. W., Galic, Z., & Zack, J. A. (2003). Controlling false-negative errors in microarray differential expression analysis: a PRIM approach. *Bioinformatics, 19*(14), 1808–1816.

Cole, S. W., Hawkley, L. C., Arevalo, J. M., Sung, C. Y., Rose, R. M., & Cacioppo, J. T. (2007). Social regulation of gene expression in human leukocytes. *Genome Biology, 8*(9), R189.

Cole, S. W., Kemeny, M. E., Fahey, J. L., Zack, J. A., & Naliboff, B. D. (2003). Psychological risk factors for HIV pathogenesis: Mediation by the autonomic nervous system. *Biological Psychiatry, 54*, 1444–1456.

Cole, S. W., Kemeny, M. E., & Taylor, S. E. (1997). Social identity and physical health: Accelerated HIV progression in rejection-sensitive gay men. *Journal of Personality and Social Psychology, 72*, 320–336.

Cole, S. W., Mendoza, S. P., & Capitanio, J. P. (2009). Social stress desensitizes lymphocytes to regulation by endogenous glucocorticoids: insights from in vivo cell trafficking dynamics in rhesus macaques. *Psychosomatic Medicine, 71*(6), 591–597. Epub 2009 Jun 2024.

Cole, S. W., Sloan, E. K., Arevalo, J., & Capitanio, J. P. (2010). Socio-environmental regulation of systemic gene expression. Under review.

Cole, S. W., Yan, W., Galic, Z., Arevalo, J., & Zack, J. A. (2005). Expression-based monitoring of transcription factor activity: The TELiS database. *Bioinformatics, 21*(6), 803–810.

Collado-Hidalgo, A., Sung, C., & Cole, S. (2006). Adrenergic inhibition of innate anti-viral response: PKA blockade of Type I interferon gene transcription mediates catecholamine support for HIV-1 replication. *Brain, Behavior, and Immunity, 20*(6), 552–563. Epub 2006 Feb 28.

Dickerson, S. S., & Kemeny, M. E. (2004). Acute stressors and cortisol responses: a theoretical integration and synthesis of laboratory research. *Psychological Bulletin, 130*(3), 355–391.

Dimsdale, J. E., & Moss, J. (1980). Plasma catecholamines in stress and exercise. *Journal of the American Medical Association, 243*(4), 340–342.

Dusek, J. A., Otu, H. H., Wohlhueter, A. L., Bhasin, M., Zerbini, L. F., Joseph, M. G., et al. (2008). Genomic counter-stress changes induced by the relaxation response. *PLoS ONE., 3*(7), e2576.

Fraga, M. F., Ballestar, E., Paz, M. F., Ropero, S., Setien, F., Ballestar, M. L., et al. (2005). Epigenetic differences arise during the lifetime of monozygotic twins. *Proceedings of the National Academy of Sciences of the U.S.A., 102*(30), 10604–10609. Epub 2005 Jul 11.

Frankenhauser, M. (1975). Experimental approaches to the study of catecholamines and emotion. In L. Levi (Ed.), *Emotions - Their parameters and measurement* (pp. 209–234). New York: Raven Press.

Heckman, J. J. (2007). The economics, technology, and neuroscience of human capability formation. *Proceedings of the National Academy of Sciences of the U.S.A., 104*(33), 13250–13255 Epub 2007 Aug 8.

Irwin, M. R., Wang, M., Campomayor, C. O., Collado-Hidalgo, A., & Cole, S. (2006). Sleep deprivation and activation of morning levels of cellular and genomic markers of inflammation. *Archives of Internal Medicine, 166*(16), 1756–1762.

Kagan, J. (1994). *Galen's prophecy: temperament in human nature.* New York: Basic Books.

Kagan, J., Reznick, J. S., & Snidman, N. (1988). Biological bases of childhood shyness. *Science, 240,* 167–171.

Karssen, A. M., Her, S., Li, J. Z., Patel, P. D., Meng, F., Bunney, W. E., Jr., et al. (2007). Stress-induced changes in primate prefrontal profiles of gene expression. *Molecular Psychiatry, 12*(12), 1089–1102. Epub 2007 Sep 25.

Kauffman, S. (1993). *The origins of order: self-organization and selection in evolution.* Oxford: Oxford University Press.

Kim, H. D., Shay, T., O'Shea, E. K., & Regev, A. (2009). Transcriptional regulatory circuits: predicting numbers from alphabets. *Science, 325*(5939), 429–432.

Koch, J. M., Hinze-Selch, D., Stingele, K., Huchzermeier, C., Goder, R., Seeck-Hirschner, M., et al. (2009). Changes in CREB phosphorylation and BDNF plasma levels during psychotherapy of depression. *Acta Psychotherapeutica et Psychosomatica, 78*(3), 187–192. Epub 2009 Mar 24.

Koch, J. M., Kell, S., & Aldenhoff, J. B. (2003). Differential effects of fluoxetine and imipramine on the phosphorylation of the transcription factor CREB and cell-viability. *Journal of Psychiatric Research, 37*(1), 53–59.

Koch, J. M., Kell, S., Hinze-Selch, D., & Aldenhoff, J. B. (2002). Changes in CREB-phosphorylation during recovery from major depression. *Journal of Psychiatric Research, 36*(6), 369–375.

Kramer, P. D. (1992). *Listening to Prozac: A psychiatrist explores antidepressant drugs and the remaking of the self.* New York: Penguin Books.

Lutgendorf, S. K., Lamkin, D. M., Jennings, N. B., Arevalo, J. M., Penedo, F., DeGeest, K., et al. (2008). Biobehavioral influences on matrix metalloproteinase expression in ovarian carcinoma. *Clinical Cancer Research, 14*(21), 6839–6846.

Lykken, D. T. (2006). The mechanism of emergenesis. *Genes, Brain, and Behavior, 5*(4), 306–310.

Mamounas, E. P., Tang, G., Fisher, B., Paik, S., Shak, S., Costantino, J. P., et al. (2010). Association between the 21-gene recurrence score assay and risk of locoregional recurrence in node-negative, estrogen receptor-positive breast cancer: results from NSABP B-14 and NSABP B-20. *Journal of Clinical Oncology, 28*(25), 3937–3944. Epub 2010 Aug 2.

Meyer-Lindenberg, A., & Weinberger, D. R. (2006). Intermediate phenotypes and genetic mechanisms of psychiatric disorders. *Nature Reviews. Neuroscience, 7*(10), 818–827.

Miller, G., Chen, E., & Cole, S. W. (2009). Health Psychology: Developing Biologically Plausible Models Linking the Social World and Physical Health. *Annual Review of Clinical Psychology, 60*, 501–524.

Miller, G. E., Chen, E., Fok, A. K., Walker, H., Lim, A., Nicholls, E. F., et al. (2009). Low early-life social class leaves a biological residue manifested by decreased glucocorticoid and increased proinflammatory signaling. *Proceedings of the National Academy of Sciences of the U.S.A., 106*(34), 14716–14721. Epub 2009 Jul 14.

Miller, G. E., Chen, E., Sze, J., Marin, T., Arevalo, J. M., Doll, R., et al. (2008). A functional genomic fingerprint of chronic stress in humans: blunted glucocorticoid and increased NF-kappaB signaling. *Biological Psychiatry, 64*(4), 266–272. Epub 2008 Apr 28.

Miller, G. E., Cohen, S., Rabin, B. S., Skoner, D. P., & Doyle, W. J. (1999). Personality and tonic cardiovascular, neuroendocrine, and immune parameters. *Brain, Behavior, and Immunity, 13*, 109–123.

Morita, K., Saito, T., Ohta, M., Ohmori, T., Kawai, K., Teshima-Kondo, S., et al. (2005). Expression analysis of psychological stress-associated genes in peripheral blood leukocytes. *Neuroscience Letters, 381*(1-2), 57–62. Epub 2005 Feb 16.

Nater, U. M., Whistler, T., Lonergan, W., Mletzko, T., Vernon, S. D., & Heim, C. (2009). Impact of acute psychosocial stress on peripheral blood gene expression pathways in healthy men. *Biological Psychology, 82*(2), 125–132. Epub 2009 Jul 03.

Pace, T. W., Hu, F., & Miller, A. H. (2007). Cytokine-effects on glucocorticoid receptor function: Relevance to glucocorticoid resistance and the pathophysiology and treatment of major depression. *Brain, Behavior, and Immunity, 21*(1), 9–19. Epub 2006 Oct 27.

Padgett, D. A., Sheridan, J. F., Dorne, J., Berntson, G. G., Candelora, J., & Glaser, R. (1998). Social stress and the reactivation of latent herpes simplex virus type 1. *Proceedings of the National Academy of Sciences of the U.S.A., 95*(12), 7231–7235.

Paik, S., Shak, S., Tang, G., Kim, C., Baker, J., Cronin, M., et al. (2004). A multigene assay to predict recurrence of tamoxifen-treated, node-negative breast cancer. *New England Journal of Medicine, 351*(27), 2817–2826. Epub 2004 Dec 10.

Palmer, C., Diehn, M., Alizadeh, A. A., & Brown, P. O. (2006). Cell-type specific gene expression profiles of leukocytes in human peripheral blood. *BMC Genomics, 7*, 115–130.

Pressman, S. D., Cohen, S., Miller, G. E., Barkin, A., Rabin, B. S., & Treanor, J. J. (2005). Loneliness, social network size, and immune response to influenza vaccination in college freshmen. *British Journal of Health Psychology, 24*(3), 297–306.

Sapolsky, R. M. (1994). *Why zebras don't get ulcers: a guide to stress, stress-related diseases, and coping.* New York: Freeman.

Schwartz, C. E., Wright, C. I., Shin, L. M., Kagan, J., & Rauch, S. L. (2003). Inhibited and uninhibited infants "grown up": adult amygdalar response to novelty. *Science, 300*(5627), 1952–1953.

Seeman, T. E. (1996). Social ties and health: the benefits of social integration. *Annals of Epidemiology, 6*(5), 442–451.

Sheridan, J. F., Stark, J. L., Avitsur, R., & Padgett, D. A. (2000). Social disruption, immunity, and susceptibility to viral infection. Role of glucocorticoid insensitivity and NGF. *Annals of the New York Academy of Sciences, 917*, 894–905.

Sloan, E. K., Capitanio, J. P., & Cole, S. W. (2008). Stress-induced remodeling of lymphoid innervation. *Brain, Behavior, and Immunity, 22*(1), 15–21. Epub 2007 Aug 13.

Sloan, E. K., Capitanio, J. P., Tarara, R. P., & Cole, S. W. (2007). Social temperament and lymph node innervation. *Brain, Behavior, and Immunity, 22*(5), 717–726. Epub 2007 Dec 18.

Sloan, E. K., Capitanio, J. P., Tarara, R. P., Mendoza, S. P., Mason, W. A., & Cole, S. W. (2007). Social stress enhances sympathetic innervation of primate lymph nodes: mechanisms and implications for viral pathogenesis. *Journal of Neuroscience, 27*(33), 8857–8865.

Sloan, E. K., Nguyen, C. T., Cox, B. F., Tarara, R. P., Capitanio, J. P., & Cole, S. W. (2007). SIV infection decreases sympathetic innervation of lymph nodes: The role of neurotrophins. *Brain, Behavior, and Immunity*, in press.

Stark, J., Avitsur, R., Padgett, D. A., & Sheridan, J. F. (2001). Social stress induces glucocorticoid resistance in macrophages. *American Journal of Physiology. Regulatory, Integrative and Comparative Physiology, 280*, R1799–R1805.

Stevens, H., Leckman, J., Coplan, J., & Suomi, S. (2009). Risk and resilience: early manipulation of macaque social experience and persistent behavioral and neurophysiological outcomes. *Journal of the American Academy of Child and Adolescent Psychiatry, 48*(2), 114–127.

Sullivan, P. F., Neale, B. M., van den Oord, E., Miles, M. F., Neale, M. C., Bulik, C. M., et al. (2004). Candidate genes for nicotine dependence via linkage, epistasis, and bioinformatics. *American Journal of Medical Genetics. Part B, Neuropsychiatric Genetics, 126B*(1), 23–36.

Thaker, P. H., Han, L. Y., Kamat, A. A., Arevalo, J. M., Takahashi, R., Lu, C., et al. (2006). Chronic stress promotes tumor growth and angiogenesis in a mouse model of ovarian carcinoma. *Nature Medicine, 12*(8), 939–944. Epub 2006 Jul 23.

Weaver, I. C., Meaney, M. J., & Szyf, M. (2006). Maternal care effects on the hippocampal transcriptome and anxiety-mediated behaviors in the offspring that are reversible in adulthood. *Proceedings of the National Academy of Sciences of the U.S.A., 103*(9), 3480–3485. Epub 2006 Feb 16.

Weiner, H. (1992). *Perturbing the organism: the biology of stressful experience.* Chicago: University of Chicago Press.

Whitney, A. R., Diehn, M., Popper, S. J., Alizadeh, A. A., Boldrick, J., Relman, D. A., et al. (2003). Individuality and variation in gene expression patterns in human blood. *Proceedings of the National Academy of Sciences of the U.S.A., 100*(4), 1896–1901.

York, T. P., Miles, M. F., Kendler, K. S., Jackson-Cook, C., Bowman, M. L., & Eaves, L. J. (2005). Epistatic and environmental control of genome-wide gene expression. *Twin Research and Human Genetics, 8*(1), 5–15.

Zhang, T. Y., Bagot, R., Parent, C., Nesbitt, C., Bredy, T. W., Caldji, C., et al. (2006). Maternal programming of defensive responses through sustained effects on gene expression. *Biological Psychology, 73*(1), 72–89. Epub 2006 Feb 28.

Zhang, X., Odom, D. T., Koo, S. H., Conkright, M. D., Canettieri, G., Best, J., et al. (2005). Genome-wide analysis of cAMP-response element binding protein occupancy, phosphorylation, and target gene activation in human tissues. *Proceedings of the National Academy of Sciences of the U.S.A., 102*(12), 4459–4464. Epub 2005 Mar 7.

Part II

Dynamics of Genes and Environments in
Behavioral and Mental Health Conditions

10

The Interplay Between Genes and Environment in the Development of Anxiety and Depression

THALIA C. ELEY

Anxiety disorders and depressive disorders are surprisingly common in young people, affecting approximately 5%–10% of children and adolescents at any one point in time (Costello, Egger, & Angold, 2005a,b; Meltzer, Gatward, Goodman, & Ford, 2000) and up to 25% of individuals by the age of 18. Both types of disorder commonly continue into adulthood (Fombonne, Wostear, Cooper, Harrington, & Rutter, 2001a; Gregory et al., 2007), are associated with a variety of difficulties from interpersonal relationships to academic attainments (Fombonne, Wostear, Cooper, Harrington, & Rutter, 2001b; Van Ameringen, Mancini, & Farvolden, 2003), and come with a heavy financial burden (Meltzer et al., 2000). Furthermore, anxiety and depressive disorders frequently co-occur (Angold, Costello, & Erkanli, 1999).

Not all that long ago, a fierce debate raged over whether genes or the environment were the key influence on emotional development. Fortunately, that time has passed, and it is now widely accepted that genetic and environmental influences operate in conjunction with one another in complicated ways that move well beyond the simple additive factors traditionally modeled. Although this is incredibly exciting in terms of the sophistication of our models of development, it also requires new approaches and methods that reflect our growing understanding of how genes and the environment operate. Debate remains fierce, but is now focussed on the best approaches to take to understanding these complex phenomena. In this chapter, I present some of the factors relevant to understanding gene–environment interplay, with respect to the development of anxiety and depression. I focus largely on findings from studies using child and adolescent populations, but include selected studies of adults where particularly relevant.

The literature pertinent to this issue falls into two very distinct categories. Quantitative genetic studies use approaches such as the twin design, which estimate the level of genetic influence (or heritability) within a sample. In contrast, molecular genetic studies analyze specific markers within genes. Heritability estimates for complex traits tend to be around 40%, whereas the effects of any one marker are likely to be less than 1%. Thus, molecular genetic studies are necessarily more specific, but they also require huge samples to have adequate power, and the literature is full of non- and partial replications in studies that are substantially underpowered. A particularly pressing issue currently is that, even where replicated associations between genetic markers and complex traits have been found, the sum of the influence of these markers comes nowhere close to the total heritability estimates obtained from twin studies. This phenomena, recently termed "missing heritability" is now the subject of considerable discussion (Manolio et al., 2009). One implication of this is that it is likely that the effects of individual markers are even smaller than thought. Furthermore, genes almost certainly act in conjunction with other factors, making it particularly difficult to identify them. From this perspective, the interplay between genetic and environment factors is an area that is likely to attract increasing levels of attention.

Before turning to the literature on gene–environment interplay, it is useful to summarize what we know about the main effects of genes and the environment on the development of anxiety and depression in young people. First, although estimates vary depending on the precise phenotype being assessed, and the rater used (e.g. parent- versus child-report), both anxiety and depression have been shown to be moderately heritable, with estimates of around 30%–40% (Gregory & Eley, in press; Rice, 2009). Second, the level of genetic influence seems to be similar for both measures of symptoms and for disorders, and extremes analyses indicate that the same genes influence both normal variation and high levels of symptoms (Eley, 1997; Rende, Plomin, Reiss, & Hetherington, 1993; Stevenson, Batten, & Cherner, 1992). Third, genes appear to be more important in continuity of symptoms than in their change, whereas environmental influences are critical to change in symptoms over time (Lau & Eley, 2006). Finally, genetic influences on anxiety and depression show considerable overlap, whereas environmental influences on anxiety and depression are fairly specific (Eley & Stevenson, 1999; Silberg, Rutter, & Eaves, 2001; Thapar & McGuffin, 1997). Thus, for example, whereas loss events are commonly found to be specifically associated with depression, threat or danger events are specifically associated with anxiety (Eley & Stevenson, 2000; Finlay-Jones & Brown, 1981; Kendler, Hettema, Butera, Gardner, & Prescott, 2003).

These findings indicate that when examining gene–environment correlations and interactions on anxiety and depression during development, it is useful to keep in mind that, although they may share much of their genetic influence, environmental influences are largely disorder- and time-specific. This suggests that there is some rather general genetic vulnerability to anxiety and depression that manifests as specific symptoms at a certain point in time as a result of environmental experiences. This genetic vulnerability could also be seen in the context of a diathesis–stress model, with genetic vulnerability being the diathesis, making the exploration of gene–environment interactions particularly pertinent.

It is also critical to consider the way in which the environment is assessed. There are three key points here: the *type* of environmental influence being assessed, *timing* considerations, and the *method* of assessment. With respect to the type of influence, three main groups of factors are relevant to the development of anxiety and depression: background factors, such as socioeconomic status; social factors, such as parent–child interaction or relationships with peers or friends; and specific life events. Timing of certain experiences is also critical. Thus, with life events, it is most useful to know about events experienced in the weeks or months prior to the assessment of anxiety/depression, or to the onset of disorder, when this is the focus. Other factors such as parent–child interaction need to take into account the age of the child. Furthermore, almost all these influences are best assessed prospectively rather than retrospectively wherever possible. Finally, although all of these influences can be assessed by questionnaire, and indeed usually are in genetic studies, in which large samples are needed, they are without doubt most accurately assessed by more detailed means. Thus, for example, life events are best assessed by use of detailed standardized interviews taking into account the circumstances surrounding the events (Sandberg et al., 1993; see Uher & McGuffin, 2008 and Chapter 2 by Uher for a discussion of this issue with respect to gene–environment interaction studies). Parent–child interaction, on the other hand, can be usefully assessed by observation both in structured and unstructured activities. Given all this variation, not only in the type of environmental influences assessed but in their mode of assessment, it is particularly important to establish the extent to which different studies have used similar or different approaches.

Having discussed the background to this area, we can now turn to the main subject of this chapter, the role of gene–environment correlations and interactions in the development of anxiety and depression in young people. The chapter is broken into two main sections. First, gene–environment correlation studies are reviewed; second, gene–environment interactions are discussed. Within both of these sections, work taking a quantitative genetic approach is described first, followed by molecular genetic findings. The chapter ends with thoughts about future directions for this field.

GENE–ENVIRONMENT CORRELATIONS

Gene–environment correlations refer to the fact that genetic and environmental influences on a phenotype are sometimes correlated. Three types of gene–environment correlation have been outlined in the literature (e.g. Scarr & McCartney, 1983) and are discussed in detail by Jaffee in Chapter 4 of this volume. *Passive* gene–environment correlations come about due to the sharing in biological families of both genes and environment. Thus, for example, offspring of depressed mothers are likely to receive both a genetic predisposition for this condition and the environmental effects of a depressogenic parenting style, which may characterize the interpersonal styles of these mothers. *Evocative* gene–environment correlations refer to the genetic propensity of some individuals to elicit or evoke certain reactions from others. Intermediate factors, such as temperament or cognitive style factors, may mediate these effects. Thus, infants who cry easily or show irritability may be more likely to elicit negative reactions from caregivers, which may then impact on parenting style. Finally, *active* gene–environment correlations occur when individuals select, create, and modify their environmental experiences based on particular genetically mediated dispositions. Behaviorally inhibited or shy individuals may be less likely to seek out friends and may instead choose to engage in solitary play, thus ultimately influencing socioemotional development. We examine these effects based on both classical quantitative methods and newer molecular approaches.

Quantitative Genetic Approaches

The core requirement for identifying gene–environment correlation is to demonstrate genetic influence on a measure of the environment (see Chapters 1 and 4, by Kendler and Jaffee, respectively). Perhaps the most widely explored aspect of the environment from this perspective is life events, which have been shown not only to aggregate in families (Rijsdijk et al., 2001), but to be heritable in both adults (Kendler & Karkowski-Shuman, 1997; Kendler, Neale, Kessler, Heath, & Eaves, 1993) and in adolescents (Lau & Eley, 2008; Thapar, Harold, & McGuffin, 1998; Thapar & McGuffin, 1996; Silberg et al., 1999). This indicates that, in some way, individuals are creating their environments around them.

Of course, our environments are much more than just the life events we experience. For children, one other very key feature of the environment is the family in which the child lives. Studies of gene–environment correlation in young people have therefore moved beyond examining just life events, and genetic influence has now been demonstrated on an impressively wide array of family-based environmental variables, including family connectedness (Jacobson & Rowe, 1999); parent–child interaction as assessed by questionnaire (Plomin, Reiss, Hetherington, & Howe, 1994), observation (O'Connor, Hetherington, Reiss, & Plomin, 1995), or both (Pike, McGuire, Hetherington, Reiss, & Plomin, 1996); sibling interactions assessed by questionnaire alone (Plomin et al., 1994) or combined with observation (Pike et al., 1996); parental discipline style retrospectively (Wade & Kendler, 2000) and prospectively (Lau, Rijsdijk, & Eley, 2006); and parental divorce (O'Connor, Caspi, DeFries, & Plomin, 2000; for review of this area see Kendler & Baker, 2007, and Chapter 6 by Horwitz, Marceau, & Neiderhiser).

What is even more fascinating is that, for many of these environmental measures, the genetic influence is *shared* with that of depression. One of the most influential findings in the literature has been that genetic influences on life events overlap with those on depression (Kendler & Karkowski-Shuman, 1997). Studies of child and adolescent samples have examined links between both life events and aspects of parenting with depression and found consistent evidence for genetic influence on these associations (Eaves, Silberg, & Erkanli, 2003; Lau & Eley, 2008; Pike et al., 1996; Rice, Harold, & Thapar, 2003; Silberg et al., 1999; Thapar et al., 1998). This finding has had particular resonance because it has forced people to reconsider how they view the environment. It affirms with great clarity the role we have on shaping and influencing our own environments. Of note, in adults, it appears that the genetic link between depression and life events is also shared with that of neuroticism (Saudino, Pedersen, Lichtenstein, McClearn, & Plomin, 1997), indicating that personality may play a crucial role in the mechanism by which genetic factors influence both life events and depressive symptoms.

One of the most thought-provoking of the papers on this topic (Rice et al., 2003) specifically addressed the potential role gene–environment correlations might have on the increase in heritability of depression that tends to be seen from childhood into adolescence (e.g., Scourfield et al., 2003). The hypothesis was that gene–environment correlations are likely to become increasingly relevant as children move into adolescence and begin to gain greater independence in their social interactions and to have greater choices over how they spend their time. As predicted, the authors found that genetic influences on life events were greater in adolescents relative to children, and accounted not only for much of their covariation, but for the increase in heritability in depression in adolescents relative to children (Rice et al., 2003, see also Chapter 5, by Dickens, Turkheimer, & Beam for similar findings regarding cognitive ability).

Although work of this kind has proliferated with respect to environmental influences on depression, much less is known about the role of gene–environment correlations on the development of anxiety. One key aspect of the family environment that has been repeatedly shown to be associated with child anxiety is maternal overcontrol (Rapee et al., 2009). However, it is unclear whether this is because the mother is responding to anxiety in the child (i.e., the child elicits the overcontrol), or whether the child becomes anxious as a result of the control. We examined this association in our study of 8-year-old twins selected for high levels of parent-reported anxiety (Eley et al., 2008). All the children completed a difficult task (one twin at a time) using an Etch-a-Sketch toy with their mothers. They had to draw a house together, and there was one strict rule. The mother was only allowed to use the dial that made the drawing line go up and down, the child was only allowed to use the dial that made the line go from side to side. In a small proportion of cases, mothers broke this rule and moved to using the child's dial as well as their own. We coded this behavior as extreme control (coded with 90% exact agreement between raters) and found it to be highly heritable ($h^2 = .63$), reflecting an association between child genotype and maternal control (Eley, Napolitano, Lau, & Gregory, 2010). Furthermore, this genetic influence was also shared with that on child anxiety, reflecting one of two different types of gene–environment (GE) correlation. Either it is a passive GE correlation, in which the same genes influence parental control that influence child anxiety, or it is an evocative GE correlation with the child's anxiety (which is itself heritable) eliciting the maternal control. To identify which of these two possibilities is the case, methods that disentangle parental and child genetic influences are needed, such as the children-of-twins design (Narusyte et al., 2008; see also Chapters 4 and 6). What is notable, however, is that this indicates that it is not as simple a situation as maternal control causing child anxiety. This affirms what some clinical psychologists have suspected for a while: that the association between child anxiety and maternal control is likely to be bidirectional, and that the child's anxiety has an important part to play in eliciting maternal controlling behavior.

Molecular Genetic Approaches

Another approach to gene–environment correlations is to look for associations between specific measured genes and aspects of the environment. Associations have been found between specific genes and a range of environmental stresses relevant to the development of anxiety and depression. These include associations between the serotonin transporter receptor 2A gene and popularity (Burt, 2009), the catechol-O-methyltransferase gene and daily hassles (van Ijzendoorn, Bakermans-Kranenburg, & Mesman, 2008), the brain-derived neurotrophic factor gene and childhood adversity

(Wichers et al., 2008), and the dopamine D_2 receptor gene and perceived parenting behaviors (Lucht et al., 2006). However, none of these has been replicated to date, so it is unclear the extent to which these reflect genuine associations. There has also been a genome-wide association scan of a measure of the home environment, but this yielded no positive associations (Butcher & Plomin, 2008). At this point, it is probably fair to say that this approach has not yet yielded any substantial contributions to our understanding of gene–environment correlations, but that is largely because so little work has been done. As more psychologists get interested in genetics, it is likely that there will be more studies of this kind, and hopefully replications will begin to be seen.

GENE–ENVIRONMENT INTERACTION

Gene–environment interaction (G×E) refers to the process whereby genetic influence on a phenotype varies as a function of the environment or vice versa. For example, G×E interaction includes situations in which genetic influence is enhanced or attenuated by aspects of the environment. It also includes situations in which the positive or negative influences of a specific environmental factor are dependent on the presence of a particular genetic profile. Thus, an environmental influence could either protect against or exacerbate the effects of a gene.

Quantitative Genetic Approaches

Although anxiety and depression are influenced by largely the same genes, G × E interaction effects are likely to be specific to either anxiety or depression because of the specificity of environmental influences on these phenotypes. To date, many more G × E investigations in this area have examined depression, whereas anxiety remains far less well explored.

One of the earliest findings in this area examined risk of onset of major depressive disorder (MDD) in adult women as a function of both genetic liability and presence of a severe stressful life event that month (Kendler et al., 1995). An innovative approach was taken to genetic liability in that the presence of MDD in a co-twin (yes vs. no) and their genetic relatedness (monozygotic [MZ] vs. dizygotic [DZ]) were used. Thus, the co-twin of an adult with MDD from an MZ pair was rated as having highest genetic liability, whereas the co-twin of an adult with MDD from a DZ twin pair was coded as moderate liability. The co-twin of an adult without MDD from a DZ pair was rated as having even lower genetic liability, whereas the co-twin without MDD from a MZ pair was coded as having lowest genetic liability. In those who had experienced a severely stressful life event in the preceding month, risk of depression was found to increase systematically with genetic liability, a "fan-shaped interaction" (see Chapter 1, by Kendler). Similarly, in a study of risk of depression in married and unmarried women (Heath, Eaves, & Martin, 1998), heritability of depression was greatest in unmarried women, particularly if they were over the age of 30 years. Furthermore, in our own study of adolescents, risk of depression was greatest in those who showed not only familial risk for depression but who also experienced high levels of family stress (Eley et al., 2004a).

Although these studies are informative, none of them considers the potential role for gene–environment correlation. Given the complex interplay between genes and the environment, it has become clear that studies exploring gene–environment interaction need to take gene–environment correlation into account. In an attempt to deal with this issue, an early study examined only life events, independent of the behavior of the adolescent, and for which there was no evidence of genetic influence (Silberg, Rutter, Neale, & Eaves, 2001). Heritability of both anxiety and depression varied as a function of these life events and parental emotional disorder.

More recently, new statistical models have become available that allow us to model both gene–environment correlation and interaction together. One of these is an extension of the traditional bivariate genetic model, in which an environmental moderator and outcome are considered as the two phenotypes. In a standard bivariate model, one shared set of latent genetic and environmental factors influences both the measured environmental moderator and the measured outcome. Furthermore, a second set of specific latent factors influences the outcome alone. The novel aspect of this approach is that the extent to which latent genetic and environmental factors influence both the environmental moderator and the outcome (e.g., depression), may vary as a function of the moderator (Purcell, 2002). As such, all three sets of influences (those on the environmental moderator, those from the shared factors to the outcome, and those specific to the outcome) can vary as a function of the moderator.

One of the earliest studies using this approach revealed that genetic influences on depression were greater in families in which high levels of conflict were present (Rice et al., 2006). We examined the heritability of adolescent depression as a function of dependent negative life events (those that arise from the behavior of the individual, such as a relationship break-up) and negative parenting (Lau & Eley, 2008). We found that overall variance and mean scores for depression increased with increasing levels of both negative life events and maternal punitive discipline, and that both were heritable. Furthermore, both variables moderated genetic influence on depressive symptoms, such that the genetic influence increased with increasing levels of the environmental moderator (see Figures 10.1A and B). Note that, in this figure, we see a change in genetic variance *for the whole sample*, as a function of differing levels of environmental stress. This is in contrast to the more common representation of G × E, in which subgroups of individuals who differ depending on their level of genetic risk are represented by different lines (e.g., Figure 1.1 from Chapter 1, or Figure 10.2 below). Extending this approach to anxiety within our dataset (Lau, Gregory, Goldwin, Pine, & Eley, 2007), we found evidence for moderation by negative

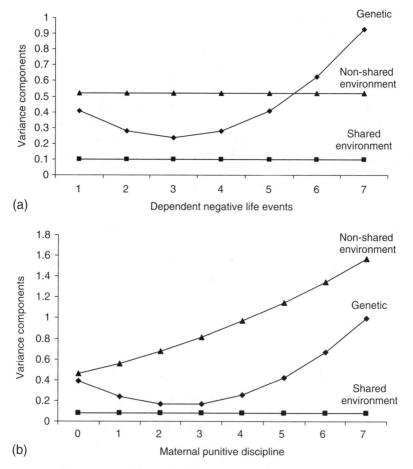

(a)

(b)

Figure 10.1. **Variance in depression increased as a function of both increasing dependent negative life events and maternal punitive discipline. (A)** This increase was entirely accounted for by increasing genetic influence. **(B)** Both genetic and nonshared environmental variance increased as a function of the measured environmental risk factor (maternal punitive discipline). Reproduced with permission from: Lau & Eley(2008).

life events of genetic events on separation anxiety in childhood and panic symptoms in adolescence. However, as we looked at several subscales of anxiety across both childhood and adolescence, these findings could be chance effects and need replicating. Most recently, this approach was used to examine the association between peer-rejection and depressive behavior in children (Brendgen et al., 2009). Moderate genetic overlap was found between peer rejection and depressive behavior, such that genetic influence on depression was also associated with increased risk of peer rejection. Furthermore, heritability of depressive behavior varied as a function of peer rejection, but in this case heritability of depression was *reduced* as levels of peer rejection rose. This may indicate that rejection overwhelms genetic effects on depression.

Finally, another group used a Markov chain Monte Carlo approach (Eaves et al., 2003) to analyze adolescent anxiety, depression, and life events. In line with their own and others' previous work, they found a genetic link between earlier anxiety and later depression. Of particular note here, however, they found that genetic risk for earlier anxiety increased both risk of and sensitivity to later depressogenic environments. Thus, the study provided evidence for both gene–environment correlation and gene–environment interaction.

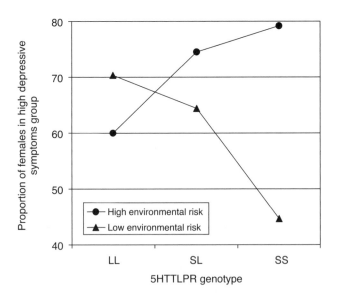

Figure 10.2. **Variation by environmental risk groups.** Environmental risk groups were those below (low) and above (high) the mean for a composite score of parent-reported social adversity, negative life events relevant to the whole family, and parental education. Reproduced with permission from: Eley et al.(2004c).

Of note, in all but one (Brendgen et al., 2009) of these studies, the familial or genetic risk for depression was exacerbated by the presence of stress. This indicates that what is inherited may be a vulnerability to environmental stress, which sits well with a diathesis–stress model. If genetic vulnerability makes people differentially vulnerable to stress, or differentially responsive to the environment more broadly, then the obvious place to look for potential mechanisms of these effects is the brain. With this in mind, we now explore genetically informative data on measures of cognitive processing and brain function.

Cognitive Processing and the Brain: Where Genes Meet the Environment?

Our cognitions or cognitive style are like prisms through which we interpret and attend to the world. They, therefore, represent one logical place to start when investigating the interplay between genes and environment. Cognitive biases relevant to the development of anxiety and depression include biases of attention, memory, and interpretation (Williams, Watts, MacLeod, & Mathews, 1997). One aspect of cognitive style important in the development of depression is attributional style (Abramson, Seligman, & Teasdale, 1978), which reflects a tendency to attribute negative experiences to internal (to do with oneself rather than someone else), global (i.e., applying generally rather than to this one specific instance), and stable (as an ongoing tendency rather than discrete event) causes. We found in both child and adolescent twin samples that depression was associated with this negative interpretational bias (Lau et al., 2006; Lau & Eley, 2009b). Interestingly, in line with the increased prevalence and heritability of depression in adolescence as compared with childhood, we found that both attributional style and its association with depression were far more strongly influenced by genetic factors in adolescence (Lau et al., 2006) than in childhood (Lau & Eley, 2009b). Similarly, in our child sample, all scales from our measure of children's expectations and perceptions of social situations showed only minimal genetic influence, and nonshared environment largely accounted for associations with depression (Gregory et al., 2007). This supports the interesting possibility that genetic influences on these types of cognitions only come online during adolescence. In contrast, when we examined the tendency to interpret ambiguous stimuli in our child sample, which was also specifically associated with depression (rather than anxiety), there was significant genetic influence on this measure and on its association with depression (Eley et al., 2008).

We also explored genetic and environmental influences on associations between anxiety and both heartbeat perception (an attentional bias) and anxiety sensitivity (an interpretational bias). Heartbeat perception or awareness of one's own heartbeat is associated with panic disorder in adults (Van der Does, Antony, Ehlers, & Barsky, 2000) and panic/somatic symptoms in children (Eley, Stirling, Ehlers, Gregory, & Clark, 2004b). Anxiety sensitivity reflects fear of physical symptoms and the belief that these are harmful (Reiss, 1986). It was initially proposed as a risk factor for panic disorder, but findings suggest a broader association with anxiety more generally (Schmidt et al., 2006). In our child twin sample, we found that both heartbeat perception and anxiety sensitivity were heritable and shared substantial genetic influence with panic/somatic symptoms (Eley et al., 2004b). Similarly, in our adolescent sample, we found evidence for genetic overlap between anxiety sensitivity and both anxiety and depression self-report scales (Zavos, Rijsdijk, Gregory, & Eley, 2010).

These results are as yet unreplicated and need to be examined in other twin samples. However, these findings imply that some aspects of cognitive processing may reflect a developmental progression from environmental influence during childhood to genetic influence during adolescence, but that other processes reflect a mixture of genetic and environmental influences at both developmental stages. Finding genetic links between aspects of cognitive functioning and outcomes is also compatible with the possibility that these processes mediate interactions between genetic vulnerabilities and response to environmental stress. Continuing to refine our measures and clarify exactly which parts reflect genetic influence and which reflect environmental influence will help in specifying the mechanisms involved.

Another useful approach to identifying mechanisms that might mediate either genetic or environmental risk is to examine brain function. As with genetically informative studies of cognitive processing, this area is very new, and findings are limited and somewhat contradictory. Thus, for example, there is evidence for involvement of the amygdala in vulnerability to depression, but studies differ as to whether this is found to reflect genetic or environmental influence (Monk et al., 2008; Wolfensberger, Veltman, Hoogendijk, Boomsma, & de Geus, 2008; see also Chapter 11, by Casey, Ruberry, and Libby).

The major difficulty with this approach is that the two methodologies being combined (genetics and functional magnetic resonance imaging [fMRI]) have widely differing methodological issues that are hard to reconcile, the most critical of which is sample size. Cost constraints mean that fMRI studies usually have fairly small samples, whereas genetic studies require far larger samples. Strategies to enrich the power to identify genetic effects in smaller samples are likely to be the best route forward.

Identifying individuals from the extremes of the distribution, and comparing members of discordant twin pairs (e.g., de Geus et al., 2007) are likely to be useful strategies.

Molecular Genetic Approaches

Association studies of candidate genes in anxiety and depression have primarily focused on the serotonin system, dysregulation of which is a core feature of both disorders (Lesch, 2003), as seen by the efficacy of selective serotonin reuptake inhibitors (SSRIs; Kronenberg et al., 2008). The most widely explored polymorphism is a common 44-base variable number tandem repeat (VNTR) (*5HTTLPR*), in the promoter region of the serotonin transporter gene, in which the short (S) allele results in decreased expression. Subjects with at least one S allele were found to have higher anxiety-related personality scores than were those with none (Lesch et al., 1996). There have been many attempts at replication for this finding, some positive (Katsuragi et al., 1999; Osher, Hamer, & Benjamin, 2000), others not (Ball et al., 1997; Jorm et al., 1998), and meta-analyses conclude that the association is of very small effect (as expected), and sensitive to measurement and sampling (Schinka, Busch, & Robichaux-Keene, 2004; Sen et al., 2004; Munafo, Clark, & Flint, 2005). One likely influence on the variability of findings across studies is sample heterogeneity, one source of which is exposure to stressful environmental influences. As is now clear from twin studies, such influences can alter the level of genetic influence on a phenotype, and therefore, by extension, presumably the influence of specific genes on outcome. This is where gene–environment interaction reaches its most specific level of analysis.

The field of gene–environment interaction has been hugely influenced by some very high-profile findings presenting interactions between specific genes and specific elements of the environment on psychiatric outcomes. Of particular note here was the identification of an interaction between the serotonin transporter promoter (*5HTTLPR*) genotype and both childhood maltreatment and stressful life events on adult depression (Caspi et al., 2003). This finding has been the subject of considerable attempts at replication, with mixed results, and these are discussed in detail by Uher, in Chapter 2. Here, I focus on reported associations between this marker and various anxiety- and depression-related phenotypes in younger samples. Specifically, the *5HTTLPR* has been found to interact with measures of stress on behavioural inhibition/shyness in children (Fox et al., 2005), and depression in adolescence (Cicchetti, Rogosch, & Sturge-Apple, 2007), sometimes in females only (e.g., Eley et al., 2004c; Sjoberg et al., 2006). There are also other studies in young samples that do not find this interaction (see Uher & McGuffin, 2009), but overall, these data suggest that the influence of *5HTTLPR* on adolescent depression is contingent on environmental stress, possibly only in females. Of particular note, in the females in our study and in several other studies, not only was the short allele associated with worse symptoms in the presence of stress, but also with *lower* levels of symptoms when in low stress (see Figure 10.2). These findings lead to the intriguing suggestion that the marker reflects environmental plasticity, or responsivity to the environment, be it negative or positive (Belsky et al., 2009), perhaps mediated by aspects of anxiety-relevant cognitive processing. Initial support for this hypothesis is provided by data from macaques, which reveal that this marker interacts with rearing environment (mother vs. peer) on temperamental affect (Champoux et al., 2002) and on hypothalamic-pituitary-adrenal (HPA) axis reactivity (Barr et al., 2004).

To test the hypothesis that *5HTTLPR* reflects a tendency to be responsive to the environment, a number of researchers have tested for associations between the serotonin transporter promoter polymorphism and aspects of stress reactivity, including cognitive processing, brain function, and HPA axis functioning. With respect to aspects of cognitive processing, an interaction has been found between the short-short (SS) genotype and childhood maltreatment on anxiety sensitivity in adulthood (Stein et al., 2008). The short allele also appears to interact with negative attentional biases in predicting the association between child and mother depression (Gibb, Benas, Grassia, & McGeary, 2009) and to be associated with increased levels of anxiety-related traits and state anxiety and enhanced amygdala response to anxiety provocation (e.g., a public speaking task, Furmark et al., 2004). Another study of shyness in children assessed visual event-related potentials (ERPs) in response to emotional facial expressions and *5HTTLPR* (Battaglia et al., 2005). As expected, they found higher shyness scores for children with two copies of the short allele as compared with those with two long copies. Furthermore, a significant interaction existed between genotype and expression type, such that a smaller amplitude was elicited by anger expressions for those children with the SS genotype compared with those with the long-long (LL). Subsequently, another group used fMRI and found an interaction between *5HTTLPR* and responses to facial emotional expression type on depression in adolescence (Lau et al., 2009a; although direction of effects remain somewhat unclear). Both these studies followed an initial report in adults that has now been replicated a number of times (Hariri et al., 2002; for a meta-analysis see Munafo et al., 2008).

The *5-HTTLPR* has also been examined with regard to fMRI response to an emotional face task in healthy subjects with "phobic-prone" personality style as compared with a group described as "eating-disorder prone"

(Bertolino et al., 2005). The phobic-prone subjects showed higher amygdala activity during the emotional faces task than did the comparison group, and activity in the amygdala was predicted by both personality style and the *5-HTTLPR* genotype. Finally, another approach is to look at genetic influences on HPA functioning. Recent data from a sample of never-depressed girls revealed that those with two copies of the short allele of the *5HTTLPR* showed greater cortisol response following a stressful laboratory task (Gotlib et al., 2008). Subsequent work by the same group extended these findings to reveal an association between the short allele of the *5HTTLPR* and diurnal cortisol variation (Chen et al., 2009). These results are particularly exciting, as they offer some insight into potential mechanisms by which genotypes can alter stress reactivity and thus sensitivity to the environment. Overall, this type of work highlights the possible interplay between a specific genetic risk factor, specific aspects of the environment, and anxiety-related cognitive, brain, or stress reactivity processes.

A number of other genes have been associated with anxiety and depression-related traits, possibly in conjunction with stress. Still within the serotonin system, the tryptophan hydroxylase 1 gene (*TPH1*) has been associated with depression (Levinson, 2006), particularly in conjunction with features of the environment, including social support (Jokela, Räikkönen, Lehtimäki, Rontu, & Keltikangas-Jarvinen, 2007) and stress (Gizatullin, Zaboli, Jonsson, Asberg, & Leopardi, 2006). There is similar evidence that socioeconomic status and maternal nurturance moderate an association between *5HT2A* and adult depression (Jokela et al., 2007). Also of relevance are genes relating to the HPA axis. Thus, an association has been found between the corticotropin-releasing hormone gene (*CRH*) and behavioural inhibition in children (Smoller et al., 2005). Furthermore, an association between *CRH1* and suicide attempts was found to be moderated by levels of stress (Wasserman, Sokolowski, Rozanov, & Wasserman, 2007). Finally, polymorphisms in the glucocorticoid receptor gene (*GR*) have been associated (weakly) with treatment response in adults (van Rossum et al., 2006), and also appear to interact with childhood adversity on depression (Bet et al., 2009).

CONCLUSION

Future Directions

I hope that I have shown in this chapter that the field of gene–environment interplay has much to say about the development of anxiety and depression in young people. To date, extensive findings have emerged with respect to genetic influence on aspects of the environment associated with anxiety and depression, reflecting the phenomenon known as gene–environment correlation. Work is now under way to try to tie down the mechanisms of these correlations by examining genetic and environmental influences on cognitive and brain processing. Specifying the individual genetic markers involved in this genetic influence on the environment is proving more difficult, but some progress has been made in this area. With respect to gene–environment interactions, findings are more evenly split across quantitative and molecular genetic approaches. Evidence from multiple twin studies suggests that genetic influences on depression, and probably on anxiety, vary as a function of the environment. The majority of this work indicates that genetic influence increases as a function of worsening environmental experiences. As noted earlier, this reflects a diathesis–stress model in which a genetic diathesis or vulnerability is present, that is only revealed in the presence of high stress. Such a diathesis is likely to be reflected in differences in cognitive and brain functioning. These differences may help explain the evolutionary role of such behaviors. For example, selective attention to threat has a role to play in maintaining security. It is only when someone succumbs from being attentive to threat to showing crippling levels of anxiety that this processing style moves from being a useful to a damaging attribute, and this may arise in response to the experience of high levels of stress. Work from molecular genetics has largely focused on the serotonin transporter receptor, and findings with respect to this marker have broadened out from the original gene–environment interaction approach to incorporate measures of cognitive and brain functioning and even stress reactivity. These innovative interdisciplinary studies are likely to be key to making progress in understanding the complexities of the development of anxiety and depression in young people.

In thinking about where this type of work might lead, two clear areas are ripe for development. The first is treatment. Taking child anxiety as an example, we know that the treatment of choice, cognitive-behavior therapy (CBT), is effective in around 60% of cases (Cartwright-Hatton, Roberts, Chitsabesan, Fothergill, & Harrington, 2004); thus for a substantial minority of children other treatment options are needed. This may include medication, and a recent highly influential study provides useful data on this point. Significant improvement was seen in 60% of children receiving CBT, 55% of those receiving sertraline, but in 80% of those receiving combined CBT and sertraline (Walkup et al., 2008). This indicates that, for those unlikely to respond to CBT alone, combining with sertraline is likely to improve the potential for positive outcome. It would therefore be useful to know at the outset which children are less likely to respond to CBT alone. Although results are somewhat inconsistent, there are indications that younger children and girls may respond better to CBT (Hudson, 2005) than do males and adolescents. Furthermore, greater symptom severity and parental psychopathology may be associated with poorer response to CBT (Bodden et al., 2008; Hudson, 2005). These latter two factors could be considered as a proxy for greater genetic influence, leading to the intriguing possibility that genetic factors may be relevant in response to CBT. This is not as unlikely as it may sound. We know that the types of cognitions that CBT targets are heritable (Eley & Zavos, 2009), but we also know that *change* in such cognitions is influenced not only by the environment (e.g., by treatment) but also by genetic factors (Zavos, Rijsdijk, & Eley, 2010). Indeed, as described above, it is possible that some genetic influences operate to make individuals more reactive to the environment, whatever that may be (Belsky et al., 2009). One possible outcome of such a hypothesis would be that a genetic risk factor that makes someone susceptible to environmental stress also makes him or her more likely to be responsive to an environmentally mediated treatment, such as CBT. Although the mechanisms are quite different, the possibility that genetic factors may alter treatment response has already begun to be investigated with respect to pharmacological treatments, and indeed there is now growing evidence that *5HTTLPR* is associated with antidepressant response (Serretti, Calati, Mandelli, & De Ronchi, 2006).

The second area in which this type of work will hopefully become useful is in the field of prevention. It is obviously useful to improve treatment, for example, to better target treatment to particular subgroups, and to offer treatment most likely to work and/or least likely to have detrimental side effects. It would, however, be even more useful to be able to prevent the development of disorders in the first place by targeting those who are vulnerable and offering some sort of psychological training package. The literature is currently awash with studies testing such approaches, and as noted in a recent review (Gladstone & Beardslee, 2009), a number of key points emerge. Of these, two are particularly relevant here. First, it is important to target specific high-risk individuals, and second, it is important to consider moderators of intervention effects. Both of these issues raise the possibility of genetic information being useful—either to help refine definitions of high-risk groups, or as potential moderators of intervention effects. To date, there have been no genetically informed studies of such prevention programs, but it is likely that such work would be beneficial in the long run.

In summary, there is much yet to be learned within the field of gene–environment interplay with respect to the development of anxiety and depression. However, given the longstanding nature of these disorders and their enormous cost in both human and financial terms, it is critical that we continue to try to understand the mechanisms by which they develop in the hope of identifying better approaches to treatment or even prevention.

REFERENCES

Abramson, L. Y., Seligman, M. E. P., & Teasdale, J. D. (1978). Learned helplessness in humans: Critique and reformulation. *Journal of Abnormal Psychology, 87,* 49–74.

Angold, A., Costello, E. J., & Erkanli, A. (1999). Comorbidity. *Journal of Child Psychology and Psychiatry, 40,* 57–87.

Ball, D. M., Hill, L., Freeman, B., Eley, T. C., Strelau, J., Riemann, R., et al. (1997). The serotonin transporter gene and peer-rated neuroticism. *NeuroReport, 8,* 1301–1304.

Barr, C. S., Newman, T. K., Shannon, C., Parker, C., Dvoskin, R. L., Becker, M. L., et al. (2004). Rearing condition and rh5-HTTLPR interact to influence limbic-hypothalamic-pituitary-adrenal axis response to stress in infant macaques. *Biological Psychiatry, 55,* 733–738.

Battaglia, M., Ogliari, A., Zanoni, A., Citterio, A., Pozzoli, U., Giorda, R., et al. (2005). Influence of the serotonin transporter promoter gene and shyness on children's cerebral responses to facial expressions. *Archives of General Psychiatry, 62,* 85–94.

Belsky, J., Jonassaint, C., Pluess, M., Stanton, M., Brummett, B., & Williams, R. (2009). Vulnerability genes or plasticity genes? *Molecular Psychiatry, 14,* 746–754.

Bertolino, A., Arciero, G., Rubino, V., Latorre, V., De Candia, M., Mazzola, V., et al. (2005). Variation of human amygdala response during threatening stimuli as a function of 5'HTTLPR genotype and personality style. *Biological Psychiatry, 57,* 1517–1525.

Bet, P. M., Penninx, B. W., Bochdanovits, Z., Uitterlinden, A. G., Beekman, A. T., van Schoor, N. M., et al. (2009). Glucocorticoid receptor gene polymorphisms and childhood adversity are associated with depression: New evidence for a gene-environment interaction. *American Journal of Medical Genetics, Part B, Neuropsychiatric Genetics,150B(5),* 660–669.

Bodden, D. H., Bogels, S. M., Nauta, M. H., De Haan, E., Ringrose, J., Appelboom, C., et al. (2008). Child versus family cognitive-behavioral therapy in clinically anxious youth: an efficacy and partial effectiveness study. *Journal of the American Academy of Child & Adolescent Psychiatry, 47,* 1384–1394.

Brendgen, M., Vitaro, F., Boivin, M., Girard, A., Bukowski, W. M., Dionne, G., et al. (2009). Gene-environment interplay between peer rejection and depressive behavior in children. *Journal of Child Psychology & Psychiatry & Allied Disciplines, 50,* 1009–1017.

Burt, A. (2009). A mechanistic explanation of popularity: genes, rule breaking, and evocative gene-environment correlations. *Journal of Personality & Social Psychology, 96,* 783–794.

Butcher, L. M., & Plomin, R. (2008). The nature of nurture: a genomewide association scan for family chaos. *Behavior Genetics, 38,* 361–371.

Cartwright-Hatton, S., Roberts, C., Chitsabesan, P., Fothergill, C., & Harrington, R. (2004). Systematic review of the efficacy of cognitive behavior therapies for childhood and adolescent anxiety disorders. *British Journal of Clinical Psychology, 43,* 421–436.

Caspi, A., Sugden, K., Moffitt, T. E., Taylor, A., Craig, I. W., Harrington, H., et al. (2003). Influence of life stress on depression: Moderation by a polymorphism in the 5-HTT gene. *Science, 301,* 386–389

Champoux, M., Bennett, A., Shannon, C., Higley, J. D., Lesch, K. P., & Suomi, S. J. (2002). Serotonin transporter gene polymorphism, differential early rearing, and behavior in rhesus monkey neonates. *Molecular Psychiatry, 7,* 1058–1063.

Chen, M. C., Joormann, J., Hallmayer, J., Gotlib, I. H., Chen, M. C., Joormann, J., et al. (2009). Serotonin transporter polymorphism predicts waking cortisol in young girls. *Psychoneuroendocrinology, 34,* 681–686.

Cicchetti, D., Rogosch, F. A., & Sturge-Apple, M. L. (2007). Interactions of child maltreatment and serotonin transporter and monoamine oxidase A polymor-phisms: depressive symptomatology among adolescents from low socioeco-nomic status backgrounds. *Development & Psychopathology, 19,* 1161–1180.

Costello, E. J., Egger, H., & Angold, A. (2005a). 10-year research update review: the epidemiology of child and adolescent psychiatric disorders: I. *Methods and public health burden. Journal of the American Academy of Child & Adolescent Psychiatry, 44,* 972–986.

Costello, E. J., Egger, H. L., & Angold, A. (2005b). The developmental epidemiology of anxiety disorders: phenomenology, prevalence, and comorbidity. *Child & Adolescent Psychiatric Clinics of North America, 14,* 631–648.

de Geus, E. J., van't Ent, D., Wolfensberger, S. P., Heutink, P., Hoogendijk, W. J., Boomsma, D. I., et al. (2007). Intrapair differences in hippocampal volume in monozygotic twins discordant for the risk for anxiety and depression. *Biological Psychiatry, 61,* 1062–1071.

Eaves, L., Silberg, J., & Erkanli, A. (2003). Resolving multiple epigenetic pathways to adolescent depression. *Journal of Child Psychology and Psychiatry, 44,* 1006–1014.

Eley, T. C. (1997). Depressive symptoms in children and adolescents: etiological links between normality and abnormality: a research note. *Journal of Child Psychology and Psychiatry, 38,* 861–866.

Eley, T. C., Gregory, A. M., Lau, J. Y. F., McGuffin, P., Napolitano, M., Rijsdijk, F. V., et al. (2008). In the face of uncertainty: A genetic analysis of ambiguous infor-mation, anxiety and depression in children. *Journal of Abnormal Child Psychology, 36*(1), 55–65.

Eley, T. C., Liang, H., Plomin, R., Sham, P., Sterne, A., Williamson, R., et al. (2004a). Parental familial vulnerability, family environment, and their interactions as predictors of depressive symptoms in adolescents. *Journal American Academic and Child and Adolescent Psychiatry, 43,* 298–306.

Eley, T. C., Napolitano, M., Lau, J. Y. F., & Gregory, A. M. (2010). Does childhood anxiety evoke maternal control? A genetically informed study. *Journal of Child Psychology and Psychiatry,51*(7), 772–779.

Eley, T. C., & Stevenson, J. (1999). Using genetic analyses to clarify the distinction between depressive and anxious symptoms in children and adolescents. *Journal of Abnormal Child Psychology, 27,* 105–114.

Eley, T. C., & Stevenson, J. (2000). Specific life events and chronic experiences differentially associated with depression and anxiety in young twins. *Journal of Abnormal Child Psychology, 28*, 383–394.

Eley, T. C., Stirling, L., Ehlers, A., Gregory, A. M., & Clark, D. M. (2004b). Heart-beat perception, panic/somatic symptoms and anxiety sensitivity in children. *Behavior Research and Therapy, 42*, 439–448.

Eley, T. C., Sugden, K., Gregory, A. M., Sterne, A., Plomin, R., & Craig, I. W. (2004c). Gene-environment interaction analysis of serotonin system markers with adolescent depression. *Molecular Psychiatry, 9*, 908–915.

Eley, T. C., & Zavos, H. M. S. (2010). Genetics. In J. A. Hadwin & A. P. Field (Eds.), *Information processing biases in child and adolescent anxiety: A developmental perspective* (1 ed., pp. 209-232). Chichester: Wiley Blackwell.

Finlay-Jones, R., & Brown, G. W. (1981). Types of stressful life events and the onset of anxiety and depressive disorders. *Psychological Medicine, 11*, 803–815.

Fombonne, E., Wostear, G., Cooper, V., Harrington, R., & Rutter, M. (2001a). The Maudsley long-term follow-up of child and adolescent depression. *1. Psychiatric outcomes in adulthood. British Journal of Psychiatry, 179*, 210–217.

Fombonne, E., Wostear, G., Cooper, V., Harrington, R., & Rutter, M. (2001b). The Maudsley long-term follow-up of child and adolescent depression. 2. Suicidality, criminality and social dysfunction in adulthood. *British Journal of Psychiatry, 179*, 218–223.

Fox, N. A., Nichols, K. E., Henderson, H. A., Rubin, K., Schmidt, L., Hamer, D., et al. (2005). Evidence for a gene-environment interaction in predicting behavioral inhibition in middle childhood. *Psychological Science, 16*, 921–926.

Furmark, T., Tillfors, M., Garpenstrand, H., Marteinsdottir, I., Langstrom, B., Oreland, L., et al. (2004). Serotonin transporter polymorphism related to amygdala excitability and symptom severity in patients with social phobia. *Neuroscience Letters, 362*, 189–192.

Gibb, B. E., Benas, J. S., Grassia, M., & McGeary, J. (2009). Children's attentional biases and 5-HTTLPR genotype: potential mechanisms linking mother and child depression. *Journal of Clinical Child & Adolescent Psychology, 38*, 415–426.

Gizatullin, R., Zaboli, G., Jonsson, E. G., Asberg, M., & Leopardi, R. (2006). Haplotype analysis reveals tryptophan hydroxylase (TPH) 1 gene variants associated with major depression. *Biological Psychiatry, 59*, 295–300.

Gladstone, T. R., & Beardslee, W. R. (2009). The prevention of depression in children and adolescents: a review. *Canadian Journal of Psychiatry, 54*, 212–221.

Gotlib, I. H., Joormann, J., Minor, K. L., Hallmayer, J., Gotlib, I. H., Joormann, J., et al. (2008). HPA axis reactivity: a mechanism underlying the associations among 5-HTTLPR, stress, and depression. *Biological Psychiatry, 63*, 847–851.

Gregory, A. M., Caspi, A., Moffitt, T. E., Koenen, K., Eley, T. C., & Poulton, R. (2007). Juvenile mental health histories of adults with anxiety disorders. *American Journal of Psychiatry, 164*, 301–308.

Gregory, A. M., & Eley, T. C. (in press). The genetic basis of child and adolescent anxiety. In W. K. Silverman, & A. P. Field (Eds.), *Anxiety disorders in children and adolescents: research, assessment and intervention* (2nd ed.). Cambridge: Cambridge University Press.

Hariri, A. R., Mattay, V. S., Tessitore, A., Kolachana, B., Fera, F., Goldman, D., et al. (2002). Serotonin transporter genetic variation and the response of the human amygdala. *Science, 297,* 400–403.

Heath, A. C., Eaves, L. J., & Martin, N. G. (1998). Interaction of marital status and genetic risk for symptoms of depression. *Twin Research,* 119–122.

Hudson, J. L. (2005). Efficacy of cognitive-behavioural therapy for children and adolescents with anxiety disorders. *Behavior Change, 22,* 55–70.

Jacobson, K. C., & Rowe, D. C. (1999). Genetic and environmental influences on the relationships between family connectedness, school connectedness, and adolescent depressed mood: sex differences. *Developmental Psychology, 35,* 926–939.

Jokela, M., Keltikangas-Jarvinen, L., Kivimaki, M., Puttonen, S., Elovainio, M., Rontu, R., et al. (2007). Serotonin receptor 2A gene and the influence of childhood maternal nurturance on adulthood depressive symptoms. *Archives of General Psychiatry, 64,* 356–360.

Jokela, M., Räikkönen, K., Lehtimäki, T., Rontu, R., & Keltikangas-Jarvinen, L. (2007). Tryptophan hydroxylase 1 gene (TPH1) moderates the influence of social support on depressive symptoms in adults. *Journal of Affective Disorders, 100,* 191–197.

Jorm, A. F., Henderson, A. S., Jacomb, P. A., Christensen, H., Korten, A. E., Rodgers, B., et al. (1998). An association study of a functional polymorphism of the serotonin transporter gene with personality and psychiatric symptoms. *Molecular Psychiatry, 3,* 449–451.

Katsuragi, S., Kunugi, H., Sano, A., Tsutsumi, T., Isogawa, K., Nanko, S., et al. (1999). Association between serotonin transporter gene polymorphism and anxiety-related traits. *Biological Psychiatry, 45,* 368–370.

Kendler, K. S., & Baker, J. H. (2007). Genetic influences on measures of the environment: a systematic review. *Psychological Medicine, 37,* 615–626.

Kendler, K. S., Hettema, J. M., Butera, F., Gardner, C. O., & Prescott, C. A. (2003). Life event dimensions of loss, humiliation, entrapment, and danger in the prediction of onsets of major depression and generalized anxiety. *Archives of General Psychiatry, 60,* 789–796.

Kendler, K. S., & Karkowski-Shuman, L. (1997). Stressful life events and genetic liability to major depression: Genetic control of exposure to the environment. *Psychological Medicine, 27,* 539–547.

Kendler, K. S., Kessler, R. C., Walters, E. E., MacLean, C. J., Neale, M. C., Heath, A. C., et al. (1995). Stressful life events, genetic liability, and onset of an episode of major depression in women. *American Journal of Psychiatry, 152,* 833–842.

Kendler, K. S., Neale, M., Kessler, R., Heath, A., & Eaves, L. (1993). A twin study of recent life events and difficulties. *Archives of General Psychiatry, 50,* 789–796.

Kronenberg, S., Frisch, A., Rotberg, B., Carmel, M., Apter, A., & Weizman, A. (2008). Pharmacogenetics of selective serotonin reuptake inhibitors in pediatric depression and anxiety. *Pharmacogenomics, 9*, 1725–1736.

Lau, J. Y., Goldman, D., Buzas, B., Fromm, S. J., Guyer, A. E., Hodgkinson, C., et al. (2009a). Amygdala function and 5-HTT gene variants in adolescent anxiety and major depressive disorder. *Biological Psychiatry, 65*, 349–355.

Lau, J. Y. F., & Eley, T. C. (2006). Changes in genetic and environmental influences on depressive symptoms across adolescence: A twin and sibling study. *British Journal of Psychiatry, 189*, 422–427.

Lau, J. Y. F., & Eley, T. C. (2008). Disentangling gene environment correlations and interactions in adolescent depression. *Journal of Child Psychology and Psychiatry, 49*, 142–150.

Lau, J. Y. F., & Eley, T. C. (2009b). Developmental behavioural genetic analysis of the relationship between attributional style and depressive symptoms in middle childhood. *In preparation.*

Lau, J. Y. F., Gregory, A. M., Goldwin, M. A., Pine, D. S., & Eley, T. C. (2007). Assessing gene-environment interactions on anxiety symptom subtypes across childhood and adolescence. *Development and Psychopathology, 19*, 1129–1146.

Lau, J. Y. F., Rijsdijk, F. V., & Eley, T. C. (2006). I think, therefore I am: a twin study of attributional style in adolescents. *Journal of Child Psychology and Psychiatry, 47*, 696–703.

Lesch, K. P. (2003). Neuroticism and serotonin: a developmental genetic perspective. In R. Plomin, J. C. DeFries, I. W. Craig, & P. McGuffin (Eds.), *Behavioral genetics in the postgenomic era* (pp. 389–423). Washington, D.C.: American Psychological Association.

Lesch, K. P., Bengel, D., Heils, A., Zhang Sabol, S., Greenburg, B. D., Petri, S., et al. (1996). Association of anxiety-related traits with a polymorphism in the serotonin transporter gene regulatory region. *Science, 274*, 1527–1531.

Levinson, D. F. (2006). The genetics of depression: a review. *Biological Psychiatry, 60*, 84–92.

Lucht, M., Barnow, S., Schroeder, W., Grabe, H. J., Finckh, U., John, U., et al. (2006). Negative perceived paternal parenting is associated with dopamine D2 receptor exon 8 and GABA(A) alpha 6 receptor variants: an explorative study. *American Journal of Medical Genetics Part B Neuropsychiatric Genetics, 141*, 167–172.

Manolio, T. A., Collins, F. S., Cox, N. J., Goldstein, D. B., Hindorff, L. A., Hunter, D. J., et al. (2009). Finding the missing heritability of complex diseases. *Nature, 461*, 747–753.

Meltzer, H., Gatward, R., Goodman, R., & Ford, T. (2000). *Mental health of children and adolescents in Great Britain.* London: The Stationery Office.

Monk, C. S., Klein, R. G., Telzer, E. H., Schroth, E. A., Mannuzza, S., Moulton, J. L., III, et al. (2008). Amygdala and nucleus accumbens activation to emotional facial expressions in children and adolescents at risk for major depression. *American Journal of Psychiatry, 165*, 90–98.

Munafo, M. R., Brown, S. M., Hariri, A. R., Munafo, M. R., Brown, S. M., & Hariri, A. R. (2008). Serotonin transporter (5-HTTLPR) genotype and amygdala activation: a meta-analysis. *Biological Psychiatry, 63*, 852–857.

Munafo, M. R., Clark, T., & Flint, J. (2005). Does measurement instrument moderate the association between the serotonin transporter gene and anxiety-related personality traits? A meta-analysis. *Molecular Psychiatry, 10,* 415–419.

Narusyte, J., Neiderhiser, J. M., D'Onofrio, B. M., Reiss, D., Spotts, E. L., Ganiban, J., et al. (2008). Testing different types of genotype-environment correlation: an extended children-of-twins model. *Developmental Psychology, 44,* 1591–1603.

O'Connor, T. G., Caspi, A., DeFries, J. C., & Plomin, R. (2000). Are associations between parental divorce in children's adjustment genetically mediated? An adoption study. *Developmental Psychology, 36,* 429–437.

O'Connor, T. G., Hetherington, E. M., Reiss, D., & Plomin, R. (1995). A twin-sibling study of observed parent-adolescent interactions. *Child Development, 66,* 812–829.

Osher, Y., Hamer, D., & Benjamin, J. (2000). Association and linkage of anxiety-related traits with a functional polymorphism of the serotonin transporter gene regulatory region in Israeli sibling pairs. *Molecular Psychiatry, 5,* 216–219.

Pike, A., McGuire, S., Hetherington, E. M., Reiss, D., & Plomin, R. (1996). Family environment and adolescent depressive symptoms and antisocial behavior: a multivariate genetic analysis. *Developmental Psychology, 32,* 590–603.

Plomin, R., Reiss, D., Hetherington, E. M., & Howe, G. W. (1994). Nature and nurture: Genetic contributions to measures of the family environment. *Developmental Psychology, 30,* 32–43.

Purcell, S. (2002). Variance components models for gene-environment interaction in twin analysis. *Twin Research, 5,* 554–571.

Rapee, R. M., Schniering, C. A., & Hudson, J. L. (2009). Anxiety disorders during childhood and adolescence: origins and treatment. *[Review]. Annual Review of Clinical Psychology, 5,* 311–341.

Reiss, S. (1986). Anxiety sensitivity, anxiety frequency and the predictions of fearfulness. *Behavior Research & Therapy, 24,* 1–8.

Rende, R. D., Plomin, R., Reiss, D., & Hetherington, E. M. (1993). Genetic and environmental influences on depressive symptomatology in adolescence: Individual differences and extreme scores. *Journal of Child Psychology and Psychiatry, 34,* 1387–1398.

Rice, F. (2009). The genetics of depression in childhood and adolescence. *Current Psychiatry Reports, 11,* 167–173.

Rice, F., Harold, G. T., Shelton, K. H., Thapar, A., Rice, F., Harold, G. T., et al. (2006). Family conflict interacts with genetic liability in predicting childhood and adolescent depression. *Journal of the American Academy of Child & Adolescent Psychiatry, 45,* 841–848.

Rice, F., Harold, G. T., & Thapar, A. (2003). Negative life events as an account of age-related differences in the genetic aetiology of depression in childhood and adolescence. *Journal of Child Psychology and Psychiatry, 44,* 977–987.

Rijsdijk, F. V., Sham, P. C., Sterne, A., Purcell, S., McGuffin, P., Farmer, A., et al. (2001). Life events and depression in a community sample of siblings. *Psychological Medicine, 31,* 401–410.

Sandberg, S., Rutter, M., Giles, S., Owen, A., Champion, L., Nicholls, J., et al. (1993). Assessment of psychosocial experiences in childhood: Methodological issues and some illustrative findings. *Journal of Child Psychology and Psychiatry, 34,* 879–897.

Saudino, K. J., Pedersen, N. L., Lichtenstein, P., McClearn, G. E., & Plomin, R. (1997). Can personality explain genetic influences on life events? *Journal of Personality and Social Psychology, 72,* 196–206.

Scarr, S., & McCartney, K. (1983). How people make their own environments: A theory of genotype greater than environmental effects. *Child Development, 54,* 424–435.

Schinka, J. A., Busch, R. M., & Robichaux-Keene, N. (2004). A meta-analysis of the association between the serotonin transporter gene polymorphism (5-HTTLPR) and trait anxiety. *Molecular Psychiatry, 9,* 197–202.

Schmidt, N. B., Zvolensky, M. J., Maner, J. K., Schmidt, N. B., Zvolensky, M. J., & Maner, J. K. (2006). Anxiety sensitivity: prospective prediction of panic attacks and Axis I pathology. *Journal of Psychiatric Research, 40,* 691–699.

Scourfield, J., Rice, F., Thapar, A., Harold, G. T., Martin, N., & McGuffin, P. (2003). Depressive symptoms in children and adolescents: changing aetiological influences with development. *Journal of Child Psychology & Psychiatry & Allied Disciplines., 44,* 968–976.

Sen, S., Burmeister, M., Ghosh, D., Sen, S., Burmeister, M., & Ghosh, D. (2004). Meta-analysis of the association between a serotonin transporter promoter polymorphism (5-HTTLPR) and anxiety-related personality traits. *American Journal of Medical Genetics, Part B, Neuropsychiatric Genetics,* 85–89.

Serretti, A., Calati, R., Mandelli, L., & De Ronchi, D. (2006). Serotonin transporter gene variants and behavior: a comprehensive review. *Current Drug Targets, 7,* 1659–1669.

Silberg, J., Pickles, A., Rutter, M., Hewitt, J., Simonoff, E., Maes, H., et al. (1999). The influence of genetic factors and life stress on depression among adolescent girls. *Archives of General Psychiatry, 56,* 225–232.

Silberg, J., Rutter, M., Neale, M., & Eaves, L. (2001). Genetic moderation of environmental risk for depression and anxiety in adolescent girls. *British Journal of Psychiatry, 179,* 116–121.

Silberg, J. L., Rutter, M., & Eaves, L. (2001). Genetic and environmental influences on the temporal association between earlier anxiety and later depression in girls. *[erratum appears in Biological Psychiatry 2001 Sep 1;50(5):393.]. Biological Psychiatry, 49,* 1040–1049.

Sjoberg, R. L., Nilsson, K. W., Nordquist, N., Ohrvik, J., Leppert, J., Lindstrom, L., et al. (2006). Development of depression: sex and the interaction between environment and a promoter polymorphism of the serotonin transporter gene. *International Journal of Neuropsychopharmacology, 9,* 443–449.

Smoller, J. W., Yamaki, L. H., Fagerness, J. A., Biederman, J., Racette, S., Laird, N. M., et al. (2005). The corticotropin-releasing hormone gene and behavioral inhibition in children at risk for panic disorder. *Biological Psychiatry, 57,* 1485–1492.

Stein, M. B., Schork, N. J., Gelernter, J., Stein, M. B., Schork, N. J., & Gelernter, J. (2008). Gene-by-environment (serotonin transporter and childhood maltreatment) interaction for anxiety sensitivity, an intermediate phenotype for anxiety disorders. *Neuropsychopharmacology, 33,* 312–319.

Stevenson, J., Batten, N., & Cherner, M. (1992). Fears and fearfulness in children and adolescents: a genetic analysis of twin data. *Journal of Child Psychology and Psychiatry, 33,* 977–985.

Thapar, A., Harold, G., & McGuffin, P. (1998). Life events and depressive symptoms in childhood—shared genes or shared adversity? A research note. *Journal of Child Psychology and Psychiatry, 39,* 1153–1158.

Thapar, A., & McGuffin, P. (1996). Genetic influences on life events in childhood. *Psychological Medicine, 26,* 813–820.

Thapar, A., & McGuffin, P. (1997). Anxiety and depressive symptoms in childhood—a genetic study of comorbidity. *Journal of Child Psychology and Psychiatry, 38,* 651–656.

Uher, R., & McGuffin, P. (2008). The moderation by the serotonin transporter gene of environmental adversity in the aetiology of mental illness: review and methodological analysis. *Molecular Psychiatry, 13,* 131–146.

Uher, R., & McGuffin, P. (2009). The moderation by the serotonin transporter gene of environmental adversity in the etiology of depression: 2009 update. *Molecular Psychiatry, 15,* 18–22.

Van Ameringen, M., Mancini, C., & Farvolden, P. (2003). The impact of anxiety disorders on educational achievement. *Journal of Anxiety Disorders, 17,* 561–571.

Van der Does, A. J. W., Antony, M. M., Ehlers, A., & Barsky, A. J. (2000). Heartbeat perception in panic disorder: a reanalysis. *Behavior Research & Therapy, 38,* 47–62.

van Ijzendoorn, M. H., Bakermans-Kranenburg, M. J., & Mesman, J. (2008). Dopamine system genes associated with parenting in the context of daily hassles. *Genes, Brain, & Behavior, 7,* 403–410.

van Rossum, E. F., Binder, E. B., Majer, M., Koper, J. W., Ising, M., Modell, S., et al. (2006). Polymorphisms of the glucocorticoid receptor gene and major depression. *Biological Psychiatry, 59,* 681–688.

Wade, T. D., & Kendler, K. S. (2000). The genetic epidemiology of parental discipline. *Psychological Medicine, 30,* 1303–1313.

Walkup, J. T., Albano, A. M., Piacentini, J., Birmaher, B., Compton, S. N., Sherrill, J. T., et al. (2008). Cognitive behavioral therapy, sertraline, or a combination in childhood anxiety. *New England Journal of Medicine, 359,* 2753–2766.

Wasserman, D., Sokolowski, M. B., Rozanov, V., & Wasserman, J. (2007). The CRHR1 gene: a marker for suicidality in depressed males exposed to low stress. *Genes, Brain and Behavior 7(1),* 14–19. doi: 10.1111/j.1601-183X.2007.00310.x

Wichers, M., Kenis, G., Jacobs, N., Mengelers, R., Derom, C., Vlietinck, R., et al. (2008). The BDNF Val(66)Met x 5-HTTLPR x child adversity interaction and depressive symptoms: An attempt at replication. *American Journal of Medical Genetics, Part B, Neuropsychiatric Genetics,* 120–123.

Williams, J. M. G., Watts, F. N., MacLeod, C., & Mathews, A. (1997). *Cognitive psychology and emotional disorders.* (2nd ed.) Guilford: John Wiley & Sons.

Wolfensberger, S. P., Veltman, D. J., Hoogendijk, W. J., Boomsma, D. I., & de Geus, E. J. (2008). Amygdala responses to emotional faces in twins discordant or concordant for the risk for anxiety and depression. *Neuroimage, 41*, 544–452.

Zavos, H. M. S., Rijsdijk, F. V., & Eley, T. C. (2010). Longitudinal associations between Anxiety Sensitivity, Anxiety and Depression. *Submitted for publication.*

Zavos, H. M. S., Rijsdijk, F. V., Gregory, A. M., & Eley, T. C. (2010) Genetic influences on the cognitive biases associated with anxiety and depression symptoms in adolescents. *Journal of Affective Disorders, 124*(1–2), 45–53.

11

Adolescence and Risk for Anxiety and Depression

Insights from Human Imaging to Mouse Genetics

B.J. CASEY, ERIKA RUBERRY, AND VICTORIA LIBBY

Adolescence is described as a period of heightened stress due to the many changes experienced concomitantly, including physical maturation, drive for independence, increased salience of social and peer interaction, and brain development. Although new-found independence can be stimulating, it may also lead to feelings of being overwhelmed by change, which has historically led some researchers to characterize adolescence as being ridden with "storm and stress" (Hall, 1904). The controversial "storm and stress" viewpoint is supported by the sharp increase in onset of many psychiatric illnesses from childhood to adolescence and by the alarming U.S. health statistics on mortality associated with this developmental period (Casey et al., 2010). However, the majority of teens experience and emerge from this period in a healthy, positive manner (Lerner, 2008). In this chapter, we highlight recent empirical behavioral, imaging, and genetic findings that may explain why some teens are at greater risk for storm and stress during this developmental period than others.

A number of cognitive and neurobiological hypotheses have been postulated for why adolescence may be a period of heightened emotion. In a review of the literature on human adolescent brain development, Yurgelun-Todd (2007) suggests that development through the adolescent years is associated with progressively greater efficiency of cognitive control

capacities. This efficiency is described as being dependent on linear maturation of the prefrontal cortex from childhood to adulthood. Yet the emotive
behaviors observed during adolescence represent a nonlinear change that can
be distinguished from childhood and adulthood, as evidenced by the National
Center for Health Statistics on adolescent behavior and mortality. Thus, if
immature prefrontal cortex were the basis for teen behavior, then children
should look remarkably similar or even worse than adolescents, given their
less developed prefrontal cortex and cognitive abilities. Thus, immature prefrontal function alone cannot account for adolescent behavior.

To understand this developmental period, transitions into and out of
adolescence are necessary to distinguish distinct attributes of this period of
development. A theoretical model of adolescence must account for *nonlinear changes*, such as deflections or inflections during adolescence relative to
both childhood and adulthood. Recently, we have developed a testable
model of brain development that may account for the unique affective
behavioral changes that arise during adolescence (Casey, Getz & Galvan,
2008; Casey, Jones & Hare, 2008).

Our characterization of adolescence in this model goes beyond exclusive association of teen behavior with immaturity of the prefrontal cortex.
Instead, we suggest that the development of limbic subcortical regions
involved in desire, fight, and flight—which are reminiscent of teens'
heightened emotional reactions—must be considered together with prefrontal top-down control regions. The cartoon in Figure 11.1 illustrates different
developmental trajectories for these systems, with limbic systems developing
earlier than prefrontal control regions. This "imbalance" model proposes that
during adolescence, differential timing of brain development induces a dis-
parity between the structural and functional maturity of brain systems critical to affective processing (e.g., subcortical regions including the amygdala),
relative to cortical regions of the brain important in control over emotional
responses (e.g., the prefrontal cortex). Differential developmental timing of
these regions is consistent with nonhuman primate and human postmortem
studies showing that the prefrontal cortex is one of the last brain regions to
mature (Bourgeois, Goldman-Rakic & Rakic, 1994; Huttenlocher, 1979),
while subcortical and sensorimotor regions develop sooner.

In this chapter, we provide empirical support for our imbalance model
from recent behavioral and human imaging studies on the development of
emotion regulation. We then provide examples of environmental factors
that may exacerbate imbalances in emotion-related brain circuitry. Finally,
we present data from human imaging and mouse studies to illustrate how
genetic factors may exacerbate or diminish this risk. Together, these studies
provide a converging methods approach for understanding the highly
variable experience of adolescence among our teens.

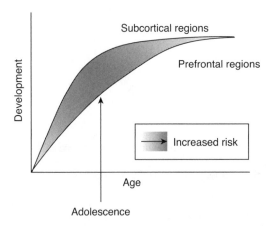

Figure 11.1. **Neurobiological model depicting later development of top down prefrontal regions relative to subcortical limbic regions involved in emotional processes.** This imbalance in the development of these systems is suggested to be at the core of aberrant teen behavior, in contrast to the popular view of adolescent behavior as being due to the protracted development of the prefrontal cortex alone. Adapted from Somerville, L. H., Jones, R. M., & Casey, B. J. (2010). A time of change: behavioral and neural correlates of adolescent sensitivity to appetitive and aversive environmental cues. *Brain and Cognition*, 72, 124–133, with permission.

DEVELOPMENT OF FRONTOLIMBIC CIRCUITRY

The development of the prefrontal cortex is believed to play an important role in the maturation of higher cognitive abilities and goal-oriented behavior. Many paradigms have been used, together with functional magnetic resonance imaging (fMRI), to assess the neurobiological basis of these abilities. Collectively, these studies suggest a fine-tuning of brain regions important for behavior regulation. In other words, there appears to be a shift from diffuse recruitment of several brain areas to more focal recruitment of a refined pathway. Although imaging studies cannot definitively characterize the mechanism of these developmental changes (e.g., dendritic arborization, synaptic pruning), postmortem data provides converging evidence for the continued pruning of prefrontal cortical synapses throughout adolescence (Rakic et al., 1986; Huttenlocher & Dabholkar, 1997). This pattern of development occurs earlier in sensorimotor and subcortical areas.

For the purposes of this chapter, we focus on the development of and imbalance in frontoamygdala circuitry involved in emotion regulation and anxiety. Figure 11.2 is an oversimplified diagram of this circuit, which has been delineated in the human and rodent using fear conditioning and extinction. This diagram illustrates how top-down prefrontal input to the amygdala,

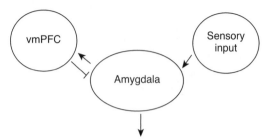

Figure 11.2. **Cartoon of frontoamygdala circuit**

a region important for detecting cues of safety and threat, can reduce the fear response. Basically, sensory input is received by the amygdala mainly through the lateral nuclei of the amygdala. The lateral nuclei project to the basal and central nuclei of the amygdala. The basal nuclei project predominantly to cortical regions (e.g., ventromedial prefrontal cortex) and striatum, whereas the central nuclei project mainly to subcortical regions involved in the fear response, including neuromodulatory systems, the hypothalamus, periaqueductal gray, and vagus. The ventromedial prefrontal cortex (vmPFC) projects to the amygdala and dampens its response via projections to inhibitory intercalated cells at the level of the central and basal nuclei. Thus, the amygdala can both signal the vmPFC of the significance of a cue and receive input from the vmPFC that dampens the fear response by suppressing output from the amygdala to neuromodulatory systems, the hypothalamus, periaqueductal gray, and vagus (LeDoux & Schiller, 2009; Phelps, 2004).

To test the hypothesis of our model for differential development of subcortical regions (amygdala) relative to prefrontal cortex (the vmPFC), we examined emotion regulation in 60 children, adolescents, and adults using fMRI. We went beyond examining the magnitude of brain activity that has been shown by several groups to be higher in adolescents than in children or adults (Ernst et al., 2005; Guyer et al., 2008; Guyer, McClure-Tone, Shiffrin, Pine, & Nelson, 2009; Monk et al., 2003; Rich et al., 2006; Williams et al., 2006). Instead, we focused on neural adaptation within frontoamygdala circuitry across time to capture both developmental and individual differences (Hare et al., 2008). Individual differences in emotional reactivity might put some teens at greater risk during this sensitive transition in development. We assessed everyday anxiety using the Spielberger Trait Anxiety Inventory to test this hypothesis.

Our imaging results showed that adolescents have an initial exaggerated amygdala response to cues that signal threat (fearful faces) relative to children and adults (see Figure 11.3). This initial heightened response in

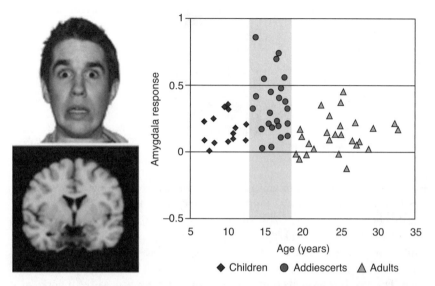

Figure 11.3. **Amygdala response to empty threat (fearful faces) as a function of age.** Adapted from Hare et al., 2008. (See Plate 3 for a color version of this figure.)

amygdala activity is age-dependent and does not correlate with trait anxiety. Moreover, although several groups have shown elevated amygdala activity to emotional pictures in adolescents relative to adults (Guyer at al., 2008; Monk et al., 2003), these data showed a pattern in adolescents that is distinct from that in both children and adults.

The age-related difference of elevated amygdala activity during adolescence diminishes with repeated exposure of empty threat (fearful faces) across the experiment. Anonymous self-report ratings of everyday anxiety predicted the extent of habituation in the amygdala to empty threat. Habituation was measured by subtracting average MR signal in late trials of the experiment from early trials of the experiment (i.e., the larger the value, the greater the habituation of the amygdala response). Cases in which the value is negative indicate an actual increase in amygdala activity over time. Individuals with higher trait anxiety showed less habituation of the amygdala response over time, with some actually showing increases (Fig. 11.4a). This failure of the amygdala response to habituate over time was associated with frontolimbic activity, specifically between the ventromedial prefrontal cortex (vmPFC) and the amygdala (refer back to Figure 11.2). Inverse functional coupling of these regions, consistent with greater top-down regulation (higher vmPFC activity) of the amygdala, was correlated with more habituation (i.e., negative values indicate inverse coupling, whereas a lack of coupling was indicated by values of 0 or slightly positive [see Figure 11.4]).

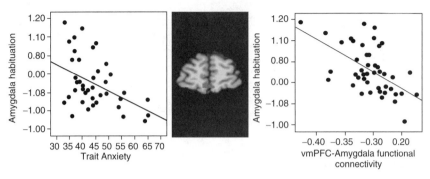

Figure 11.4. **Habituation of amygdala response is associated with anxiety and less ventral prefrontal cortex (vmPFC) activity.** Trait anxiety scores were negatively correlated with habituation (decrease from early to late trials) of amygdala activity ($r = -.447$; $p < .001$). **Left panel:** Scatter plot of the correlation between trait anxiety and amygdala habituation. The y-axis represents magnetic resonance (MR) signal in the left amygdala for early-late trials. The x-axis represents trait anxiety score. There was negative functional coupling between the amygdala and the ventral prefrontal cortex (vmPFC). The magnitude of activity in vmPFC (*middle panel*) and the strength of the connectivity between vmPFC and the amygdala were negatively correlated with amygdala habituation ($r = -.56$ $p < .001$). Right panel: Scatterplot of vmPFC–amygdala connectivity values versus amygdala habituation. The y-axis represents MR signal in the left amygdala for early-late trials. The x-axis represents Z-scored vmPFC–amygdala connectivity values. Adapted from Hare, T. A., Tottenham, N., Galvan, A., Voss, H. U., Glover, G. H., & Casey, B. J. (2008). Biological substrates of emotional reactivity and regulation in adolescence during an emotional go-nogo task. *Biological Psychiatry*, 63(10), 927–934, with permission. (See Plate 4 for a color version of this figure.)

The findings suggest that initial emotional reactivity as indexed by elevated amygdala activity is typical of adolescence, but that failure of this response to subside over time with no impending threat is atypical and may be indicative of trait anxiety. Individual data are shown, rather than grouping adolescents to show the large variability that may underlie why some adolescents experience and emerge from adolescence with healthy positive outcomes whereas others may not. Future studies of populations at risk for anxiety will need to examine carefully not only what triggers a heightened amygdala response, but what causes the sustained response in other brain regions over time.

The observation of imbalanced activity in the amygdala–vmPFC network, as shown by elevated amygdala and less vmPFC activity in highly anxious individuals, is consistent with a variety of work in animals (Baxter et al., 2000; Milad & Quirk, 2002) and humans (Delgado et al., 2006; Etkin et al., 2006; Haas et al., 2007; Johnstone et al., 2007; Urry et al., 2006), thus implicating an inverse relationship between these structures that governs affective output. In particular, increased response in the vmPFC is inversely correlated with response in the amygdala, and predicts behavioral outcomes such as fear extinction, downregulation of autonomic responses (Phelps, Delgado, Nearing, & LeDoux, 2004), and more positive interpretations of emotionally ambiguous information (Kim et al., 2004). Thus, inverse functional coupling of these structures is key to the down-regulation of heightened emotional responses. During adolescence, when the amygdala response is heightened relative to that observed in children and adults (imbalance), more top-down control is needed. A lack of coupling between these regions with development to help provide that down-regulation may lead to symptoms and ultimately diagnosis of anxiety.

TIPPING THE IMBALANCE: ENVIRONMENTAL FACTORS

A number of studies have shown the significance of environmental factors, such as stress and early adversity, on brain and behavior (Liston, McEwen, & Casey, 2009; Liston et al., 2006; Tottenham et al., 2010) and risk for psychopathology. Trauma exposure is a particularly potent environmental risk factor for anxiety and depression. A recent study by our group examined the effects of a naturally occurring disaster on affective processing of cues of threat (Ganzel, Casey, Voss, Glover, & Temple, 2007). Specifically, we used fMRI to assess the impact of proximity to the disaster of September 11, 2001, on amygdala function in 22 healthy young adults.

Our findings suggest that more than 3 years after the terrorist attacks, bilateral amygdala activity in response to viewing fearful faces compared to calm ones was higher in individuals who were within 1.5 miles of the World Trade Center on 9/11, relative to those who were living more than 200 miles away (all were living in the New York metropolitan area at time of scan). This effect was statistically driven by time since worst lifetime trauma, other than 9/11, and intensity of that most severe trauma, as indicated by reported number of symptoms at the time of the prior trauma (see Figure 11.5). These data are consistent with a model of heightened amygdala reactivity following high-intensity trauma exposure, with relatively slow recovery, similar to that reported in animal studies (Adamec, 2005; Vyas, Mitra, Shankaranarayana Rao & Chattarji, 2002).

In the context of our neurobiological model of adolescence, individuals who experience trauma during this period or have experienced multiple traumas may be especially vulnerable for developing symptoms of anxiety and depression as teens or adults. A large epidemiological and clinical literature supports this claim (Brown, Cohen, Johnson & Smailes, 1999;

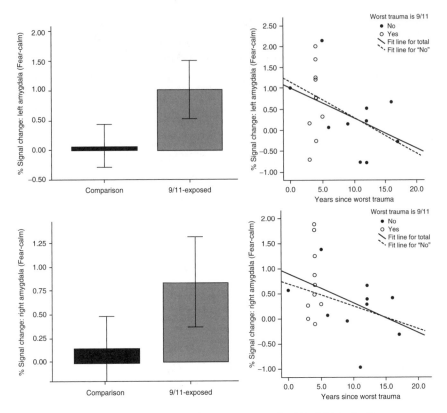

Figure 11.5. **Amygdala activity and proximity to the World Trade Center on September 11th, 2001.** Fearful emotional faces elicited greater left (top panel) and right (bottom panel) amygdala activity in the 9/11-exposed group relative to a comparison group ($p < 05$). Amygdala activity was negatively correlated with time since worst trauma in lifetime in left ($r = -.46$, $p < .05$) and right amygdala ($r = -.45$, $p < .06$). Adapted from Ganzel, B., Casey, B. J., Voss, H. U., Glover, G., & Temple, E. (2007). The aftermath of 9/11: effect of intensity and recency of trauma on outcome. *Emotion*, 7(2), 227–238, with permission.

Heim & Nemeroff, 2001; Pine & Cohen, 2002; Yule et al., 2000). Our model simply provides a biological basis for this vulnerability. In other words, although heightened emotional reactivity is typical during the period of adolescence, failure to suppress that emotional reactivity over time when there is no impending threat is associated with symptoms of anxiety.

The large variability observed in our developmental studies of emotion regulation may in part be due to variation in individuals' experiences. An imbalance in amygdala–vmPFC coupling has been implicated in the pathophysiology of childhood and adolescent mood and anxiety disorders (Guyer et al., 2009; Leibenluft, Blair, Charney & Pine, 2003; Monk et al.,

2008) showing greater amygdala relative to prefrontal activity. As such, improving our understanding of the development of these circuits and the source of biased responding in some adolescents over others will facilitate our understanding of the most commonly experienced psychiatric illnesses of this developmental period (i.e., anxiety and depression).

GENETIC FACTORS

A number of human genetic studies have begun to identify candidate genes that may play a role in increased risk for anxiety and depression. The main avenues for understanding gene function in these disorders have been in behavioral genetics on one end and on the other end, molecular mouse models. Attempts to bridge these approaches have used brain imaging to conveniently link anatomical abnormalities seen in knock-out/transgenic mouse models and abnormal patterns of brain activity seen in humans. Recently, we completed a study using human and mouse behavioral genetics together with imaging genetics. Each of these approaches alone provides limited information on gene function in complex human behavior, such as emotion regulation and its dysregulation in psychopathology, but together they are forming bridges between animal models and human psychiatric disorders.

In this study, we utilized a common single nucleotide polymorphism (SNP) in the brain-derived neurotrophic factor (*BDNF*) gene that leads to a valine (Val) to methionine (Met) substitution at codon 66 (Val66Met). This polymorphism leads to decreased trafficking of *BDNF* into the regulated secretory pathway. This deficit leads to impaired activity-dependent release of *BDNF*. In an inbred genetic knock-in mouse strain that expresses the variant *BDNF* allele to recapitulate the specific phenotypic properties of the human polymorphism in vivo, we found the *BDNF* Val66Met genotype was associated with treatment-resistant forms of anxiety-like behavior (Chen et al., 2006). A key feature of anxiety is difficulty in learning of cues that signal safety versus threat and in learning new associations when a previous threat cue is no longer associated with that threat (i.e., extinction). Thus, the objective of our study was to test if the Val66Met genotype could impact extinction learning in our mouse model, and if such findings could be generalized to human populations.

We examined the impact of the variant *BDNF* on fear conditioning and extinction paradigms (Soliman et al., 2010). Approximately 70 mice and 70 humans were tested. The mice included 17 $BDNF^{Val/Val}$, 33 $BDNF^{Val/Met}$, and 18 $BDNF^{Met/Met}$. The human sample included 36 Met allele carriers (31 $BDNF^{Val/Met}$ and 5 $BDNF^{Met/Met}$) and 36 nonMet allele carriers group-matched on age, gender and ethnic background. Fear conditioning consisted of pairing a neutral cue with an unconditioned aversive stimulus until the cue itself took on properties of the unconditioned stimulus (US) of an impending aversive event. The extinction procedure consisted of repeated presentations of the cue (i.e., conditioned stimulus or CS) alone. Behavioral responses of percentage of time freezing in the mouse and amplitude of the galvanic skin response in the human were the dependent measures. In addition, we collected brain imaging data using fMRI with the human sample.

Our findings showed no effects of *BDNF* genotype on fear conditioning in the mice or humans as measured by freezing behavior to the conditioned stimulus in the mice ($F(2,65) = 1.58$, $p < 0.22$) and by skin conductance response in humans to the cue predicting the aversive stimulus relative to a neutral cue ($F(1,70) = 0.67$, $p < 0.42$). However, both the mice and humans showed slower extinction in Met allele carriers than in nonMet allele carriers, as shown in Figure 11.6A and B. Specifically, Met allele carriers showed slower extinction than nonMet allele carriers. Moreover, human fMRI data provide neuroanatomical validation of the cross-species translation. Specifically, we show alterations in frontoamygdala circuitry, shown to support fear conditioning and extinction in previous rodent (LeDoux, 2000; Myers & Davis, 2002) and human (Delgado, Nearing, LeDoux, & Phelps, 2008; Gottfried & Dolan 2004; LaBar, Gatenby, Gore, LeDoux, & Phelps, 1998; Phelps et al., 2004) studies, as a function of *BDNF* genotype. Met allele carriers show less vmPFC activity during extinction relative to nonMet allele carriers (Fig. 11.6C), but greater amygdala activity relative to nonMet allele carriers (Fig. 11.6D). These findings suggest that cortical regions essential for extinction in animals and humans are less responsive in Met allele carriers. Moreover, amygdala recruitment that should show diminished activity during the extinction was elevated in Met allele carriers, suggesting less dampening by vmPFC and more fear response as generated by amygdala output to neuromodulatory systems, the hypothalamus, periaqueductal gray, and vagus (LeDoux & Schiller, 2009; Phelps, 2009).

In the current study, we hypothesized that extinction would be slower in human Met allele carriers than nonMet allele carriers. Replication of the effect in both mice and humans is a convenient first step in showing this association. However, given that the genetic findings are based on fewer than 100 participants for only a modest effect size, and neural anatomical evidence for extinction was examined across two regions of interest and confirmed in a whole-brain analysis across multiple voxels, the human findings will need to be replicated.

Nonetheless, the findings are provocative as they provide an example of bridging human behavioral and imaging genetics with a molecular mouse

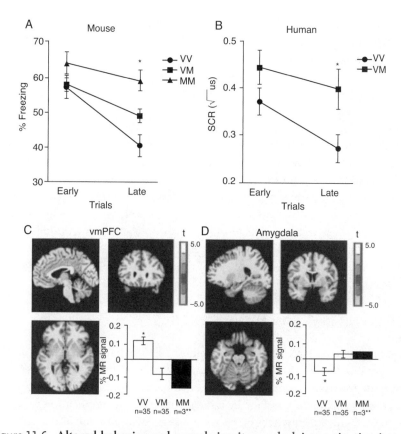

Figure 11.6. **Altered behavior and neural circuitry underlying extinction in mice and humans with brain-derived neurotrophic factor (*BDNF*) Val66Met.** Impaired extinction in Met allele carriers (Val/Met and Met/Met) as a function of time in 68 mice (**A**) and 72 humans (**B**) as indexed by percent time freezing in mice and skin conductance response (SCR) in humans to the conditioned stimulus when it was no longer paired with the aversive stimulus. (**C**) Brain activity as indexed by percent change in MR signal during extinction in the ventromedial prefrontal cortex (vmPFC) by genotype (*xyz* = –4, 24, 3), with Met allele carriers having significantly less activity than Val/Val homozygotes (VM < VV = blue), image threshold $p < 0.05$, corrected. (**D**) Genotypic differences in left amygdala activity during extinction (*xyz* = –25, 2, – 20) in 70 humans, with Met allele carriers having significantly greater activity than Val/Val homozygotes (VM > VV = orange), image threshold $p < 0.05$, corrected. $*p < 0.05$. **MM were included in the analysis with VM, but plotted separately to see dose response. All results are presented as a mean ± SEM. VV = Val/Val; VM = Val/Met; MM = Met/Met. From Soliman, F., Glatt, C. E., Bath, K. G., Levita, L., Jones, R. M., Pattwell, S. S., Jing, D., et al. (2010). A genetic variant *BDNF* polymorphism alters extinction learning in both mouse & human. *Science, 327*(5967), 863–866, with permission. (See Plate 5 for a color version of this figure.)

model to suggest a role for *BDNF* in anxiety disorders. In the context of our neurobiological model of adolescence, individuals with the *BDNF* Met allele may be more vulnerable for developing symptoms of anxiety as teens, in that they show higher and prolonged patterns of amygdala activity and less vmPFC activity to emotional cues. During a period when evaluating social cues from peers is essential in forming and maintaining healthy peer relationships, the failure to suppress heightened emotional responses to empty threat in these interactions (e.g., failure of a peer to notice or smile at a teenager without any negative intent) could lead to overinterpretation and ruminations of self-doubt. These genetic data provide an example then of how an imbalance in amygdala–vmPFC coupling during typical development could be exacerbated and lead to clinical symptoms of anxiety. Moreover, these data may have important implications for the efficacy of treatments for anxiety disorders that rely on extinction mechanisms, such as exposure therapy.

CHALLENGES FOR IMAGING GENETICS

The previous example uses brain imaging to link behavioral and anatomical abnormalities seen in a genetic mouse model with abnormal patterns of behavior and brain activity in humans. Although a rapidly growing approach, there are many challenges for imaging genetics and its application to development. Genetic influences on behavior are mediated by the impact of their molecular and cellular effects on brain development and function. As such, genetic imagers have argued that the functional effects of candidate variants on brain structure and function may be more readily measurable than are the cognitive and emotional processes supported by these substrates, and thus, functional polymorphisms in genes weakly related to behaviors or psychiatric syndromes may be more strongly related to the function of specific neural systems (Casey & Durston, 2006; Fossella & Casey, 2006). With 25,000 known genes in the human/mouse mammalian genome and up to 100 variable sites within each of these genes in an individual with a brain that can be measured at the 1–3 mm^3 voxel level and whose dynamic networks are capable of an infinite variety of computational states, the one-to-one statistical mappings of genetic variants to brain function is a challenging yet convenient first step.

The challenge of linking genes, brain, and behavior becomes more daunting when we consider gene function in a dynamically changing system, such as in the developing brain. In the excitement in using the Human Genome Project to uncover the functions of specific genes, researchers often ignore a fundamental factor: the gradual process of ontogenetic development.

Claims about gene–behavior relations are typically based on a phenotypic end-state and grounded in the neuropsychological tradition of studying adults whose brains were fully and normally developed until a brain insult later in life. The developing brain, of course, is different in that brain regions and circuitry are not specialized at birth. Many years are required for the specialization of neural networks as a result of the complex interaction with the environment in gene expression and the resulting phenotype. Karmiloff-Smith (2006) has argued that "because ontogenetic development and timing play such important roles in development, a tiny impairment in the start state of the brain of a child with a genetic disorder may affect several brain regions, some more profoundly and others more subtly, giving rise in the phenotypic endstate to what appears to be a domain-specific outcome." This neuroconstructivist perspective is an important one to keep in mind when interpreting simple gene–behavior associations in typical develop-ment and in developmental disorders.

Perhaps most challenging for genetic studies is the multiple-comparisons problem. In its most severe form, multiple endpoints or measures are exam-ined and then an unconstrained search for some statistically significant dif-ference between genotypic groups is performed without a specific hypothesis or a priori determined endpoint. When these samples are small or there is uneven representation of ethnicities, gender or age in the genotypic groups, it can lead to "genotypic group differences" that have nothing to do with the genotype at all (see review of limitations in Casey, Soliman, Bath & Glatt, 2010). Using a significance level of $p = 0.05$ to decide which of the differ-ences is reliable only increases the chances far beyond 0.05 of finding differ-ences that are not real. The behavioral and imaging genetic literature is filled with such examples, and several meta-analyses in the field underscore the inconsistencies in findings from such studies (Risch et al., 2009).

CONCLUSION

Taken together, the findings synthesized here indicate that increased risk in adolescence for storm and stress is associated with different developmental trajectories of subcortical emotional systems and cortical control regions. This differential development may lead to an imbalance in control by sub-cortical regions over prefrontal ones and heightened emotional reactivity. Although elevated emotional reactivity is typical during the period of ado-lescence, failure to suppress that emotional reactivity with time is associated with symptoms of anxiety. The large variability observed in our develop-mental studies of emotion regulation may in part be due to variation in individuals' environmental and genetic background. Both environmental and genetic factors can enhance or minimize the imbalance between limbic and control regions and, in turn, increase or decrease risk for anxiety and depression in some individuals over others.

Together, our studies provide a converging methods approach for understanding the highly variable experiences of adolescence. Important future directions will be to consider how genetic, environmental, and developmental factors interrelate in sufficiently large samples to directly test these effects from a developmental perspective. Such genetic studies will need to entertain dynamic models that capture the effects of changing environmental and developmental conditions or contexts, rather than relying on static models.

ACKNOWLEDGMENTS

This work was supported in part by NIMH 1R01 MH73175, NIDA R01, DA018879, NIMH P50 MH62196, the Mortimer D. Sackler family, the Dewitt-Wallace fund, and by the Weill Cornell Medical College Citigroup Biomedical Imaging Center and Imaging Core.

REFERENCES

Adamec, R. E., Blundell, J., & Burton, P. (2005). Neural circuit changes mediating lasting brain and behavioral response to predator stress. *Neuroscience and Biobehavioral Reviews, 29*, 1225–1241.

Baxter, M. G., Parker, A., Lindner, C. C., Izquierdo, A. D. & Murray, E. A. (2000). Control of response selection by reinforcer value requires interaction of amygdala and orbital prefrontal cortex. *Journal of Neuroscience, 20*(11), 4311–4319.

Bourgeois, J. P., Goldman-Rakic, P. S. & Rakic, P. (1994). Synaptogenesis in the prefrontal cortex of rhesus monkeys. *Cerebral Cortex, 4*, 78–96.

Brown, J., Cohen, P., Johnson, J. G. & Smailes, E. M. (1999). Childhood abuse and neglect: Specificity of effects on adolescent and young adult depression and suicidality. *Journal of the American Academy of Child and Adolescent Psychiatry, 38*, 1490–1496.

Casey B. J., & Durston, S. (2006). From behavior to cognition to the brain and back: what have we learned from functional imaging studies of attention deficit hyperactivity disorder? *American Journal of Psychiatry, 163*(6), 957–960.

Casey, B. J., Getz, S., & Galvan, A. (2008). The adolescent brain and risky decisions. *Developmental Reviews, 28*(1), 62–77.

Casey, B. J., Jones, R., & Hare, T. A. (2008). The adolescent brain. *The Year in Cognitive Neuroscience, 1124*, 111–126.

Casey, B. J., Jones, R., Levita, L., Libby, V., Pattwell, S. S., Ruberry, E., Soliman, F., et al. (2010). The storm and stress of adolescence: insights from human imaging and mouse genetics. *Developmental Psychobiology, 52*(3), 225–235.

Casey, B. J., Soliman, F., Bath, K. G., & Glatt, C. E. (2010). Imaging genetics and development: challenges and promises. *Human Brain Mapping, 31*(6), 838–851.

Chen, Z. Y., Jing, D., Bath, K. G., Ieraci, A., Khan, T., Siao, C. J., et al. (2006). Genetic variant BDNF (Val66Met) polymorphism alters anxiety-related behavior. *Science, 314*(5796), 140–143.

Delgado, M. R., Nearing, K. I., LeDoux, J. E., & Phelps, E. A. (2008). Neural circuitry underlying the regulation of conditioned fear and its relation to extinction. *Neuron, 59*(5), 829–838.

Delgado, M. R., Olsson, A., & Phelps, E. A. (2006). Extending animal models of fear conditioning to humans. *Biological Psychology, 73*(1), 39–48.

Ernst, M., Nelson, E. E., Jazbec, S., McClure, E. B., Monk, C. S., Leibenluft, E., et al. (2005). Amygdala and nucleus accumbens in responses to receipt and omission of gains in adults and adolescents. *Neuroimage, 25,* 1279–1291.

Etkin, A., Egner, T., Peraza, D. M., Kandel, E. R., & Hirsch, J. (2006). Resolving emotional conflict: a role for the rostral anterior cingulate cortex in modulating activity in the amygdala. *Neuron 51*(6), 871–882.

Fossella, J. A., & Casey, B. J. (2006). Genes, brain, and behavior: bridging disciplines. *Cognitive, Affective & Behavioral Neuroscience, 6*(1), 1–8.

Ganzel, B., Casey, B. J., Voss, H. U., Glover, G., & Temple, E. (2007). The aftermath of 9/11: effect of intensity and recency of trauma on outcome. *Emotion, 7*(2), 227–238.

Gottfried, J. A., & Dolan, R. J. (2004). Human orbitofrontal cortex mediates extinction learning while accessing conditioned representations of value. *Nature. Neuroscience, 7*(10), 1144–1152.

Guyer, A. E., Monk, C. S., McClure-Tone, E. B., Nelson, E. E., Roberson-Nay, R., Adler, A., et al. (2008). A developmental examination of amygdala response to facial expressions. *Journal of Cognitive Neuroscience, 20*(9), 1565–1582.

Guyer, A. E., McClure-Tone, E. B., Shiffrin, N. D., Pine, D. S., & Nelson, E. E. (2009). Probing the neural correlates of anticipated peer evaluation in adolescence. *Child Development, 80,* 1000–1015.

Haas, B. W., Omura, K., Constable, R. T., & Canli, T. (2007). Emotional conflict and neuroticism: personality-dependent activation in the amygdala and subgenual anterior cingulate. *Behavioral Neuroscience, 121*(2), 249–256.

Hall, G. S. (1904). Adolescence: its psychology and its relation to physiology, anthropology, sociology, sex, crime, religion, and education (vols. I & II). Englewood Cliffs, NJ: Prentice- Hall.

Hare, T. A., Tottenham, N., Galvan, A., Voss, H. U., Glover, G. H., & Casey, B. J. (2008). Biological substrates of emotional reactivity and regulation in adolescence during an emotional go-nogo task. *Biological Psychiatry, 63*(10), 927–934.

Heim, C., & Nemeroff, C. B. (2001). The role of childhood trauma in the neurobiology of mood and anxiety disorders: preclinical and clinical studies. *Biological Psychiatry, 49,* 1023–1039.

Huttenlocher, P. R. (1979). Synaptic density in human frontal cortex—developmental changes and effects of aging. *Brain Research, 163,* 195–205.

Huttenlocher, P. R, & Dabholkar, A. S. (1997). Regional differences in synaptogenesis in human cerebral cortex. *The Journal of comparative neurology*, *387*(2), 167–178.

Johnstone, T., van Reekum, C. M., Urry, H. L., Kalin, N. H. & Davidson, R. J. (2007). Failure to regulate: counterproductive recruitment of top-down prefrontal-subcortical circuitry in major depression. *Journal of Neuroscience*, *27*(33), 8877–8884.

Karmiloff-Smith, A. (2006). The tortuous route from genes to behavior: a neuroconstructivist approach. *Cognitive, Affective, & Behavioral Neuroscience*, 6 (*1*), 9–17.

Kim H., Somerville L. H., Johnstone T., Polis S., Alexander A. L., Shin L. M., & Whalen P. J. (2004). Contextual modulation of amygdala responsivity to surprised faces. *Journal of Cognitive Neuroscience*, *16*(10): 1730-1745.

LaBar, K. S., Gatenby, J. C., Gore, J. C., LeDoux, J. E., & Phelps, E. A. (1998). Human amygdala activation during conditioned fear acquisition and extinction: a mixed-trial fMRI study. *Neuron*, *20*(5), 937–945.

LeDoux, J. E. (2000.) Emotion circuits in the brain. *Annual Review of Neuroscience*, *23*, 155–184.

LeDoux, J. E. & Schiller, D. (2009) What animal fear models have taught us about human amygdala function. In P. J. Whalen & E. A. Phelps, EA (Eds.). *The Human Amygdala* (43–60). New York, NY: Guilford.

Leibenluft, E., Blair, R. J., Charney, D. S., & Pine, D. S. (2003). Irritability in pediatric mania and other childhood psychopathology. *Annals of the New York Academy of Sciences*, *1008*, 201–218.

Lerner, R. M. (2008). *The good teen: rescuing adolescence from the myths of the storm and stress years*. New York: Random House.

Liston, C., McEwen, B., & Casey, B. J. (2009). Psychosocial stress reversibly disrupts prefrontal processing and attentional control. *Proceedings of the National Academy of Sciences of the United States of America*, *106*, 912–917.

Liston, C., Miller, M. M., Goldwater, D. S., Radley, J. J., Rocher, A. B., Hof, P. R., et al. (2006). Stress-induced alterations in prefrontal cortical dendritic morphology predict selective impairments in perceptual attentional set-shifting. *Journal of Neuroscience*, *26*(30), 7870–7874.

Milad, M. R. & Quirk, G. J. (2002). Neurons in medial prefrontal cortex signal memory for fear extinction. *Nature*, *420*(6911), 70–74.

Monk, C. S., McClure, E. B., Nelson, E. E., Zarahn, E., Bilder, R. M., Leibenluft, E., et al. (2003). Adolescent immaturity in attention-related brain engagement to emotional facial expressions. *Neuroimage*, *20*, 420–428.

Monk, C. S., Telzer, E. H., Mogg, K., Bradley, B. P., Mai, X., Louro H. M., et al. (2008). Amygdala and ventrolateral prefrontal cortex activation to masked angry faces in children and adolescents with generalized anxiety disorder. *Archives of general psychiatry*, *65*(5), 568–576.

Myers, K. M. & Davis, M. (2002). Behavioral and neural analysis of extinction. *Neuron 36*(4), 567–584.

Phelps, E. A., Delgado, M. R., Nearing, K. I., & LeDoux, J. E. (2004). Extinction learning in humans: role of the amygdala and vmPFC. *Neuron, 43*(6), 897–905.

Pine, D. S., & Cohen, J. A. (2002). Trauma in children and adolescents: risk and treatment of psychiatric sequelae. *Biological Psychiatry, 51,* 519–531

Rakic, P., Bourgeois, J. P., Eckenhoff, M. F., Zecevic, N., & Goldman-Rakic, P. S. (1986). Concurrent overproduction of synapses in diverse regions of the primate cerebral cortex. *Science, 232,* 232–235.

Rich, B. A., Vinton, D. T., Roberson-Nay, R., Hommer, R. E., Berghorst, L. H., McClure, E. B., et al. (2006). Limbic hyperactivation during processing of neutral facial expressions in children with bipolar disorder. *Proceedings of the National Academy of Sciences USA, 103*(23), 8900–8905.

Risch, N., Herrell, R., Lehner, T., Liang, K.Y., Eaves, L., Hoh, J., et al. (2009). Interaction between the serotonin transporter gene (5-HTTLPR), stressful life events, and risk of depression: a meta-analysis. *The journal of the American Medical Association, 301*(23), 2462–2471.

Soliman, F., Glatt, C. E., Bath, K. G., Levita, L., Jones, R. M., Pattwell, S. S., Jing, D., et al. (2010). A genetic variant BDNF polymorphism alters extinction learning in both mouse and human. *Science, 327*(5967), 863–866.

Somerville, L. H., Jones, R. M., & Casey, B. J. (2010). A time of change: behavioral and neural correlates of adolescent sensitivity to appetitive and aversive environmental cues. *Brain and Cognition, 72,* 124–133.

Tottenham, N., Hare, T. A., Quinn, B. T., McCarry, T. W., Nurse, M., Gilhooly, T., et al. (2010). Prolonged institutional rearing is associated with atypically larger amygdala volume and difficulties in emotion regulation. *Developmental Science, 13*(1), 46–61.

Urry, H. L., van Reekum, C. M., Johnstone, T., Kalin, N. H., Thurow, M. E., Schaefer, H. S., et al. (2006). Amygdala and ventromedial prefrontal cortex are inversely coupled during regulation of negative affect and predict the diurnal pattern of cortisol secretion among older adults. *Journal of Neuroscience, 26*(16), 4415–4425.

Vyas, A., Mitra, R., Shankaranarayana Rao, B. S., & Chattarji, S. (2002). Chronic stress induces contrasting patterns of dendritic remodeling in hippocampal and amygdalaloid neurons. *Journal of Neuroscience, 22,* 6810–6818.

Williams, L. M., Brown, K. J., Palmer, D., Liddell, B. J., Kemp, A. H., Olivieri, G., et al. (2006). The mellow years? Neural basis of improving emotional stability with age. *Journal of Neuroscience, 26*(24), 6422–6430.

Yule, W., Bolton, D., Udwin, O., Boyle, S., O'Ryan, D., & Nurrish, J. (2000). The long-term psychological effects of a disaster experienced in adolescence: I: The incidence and course of PTSD. *Journal of child psychology and psychiatry, 41,* 503–511.

Yurgelun-Todd, D. (2007). Emotional and cognitive changes during adolescence. *Current Opinion in Neurobiology, 17,* 251–257.

12

Genetic and Environmental Predictors of Depression

FRANCHESKA PEREPLETCHIKOVA AND JOAN KAUFMAN

By all standards of measurement, the problem of child maltreatment is enormous in terms of both its cost to the individual, and its cost to society (Zigler, 1980). Child abuse occurs at epidemic rates, with victims of abuse comprising a significant proportion of all child psychiatric admissions. The lifetime incidence of physical and sexual abuse is estimated at 30% among child and adolescent outpatients (Lanktree, Briere, & Zaidi, 1991), and approximately 55% among psychiatric inpatients (McClellan, Adams, Douglas, McCurry, & Storck, 1995).

Although not all abused children develop difficulties, many experience a chronic course of psychopathology (Molnar, Buka, & Kessler, 2001). Maltreatment is associated with elevated rates of conduct and antisocial personality disorders (Garland et al., 2001; Kolko, 2002; Rutter, Giller, & Hagell, 1998; Widom, 1989; Widom & Ames, 1994); posttraumatic stress disorder (Famularo, Fenton, Kinscherff, & Augustyn, 1996; Famularo, Kinscherff, & Fenton, 1992; Kilpatrick et al., 2003; Ruggiero, McLeer, & Dixon, 2000); major depression (Kaufman, 1991; Pelcovitz, Kaplan, DeRosa, Mandel, & Salzinger, 2000); and drug and alcohol problems (Molnar et al., 2001; Schuck & Widom, 2003; Widom, Ireland, & Glynn, 1995).

In the late 1970s, an influential study examining the outcomes of physically abused children suggested that the effects on child development of lower-class membership may be as powerful as abuse (Elmer, 1977). In Chapter 3 by Shanahan and Bouldry (this volume), it is noted that it is often difficult to show causal influence of maltreatment on child outcome in epidemiological research, as severe maltreatment is an uncommon event, it is rarely optimally assessed in large-scale studies, and it co-occurs with other risk factors (e.g., poverty, community violence) that may account for associations between maltreatment and the outcomes under investigation. Three decades of research with carefully characterized cases and controls (Cicchetti & Toth, 2005), however, have clearly demonstrated that when maltreated children are compared with demographically matched comparison children, maltreatment is associated with increased risk for a whole host of negative outcomes. The types of experimental controls that would be necessary to definitively make causal attributions would not be ethical to impose on children, but as discussed later in this chapter, have been elegantly modeled in animal studies. Adverse early rearing conditions are associated with widespread biological and maladaptive behavioral changes.

Genetic and environmental factors have been shown to contribute to the etiology of depression and other stress-related psychiatric problems in maltreated children, but there is currently controversy in the field regarding the strength of associations reported in prior research examining gene by environment (G × E) interactions. The first section of this chapter discusses a recent meta-analysis that has fueled debate regarding the utility of candidate gene G × E research. The second section then provides an overview of the population and methods utilized in our G × E studies, and the third section reviews a representative sample of our work. Future directions that highlight the value of translational multidisciplinary research approaches are then delineated in the closing section of this chapter. The lines between "genes" and "environment" have blurred over the last decade, and old debates that pitted nature against nurture have been replaced with more refined arguments involving specific genes, types of life events, the magnitude of the influences of both factors, and the mechanisms responsible for interactive effects. These themes are also articulated in Chapter 7, by Mill, with the focus of this chapter on understanding risk and resiliency in maltreated children.

META-ANALYSIS OF G × E DEPRESSION RESEARCH

As discussed in other chapters in this book, Caspi, Moffitt, and colleagues (Caspi et al., 2002) were the first to examine the role of genetic factors in moderating the outcome of individuals with a history of child maltreatment. They studied a large epidemiological sample of 1,037 males from birth to adulthood. They used the dataset to examine why some children who are maltreated grow up to develop psychiatric problems and others do not. They identified a functional polymorphism in the promoter region of the serotonin transporter gene (*5-HTTLPR*) that moderated the influence of early child maltreatment and stressful life events (SLE) on the development of depression (Caspi et al., 2003). The serotonin transporter is a protein critical to the regulation of serotonin function in the brain, because it terminates the action of serotonin in the synapse via reuptake. This gene has a well-studied functional variable number tandem repeat (VNTR) polymorphism in the promoter region. There are two common functional alleles of *5-HTTLPR*, the short (S) allele and the long (L) allele. The S allele encodes an attenuated promoter segment, and is associated with in vitro reduced transcription and functional capacity of the serotonin transporter relative to the L allele (Lesch et al., 1995). Capsi and colleagues (2003) reported that individuals with a history of abuse or multiple recent SLE, and with one or two copies of the S allele of *5-HTTLPR*, exhibited more depressive symptoms, diagnosable depression, and suicidality than did individuals homozygous for the L allele.

Risch, Merikangas, and colleagues recently published a highly publicized meta-analysis of studies examining the association between *5-HTTLPR*, SLE, and depression (Risch et al., 2009). The authors of this article concluded that "…This meta-analysis yielded no evidence that the serotonin transporter genotype alone or in interaction with stressful life events is associated with an elevated risk of depression in men alone, women alone, or in both sexes combined." After the publication of this paper, the *New York Times* featured an article with the headline "Report on Gene for Depression Is Now Faulted" (Carey, 2009). A smaller-scale meta-analysis published earlier in the year was less noticed, but similarly concluded that "… the literature leads us to suggest that the positive results for the *5-HTTLPR* × SLE interactions in logistic regression models are compatible with chance findings" (Munafo, Durrant, Lewis, & Flint, 2009). These conclusions, however, were based on assumptions in the simulation model that are inconsistent with prior research findings, and a bias in sampling strategy that call into question the authors' conclusions.

We have published a correspondence in *Biological Psychiatry* that delineates our concerns with the two meta-analyses (Kaufman, Gelernter, Kaffman, Caspi, & Moffitt, 2010). The nonrepresentative nature of the sample of studies included in both meta-analyses is a significant limitation of the reports. In the first published meta-analysis (Munafo et al., 2009), 14 studies out of 33 reports met the criteria for inclusion in the meta-analysis, but the authors were only able to get data from five of the 14 studies. The meta-analysis was conducted with data from less than half the studies that met the stated inclusion criteria, with a trend for negative studies to be over-represented. Of the 14 studies meeting the inclusion criteria for the meta-analysis, 75% (3/4) of the negative studies and only 20% (2/10) of the positive studies were included in the analysis (Fisher's Exact Test: $p < .10$). A similar problem exists with the second larger-scale meta-analysis (Risch et al., 2009). Although a larger number of studies were included in the second meta-analysis, 12 studies were published after the dataset for this second meta-analysis was closed and analyses were initiated, with only three of the 12 studies including adequate data in the published reports to be examined in the second meta-analysis (Merikangas, 2009). Seven of the nine studies excluded because they were published after the dataset was closed were full or partial replications, as were four other earlier studies that were not included in the second meta-analysis because the data were either not received or received in an incompatible format (Kaufman et al., 2010). Like the first meta-analysis, the second meta-analysis also included an excess representation of nonreplications. It was conducted with data from half of the 26 studies that met the inclusion criteria for the meta-analysis (Risch et al., 2009), with a trend again for negative studies to be over-represented. Of the 26 studies meeting the inclusion criteria for the second meta-analysis, 78% (7/9) of the negative studies and only 35% (6/17) of the positive studies were included in the analysis (Fisher's Exact Test: $p < .10$). In the six months following the publication of the *JAMA* meta-analysis, 12 additional papers were published that examined 5-*HTTLPR* × SLE interactions, leading us to believe that the verdict is still out on whether or not this association is real.

We agree with the arguments of Shanahan and Bauldry (Chapter 3, this volume) that the distribution of severity of life events in the population and the sample size jointly determine the likelihood of replicating G × E associations, and that the conclusions of recent published meta-analytic studies that suggest a significant 5-*HTTLPR* × SLE interaction does not exist are premature. A comprehensive meta-analysis that includes the majority of completed studies was accepted for publication in the *Archives of General Psychiatry* when copy edited versions of this chapter were under review (Karg et al., in press). A total of 54 studies were included in the meta-analysis, and the 5-*HTTLPR* S allele was found to confer a significant increased risk of developing depression in individuals with histories of significant life stress (p=0.00002). When analyses were conducted using only the studies included in the previous published meta-analyses, there was no evidence of this association (Munafo studies p=0.16; Risch studies p=0.11). This suggests that the difference in results between meta-analyses was due to the different set of included studies, not the techniques used to conduct the meta-analyses. Contrary to the results of the smaller earlier published meta-analyses, this new comprehensive meta-analysis found strong evidence in support of the hypothesis that 5-*HTTLPR* moderates the relationship between stress and depression. More research is needed, however, to understand the mechanism by which the S allele of *5-HTTLPR* confers risk.

PROGRAM BACKGROUND

Participants in our research studies were drawn from a larger study examining the efficacy of a State intervention for maltreated children removed from their parents' care due to allegations of abuse or neglect (Permanency Planning: Service Use and Child Outcomes; Principal Investigator, Joan Kaufman, Ph.D.). The maltreated children were recruited within 6 months of their initial out-of-home placement. Approximately equal numbers of maltreated and demographically matched community controls with no history of maltreatment or exposure to intrafamilial violence were recruited for the research. Multiple data sources (e.g., protective service records, parent-report, child-report) were used to quantify children's maltreatment experiences and verify the absence of abuse and domestic violence within the control children's families. Children ranged in age from 5 to 15 years, with a mean age of 9.2. The sample was also approximately evenly divided by sex, and of mixed ethnic origin. Although the maltreated and comparison groups were comparable in terms of racial composition, to prevent spurious associations that can result from variation in allele frequency and prevalence of trait by population, ancestry proportion scores were generated and included as covariates in all analyses. Participants underwent baseline interviews at their place of residence, and the baseline data were collected in one session with the child, and two sessions with the parent or guardian.

Approximately 1 month following the baseline interview, all children attended a 1-week summer day-camp program established specifically for our research purposes. The camp, free of charge to all participants, included 1 to 2 hours of research assessments per day. In the remaining time, children engaged in recreational activities including art, sports, music, and outdoor water games. This data collection procedure allowed for naturalistic observation and comprehensive assessments without overburdening the children. The camps were fun for the children, cost-effective, and promoted strong collaboration with birth parents and protective service workers. DNA specimens were collected from buccal (e.g., saliva) cell specimens while the children were at camp, and a range of standard and well-validated research instruments were completed. At study entry, 67% of the maltreated children met criteria for one or more diagnoses, compared to 18% of the controls. Post-traumatic stress disorder (PTSD) was the most common diagnosis (55%), with a large proportion of the maltreated children also meeting criteria for a depressive (34%) or behavioral (25%)

disorder, and many (41%) expressing suicidal ideation at baseline. Annual follow-ups were conducted in the 2 years following attendance in the research camp program.

The Role of Genetics in Modifying the Outcomes of Maltreated Children

This section reviews studies examining predictors of depression utilizing G × E and gene-by-gene-by-environment (G × G × E) models. Preliminary imaging genomics and translational studies of the effects of early stress that incorporate a G × E framework are also presented.

5-HTTLPR × BDNF × MALTREATMENT PREDICTORS OF DEPRESSION

In an initial study (Kaufman et al., 2004), we replicated the *5-HTTLPR* × maltreatment finding discussed in the introduction that was originally reported by Caspi and colleagues (2003). Given that gene-by-gene interactions have been theorized to contribute to the etiology of depression (Holmans et al., 2004; Kendler & Karkowski-Shuman, 1997), we hypothesized that a polymorphism in the brain-derived neurotrophic factor (*BDNF*) gene might interact with *5-HTTLPR* to further increase the risk for depression in maltreated children. Both *BDNF* (e.g., the protein product of the *BDNF* gene) and serotonin have been implicated in the etiology of depression, and they are known to interact at multiple intra- and intercellular levels (Duman, Heninger, & Nestler, 1997; Malberg, Eisch, Nestler, & Duman, 2000). In addition, *BDNF* genetic variation has been associated with child-onset depression in two independent samples (Strauss, Barr, George et al., 2004; Strauss, Barr, Vetro et al., 2004), although genetic association studies of *BDNF* have produced inconsistent results in studies examining adult-onset major depression (MDD), with results of a recent meta-analysis suggesting that *BDNF* is of greater importance in the development of adult-onset MDD in men than in women (Verhagen et al., 2008). As there are no gender differences in the prevalence of MDD among preadolescents (Birmaher et al., 1996), to date there is no evidence of gender-specific associations with *BDNF* in child samples.

As predicted, we were able to document a significant three-way interaction between *BDNF* genotype, *5-HTTLPR*, and maltreatment history in predicting depression (N = 196). As depicted in Figure 12.1, children with the met allele of the *BDNF* gene and two S alleles of *5-HTTLPR* had the highest depression scores, but the vulnerability associated with these two genotypes was only evident in the maltreated children (Kaufman et al., 2006). Since the publication of our original report, this three-way association has likewise been reported by other investigators (Wichers et al., 2008).

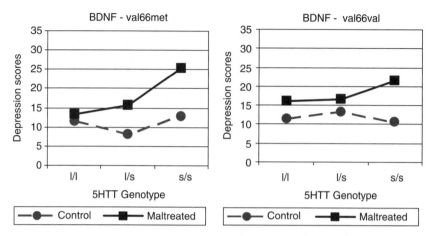

Figure 12.1. **Three-way interaction among maltreatment history, *BDNF*, and *5-HTTLPR* genotype.** The graphs depict the data of the maltreated and control children. Significant three-way interaction existed between *BDNF* genotype, *5-HTTLPR* genotype, and maltreatment history in predicting children's depression scores. Children with the *BDNF* gene Val66Met polymorphism and the SS *5-HTTLPR* genotype had the highest depression scores, with the vulnerability associated with these two genotypes only elevated in the maltreated children. Adapted from Kaufman, J., Yang, B. Z., Douglas-Palumberi, H., Grasso, D., Lipschitz, D., Houshyar, S., et al. (2006). Brain-derived neurotrophic factor-*5-HTTLPR* gene interactions and environmental modifiers of depression in children. *Biological Psychiatry, 59*, 673–680, with permission.

ROBUSTNESS OF THREE-WAY INTERACTION

Our finding of a *BDNF–5-HTTLPR* interaction is consistent with preclinical studies of double mutant mice generated by breeding serotonin transporter protein gene knock-out mice with *BDNF* heterozygous knockout mice. Compared with mice with knock-outs in only one of these systems, double mutant mice show enhanced emotional distress, elevations in stress hormones, and exaggerated deficits in serotonergic availability in the hypothalamus and hippocampus—key brain regions involved in stress reactivity and affect regulation (Ren-Patterson et al., 2005).

To examine the robustness of our findings, secondary analyses were conducted with each population group: European American ($N = 57$), African American ($N = 64$), and Hispanic ($N = 49$) children. The three-way interaction was replicated in the European American and Hispanic cohorts. Because of the lower frequency of the risk alleles in the African American cohort, the sample size was insufficient to examine the three-way interactions.

However, the two-way interaction between 5-*HTTLPR* and maltreatment status was also significant among the African American subjects.

As an additional check of the robustness of the detected three-way interaction between *BDNF* genotype, 5-*HTTLPR* genotype, and maltreatment history, we analyzed the data collected during our two waves of recruitment separately. Cohort One includes the 101 subjects who were available at the time of our earlier publication (Kaufman et al., 2004), and Cohort Two represents the additional 95 subjects recruited afterward. The three-way interaction identified in the entire sample was also statistically significant in Cohort One ($N = 101; \chi^2 = 11.6, p = .02$), and showed a trend toward significance in Cohort Two ($N = 95; \chi^2 = 7.8, p = .07$).

We believe our power to detect $G \times E$ and $G \times G \times E$ associations was enhanced by the extreme severity of the verified maltreatment experiences of the children in our sample, the careful characterization of controls matched on demographic factors but distinct on family functioning measures, and the assessment of depression in close proximity to the children's maltreatment experiences and shortly after the significant stressor of an out-of-home placement.

5-*HTTLPR* × *BDNF* × MALTREATMENT × SOCIAL SUPPORT PREDICTORS OF DEPRESSION

Clinical studies of individuals with a history of abuse suggest that the availability of a caring and stable parent or alternate guardian is one of the most important factors that distinguish abused individuals with good developmental outcomes from those with more deleterious outcomes (Kaufman & Henrich, 2000). Consequently, we examined the effect of social supports in our two $G \times E$ studies predicting depression in maltreated children (Kaufman et al., 2006). Children were asked to name people they (1) talk to about personal things; (2) count on to buy the things they need; (3) share good news with; (4) get together with to have fun; and (5) go to if they need advice. The summary social support measure used was the number of positive support categories listed for the child's top support.

As depicted in Figure 12.2, maltreated children with positive supports had depression scores that were only slightly greater than controls, regardless of genotype. The quality and availability of social supports was extremely potent in reducing risk for depression in maltreated children—with the effect greatest for those maltreated children with the most vulnerable genotypes. Negative sequelae associated with abuse are not inevitable; they can be modified by both genetic and environmental factors. The ability of social supports to modify genetic and environmental risk for depression has likewise been replicated in several independent studies (Kilpatrick et al., 2007; Koenen et al., 2009).

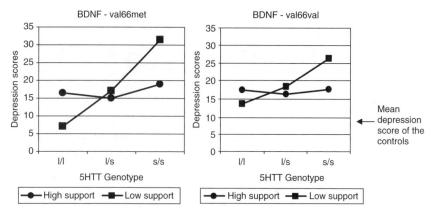

Figure 12.2. **Four-way interaction among maltreatment history,** *BDNF* **genotype,** *5-HTTLPR* **genotype, and social supports: maltreated children's data.** The graphs depict only the data of the maltreated children. The mean score of the controls is indicated on the right, as a frame of reference. The depression scores of the maltreated children with high social supports were close to the mean depression score of the controls, regardless of genotype. The SS genotype was associated with an increase in maltreated children's depression scores, which was greatest for the children without positive supports and the additional presence of the met allele of the *BDNF* polymorphism. Adapted from Kaufman, J., Yang, B. Z., Douglas-Palumberi, H., Grasso, D., Lipschitz, D., Houshyar, S., et al. (2006). Brain-derived neurotrophic factor-*5-HTTLPR* gene interactions and environmental modifiers of depression in children. *Biological Psychiatry, 59*, 673–680, with permission.

GENE–ENVIRONMENT CORRELATIONS

The publication of this study elicited a commentary questioning whether the G × E and G × G × E effects might better be understood by gene–environment correlations (Kaufman, Massey, Krystal, & Gelernter, 2007). As indicated in Figure 12.3, there were no associations between the genotypes we examined and group status. Maltreated children and community controls had comparable proportion of high-risk *5-HTTLPR* and *BDNF* alleles, and if anything, there was a nonsignificant trend for the controls to have a slightly higher proportion of the high-risk val66met allele of the *BDNF* gene. There was also no association between these two genotypes and children's ratings of the quality of their relationship with their primary social support (Social Support Scores by Genotype: L/L: 4.1 ± 1.0; S/L: 4.0 ± 1.1; S/S: 3.8 ± 1.1; val66met: 4.0 ± 1.0; val66val: 4.0 ± 1.0; Wald Chi Square = ns, all contrasts).

We were also asked whether the findings could be explained by sociability characteristics of the child. As part of the larger study, we obtained

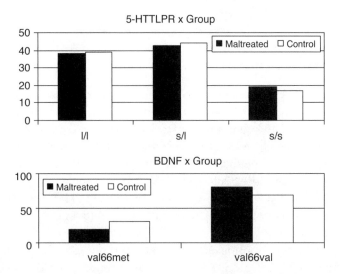

Figure 12.3. **No differences in allele frequencies in maltreated and control children.**

teachers' ratings of children's social competence. One of the measures we collected, the Teacher Child Rating Scale (TCRS; Hightower, 1986), generates a sociability factor score. Depression, social supports, and teacher sociability ratings were available for 153 children. As there was no association between children's depression ratings and the sociability index ($P = 0.06$, ns), and no association between children's ratings of the quality of their relationship with primary social support and the sociability index ($P = -0.09$, ns), this proposed alternative explanation is not supported.

The children in our study were most likely to name an adult as their primary support. Sixty-one percent of the maltreated children and 83% of the controls listed their mothers as their top support, and 30% of the maltreated children and 10% of the controls listed alternative parental figures (e.g., father, stepfather, foster mother), grandparents, or other adult relatives as their primary support. The authors of the commentary, however, were right to suggest that cross-sectional associations between social supports and depressive symptomatology must be interpreted with caution. Prospective longitudinal investigations of the ameliorating effect of social supports on genetic and environmental risk for depression are needed. Egeland and colleagues have conducted a prospective longitudinal study of high-risk children from the third trimester of pregnancy through to young adulthood (Duggal, Carlson, Sroufe, & Egeland, 2001). In their study, measures of supportive early care obtained via observation, and reports of abuse verified with child protective services, independently predicted depression in childhood and adolescence. Additional longitudinal work in this area is needed.

IMAGING GENOMICS RESEARCH

There is growing interest in imaging genomics in clinical and nonclinical samples—examining the relationship between genetic variants and alterations in brain structure and function in order to understand potential mechanisms by which high-risk genes may confer risk for psychopathology. Hariri and colleagues were the first to use functional magnetic resonance imaging (fMRI) to show that the S allele of 5-HTTLPR is associated with increased amygdala activation in response to aversive stimuli (Hariri, et al., 2002). The amygdala is a key structure involved in the brain's orchestrated response to stress (Kaufman, Plotsky, Nemeroff, & Charney, 2000), and is a critical component of the emotion and reward processing circuits in the brain, interconnected neural circuits implicated in the etiology of MDD (Drevets, Price, & Furey, 2008; Krishnan & Nestler, 2008). Since Hariri's initial publication, 17 studies have investigated this association, and results of a recent meta-analysis supports the conclusion that 5-HTTLPR variation moderates amygdala activation in response to a range of negative stimuli (Munafo, Brown, & Hariri, 2008).

In our imaging genomics work, we chose to use a dichotic listening prosody task developed by Vuilleumier and colleagues (Grandjean et al., 2005; Sander et al., 2005). Male and female actors created the stimuli for the prosody task using nonsense words that were spoken in either an angry or neutral tone of voice (e.g., "goster" and "niuvenci"). Participants simultaneously heard two stimuli, one spoken by a male in one ear, and one spoken by a female in the other ear. Children were asked to attend to either the left or right ear and indicate whether a man or woman spoke the word in that ear. Across blocks of trials, the to-be-attended-to ear was indicated via the presentation of the letters L and R, for left and right ear, respectively.

This implicit emotion-processing task, like other emotion-processing tasks used in previous studies, elicits amygdala activation when attention is focused on the angry prosody (Grandjean et al., 2005; Sander et al., 2005). In addition, performance of the dichotic listening task requires interhemispheric transfer through the posterior portion of the corpus callosum (Westerhausen et al., 2006), a region that has been shown in several independent samples to be altered in maltreated children and adolescents (De Bellis et al., 1999, 2002; Jackowski et al., 2008; Teicher, Andersen, Polcari, Anderson, & Navalta, 2002).

The children ($N = 18$) for this pilot project were recruited from our prior G × E studies. Participants were stratified in a 2 × 2 factorial design according to maltreatment history (e.g., maltreated vs. control) and genetic risk (e.g., SS vs. LL) (Herrington et al., 2007). To test for differences in activation during emotion processing of negative stimuli, an analysis was conducted comparing areas of activation during the attended angry versus nonattended angry conditions. The results of the voxel-based random effects analyses showed significant genotype effects in the left amygdala, with significantly greater activation in SS compared to LL subjects. No main effect for group was evident in the maltreated versus control contrast, which may be due to the fact that the scans were completed 2 years after recruitment for the G × E studies, and 42% of the maltreated children included in this pilot study no longer met full criteria for PTSD or any other psychiatric diagnosis.

In terms of the interhemispheric transfer of information, during the dichotic listening task, auditory information sent along ipsilateral pathways seems to be blocked or suppressed by information from the contralateral pathways (Plessen et al., 2007; Springer & Gazzaniga, 1975), such that auditory input received via the right ear in the forced right-ear-attend condition is initially processed in the left hemisphere. Processing of paralinguistic information (e.g., gender determination, prosody), however, is lateralized to the right hemisphere in the superior temporal gyrus (STG). Therefore, completion of this task in the right-ear-attend condition requires transfer of stimulus information from the left hemisphere to the right STG via the posterior region of the corpus callosum. Activation in the right STG in the right-forced-ear attend condition provides a test of interhemispheric transfer of information and corpus callosum integrity. During the right-ear-attend condition, when compared with maltreated children, control children showed significantly greater activation of the right STG. Group differences in this same region were also evident when comparing children with two S alleles to children with two L alleles of 5-HTTLPR. As predicted, however, there were no differences in the activation of the right STG in the left-ear-attend condition when interhemispheric transfer of information is not necessary for processing the prosody stimuli. Imaging genomics research, especially when conducted in conjunction with translational research studies, holds significant promise in unraveling the mechanisms by which alleles of certain genes and experiences of maltreatment confer vulnerability to psychopathology.

TRANSLATIONAL RESEARCH STUDIES

Preclinical (e.g., animal) studies of the effects of stress provide a valuable heuristic in understanding the pathophysiology of depression and other stress-related psychiatric illnesses (Gorman, Mathew, & Coplan, 2002; Heim, Owens, Plotsky, & Nemeroff, 1997), with many of the biological alterations associated with early stress in preclinical studies reported in adults with MDD. Preclinical studies also allow for experimental manipulation of the environment and examination of molecular mechanisms that are simply not ethically or physically possible in living clinical populations.

To further our understanding on the effects of stress on corpus callosum development, we formed a collaboration with Dr. Jeremy Coplan and colleagues. Coplan and colleagues developed an experimental model of early stress in nonhuman primates called *variable foraging demand* (VFD). Variable foraging demand rearing is an early-life stress paradigm in which infant bonnet macaques are reared by mothers undergoing an experimentally induced "perception" of food unavailability (Rosenblum et al., 1994). Although adequate food is always available, the ease with which food is accessed is variable and unpredictable for 12–16 weeks within the first year of the infant's life. At times, food is easily accessed; at other times, mothers must forage for food in a bin of wood chips.

Infants are assigned randomly to VFD or non-VFD rearing conditions, a manipulation that would never be ethical to pursue in human populations. There are no differences in weight between VFD and non-VFD mothers or infants, but VFD conditions are stressful for mothers. Stress is transmitted to the offspring via disruption of maternal attachment and affective reciprocity, with increased stress hormones observed in both mothers and infants (Coplan, Kaufman, Shorman, et al., 2005). Variable foraging demand rearing is associated with pervasive and sustained biobehavioral effects throughout the developmental trajectory of the offspring and into adulthood that closely resemble the biobehavioral abnormalities associated with human anxiety and depressive disorders (Coplan et al., 2005; Gorman et al., 2002).

Coplan and colleagues had data from 23 young adult male subjects who underwent MRI scans: 13 were reared under the VFD condition (eight LL and five SL or SS), and nine age-matched male subjects were normally reared (seven LL and two SL subjects). Corpus callosum area measurements were obtained at the midsagittal slice using validated procedures (Witelson, 1976). Exploratory analyses revealed a significant effect for rearing condition and a gene × rearing effect in total corpus callosum area, with VFD-reared subjects with an S allele having the smallest corpus callosum area (Jackowski et al., in press).

Postmortem microarray studies are planned to look at changes in gene expression in key cortical areas. We are especially interested in examining the expression of myelin-related genes, given the prominence of white matter changes reported in neuroimaging studies of maltreated children and adolescents (De Bellis et al., 1999, 2002; Jackowski et al., 2008; Jackowski et al., in press; Teicher et al., 2004), and results of postmortem studies of adults with depression documenting oligodendrocyte reduction in the amygdala (Hamidi,

Drevets, & Price, 2004) and decreased expression of 17 genes related to oligodendrocyte function in the temporal cortex (Barley, Dracheva, & Byne, 2009). Oligodendrocytes are the myelinating cells of the central nervous system (CNS) that arise from progenitors in the subventricular zone and undergo a well-regulated process of proliferation, migration, and differentiation (Miller, 1996). One hypothesis emerging from our research is that genes involved in oligodendrocyte proliferation, migration, and differentiation may have a role in moderating the effects of stress.

It is the hope that these translational studies will identify novel candidate genes that can be examined in our clinical samples. Unfortunately, at the time of the writing of this chapter, we were 4 years and 3 months into a moratorium on our G × E work. With a change in leadership of the State Institutional Review Board, a moratorium was imposed on our work that letters of support from the Directors of the National Institute of Mental Health and National Institute on Alcohol Abuse and Alcoholism, the Deans of the two medical schools in our state, the Child Advocate, and many others failed to affect. We have been forbidden to recruit new maltreated children, further characterize stored DNA specimens, or freely analyze data we had permission to collect. We are currently pursuing permission to reinitiate our studies in another jurisdiction.

FUTURE DIRECTIONS: CLINICAL AND RESEARCH IMPLICATIONS

An important take-home message from the G × E studies conducted to date is that neither high-risk genes, nor high-risk environments guarantee deleterious outcomes. Positive factors in the environment can reduce risk, and the more we understand about the relevant risk, protective factors, and mechanisms involved, the more opportunity there will be to optimize child outcomes. It is our firm belief that identifying susceptibility genes and how environmental factors may interact with them will increase our capacity to identify those at risk for depression and other stress-related psychiatric illnesses, identify targets for developing interventions to preempt the onset of these problems, and guide efforts to tailor and personalize treatments for individuals with these disorders.

Medical treatments will not necessarily dominate the future of care for these problems, as in many areas of medicine understanding risk has led to nonmedical interventions to prevent disease onset. Phenylketonuria (PKU) is a good illustration of this point. Phenylketonuria is a genetic disorder caused by an enzyme deficiency that affects the metabolism of the amino acid phenylalanine, causing toxic byproducts to build up, damage to the infant brain, and mental retardation. In PKU, mental retardation can be prevented by controlling the environment, by eliminating phenylalanine from the diet. Unlike in PKU, depression is not caused by a single gene, and there is not one environmental factor that can eliminate risk for depression, but we are beginning to unravel the genetic and environmental factors associated with risk and resiliency in maltreated children.

As discussed previously, in our studies, the availability of positive social supports ameliorated risk for depression in maltreated children, with this effect greatest for maltreated children at highest genetic risk for depression. There are probably multiple mechanisms by which social supports and the availability of a positive stable adult relationship may ameliorate risk for depression in maltreated children. Data from preclinical (e.g., animal) studies suggest that maternal behavior can produce stable changes in DNA methylation and chromatin structure of the glucocorticoid receptor gene promoter in the hippocampus (Weaver et al., 2004). These epigenetic changes and the subsequent alteration in hypothalamic-pituitary-adrenal (HPA) axis response to stress may be one important mechanism by which variations in maternal behavior/social supports alter risk for stress-related disorders.

Epigenetic changes associated with maternal care have been studied most extensively in two breeds of rats—one that displays optimal parenting, evidenced by high licking and grooming (high-LG) behavior; and the other that evidences low-LG behavior (Weaver et al., 2004; Weaver, Meaney, & Szyf, 2006). Low-LG is associated with depressive and anxious behavior phenotype in the offspring, increased stress reactivity, and methylation and decreased expression of the glucocorticoid receptor gene in the hippocampus. This gene is always methylated in the offspring of low-LG mothers, and rarely methylated in those of high-LG dams. Cross-fostering low-LG offspring with high-LG dams reverses the differences in the methylation and suggests a direct relationship between maternal behavior and changes in DNA methylation and gene expression.

Evidence that variation in early care may also lead to epigenetic changes in humans is derived from a recent study of postmortem hippocampus tissue obtained from suicide victims with a history of childhood abuse, suicide victims with no childhood abuse, and nonabused nonpsychiatric controls (McGowan et al., 2009). Consistent with preclinical studies, samples from abused suicide victims showed increased cytosine methylation of the glucocorticoid receptor promoter, decreased transcription factor binding, and decreased gene transcription compared to nonabused suicide victims and nonsuicide controls. These data add to a growing body of evidence that suggests the neurobiological mechanisms responsible for depression in individuals with a history of early trauma may be different from the mechanisms responsible for depression in individuals without adverse early experiences.

Better understanding of the mechanisms by which early adversity confers risk for depression and other stress-related psychiatric disorders may lead to novel foci for intervention efforts and more personalized approaches to treatment selection. Both pharmacological and environmental interventions may be indicated. For example, recent data suggest that the epigenetic changes associated with variation in maternal care can be reversed in adult animals via infusion of the histone deacetylase inhibitor l-methionine or trichostatin A (Weaver et al., 2006). As a result of this research, histone deacetylase inhibitors are currently under investigation as a novel treatment strategy for depression (Covington et al., 2009).

Environmental enrichment in adolescent animals has also been found to reverse HPA axis hyperactivity, memory deficits, and other behavioral alterations associated with early nonoptimal rearing, with the reversal of many of these effects apparently mediated by enhancement of glutamate N-methyl-D-aspartate (NMDA) and α-amino-3hydroxy-5methly-4isoxazolepropionic acid (AMPA) receptor subunit gene expression (Bredy et al., 2003, 2004). Improvement in stress and behavioral measures is not associated with reversal of glucocorticoid receptor changes, but compensatory changes in the glutamate system. It is becoming increasingly evident that multiple neurochemical systems can be altered to modify the behavioral and biological sequelae associated with early stress, and that environmental manipulations can lead to neurobiological changes.

What are the clinical implications of the emerging findings for intervention research? We are not at the point where we can prioritize children for interventions based on genotype, but we'll never get there if genetic measures are not included in intervention studies. Likewise, to identify those at highest risk for disorder, it is necessary that environmental measures be included in genetic studies, and that epigenetic mechanisms be examined in clinical and basic research.

CONCLUSION

As noted previously, the line between genes and environment has blurred in the past decade. It is our firm belief that identifying susceptibility genes and how environmental factors may interact with them will increase our capacity to identify those at risk for stress-related psychiatric illnesses, identify targets for developing interventions to prevent the onset of these problems, and guide efforts to tailor and personalize treatments for individuals with these disorders. Multidisciplinary efforts that allow for cross-fertilization between basic and clinical investigators hold significant promise for delineating mechanism of risk and identifying novel foci for future clinical research efforts. We are committed to ongoing work with this population—with foci of research that spans neurobiology to social policy. There is much left to learn to optimize the outcomes of maltreated children, and we believe G × E clinical and translational research approaches hold much promise.

REFERENCES

Barley, K., Dracheva, S., & Byne, W. (2009). Subcortical oligodendrocyte- and astrocyte-associated gene expression in subjects with schizophrenia, major depression and bipolar disorder. *Schizophrenia Research, 112*(1–3), 54–64.

Birmaher, B., Ryan, N. D., Williamson, D. E., Brent, D. A., Kaufman, J., Dahl, R. E., et al. (1996). Childhood and adolescent depression: a review of the past 10 years. Part I. *Journal of the American Academy of Child and Adolescent Psychiatry, 35*(11), 1427–1439.

Bredy, T. W., Humpartzoomian, R. A., Cain, D. P., Meaney, M. J., Morley-Fletcher, S., Rea, M., et al. (2003). Partial reversal of the effect of maternal care on cognitive function through environmental enrichment. *European Journal of Neuroscience, 18*(12), 571–576.

Bredy, T. W., Zhang, T. Y., Grant, R. J., Diorio, J., Meaney, M. J., Humpartzoomian, R. A., et al. (2004). Peripubertal environmental enrichment reverses the effects of maternal care on hippocampal development and glutamate receptor subunit expression. *European Journal of Neuroscience, 20*(5), 1355–1362.

Carey, B. (2009). Report on gene for depression is now faulted. *New York Times, 2009 Jun 17.*

Caspi, A., McClay, J., Moffitt, T. E., Mill, J., Martin, J., Craig, I. W., et al. (2002). Role of genotype in the cycle of violence in maltreated children. *Science, 297*(5582), 851–854.

Caspi, A., Sugden, K., Moffitt, T. E., Taylor, A., Craig, I. W., Harrington, H., et al. (2003). Influence of life stress on depression: moderation by a polymorphism in the 5-HTT gene. *Science, 301,* 386–389.

Cicchetti, D., & Toth, S. L. (2005). Child maltreatment. *Annual Review of Clinical Psychology, 1,* 409–438.

Coplan, J. D., Altemus, M., Mathew, S. J., Smith, E. L., Sharf, B., Coplan, P. M., et al. (2005). Synchronized maternal-infant elevations of primate CSF CRF concentrations in response to variable foraging demand. *CNS Spectrums, 10*(7), 530–536.

Coplan, J. D., Kaufman, D., Shorman, I., Smith, E. L., Owens, M. J. Nemeroff, C. B., et al. (2005). *Variable foraging demand (VFD) exposure of primate maternal-infant dyads and impaired insulin action in juvenile offspring.* Paper presented at the American College of Neuropsychopharmacology Annual Meeting, 2005 Dec 13, Waikoloa, Hawaii.

Covington, H. E., 3rd, Maze, I., LaPlant, Q. C., Vialou, V. F., Ohnishi, Y. N., Berton, O., et al. (2009). Antidepressant actions of histone deacetylase inhibitors. *Journal of Neuroscience, 29*(37), 11451–11460.

De Bellis, M. D., Keshavan, M. S., Clark, D. B., Casey, B. J., Giedd, J. N., Boring, A. M., et al. (1999). Developmental traumatology. *Part II: brain development. Biological Psychiatry, 45*(10), 1271–1284.

De Bellis, M. D., Keshavan, M. S., Shifflett, H., Iyengar, S., Beers, S. R., Hall, J., et al. (2002). Brain structures in pediatric maltreatment-related posttraumatic stress disorder: a sociodemographically matched study. *Biological Psychiatry, 52*(11), 1066–1078.

Drevets, W. C., Price, J. L., & Furey, M. L. (2008). Brain structural and functional abnormalities in mood disorders: implications for neurocircuitry models of depression. *Brain Structure and Function, 213*(1-2), 93–118.

Duggal, S., Carlson, E. A., Sroufe, L. A., & Egeland, B. (2001). Depressive symptomatology in childhood and adolescence. *Developmental Psychopathology, 13*(1), 143–164.

Duman, R. S., Heninger, G. R., & Nestler, E. J. (1997). A molecular and cellular theory of depression [see comments]. *Archives of General Psychiatry, 54*(7), 597–606.

Elmer, E. (1977). A follow-up study of traumatized children. *Pediatrics, 59*(2), 273–279.

Famularo, R., Fenton, T., Kinscherff, R., & Augustyn, M. (1996). Psychiatric comorbidity in childhood post traumatic stress disorder. *Child Abuse and Neglect, 20*(10), 953–961.

Famularo, R., Kinscherff, R., & Fenton, T. (1992). Psychiatric diagnoses of maltreated children: preliminary findings. *Journal of the American Academy of Child and Adolescent Psychiatry, 31*, 863–867.

Garland, A., Hough, R., McCabe, K., Yeh, M., Wood, P., & Aarons, G. (2001). Prevalence of psychiatric disorders in youths across five sectors of care. *Journal of the American Academy of Child and Adolescent Psychiatry, 40*(4), 409–418.

Gorman, J. M., Mathew, S., & Coplan, J. (2002). Neurobiology of early life stress: nonhuman primate models. *Seminars in Clinical Neuropsychiatry, 7*(2), 96–103.

Grandjean, D., Sander, D., Pourtois, G., Schwartz, S., Seghier, M. L., Scherer, K. R., et al. (2005). The voices of wrath: brain responses to angry prosody in meaningless speech. *Nature. Neuroscience, 8*(2), 145–146.

Hamidi, M., Drevets, W. C., & Price, J. L. (2004). Glial reduction in amygdala in major depressive disorder is due to oligodendrocytes. *Biological Psychiatry, 55*(6), 563–569.

Hariri, A. R., Mattay, V. S., Tessitore, A., Kolachana, B., Fera, F., Goldman, D., et al. (2002). Serotonin transporter genetic variation and the response of the human amygdala. *Science, 297*(5580), 400–403.

Heim, C., Owens, M. J., Plotsky, P. M., & Nemeroff, C. B. (1997). The role of early adverse life events in the etiology of depression and posttraumatic stress disorder. *Focus on corticotropin-releasing factor. Annals of the New York Academy of Science, 821*, 194–207.

Herrington, J., Douglas-Palumberi, H., Meadows, A., Gelernter, J., Kaffman, A., Coplan, J., et al. (2007). *fMRI Study of Prosody Processing in Maltreated Children.*

Paper presented at the Annual Meeting of the Society of Biological Psychiatry, 2007, May, San-Diego, CA.

Hightower, A. (1986). The teacher-child rating scale: a brief objective measure of elementary children's school problem behaviors and competencies. *School Psychology Review, 15*(3), 393–409.

Holmans, P., Zubenko, G. S., Crowe, R. R., DePaulo, J. R., Jr., Scheftner, W. A., Weissman, M. M., et al. (2004). Genomewide significant linkage to recurrent, early-onset major depressive disorder on chromosome 15q. *American Journal of Human Genetics, 74*(6), 1154–1167.

Jackowski, A., Douglas-Palumberi, H., Jackowski, M., Win, L., Schultz, R. T., Staib, L. H., et al. (2008). Corpus callosum in maltreated children with PTSD: a diffusion tensor imaging study. *Psychiatry Research: Neuroimaging, 162*(3), 256–261.

Jackowski, A., Perera, T., Garrido, G., Tang, C. Y., Martinez, J., Sanjay, M., et al. (in press). Early life stress, corpus callosum development, and anxious behavior in nonhuman primates. *Psychiatry Research Neuroimaging.*

Karg, K., Shedden, K., Burmeister, M., & Sen, S. (in press). The serotonin transporter promoter variant (5-HTTLPR), stress, and depression meta-analysis revisited: Evidence of genetic moderation. *Archives of General Psychiatry.*

Kaufman, J. (1991). Depressive disorders in maltreated children. *Journal of the American Academy of Child and Adolescent Psychiatry, 30*(2), 257–265.

Kaufman, J., Gelernter, J., Kaffman, A., Caspi, A., & Moffitt, T. E. (2010). Arguable assumptions, questionable conclusions. *Biological Psychiatry 67*(4): 19–20.

Kaufman, J., & Henrich, C. (2000). Exposure to violence and early childhood trauma. In C. Zeanah Jr. (Ed.), *Handbook of infant mental health* (pp. 195–207). New York: Guilford Press.

Kaufman, J., Massey, J., Krystal, J., & Gelernter, J. (2007). BDNF, serotonin transporter and depression: response to Kalueff and colleagues: reply. *Biological Psychiatry,* 1112–1113.

Kaufman, J., Plotsky, P., Nemeroff, C., & Charney, D. (2000). Effects of early adverse experience on brain structure and function: clinical implications. *Biological Psychiatry, 48*(8), 778–790.

Kaufman, J., Yang, B. Z., Douglas-Palumberi, H., Grasso, D., Lipschitz, D., Houshyar, S., et al. (2006). Brain-derived neurotrophic factor-5-HTTLPR gene interactions and environmental modifiers of depression in children. *Biological Psychiatry, 59,* 673–680.

Kaufman, J., Yang, B. Z., Douglas-Palumberi, H., Houshyar, S., Lipschitz, D., Krystal, J., et al. (2004). Social supports and serotonin transporter gene moderate depression in maltreated children. *Proceedings of the National Academy of Sciences of the U.S.A., 101*(49), 17316–17321.

Kendler, K. S., & Karkowski-Shuman, L. (1997). Stressful life events and genetic liability to major depression: genetic control of exposure to the environment? *Psychological Medicine, 27*(3), 539–547.

Kilpatrick, D., Ruggiero, K., Acierno, R., Saunders, B., Resnick, H., & Best, C. (2003). Violence and risk of PTSD, major depression, substance abuse/dependence, and comorbidity: results from the National Survey of Adolescents. *Journal of Consulting and Clinical Psychology, 71*(4), 692–700.

Kilpatrick, D. G., Koenen, K. C., Ruggiero, K. J., Acierno, R., Galea, S., Resnick, H. S., et al. (2007). The serotonin transporter genotype and social support and moderation of posttraumatic stress disorder and depression in hurricane-exposed adults. *American Journal of Psychiatry, 164*(11), 1693–1699.

Koenen, K. C., Aiello, A. E., Bakshis, E., Amstadter, A. B., Ruggiero, K. J., Acierno, R., et al. (2009). Modification of the association between serotonin transporter genotype and risk of posttraumatic stress disorder in adults by county-level social environment. The serotonin transporter genotype and social support and moderation of posttraumatic stress disorder and depression in hurricane-exposed adults. *American Journal of Epidemiology, 169*(6), 704–711.

Kolko, D. (2002). Child Physical Abuse. In J. Myers, L. Berliner, J. Briere, C. T. Hendrix, C. Jenny, & T. A. Reid (Eds.), *The APSAC handbook on child maltreatment* (2nd ed., pp. 21–54). Thousand Oaks, CA: Sage Publications, Inc.

Krishnan, V., & Nestler, E. J. (2008). The molecular neurobiology of depression. *Nature, 455*(7215), 894–902.

Lanktree, C., Briere, J., & Zaidi, L. (1991). Incidence and impact of sexual abuse in a child outpatient sample: the role of direct inquiry. *Child Abuse and Neglect, 15*(4), 447–453.

Lesch, K. P., Gross, J., Franzek, E., Wolozin, B. L., Riederer, P., & Murphy, D. L. (1995). Primary structure of the serotonin transporter in unipolar depression and bipolar disorder. *Biological Psychiatry, 37*(4), 215–223.

Malberg, J. E., Eisch, A. J., Nestler, E. J., & Duman, R. S. (2000). Chronic antidepressant treatment increases neurogenesis in adult rat hippocampus. *Journal of Neuroscience, 20*(24), 9104–9110.

McClellan, J., Adams, J., Douglas, D., McCurry, C., & Storck, M. (1995). Clinical characteristics related to severity of sexual abuse: a study of seriously mentally ill youth. *Child Abuse and Neglect, 19*(10), 1245–1254.

McGowan, P. O., Sasaki, A., D'Alessio, A. C., Dymov, S., Labonte, B., Szyf, M., et al. (2009). Epigenetic regulation of the glucocorticoid receptor in human brain associates with childhood abuse. *Nature. Neuroscience, 12*(3), 342–348.

Merikangas, K. R. (2009). Meta-Analysis Question. In J. Kaufman (Ed.). New Haven, CT.

Miller, R. H. (1996). Oligodendrocyte origins. *Trends in Neuroscience, 19*(3), 92–96.

Molnar, B. E., Buka, S. L., & Kessler, R. C. (2001). Child sexual abuse and subsequent psychopathology: results from the National Comorbidity Survey. *American Journal of Public Health, 91*(5), 753–760.

Munafo, M. R., Brown, S. M., & Hariri, A. R. (2008). Serotonin transporter (5-HT-TLPR) genotype and amygdala activation: a meta-analysis. *Biological Psychiatry, 63*(9), 852–857.

Munafo, M. R., Durrant, C., Lewis, G., & Flint, J. (2009). Gene x environment interactions at the serotonin transporter locus. *Biological Psychiatry, 63*(3), 211–219.

Pelcovitz, D., Kaplan, S. J., DeRosa, R. R., Mandel, F. S., & Salzinger, S. (2000). Psychiatric disorders in adolescents exposed to domestic violence and physical abuse. *American Journal of Orthopsychiatry, 70*(3), 360–369.

Plessen, K. J., Lundervold, A., Gruner, R., Hammar, A., Peterson, B. S., & Hugdahl, K. (2007). Functional brain asymmetry, attentional modulation, and interhemispheric transfer in boys with Tourette syndrome. *Neuropsychologia, 45*(4), 767–774. Epub 2006 Oct 11.

Ren-Patterson, R. F., Cochran, L. W., Holmes, A., Sherrill, S., Huang, S. J., Tolliver, T., et al. (2005). Loss of brain-derived neurotrophic factor gene allele exacerbates brain monoamine deficiencies and increases stress abnormalities of serotonin transporter knockout mice. *Journal of Neuroscience Research, 79*(6), 756–771.

Risch, N., Herrell, R., Lehner, T., Liang, K. Y., Eaves, L., Hoh, J., et al. (2009). Interaction between the serotonin transporter gene (5-HTTLPR), stressful life events, and risk of depression: a meta-analysis. [Meta-Analysis Research Support, N.I.H., Extramural Research Support, N.I.H., Intramural]. *Journal of the American Medical Association, 301*(23), 2462–2471.

Rosenblum, L. A., Coplan, J. D., Friedman, S., Bassoff, T., Gorman, J. M., & Andrews, M. W. (1994). Adverse early experiences affect noradrenergic and serotonergic functioning in adult primates. *Biological Psychiatry, 35*(4), 221–227.

Ruggiero, K., McLeer, S., & Dixon, J. (2000). Sexual abuse characteristics associated with survivor psychopathology. *Child Abuse and Neglect, 24*(7), 951–964.

Rutter, M., Giller, H., & Hagell, A. (1998). *Antisocial behavior by young people.* Cambridge: Cambridge University Press.

Sander, D., Grandjean, D., Pourtois, G., Schwartz, S., Seghier, M. L., Scherer, K. R., et al. (2005). Emotion and attention interactions in social cognition: brain regions involved in processing anger prosody. *Neuroimage, 28*(4), 848–858.

Schuck, A. M., & Widom, C. S. (2003). Childhood victimization and alcohol symptoms in women: an examination of protective factors. *Journal of Studies on Alcohol, 64*(2), 247–256.

Springer, S. P., & Gazzaniga, M. S. (1975). Dichotic testing of partial and complete split brain subjects. *Neuropsychologia, 13*(3), 341–346.

Strauss, J., Barr, C. L., George, C. J., King, N., Shaikh, S., Devlin, B., et al. (2004). Association study of brain-derived neurotrophic factor in adults with a history of childhood onset mood disorder. *American Journal of Medical Genetics. Part B, Neuropsychiatric Genetics, 131*(1), 16–19.

Strauss, J., Barr, C. L., Vetro, A., King, N., Shaikh, S., Brathwaite, J., et al. (2004). *Brain derived neurotrophic factor gene and childhood-onset depressive disorder: results from a Hungarian sample.* Paper presented at the Collegium International Neuro-Psychopharmacologicum (CINP) Congress, Paris, 2004 Jun 21–22.

Teicher, M. H., Andersen, S. L., Polcari, A., Anderson, C. M., & Navalta, C. P. (2002). Developmental neurobiology of childhood stress and trauma. *The Psychiatric Clinics of North America, 25*(2), 397–426, vii-viii.

Teicher, M. H., Dumont, N. L., Ito, Y., Vaituzis, C., Giedd, J. N., & Andersen, S. L. (2004). Childhood neglect is associated with reduced corpus callosum area. *Biological Psychiatry, 56*(2), 80–85.

Verhagen, M., van der Meij, A., van Deurzen, P. A., Janzing, J. G., Arias-Vasquez, A., Buitelaar, J. K., et al. (2008). Meta-analysis of the BDNF Val66Met polymorphism in major depressive disorder: effects of gender and ethnicity. *Molecular Psychiatry, 14*, 14.

Weaver, I. C., Cervoni, N., Champagne, F. A., D'Alessio, A. C., Sharma, S., Seckl, J. R., et al. (2004). Epigenetic programming by maternal behavior. *Nature Neuroscience, 7*(8), 847–854.

Weaver, I. C., Meaney, M. J., & Szyf, M. (2006). Maternal care effects on the hippocampal transcriptome and anxiety-mediated behaviors in the offspring that are reversible in adulthood. *Proceedings of the National Academy of Sciences of the U.S.A., 103*(9), 3480–3485. Epub 2006 Feb 16.

Westerhausen, R., Woerner, W., Kreuder, F., Schweiger, E., Hugdahl, K., & Wittling, W. (2006). The role of the corpus callosum in dichotic listening: a combined morphological and diffusion tensor imaging study. *Neuropsychology, 20*(3), 272–279.

Wichers, M., Kenis, G., Jacobs, N., Mengelers, R., Derom, C., Vlietinck, R., et al. (2008). The BDNF Val(66)Met x 5-HTTLPR x child adversity interaction and depressive symptoms: An attempt at replication. *American Journal of Medical Genetics. Part B, Neuropsychiatric Genetics, 147B*(1), 120–123.

Widom, C. (1989). The cycle of violence. *Science, 244*(4901), 160–166.

Widom, C., & Ames, M. A. (1994). Criminal consequences of childhood sexual victimization. *Child Abuse and Neglect, 18*(4), 303–318.

Widom, C. S., Ireland, T., & Glynn, P. J. (1995). Alcohol abuse in abused and neglected children followed-up: are they at increased risk? *Journal of Studies on Alcohol, 56*(2), 207–217.

Witelson, S. F. (1976). Sex and the single hemisphere: specialization of the right hemisphere for spatial processing. *Science, 193*(4251), 425–427.

Zigler, E. (1980). Controlling child abuse: Do we have the knowledge and/or the will? In C. R. G. Gerbner, & E. Zigler (Eds.), *Child abuse: an agenda for action* (pp. 293–304). New York: Oxford Press.

13

Genes, Environment, and Adolescent Smoking: Implications for Prevention

JANET AUDRAIN-MCGOVERN AND KENNETH P. TERCYAK

Adolescent tobacco use is a significant public health problem. It is estimated that 1,000 adolescents start smoking regularly each day. This translates to 400,000 new adolescent daily smokers in the United States every year—many of whom will eventually die from their addiction (SAMHSA, 2008). Approximately 90% of adults who have regularly smoked cigarettes began smoking during adolescence (SAMHSA, 2008). The prevention of smoking initiation (i.e., beginning to experiment with cigarettes) and the transition to regular smoking is critical to decrease the morbidity and mortality associated with cigarette smoking (including heart disease, cancer, and stroke). According to national survey data, approximately 50% of adolescents report ever having smoked cigarettes, 20% report smoking cigarettes regularly (at least one cigarette in the past 30 days), and over 8% report smoking cigarettes frequently (at least 20 of the past 30 days) (CDC, 2008). Importantly, the percentage of adolescent regular smokers doubles, and the percentage of adolescent frequent or daily smokers triples from mid to late adolescence, making this the most vulnerable period in young people's lives for the adoption of a dangerous and lifelong habit (CDC, 2008; Johnston, O'Malley, Bachman, & Schulenberg, 2008).

The mid to late adolescent developmental stage is important for smoking initiation and progression. Among adolescents who experiment with cigarette smoking, a portion will progress to regular smoking, whereas others will not progress beyond experimentation. From a public health standpoint, it is critical to identify adolescents early who are most likely to progress to regular smoking and the factors that facilitate smoking uptake, so that targeted prevention can take place. Valuable information may also be derived from those adolescents who do not progress along the uptake continuum, offering new insights into factors that protect against smoking progression. Research indicates that there is significant heterogeneity in the timing of onset, rate of progression, and magnitude of smoking, as well as in the emergence of nicotine dependence among adolescents (Audrain-McGovern, Rodriguez et al., 2004; Colder et al., 2001; Hu, Muthen, Schaffran, Griesler, & Kandel, 2008; Soldz & Cui, 2002).

To continue to develop and refine more effective adolescent smoking prevention programs, a greater understanding of the risk and protective factors for smoking uptake is needed. Although a large body of research describes environmental influences on adolescent smoking uptake (Mayhew, Flay, & Mott, 2000), far less is known about biological influences and the interplay between biology and the environment. Research suggests that genetic factors play an important role in determining smoking behavior. Heritability studies of adolescent twins estimate that about 33% of the variance in smoking initiation (ever smoking), 80% of the variance in smoking rate, 44% in the frequency of smoking (days of smoking in the past 30 days), and 44% in nicotine dependence may be attributable to genetic factors (Boomsma, Koopmans, Van Doornen, & Orlebeke, 1994; Han, McGue, & Iacono, 1999; Koopmans, Slutske, Heath, Neale, & Boomsma, 1999; McGue, Elkins, & Iacono, 2000; Slomkowski, Rende, Novak, Lloyd-Richardson, & Niaura, 2005). Whereas environment plays a stronger role in smoking initiation, genetic factors appear to play a greater role in determining whether an adolescent progresses to regular smoking and nicotine dependence than whether an adolescent experiments with cigarettes but does not progress (Koopmans et al., 1999; McGue et al., 2000). This suggests that both the environment and biology have important roles in the adolescent smoking spectrum. Moreover, specific susceptibility genes may also interact with environmental factors to influence the rate of onset of adolescent smoking behavior. These interactions may serve to facilitate (increase risk for) or buffer (confer protection against) smoking progression among those who have been exposed to nicotine.

This chapter summarizes the evidence base supporting the role of genetic factors in adolescent cigarette smoking, and the interplay between genetic and environmental factors in smoking initiation and maintenance early in the lifespan. It also describes how the identification of social and behavioral (environmental) processes, and the interventions applied to these processes, might alter smoking trajectories from adolescence to young adulthood. This chapter includes a description of a biobehavioral model of adolescent smoking that serves as an overarching framework for integrating research on genetic and environmental smoking influences on adolescent smoking behavior. Research on genetic influences on smoking and the evidence for gene–environment interactions in adolescent smoking trajectories is then summarized in light of this framework. Finally, we discuss how insights from biobehavioral research might be used to develop more effective adolescent smoking prevention and intervention approaches, and the opportunities and challenges of applying such approaches with adolescent populations.

BIOBEHAVIORAL MODEL OF ADOLESCENT SMOKING

We begin by presenting a biobehavioral model of adolescent smoking. This model is presented to provide a broad conceptual framework for integrating research on genetic influences and environmental influences on adolescent smoking behavior, and the impact of their interactions on adolescent smoking uptake. The model is an adaptation and integration of a biobehavioral model of smoking progression and persistence (Audrain-McGovern, Nigg, & Perkins, 2009; Lerman & Niaura, 2002).

Adolescence is a developmental period during which a number of important changes in brain structure and function are occurring. The prefrontal cortex (planning and executive function) and mesolimbic regions (emotional regulation and learning) of the forebrain, and the dopamine input to these regions, undergo prominent alterations (Spear, 2000, 2002). These areas of the brain are thought to be important in modulating the reinforcing effects of drugs and may make some adolescents more vulnerable to addictive substances, such as nicotine (Chambers, Taylor & Potenza, 2003).

As depicted in Figure 13.1, this model includes three tenets. First, there are multiple pathways by which genetic factors could influence adolescent smoking uptake (Audrain-McGovern, Nigg, & Perkins, 2009). Genetic variation, coupled with early environmental influences, contributes to brain structure and function. Genetic variation can occur in pathways that are important in nicotine metabolism and pathways involved in substance use reward more generally and nicotine reinforcement specifically. Specific genetic variants may give rise to alterations in neurotransmitter and receptor systems involved in reward, emotional regulation, or cognitive deficit management. Self-medication or adaptive theories of substance use disorders postulate that substance use is motivated, at least in part, by attempts to correct these inherited deficits in neurotransmitter and receptor systems (Bardo, Donohew, & Harrington, 1996; Khantzian, 1997). It is possible that substance use is maintained by neuro-adaptations in brain regions responsible for specific functions. Comorbidity between

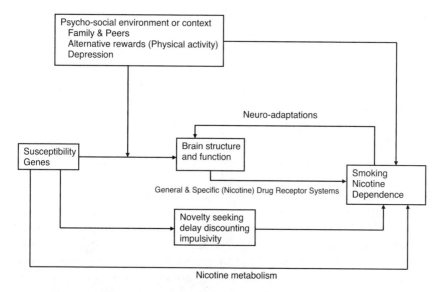

Figure 13.1. **Biobehavioral model of adolescent smoking**.

depression, attention-deficit hyperactivity disorder (ADHD), novelty-seeking, and adolescent smoking lend support to this notion (Audrain-McGovern, Rodriguez, & Kassel, 2009; Audrain-McGovern, Rodriguez, Patel et al., 2006; Masse & Tremblay, 1997; McClernon & Kollins, 2008; Rodriguez, Tercyak, & Audrain-McGovern, 2008).

A second assumption of this model is that the neurobiological influences on adolescent smoking can also be indirect and mediated by variables involved in reward, including dispositional traits. Personality characteristics such as novelty- or sensation-seeking, impulsive behavior, and impulsive decision making have been shown to predict adolescent smoking (Audrain-McGovern, Rodriguez, Epstein et al., 2009; Audrain-McGovern, Rodriguez, Patel et al., 2006; Krishnan-Sarin et al., 2007; Lewinsohn, Brown, Seeley, & Ramsey, 2000). These complex behavioral traits also have a significant heritable component (Eaves & Eysenck, 1975; Koopmans, Boomsma, Heath, & van Doornen, 1995). In fact, evidence suggests that exposure to novelty activates the same neural structures that are responsible for the rewarding effects of substances of abuse (Bardo et al., 1996). For example, the dopamine D_4 receptor gene (*DRD4*) 7-repeat allele is linked to higher novelty-seeking which, in turn, is linked to greater smoking among male adolescents (Laucht, Becker, El-Faddagh, Hohm, & Schmidt, 2005).

Third, this model suggests that genetic factors may interact with environmental or contextual factors (social, behavioral, and psychological) to influence adolescent smoking. These interactions may serve to increase the likelihood or decrease the likelihood that an adolescent will progress along the smoking uptake continuum (Audrain-McGovern, Rodriguez, Wileyto et al., 2006; Audrain-McGovern, Lerman et al., 2004; Karp, O'Loughlin, Hanley, Tyndale, & Paradis, 2006). Genetic predisposition to becoming a regular smoker may be moderated by environmental influences, which could offer intervention targets to help prevent smoking uptake in adolescents at high risk for smoking adoption. These issues are taken up later in this chapter.

Genetic Contributions to Adolescent Smoking

Evidence for the heritability of smoking has stimulated research evaluating the role of candidate genes involved in nicotine metabolism and the dopamine reward pathway in adolescent smoking and nicotine dependence. Relatively few studies have evaluated the specific genetic contributions to becoming a regular smoker or becoming nicotine dependent, which are longitudinal phenotypes. Similar to adult tobacco studies involving different phenotypes (e.g., smoking status, smoking cessation), findings have not always been replicated. Lack of replication in case-control studies have plagued the adult smoking literature (Lerman, Perkins, & Gould, 2009). We discuss the implications of few adolescent studies, small effects, and lack of replication in greater detail later in the chapter.

Cytochrome P4502A6 (CYP2A6) is involved in the metabolic inactivation of nicotine. About 80% of nicotine consumed through cigarette smoking is removed from the body via inactivation to cotinine; the CYP2A6 hepatic enzyme mediates about 90% of the metabolism of nicotine to cotinine (Benowitz & Jacob, 1994; Messina, Tyndale, & Sellers, 1997; Nakajima et al., 1996; Xu, Goodz, Sellers, & Tyndale, 2002). Genetic variation in CYP2A6 appears to be important for adult and adolescent smoking phenotypes. Evidence suggests that reduced CYP2A6 activity and null alleles (slower nicotine metabolism) are protective against smoking behavior and associated with lower cigarette consumption (Ando et al., 2003; Schoedel, Hoffmann, Rao, Sellers, & Tyndale, 2004; Tyndale & Sellers, 2001). CYP2A6 also influences the likelihood of becoming nicotine dependent and the rate of emergence of nicotine dependence symptoms in adolescents (Audrain-McGovern et al., 2007; O'Loughlin et al., 2004). Variations in the rate at which nicotine is converted to cotinine may further impact the reinforcing effects of nicotine, as well as the number of cigarettes smoked each day in order to manage craving.

Genetic variations (polymorphisms) in the dopamine reward pathway have received most attention with respect to adult and adolescent smoking behavior. The majority of these studies have investigated the dopamine receptor D_2 gene (*DRD2*)(Comings et al., 1996; Noble et al., 1994; Spitz et al., 1998). The A1 allele of *DRD2* has been associated with altered receptor density and binding characteristics and therefore less endogenous dopamine (Fuke et al., 2001; Noble, Blum, Ritchie, Montgomery, & Sheridan, 1991; Ritchie & Noble, 1996; Thompson et al., 1997).

Studies have also focused on the dopamine transporter gene *SLC6A3*, which is found on presynaptic dopaminergic neurons. *SLC6A3* regulates the reuptake of dopamine into presynaptic terminals. The gene has a polymorphic 40-base pair variable number tandem repeat (VNTR), with repeats varying between 3 and 11 copies, with the 9-repeat and 10-repeat alleles being the most common (King, Xiong, Douglas, Lee, & Ellinwood, 1999). The 10-repeat allele of *SLC6A3* has been associated with greater gene expression (Fuke et al., 2001) and dopamine transporter reuptake protein levels, which result in greater clearance and less bioavailability of dopamine (Heinz et al., 2000). Studies have linked smoking status, rate, and latency to relapse in adults with the 10-repeat allele dopamine reuptake transporter (Lerman et al., 1999; Lerman et al., 2003; Sabol et al., 1999).

It has been speculated that adolescents who carry the A1 allele of *DRD2* or the 10-repeat allele of *SLC6A3* may achieve greater reward from nicotine's effects on dopamine activity because these individuals have less naturally occurring dopamine. Greater reward derived from nicotine in cigarettes may promote further experimentation and progression to regular smoking. One prospective cohort study of the genetic, social, and behavioral predictors of adolescent smoking progression found support for the role of *DRD2*, but did not find evidence for an effect of *SLC6A3*. Specifically, the likelihood of an adolescent progressing to a higher level of smoking across years increased almost two-fold with each additional *DRD2* A1 allele (Audrain-McGovern, Lerman, Wileyto, Rodriguez, & Shields, 2004). Recent data have also linked the dopamine D_2 receptor gene to smoking progression as well as to nicotine dependence, finding that *SLC6A3* may be more strongly linked to smoking cessation, as intention to quit smoking was significantly lower in adolescents with two 10-repeat alleles (19% 36%) (Laucht et al., 2008). *SLC6A3* 10-repeat alleles were also associated with a younger age of smoking onset (12.8 13.9 years) (Laucht et al., 2008).

It is important to point out that the functional significance of these two dopamine gene polymorphisms (*DRD2* and *SLC6A3*) is not known. This raises the possibility that these polymorphisms are simply linked with other functional variants or that variants downstream of the *DRD2* and *SLC6A3* loci could be affected by these polymorphisms (Fossella, Green, & Fan, 2006; Neville, Johnstone, & Walton, 2004). The *DRD2* A1 allele is located downstream of *DRD2* and within a coding region of the ANKK1 kinase gene and *ANKK1* is expressed in the central nervous system (Hoenicka et al., 2010).

Studies have also evaluated polymorphisms in the D_4 dopamine receptor gene with respect to adolescent smoking. *DRD4* contains a 48-base pair VNTR (Lichter et al., 1993). This polymorphism has been found to vary between 2 and 11 repeats, with the 4 repeat and 7 repeat alleles being the most common. The *DRD4-7* repeat allele has been linked to a blunted intracellular response to dopamine and to a lower affinity to antagonists in vitro (Asghari et al., 1995; Van Tol et al., 1992), although its functional role remains unclear (Wong & Van Tol, 2003). Research with adults indicates that individuals with at least one long allele (6–8 repeats) had an earlier age of smoking onset and smoked more frequently than did individuals homozygous for the short alleles (2–5 repeats) (Shields et al., 1998). A cross-sectional evaluation of this variant found a link between it, adolescent smoking status, and smoking frequency (daily versus nondaily) in males, but not females (Laucht et al., 2005). This suggests an interaction effect between gender and this particular polymorphism. The moderating aspects of gender may include individual (e.g., hormones) and environmental characteristics, but this explanation remains speculative.

Tyrosine hydroxylase (TH) also is a rate-limiting enzyme involved in dopamine biosynthesis. Studies have demonstrated that nicotine increases *TH* expression and dopamine synthesis (Hiremagalur, Nankova, Nitahara, Zeman, & Sabban, 1993; Hiremagalur & Sabban, 1995; Serova, Danailov, Chamas, & Sabban, 1999). Variation in the number of sequence repeats in *TH* has been shown to affect the activity of the enzyme and is associated with brain dopamine metabolism (Johnson et al., 1996). In a study of adult smokers, the 4-repeat allele (K4) was associated with a lower smoking rate and the 1-repeat allele (K1) was associated with a higher smoking rate (Lerman et al., 1997). These findings led to investigations of whether the K4 allele was protective against nicotine dependence in adolescents. In a small group of adolescents participating in a longitudinal cohort study, adolescents with the K4 allele were over 70% less likely to smoke within an hour of waking, although a link between the K4 allele and regular smoking (>10 cigarettes a day) was not found (Olsson et al., 2004). In a replication and extension of that work, adolescents with the K4 allele were about half as likely to regularly smoke more than 10 cigarettes a day (Anney, Olsson, Lotfi-Miri, Patton, & Williamson, 2004).

The serotonin pathway has received much less attention. The human serotonin transporter gene (*5-HTT*)is responsible for the reuptake of serotonin from the synaptic cleft. The *5-HTTLPR* polymorphism consists of a 44 base-pair insertion (long allele L) or deletion (short allele S) involving 6–8 repeat sequences. It is thought that the LL variant is associated with reduced levels of intrasynaptic serotonin compared to the SL and SS variants; the short variant is associated with reduced transcription, resulting in less serotonin uptake (i.e., more available serotonin) (Heils et al., 1996; Johnson, 2000; Lesch et al., 1996). This polymorphism has been shown to interact with the personality trait of neuroticism to predict nicotine intake, level of nicotine dependence, and smoking to reduce negative affect among adult smokers with at least one short allele (i.e., SS or SL) (Lerman et al., 2000). The short allele (SS genotype) of the serotonin transporter *5-HTTLPR* has been linked to heavier smoking with earlier onset, compared to moderate smoking with later onset among adolescents (Gerra et al., 2005). However, a higher level of smoking was seen among girls who were homozygous for the long allele of *5-HTTLPR* and who lacked the dopamine receptor *DRD4*7*-repeat allele (Skowronek, Laucht, Hohm, Becker, & Schmidt, 2006).

Gene × Environment Interactions

Like most complex behavioral traits, smoking is the result of genetic and environmental influences (Hopfer, Crowley, & Hewitt, 2003). A range of environmental factors likely interact with genetic vulnerability to promote or deter smoking progression and the development of nicotine dependence. Here, as we review the available literature that has considered gene by environment interactions (G × E) and the impact on adolescent smoking, we will consider a broad definition of environment. A discussion of the environment can be found in Kendler's introductory chapter (Chapter 1). We recognize that many of the G × E interactions involving adolescent smoking behavior published to date do not reflect environmental variables per se, but rather environmental variables that reflect both genetic and environmental influences. This literature may provide some insights about the social, behavioral, and psychological contexts that enhance genetic liability to smoke and those in which genetic influences are suppressed.

Recent examples of G × E interactions in smoking come from studies utilizing the National Longitudinal Study of Adolescent Health (Add Health) dataset. For example, Boardman and colleagues showed that school context moderates the heritability of daily smoking. Heritability of daily smoking was significantly higher within schools where the most popular students smoked compared to schools where students who smoked were less salient (Boardman, Saint Onge, Haberstick, Timberlake, & Hewitt, 2008). Similarly, genetic influences on daily smoking were lower in states with higher cigarette taxes and greater restrictions on vending machines and cigarette advertising (Boardman, 2009). Likewise, a recent study found that genetic effects on adolescent smoking level at age 14 were moderated by parental monitoring (Dick et al., 2007). The significance of adolescents' genetic predispositions decreased (about 60% of the variance to 15% of the variance in smoking behavior), whereas common environmental influences increased with higher levels of parental monitoring (about 20% of the variance to 80% of the variance in smoking behavior). The findings that genetic factors interact with the environment to influence the heritability of smoking behavior underscores the importance of genetic factors within certain environments (Shanahan & Hofer, 2005). Environments that place constraints on smoking (e.g., social forces including parents, state tobacco access, and promotion policies) may not support the expression of a genetic propensity to smoke, whereas environments that promote smoking (e.g., popular peers who smoke) permit genetically-based responses to nicotine to drive smoking behavior.

Also using the data from Add Health, Timberlake and colleagues (2006) evaluated the moderating effects of religiosity on the genetic and environmental determinants of smoking initiation, which was defined as having ever smoked one whole cigarette (Timberlake et al., 2006). The heritability of smoking initiation was approximately 80% among those who reported low levels of religiosity, but less than 10% among those who reported high levels of religiosity. As this study evaluated the different influences on who would try cigarettes, rather than tapping influences on more regular use, the results highlight the potential role of individual characteristics that promote risk taking rather than influences specific to nicotine. To the extent that these individual or personality characteristics serve as endophenotypes or intermediate phenotypes for the progression to regular smoking and nicotine dependence after initial smoking exposure, they could serve as markers to target smoking prevention interventions. This point is discussed in greater detail later in this chapter when we consider smoking prevention implications.

Studies have also evaluated the interacting effects of specific genes with other factors that have been shown to increase adolescent vulnerability to smoking, such as depression (Audrain-McGovern, Rodriguez, & Kassel, 2009). A prospective cohort study of adolescent smoking adoption found that adolescents with at least one *DRD2* A1 allele and higher levels of depression symptoms were almost two times more likely to progress along the smoking uptake continuum across 2 years (Audrain-McGovern, Lerman et al., 2004). In this case, the gene interacted with depression—a well-known smoking vulnerability factor—to potentiate adolescent smoking progression (Audrain-McGovern, Lerman et al., 2004). Preliminary findings also suggest that depression and the rate of nicotine metabolism may interact to promote tobacco dependence among adolescents (Karp et al., 2006).

Adolescents with higher levels of depression symptoms had odds of being tobacco dependent that were two times greater (48% versus 22%) if they were faster nicotine metabolizers (i.e., CYP2A6 normal and intermediate metabolizer) than did slower nicotine metabolizers. This relationship may be due, in part, to genetic influences that are common to both smoking and depression (Kendler, Jacobson, Prescott, & Neale, 2003; Lyons et al., 2008) and/or to a dysfunctional dopaminergic reward pathway among depressed individuals, such that they are more responsive to substances that increase dopamine levels, such as nicotine (Cardenas et al., 2002). Indeed, recent research supports a bidirectional "self-medication" hypothesis of smoking and depression, in that adolescents with greater symptoms of depression are more likely to progress to regular smoking and adolescents who progress along the smoking uptake continuum show a deceleration in depression symptoms (Audrain-McGovern, Rodriguez, & Kassel, 2009).

In contrast to studies assessing risk-promoting interactions on adolescent smoking uptake, there are also data suggesting that specific genes interact with other factors that buffer or protect adolescents against smoking progression, such as physical activity and sports participation (Audrain-McGovern, Rodriguez, & Moss, 2003; Rodriguez & Audrain-McGovern, 2004). For example, having one or more smoking risk genotypes (i.e., DRD2 A1/* and/or dopamine transporter SLC6A3 10/10) was related to higher levels of physical activity which, in turn, was related to lower levels of smoking progression for adolescents participating in at least one team sport, but not for adolescents with no team sport participation. Team sport participation appears to interact with genes in the dopamine reward pathway (i.e., DRD2 and/or dopamine transporter SLC6A3) to mitigate adolescent smoking progression (Audrain-McGovern, Rodriguez, Wileyto, Schmitz, & Shields, 2006). The physical activity involved in team sports or the physical activity habits of those involved in sports may increase dopamine levels in the brains of adolescents with genotypes associated with less available dopamine, making smoking less likely or less rewarding. Alternatively, this buffering effect may be the product of nonsmoking norms among coaches and fellow nonsmoking teammates.

CHALLENGES IN APPLYING G × E FINDINGS IN ADOLESCENT SMOKING PREVENTION INTERVENTIONS

Although many of the above-noted findings are certainly intriguing and lend some new opportunities for breakthroughs in adolescent smoking prevention, there remain significant challenges to doing so. Here, we review some of these challenges and discuss how their resolution could ultimately lead to targeted interventions for young people.

Limitations in Knowledge

The possibility of using genetic information to help develop smoking prevention efforts is attractive, although the state of the science is early to inform intervention efforts. Several cautions apply here. First, only a handful of genes have been investigated. Of those investigated, effects tend to be small and are not always replicated (Audrain-McGovern, Nigg, & Perkins., 2009). Replication has been a problem in candidate gene association studies (Ioannidis, Trikalinos, & Khoury, 2006). Any individual genetic variant is likely to account for only a small proportion of the overall variance in smoking behavior. It is possible that a greater number of genetic variants important to adolescent smoking uptake may be identified in the future, such as those identified through genome-wide association studies (Caporaso et al., 2009). It is still unclear whether the results of numerous genotypes could be combined to predict future smoking behavior with precision, and this work remains in progress.

Second, many studies involve genes with unidentified function. The lack of clarity in the specific role that a polymorphism has in smoking progression and the emergence of nicotine dependence may limit its usefulness (Swan, 1999). Third, a consistent smoking phenotype has not been evaluated. In many instances, adolescent studies that have evaluated genetic contributions to smoking have been cross-sectional or longitudinal studies that have relied on cross-sectional phenotypes. Cross-sectional phenotypes do not adequately capture the developmental nature of smoking adoption and the emergence of nicotine dependence. Different genes may have specific effects (or no effects) depending on the smoking phenotype (e.g., initiation of smoking experimentation, smoking progression, nicotine dependence, smoking cessation).

Finally, there have been very few studies of G × E interactions on smoking acquisition and the development of nicotine dependence. Given that smoking is a product of both of these influences (Hopfer et al., 2003), adolescent smoking prevention and cessation approaches may need to consider genetic and environmental factors as etiological mechanisms underlying smoking behavior. We currently lack information on the genes that contribute to smoking uptake, and we know even less about the environmental or contextual variables that are most important for promoting and suppressing genetic liability to smoke. Until these issues are more completely resolved in the research literature, the translation of these findings to inform adolescent smoking prevention interventions remains limited.

Ethical and Social Challenges

Like all forms of clinical and translational research involving an examination of genes, behavior, and health, there are sensitivities and challenges relating to the potential use and misuse of this information (Cullen & Marshall, 2006). For example, some have expressed concern that knowledge of gene–behavior relationships in the context of cigarette smoking could undermine individuals' motivation to stop smoking or their perceived control over smoking (Shields, Lerman, & Sullivan, 2004; Wright, Weinman, & Marteau, 2003). It remains to be seen if or how such information could potentially affect adolescents. Given the gaps in our current understanding of genetic contributions to smoking and G × E effects, adolescent smoking prevention interventions based on genetic data would be premature.

For such genetically informed efforts to move forward, several important aspects of planning for such interventions would need to be resolved. These challenges include the following, and are offered as additional areas for future explorative research: (1) what to communicate (we do not presently know enough to inform risk communication), (2) to whom this information should be communicated (all adolescents, or those who are members of certain high risk subgroups?), and (3) how this information should be communicated. Again, the resolution of these issues would also require that larger social and ethical challenges be remedied as well. For example, many of the genes linked to smoking are also associated with other substances of abuse and neuropsychiatric conditions. This is termed *genetic pleiotropy*, and it raises confidentiality, privacy, and adolescent self-identity concerns (Moolchan & Mermelstein, 2002; Wilfond, Geller, Lerman, Audrain-McGovern, & Shields, 2002). As the balance is tipped toward greater risk without clear long-term benefit (i.e., prevention of smoking), the need for parental involvement in the informed consent process also becomes more pronounced. Over half of adolescents express interest in genetic testing for nicotine dependence susceptibility (Tercyak, Peshkin, Wine, & Walker, 2006). Whether an adolescent, a parent, or both provide consent will require careful consideration in ongoing examinations of the risk-to-benefit ratio of preventive use of genetic information in adolescent smoking.

Implications for Adolescent Smoking Prevention

As suggested by the biobehavioral model of adolescent smoking, the heterogeneity in smoking uptake is influenced by developmental stage, genetic factors, and the environment that inhibits or promotes genetic contributions to an adolescent's response to nicotine. To date, adolescent smoking prevention interventions have not considered these individual differences in smoking acquisition (e.g., timing, content). Heterogeneity in smoking acquisition and the emergence of nicotine dependence reflects individual genetic variation in combination with a loss of environmental protective factors and an increase in vulnerability-promoting factors (Audrain-McGovern, Rodriguez, Wileyto et al., 2006; Hu et al., 2008). The questions that remain for prevention science are: How do we identify factors in adolescents' social environments that protect them from smoking? How do we bolster/maintain those influences among adolescents in which they are present? And, how do we add them to the environments of adolescents when they are absent? Again, answers to these important questions would be necessary prior to considering how to influence G × E effects in smoking adoption and maintenance among youth.

The Role of Behavioral Traits in Adolescent Smoking Prevention

Research suggests that genetic predisposition to smoking may be mediated, in part, by individual differences in behavioral traits. Prevention programs designed to meet the needs of those with specific traits linked to smoking (i.e., high-risk subgroups) may be most effective in reducing liability to smoking and subsequent nicotine dependence. A targeted approach focusing on at-risk youth also seems consistent with models of individual differences.

Current knowledge and unresolved practical challenges indicate that genetic markers may not be the most useful target for identifying youth at risk of becoming regular smokers. Several personality traits, with genetic underpinnings, may serve as liability markers for susceptibility to smoking adoption and nicotine dependence. For example, individual differences in delay discounting and sensation-seeking reflect underlying genetic variation (Dick et al., 2008; Eisenberg et al., 2007; Volkow, Fowler, Wang, & Goldstein, 2002). Consideration of individual differences in biologically-based behavioral traits may shed light on who benefits most from targeted antismoking prevention efforts and who benefits least. An important question is whether these trait markers provide greater predictive value than the underlying genetic variants.

Delay discounting is one aspect of impulsivity that influences cigarette smoking, as well as use of other substances of abuse (Kreek, Nielsen, Butelman, & LaForge, 2005; Mitchell, 1999). This phenomena describes the tendency for persons to discount the value of a reward as a function of the length of delay to its delivery (Madden, 2000). Individuals who score higher on measures of delay discounting prefer more immediate rewards at the expense of larger rewards later (Monterosso & Ainslie, 1999). Across substances of abuse, current substance abusers discount delayed rewards more rapidly than nonusers or controls (Kirby, Petry, & Bickel, 1999; Madden, Petry, Badger, & Bickel, 1997).

Delay discounting has been shown to play an etiological role in adolescent smoking and has been shown to discriminate among those who progress to a more regular pattern of smoking versus those who do not, irrespective of the pattern of smoking acquisition (Audrain-McGovern, Rodriguez, Epstein et al., 2009). Adolescents higher in delay discounting may seek out activities that have more immediate (rather than more delayed) rewards, such as smoking and substance use. If higher rates of delay discounting contribute to smoking acquisition, then delay discounting may provide a variable by which to screen for smoking vulnerability. Adolescents at higher risk of smoking due to higher delay discounting may be a subgroup to target for more intensive smoking prevention efforts that include novel behavioral components directed toward aspects of impulsivity.

Intervention research suggests that impulsive decision making may be moderated by the acquisition of self-control skills. For example, a classroom-based behavioral management intervention focused on reducing aggressive (e.g., fighting) and disruptive (e.g., shouting out of turn) behaviors among first and second graders reduced the risk of early onset smoking initiation (age 12), smoking initiation by age 14 for boys, and regular smoking in young men (19–21 years old) (Kellam & Anthony, 1998; Kellam et al., 2008; Storr, Ialongo, Kellam, & Anthony, 2002). Proscribed behaviors were met with a team of classmates losing points. Teams received tangible rewards (e.g., classroom activities, stickers, erasers) for their points when no member exhibited the proscribed behaviors during the sessions. These rewards were delivered immediately at first, and then delayed to the end of the school day, and eventually to the end of the school week. Early interventions such as these may interrupt the development of impulsive behaviors, or reduce their occurrence by bolstering self-control skill sets, including delaying gratification.

Greater delay discounting may be a marker for the early onset of smoking, a heavier smoking rate, and other issues such as substance use and poorer academic performance. Modifying this common etiologically important antecedent early, during developmentally malleable periods, may prevent the direct effect of delay discounting on smoking progression and the indirect effect via behaviors related to smoking (e.g., poorer grades, alcohol use), which may have a significant impact on smoking uptake.

Novelty- and sensation-seeking have also been linked to genetic pathways involved in reward (Bardo et al., 1996; Dick et al., 2008; Eisenberg et al., 2007; Laviola, Adriani, Terranova, & Gerra, 1999; Volkow et al., 2002). These correlated personality traits have been linked directly and indirectly to adolescent smoking progression (Audrain-McGovern, Rodriguez, Patel et al., 2006; Skara, Sussman, & Dent, 2001). An increased risk for smoking among adolescents high in sensation- and novelty-seeking is thought to be due, in part, to neurochemical factors that enhance the rewarding properties of drug use, including nicotine (Laviola et al., 1999; Perkins, Gerlach, Broge, Grobe, & Wilson, 2000). Adolescent smoking prevention programs could promote the identification and engagement in alternative (nondrug) substitute rewards to smoking. Research indicates that youth who maintain greater alternative substitute reinforcers across time (e.g., clubs, physical activity, academic involvement) are two times less likely to initiate smoking and progress along the uptake continuum (Audrain-McGovern, Lerman et al., 2004). Because adolescents high in sensation-seeking are more likely than low sensation seekers to engage in action-adventure activities, prevention approaches that incorporate these activities as substitute reinforcers may be effective (D'Silva, Harrington, Palmgreen, Donohew, & Lorch, 2001).

In addition, novelty- and sensation-seeking personality traits may increase the likelihood of smoking uptake by increasing the receptivity to pro-smoking messages in the media and peer group (Audrain-McGovern, Tercyak et al., 2003; Ellickson, Tucker, & Klein, 2008). The American Legacy Foundation's The Truth® campaign was a well-recognized national antito-bacco advertising campaign (Davis, Nonnemaker, & Farrelly, 2007). It had a significant and favorable impact on antismoking beliefs, lowered intentions to smoke, and actual smoking behavior (Davis, Farrelly, Messeri, & Duke, 2009). Truth® depicted a raw and uncensored view of the tobacco industry and its longstanding efforts to manipulate young people to start smoking but without mentioning smoking's health hazards; Truth® was specifically designed with high-sensation-seeking adolescents in mind (Healton, 2002; Vallone, Allen, Clayton, & Xiao, 2007).

Developmentally Co-occurring Conditions and Smoking Prevention

In addition to considering behavioral traits that play an etiological role in adolescent smoking, the biobehavioral model and relevant literature suggests that targeting developmentally comorbid conditions that could be linked along reward pathways may be an effective approach to prevent adolescent smoking uptake. For example, smoking prevalence tends to increase and physical activity tends to decrease across mid to late adolescence (CDC, 2008). Longitudinal research of smoking and physical activity indicate that higher levels of physical activity reduce the odds of adolescent smoking initiation and progression by almost 50% (Audrain-McGovern, Rodriguez et al., 2003). A study of Finnish youth found that consistently sedentary adolescents were more likely to have initiated regular smoking during a 6-year follow-up compared to consistently active adolescents (30% versus

2%) (Raitakari et al., 1994). More recent research also supports the role of consistent team sport participation trajectories with a 3-fold reduction in the odds of regular smoking (Rodriguez & Audrain-McGovern, 2004). Physical activity and smoking share many of the same functions (e.g., both elevate mood, decrease perceived stress, and promote weight maintenance) (Calfas & Taylor, 1994; Kassel et al., 2007; Lewis-Esquerre, Rodrigue, & Kahler, 2005; Steptoe & Butler, 1996; Wahl, Turner, Mermelstein, & Flay, 2005; Weiss, Merrill, & Gritz, 2007) and have been linked to similar genes in the dopaminergic reward pathway (Audrain-McGovern, Rodriguez, Wileyto et al., 2006).

The protective effects of physical activity and team sport participation on adolescent smoking uptake may be due to genetics as well as to environmental influences. Physical activity may increase dopamine among adolescents with genotypes associated with less bioavailability of dopamine. Physical activity may provide reward and make smoking less rewarding, or it may reduce the need for alternative methods to increase dopamine, such as smoking. Additionally, more physically active youth and those involved in team sport are exposed to social reinforcement and athletic norms (i.e., that smoking is unhealthy, athletes do not smoke, smoking is inconsistent with athletic performance) derived from physically active peers, as well as from coaches and teammates (Audrain-McGovern, Rodriguez, Wileyto, Schmitz, & Shields, 2006; Melnick, Miller, Sabo, Farrell, & Barnes, 2001). These social influences may help offset the genetic propensity to smoke with increasing frequency over time. Future research directed toward adolescent smoking prevention efforts may focus on helping adolescents identify physical activity and team sports to participate in and promote consistent participation. Although the efficacy of a physical activity intervention to prevent smoking has yet to be investigated, the promotion of physical activity would have mental and physical health benefits beyond smoking prevention.

The comorbidity between adolescent smoking and subclinical and clinical levels of depression has been well-documented (Audrain-McGovern, Rodriguez, Cuevas, & Rodgers, 2009). Similar genes may be involved in reward derived from smoking, as well as in the decision-making and hedonic capabilities of individuals with depression (Picciotto, Brunzell, & Caldarone, 2002). For example, nicotine receptor inhibition appears to mitigate mood instability and may reduce depression (Shytle et al., 2002). Adolescents may find smoking to be especially rewarding during periods of depression as brain reward systems appear to be dysfunctional in depressed individuals, such that they are more responsive to substances that activate these reward systems, including nicotine (Cardenas et al., 2002; Forbes et al., 2006; Phillips, 1984; Tremblay, Naranjo, Cardenas, Herrmann, & Busto, 2002). Animal models suggest that nicotine potentiates reward from drug and nondrug reinforcers by increasing the sensitivity of brain reward systems (Kenny & Markou, 2006). A recent study found initial support for a bidirectional "self-medication" relationship between depression symptoms and cigarette smoking: higher depression symptoms in mid-adolescence (age 14) predicted smoking progression, while smoking progression predicted a deceleration of depression symptoms from mid to late adolescence (Audrain-McGovern, Rodriguez, & Kassel, 2009). There appears to be some overlap in the neural substrates modified by smoking and antidepressant medications (Quattrocki, Baird, & Yurgelun-Todd, 2000). Based on these converging areas of research, targeting adolescent depression (subclinical and clinical) could have an important impact on smoking uptake, as well as on subsequent depression. The same is also true for attention-deficit hyperactivity disorder, which has also been related to smoking onset and progression (McClernon & Kollins, 2008).

CONCLUSION

Like most complex traits, smoking is the result of genetic and environmental influences. Evidence for specific genetic variation in adolescent smoking and nicotine dependence is emerging, but a substantial amount of information will be needed before we can determine its utility in applied settings. Environmental factors play an important role in directly affecting adolescent smoking behavior and can interact with adolescents' genetic make-up to either mitigate or promote smoking uptake. Presently, we know little about the environmental or contextual variables that are most important for suppressing genetic liability to smoke, although clues are beginning to emerge. This makes it difficult to harness information about adolescents' future risks of smoking based on personal genotype. Behavioral traits that are stable across time and contexts may serve as important phenotypic markers that can be used to identify youth at risk for smoking uptake. Understanding the links between genes, phenotypic markers, and smoking behavior may help inform the broad-based smoking prevention approaches that target the needs of high-risk subgroups of adolescents. Interactions with the environment and knowledge of the behavioral proclivities associated with such traits could provide intervention points. Comorbid conditions, such as physical inactivity and depression, may share some underlying genetic features with smoking. Targeting psychological (e.g., depression) and behavioral (e.g., declining physical activity) comorbidities may be a way to target youth more broadly to prevent smoking uptake.

The lack of practical clarity and the feasibility of using genetic data as a way to identify and target youth at risk of becoming smokers does not diminish the importance of identifying the genetic mechanisms underlying smoking acquisition and the emergence of nicotine dependence. Understanding the biobehavioral underpinnings may (a) shed light on when youth are most vulnerable to nicotine (optimize intervention timing), (b) identify underlying mechanisms that can be affected by behavioral interventions (intervention content), and (c) explain why some youth are more vulnerable to becoming regular smokers than others (youth most likely to benefit from a smoking prevention intervention). A lack of attention to the underlying genetic mechanisms that work in concert with environmental influences to affect smoking acquisition may explain, in part, why adolescent smoking prevention interventions have had limited long-term effects (Clayton, 1999; Clayton, Scutchfield, & Wyatt, 2000; Skara & Sussman, 2003). Increased knowledge of genetic variation may promote more informed smoking prevention approaches by highlighting underlying mechanisms for smoking liability and providing a greater understanding of environmental influence on expression.

REFERENCES

Ando, M., Hamajima, N., Ariyoshi, N., Kamataki, T., Matsuo, K., & Ohno, Y. (2003). Association of CYP2A6 gene deletion with cigarette smoking status in Japanese adults. *Journal of Epidemiology, 13*(3), 176–181.

Anney, R. J., Olsson, C. A., Lotfi-Miri, M., Patton, G. C., & Williamson, R. (2004). Nicotine dependence in a prospective population-based study of adolescents: the protective role of a functional tyrosine hydroxylase polymorphism. *Pharmacogenetics, 14*(2), 73–81.

Asghari, V., Sanyal, S., Buchwaldt, S., Paterson, A., Jovanovic, V., & Van Tol, H. H. (1995). Modulation of intracellular cyclic AMP levels by different human dopamine D4 receptor variants. *Journal of Neurochemistry, 65*(3), 1157–1165.

Audrain-McGovern, J., Al Koudsi, N., Rodriguez, D., Wileyto, E. P., Shields, P. G., & Tyndale, R. F. (2007). The role of CYP2A6 in the emergence of nicotine dependence in adolescents. *Pediatrics, 119*(1), e264–e274.

Audrain-McGovern, J., Lerman, C., Wileyto, E. P., Rodriguez, D., & Shields, P. G. (2004). Interacting effects of genetic predisposition and depression on adolescent smoking progression. *American Journal of Psychiatry, 161*(7), 1224–1230.

Audrain-McGovern, J., Nigg, J. T., & Perkins, K. (2009). Endophenotypes for nicotine dependence risk at or before initial nicotine exposure. In G. E. Swan, T. B. Baker, L. Chassin, D. V. Conti, C. Lerman & K. A. Perkins (Eds.), *Phenotypes and endophenotypes: foundations for genetic studies of nicotine use and dependence* (Vol. Tobacco Control Monograph No. 20). Bethesda, MD: U.S. Department of Health and Human Services, National Institutes of Health, National Cancer Institute.

Audrain-McGovern, J., Rodriguez, D., Cuevas, J., & Rodgers, K. (2009). The role of depression in adolescent smoking trajectories. *Journal of Dual Diagnosis, 5*(2), 179–190.

Audrain-McGovern, J., Rodriguez, D., Epstein, L. H., Cuevas, J., Rodgers, K., & Wileyto, E. P. (2009). Does delay discounting play an etiological role in smoking or is it a consequence of smoking? *Drug and Alcohol Dependence*, *103*(3), 99–106.

Audrain-McGovern, J., Rodriguez, D., & Kassel, J. D. (2009). Adolescent smoking and depression: evidence for self-medication and peer smoking mediation. *Addiction*, 104(10), 1743–1756.

Audrain-McGovern, J., Rodriguez, D., & Moss, H. B. (2003). Smoking progression and physical activity. *Cancer Epidemiology, Biomarkers & Prevention*, *12*(11 Pt 1), 1121–1129.

Audrain-McGovern, J., Rodriguez, D., Patel, V., Faith, M. S., Rodgers, K., & Cuevas, J. (2006). How do psychological factors influence adolescent smoking progression? The evidence for indirect effects through tobacco advertising receptivity. *Pediatrics*, *117*(4), 1216–1225.

Audrain-McGovern, J., Rodriguez, D., Tercyak, K. P., Cuevas, J., Rodgers, K., & Patterson, F. (2004). Identifying and characterizing adolescent smoking trajectories. *Cancer Epidemiology, Biomarkers & Prevention*, *13*(12), 2023–2034.

Audrain-McGovern, J., Rodriguez, D., Tercyak, K. P., Neuner, G., & Moss, H. B. (2006). The impact of self-control indices on peer smoking and adolescent smoking progression. *Journal of Pediatric Psychology*, *31*(2), 139–151.

Audrain-McGovern, J., Rodriguez, D., Wileyto, E. P., Schmitz, K. H., & Shields, P. G. (2006). Effect of team sport participation on genetic predisposition to adolescent smoking progression. *Archives of General Psychiatry*, *63*(4), 433–441.

Audrain-McGovern, J., Tercyak, K. P., Shields, A. E., Bush, A., Espinel, C. F., & Lerman, C. (2003). Which adolescents are most receptive to tobacco industry marketing? Implications for counter-advertising campaigns. *Health Communication*, *15*(4), 499–513.

Bardo, M. T., Donohew, R. L., & Harrington, N. G. (1996). Psychobiology of novelty seeking and drug seeking behavior. *Behavioral Brain Research*, *77*(1-2), 23–43.

Benowitz, N. L., & Jacob, P., 3rd. (1994). Metabolism of nicotine to cotinine studied by a dual stable isotope method. *Clinical Pharmacology and Therapeutics*, *56*(5), 483–493.

Boardman, J. D. (2009). State-level moderation of genetic tendencies to smoke. *American Journal of Public Health*, *99*(3), 480–486.

Boardman, J. D., Saint Onge, J. M., Haberstick, B. C., Timberlake, D. S., & Hewitt, J. K. (2008). Do schools moderate the genetic determinants of smoking? *Behavioral Genetics*, *38*(3), 234–246.

Boomsma, D. I., Koopmans, J. R., Van Doornen, L. J., & Orlebeke, J. F. (1994). Genetic and social influences on starting to smoke: a study of Dutch adolescent twins and their parents. *Addiction*, *89*(2), 219–226.

Calfas, K. J., & Taylor, W. C. (1994). Effects of physical activity on psychological variables in adolescents. *Pediatric Exercise Science*, *6*, 406–423.

Caporaso, N., Gu, F., Chatterjee, N., Sheng-Chih, J., Yu, K., Yeager, M., et al. (2009). Genome-wide and candidate gene association study of cigarette smoking behaviors. *PLoS One*, *4*(2), e4653.

Cardenas, L., Tremblay, L. K., Naranjo, C. A., Herrmann, N., Zack, M., & Busto, U. E. (2002). Brain reward system activity in major depression and comorbid nicotine dependence. *The Journal of Pharmacology and Experimental Therapeutics, 302*(3), 1265–1271.

Centers for Disease Control and Prevention. CDC. (2008). *Surveillance Summaries,* June 6, 2008. Atlanta, GA. MMWR; 57 (No. SS–4).

Chambers, R. A., & Potenza, M. N. (2003). Developmental neurocircuitry of motivation in adolescence: a critical period of addiction vulnerability. *The American Journal of Psychiatry, 160*(6), 1041–1052.

Clayton, R. R. (1999). Tobacco prevention research: new partnerships and paradigms of psychosocial approaches to understanding the etiology of tobacco use. *Nicotine and Tobacco Research, 1(Suppl 1)*, S57–S58.

Clayton, R. R., Scutchfield, F. D., & Wyatt, S. W. (2000). Hutchinson Smoking Prevention Project: a new gold standard in prevention science requires new transdisciplinary thinking. *Journal of the National Cancer Institute, 92*(24), 1964–1965.

Colder, C. R., Mehta, P., Balanda, K., Campbell, R. T., Mayhew, K. P., Stanton, W. R., et al. (2001). Identifying trajectories of adolescent smoking: an application of latent growth mixture modeling. *Health Psychology, 20*(2), 127–135.

Comings, D., Ferry, L., Bradshaw-Robinson, S., Burchette, R., Chiu, C., & Muhleman, D. (1996). The dopamine D2 receptor (DRD2) gene: a genetic risk factor in smoking. *Pharmacogenetics, 6*, 73–79.

Cullen, R., & Marshall, S. (2006). Genetic research and genetic information: a health information professional's perspective on the benefits and risks. *Health Information and Libraries Journal, 23*(4), 275–282.

D'Silva, M. U., Harrington, N. G., Palmgreen, P., Donohew, L., & Lorch, E. P. (2001). Drug use prevention for the high sensation seeker: the role of alternative activities. *Substance Use and Misuse, 36*(3), 373–385.

Davis, K. C., Farrelly, M. C., Messeri, P., & Duke, J. (2009). The impact of national smoking prevention campaigns on tobacco-related beliefs, intentions to smoke and smoking initiation: results from a longitudinal survey of youth in the United States. *International Journal of Environmental Research and Public Health, 6*(2), 722–740.

Davis, K. C., Nonnemaker, J. M., & Farrelly, M. C. (2007). Association between national smoking prevention campaigns and perceived smoking prevalence among youth in the United States. *Journal of Adolescent Health, 41*(5), 430–436.

Dick, D. M., Aliev, F., Wang, J. C., Grucza, R. A., Schuckit, M., Kuperman, S., et al. (2008). Using dimensional models of externalizing psychopathology to aid in gene identification. *Archives of General Psychiatry, 65*(3), 310–318.

Dick, D. M., Viken, R., Purcell, S., Kaprio, J., Pulkkinen, L., & Rose, R. J. (2007). Parental monitoring moderates the importance of genetic and environmental influences on adolescent smoking. *Journal of Abnormal Psychology, 116*(1), 213–218.

Eaves, L., & Eysenck, H. (1975). The nature of extraversion: a genetical analysis. *Journal of personality and social psychology, 32*(1), 102–112.

Eisenberg, D. T., Mackillop, J., Modi, M., Beauchemin, J., Dang, D., Lisman, S. A., et al. (2007). Examining impulsivity as an endophenotype using a behavioral approach: a DRD2 TaqI A and DRD4 48-bp VNTR association study. *Behavioral and Brain Functions, 3*, 2.

Ellickson, P. L., Tucker, J. S., & Klein, D. J. (2008). Reducing early smokers' risk for future smoking and other problem behavior: insights from a five-year longitudinal study. *Journal of Adolescent Health, 43*(4), 394–400.

Forbes, E. E., Christopher May, J., Siegle, G. J., Ladouceur, C. D., Ryan, N. D., Carter, C. S., et al. (2006). Reward-related decision-making in pediatric major depressive disorder: an fMRI study. *Journal of Child Psychology and Psychiatry, 47*(10), 1031–1040.

Fossella, J., Green, A. E., & Fan, J. (2006). Evaluation of a structural polymorphism in the ankyrin repeat and kinase domain containing 1 (ANKK1) gene and the activation of executive attention networks. *Cognitive, Affective & Behavioral Neuroscience, 6*(1), 71–78.

Fuke, S., Suo, S., Takahashi, N., Koike, H., Sasagawa, N., & Ishiura, S. (2001). The VNTR polymorphism of the human dopamine transporter (DAT1) gene affects gene expression. *Pharmacogenomics, 1*(2), 152–156.

Gerra, G., Garofano, L., Zaimovic, A., Moi, G., Branchi, B., Bussandri, M., et al. (2005). Association of the serotonin transporter promoter polymorphism with smoking behavior among adolescents. *American Journal of Medical Genetics. Part B, Neuropsychiatric Genetics, 135B*(1), 73–78.

Han, C., McGue, M. K., & Iacono, W. G. (1999). Lifetime tobacco, alcohol and other substance use in adolescent Minnesota twins: univariate and multivariate behavioral genetic analyses. *Addiction, 94*(7), 981–993.

Healton, C. (2002). Speaking truth(sm) to youth. How the American Legacy Foundation is helping teens reject tobacco. *North Carolina Medical Journal, 63*(3), 162–164.

Heils, A., Teufel, A., Petri, S., Stober, G., Riederer, P., Bengel, D., et al. (1996). Allelic variation of human serotonin transporter gene expression. *Journal of Neurochemistry, 66*(6), 2621–2624.

Heinz, A., Goldman, D., Jones, D., Palmour, R., Hommer, D., & Gorey, J., et al. (2000). Genotype influences in vivo dopamine transporter availability in human striatum. *Neuropsychopharmacology, 22*, 133–139.

Hiremagalur, B., Nankova, B., Nitahara, J., Zeman, R., & Sabban, E. L. (1993). Nicotine increases expression of tyrosine hydroxylase gene. Involvement of protein kinase A-mediated pathway. *Journal of Biological Chemistry, 268*(31), 23704–23711.

Hiremagalur, B., & Sabban, E. L. (1995). Nicotine elicits changes in expression of adrenal catecholamine biosynthetic enzymes, neuropeptide Y and immediate early genes by injection but not continuous administration. *Brain Research. Molecular Brain Research, 32*(1), 109–115.

Hoenicka, J., Quinones-Lombrana, A., Espana-Serrano, L., Alvira-Botero, X., Kremer, L., Perez-Gonzalez, R., et al. (2010). The ANKK1 gene associated with addictions is expressed in astroglial cells and upregulated by apomorphine. *Biological Psychiatry, 67*(1), 3–11.

Hopfer, C. J., Crowley, T. J., & Hewitt, J. K. (2003). Review of twin and adoption studies of adolescent substance use. *Journal of the American Academy of Child and Adolescent Psychiatry, 42*(6), 710–719.

Hu, M. C., Muthen, B., Schaffran, C., Griesler, P. C., & Kandel, D. B. (2008). Developmental trajectories of criteria of nicotine dependence in adolescence. *Drug and Alcohol Dependence, 98*(1-2), 94–104.

Ioannidis, J. P., Trikalinos, T. A., & Khoury, M. J. (2006). Implications of small effect sizes of individual genetic variants on the design and interpretation of genetic association studies of complex diseases. *American Journal of Epidemiology, 164*(7), 609–614.

Johnson, B. A. (2000). Serotonergic agents and alcoholism treatment: rebirth of the subtype concept—an hypothesis. *Alcoholism, Clinical and Experimental Research, 24*(10), 1597–1601.

Johnson, K., Strader, T., Berbaum, M., Bryant, D., Bucholtz, G., Collins, D., et al. (1996). Reducing alcohol and other drug use by strengthening community, family, and youth resiliency: an evaluation of the Creating Lasting Connections program. *Journal of Adolescent Research, 11*, 36–67.

Johnston, L. D., O'Malley, P. M., Bachman, J. G., & Schulenberg, J. E. (2008). *Monitoring the Future national survey results on drug use, 1975–2007: volume I, secondary school students (pub. no. 08-6418A).* Bethesda, MD: National Institutes of Health.

Karp, I., O'Loughlin, J., Hanley, J., Tyndale, R. F., & Paradis, G. (2006). Risk factors for tobacco dependence in adolescent smokers. *Tobacco Control, 15*(3), 199–204.

Kassel, J. D., Evatt, D. P., Greenstein, J. E., Wardle, M. C., Yates, M. C., & Veilleux, J. C. (2007). The acute effects of nicotine on positive and negative affect in adolescent smokers. *Journal of Abnormal Psychology, 116*, 543–553.

Kellam, S. G., & Anthony, J. C. (1998). Targeting early antecedents to prevent tobacco smoking: findings from an epidemiologically based randomized field trial. *American Journal of Public Health, 88*(10), 1490–1495.

Kellam, S. G., Brown, C. H., Poduska, J. M., Ialongo, N. S., Wang, W., Toyinbo, P., et al. (2008). Effects of a universal classroom behavior management program in first and second grades on young adult behavioral, psychiatric, and social outcomes. *Drug and Alcohol Dependence, 95(Suppl 1),* S5–S28.

Kendler, K. S., Jacobson, K. C., Prescott, C. A., & Neale, M. C. (2003). Specificity of genetic and environmental risk factors for use and abuse/dependence of cannabis, cocaine, hallucinogens, sedatives, stimulants, and opiates in male twins. *American Journal of Psychiatry, 160*(4), 687–695.

Kenny, P. J., & Markou, A. (2006). Nicotine self-administration acutely activates brain reward systems and induces a long-lasting increase in reward sensitivity. *Neuropsychopharmacology, 31*(6), 1203–1211.

Khantzian, E. J. (1997). The self-medication hypothesis of substance use disorders: a reconsideration and recent applications. *Harvard Review of Psychiatry, 4*(5), 231–244.

King, G. R., Xiong, Z., Douglas, S., Lee, T. H., & Ellinwood, E. H. (1999). The effects of continuous cocaine dose on the induction of behavioral tolerance and dopamine autoreceptor function. *European Journal of Pharmacology, 376*(3), 207–215.

Kirby, K. N., Petry, N. M., & Bickel, W. K. (1999). Heroin addicts have higher discount rates for delayed rewards than non-drug-using controls. *Journal of Experimental Psychology*, *128*(1), 78–87.

Koopmans, J. R., Boomsma, D. I., Heath, A. C., & van Doornen, L. J. (1995). A multivariate genetic analysis of sensation seeking. *Behavioral Genetics*, *25*(4), 349–356.

Koopmans, J. R., Slutske, W. S., Heath, A. C., Neale, M. C., & Boomsma, D. I. (1999). The genetics of smoking initiation and quantity smoked in Dutch adolescent and young adult twins. *Behavioral Genetics*, *29*(6), 383–393.

Kreek, M. J., Nielsen, D. A., Butelman, E. R., & LaForge, K. S. (2005). Genetic influences on impulsivity, risk taking, stress responsivity and vulnerability to drug abuse and addiction. *Nature Neuroscience*, *8*(11), 1450–1457.

Krishnan-Sarin, S., Reynolds, B., Duhig, A. M., Smith, A., Liss, T., McFetridge, A., et al. (2007). Behavioral impulsivity predicts treatment outcome in a smoking cessation program for adolescent smokers. *Drug and Alcohol Dependence*, *88*(1), 79–82.

Laucht, M., Becker, K., El-Faddagh, M., Hohm, E., & Schmidt, M. H. (2005). Association of the DRD4 exon III polymorphism with smoking in fifteen-year-olds: a mediating role for novelty seeking? *Journal of the American Academy of Child and Adolescent Psychiatry*, *44*(5), 477–484.

Laucht, M., Becker, K., Frank, J., Schmidt, M. H., Esser, G., Treutlein, J., et al. (2008). Genetic variation in dopamine pathways differentially associated with smoking progression in adolescence. *Journal of the American Academy of Child and Adolescent Psychiatry*, *47*(6), 673–681.

Laviola, G., Adriani, W., Terranova, M. L., & Gerra, G. (1999). Psychobiological risk factors for vulnerability to psychostimulants in human adolescents and animal models. *Neuroscience and Biobehavioral Reviews*, *23*(7), 993–1010.

Lerman, C., Audrain, J., Main, D., Boyd, N., Caporaso, N., Bowman, E., et al. (1999). Evidence suggesting the role of specific genetic factors in cigarette smoking. *Health Psychology*, *18*(1), 14–20.

Lerman, C., Caporaso, N., Audrain, J., Main, D., Boyd, N., & Shields, P. (2000). Interacting effects of the serotonin transporter gene and neuroticism in smoking practices and nicotine dependence. *Molecular Psychology*, *5*, 189–192.

Lerman, C., & Niaura, R. (2002). Applying genetic approaches to the treatment of nicotine dependence. *Oncogene*, *21*(48), 7412–7420.

Lerman, C., Perkins, K., & Gould, T. (2009). Nicotine-Dependence Endophenotypes in Chronic Smokers. In G. E. Swan, T. B. Baker, L. Chassin, D. V. Conti, C. Lerman & K. A. Perkins (Eds.), *Phenotypes and endophenotypes: foundations for genetic studies of nicotine use and dependence* (Vol. Tobacco Control Monograph No. 20). Bethesda, MD: U.S. Department of Health and Human Services, National Institutes of Health, National Cancer Institute.

Lerman, C., Shields, P. G., Main, D., Audrain, J., Roth, J., Boyd, N. R., et al. (1997). Lack of association of tyrosine hydroxylase genetic polymorphism with cigarette smoking. *Pharmacogenetics*, *7*(6), 521–524.

Lerman, C., Shields, P. G., Wileyto, E. P., Audrain, J., Hawk, L. H., Jr., Pinto, A., et al. (2003). Effects of dopamine transporter and receptor polymorphisms on smoking cessation in a bupropion clinical trial. *Health Psychology, 22*(5), 541–548.

Lesch, K. P., Bengel, D., Heils, A., Sabol, S. Z., Greenberg, B. D., Petri, S., et al. (1996). Association of anxiety-related traits with a polymorphism in the serotonin transporter gene regulatory region. *Science, 274*(5292), 1527–1531.

Lewinsohn, P. M., Brown, R. A., Seeley, J. R., & Ramsey, S. E. (2000). Psychosocial correlates of cigarette smoking abstinence, experimentation, persistence and frequency during adolescence. *Nicotine and Tobacco Research, 2*(2), 121–131.

Lewis-Esquerre, J. M., Rodrigue, J. R., & Kahler, C. W. (2005). Development and validation of an adolescent smoking consequences questionnaire. *Nicotine and Tobacco Research, 7*(1), 81–90.

Lichter, J. B., Barr, C. L., Kennedy, J. L., Van Tol, H. H., Kidd, K. K., & Livak, K. J. (1993). A hypervariable segment in the human dopamine receptor D4 (DRD4) gene. *Human Molecular Genetics, 2*(6), 767–773.

Lyons, M., Hitsman, B., Xian, H., Panizzon, M. S., Jerskey, B. A., Santangelo, S., et al. (2008). A twin study of smoking, nicotine dependence, and major depression in men. *Nicotine and Tobacco Research, 10*(1), 97–108.

Madden, G. J. (2000). A behavioral economics primer. In W. K. Bickel & R. E. Vuchinich (Eds.), *Reframing health behavior change with behavior economics* (pp. 3–25). Mahwah, NJ: Lawrence Erlbaum Associates.

Madden, G. J., Petry, N. M., Badger, G. J., & Bickel, W. K. (1997). Impulsive and self-control choices in opioid-dependent patients and non-drug-using control participants: drug and monetary rewards. *Experimental and Clinical Psychopharmacology, 5*(3), 256–262.

Masse, L. C., & Tremblay, R. E. (1997). Behavior of boys in kindergarten and the onset of substance use during adolescence. *Archives of General Psychiatry, 54*(1), 62–68.

Mayhew, K., Flay, B., & Mott, J. (2000). Stages in the development of adolescent smoking. *Drug and Alcohol Dependence, 59*, S61–S81.

McClernon, F. J., & Kollins, S. H. (2008). ADHD and smoking: from genes to brain to behavior. *Annuls of the New York Academy of Science, 1141*, 131–147.

McGue, M., Elkins, I., & Iacono, W. G. (2000). Genetic and environmental influences on adolescent substance use and abuse. *American Journal of Medical Genetics, 96*(5), 671–677.

Melnick, M. J., Miller, K. E., Sabo, D. F., Farrell, M. P., & Barnes, G. M. (2001). Tobacco use among high school athletes and nonatheletes: results of the 1997 youth risk behavior survey. *Adolescence, 36*(144), 727–747.

Messina, E. S., Tyndale, R. F., & Sellers, E. M. (1997). A major role for CYP2A6 in nicotine C-oxidation by human liver microsomes. *The Journal of Pharmacology and Experimental Therapeutics, 282*(3), 1608–1614.

Mitchell, S. H. (1999). Measures of impulsivity in cigarette smokers and non-smokers. *Psychopharmacology (Berlin), 146*(4), 455–464.

Monterosso, J., & Ainslie, G. (1999). Beyond discounting: possible experimental models of impulse control. *Psychopharmacology (Berlin), 146*(4), 339–347.

Moolchan, E. T., & Mermelstein, R. (2002). Research on tobacco use among teenagers: ethical challenges. *Journal of Adolescent Health, 30*(6), 409–417.

Nakajima, M., Yamamoto, T., Nunoya, K., Yokoi, T., Nagashima, K., Inoue, K., et al. (1996). Characterization of CYP2A6 involved in 3'-hydroxylation of cotinine in human liver microsomes. *The Journal of Pharmacology and Experimental Therapeutics, 277*(2), 1010–1015.

Neville, M. J., Johnstone, E. C., & Walton, R. T. (2004). Identification and characterization of ANKK1: a novel kinase gene closely linked to DRD2 on chromosome band 11q23.1. *Human Mutation, 23*(6), 540–545.

Noble, E., St. Jeor, S., Ritchie, T., Syndulko, K., St. Jeor, S., Fitch, R., et al. (1994). D2 dopamine receptor gene and cigarette smoking: a reward gene? *Medical Hypotheses, 42*, 257–260.

Noble, E. P., Blum, K., Ritchie, T., Montgomery, A., & Sheridan, P. J. (1991). Allelic association of the D2 dopamine receptor gene with receptor-binding characteristics in alcoholism. *Archives of General Psychiatry, 48*(7), 648–654.

O'Loughlin, J., Paradis, G., Kim, W., DiFranza, J., Meshefedjian, G., McMillan-Davey, E., et al. (2004). Genetically decreased CYP2A6 and the risk of tobacco dependence: a prospective study of novice smokers. *Tobacco Control, 13*(4), 422–428.

Olsson, C., Anney, R., Forrest, S., Patton, G., Coffey, C., Cameron, T., et al. (2004). Association between dependent smoking and a polymorphism in the tyrosine hydroxylase gene in a prospective population-based study of adolescent health. *Behavioral Genetics, 34*(1), 85–91.

Perkins, K., Gerlach, D., Broge, M., Grobe, J., & Wilson, A. (2000). Greater sensitivity to subjective effects of nicotine in nonsmokers high in sensation seeking. *Experimental and Clinical Psychopharmacology, 8*(4), 462–471.

Phillips, A. G. (1984). Brain reward circuitry: a case for separate systems. *Brain Research Bulletin, 12*(2), 195–201.

Picciotto, M. R., Brunzell, D. H., & Caldarone, B. J. (2002). Effect of nicotine and nicotinic receptors on anxiety and depression. *Neuroreport, 13*(9), 1097–1106.

Quattrocki, E., Baird, A., & Yurgelun-Todd, D. (2000). Biological aspects of the link between smoking and depression. *Harvard Review of Psychiatry, 8*(3), 99–110.

Raitakari, O. T., Porkka, K. V., Taimela, S., Telama, R., Rasanen, L., & Viikari, J. S. (1994). Effects of persistent physical activity and inactivity on coronary risk factors in children and young adults. The Cardiovascular Risk in Young Finns Study. *American Journal of Epidemiology, 140*(3), 195–205.

Ritchie, T., & Noble, E. P. (1996). [3H]naloxone binding in the human brain: alcoholism and the TaqI A D2 dopamine receptor polymorphism. *Brain Research, 718*(1–2), 193–197.

Rodriguez, D., & Audrain-McGovern, J. (2004). Team sport participation and smoking: analysis with general growth mixture modeling. *Journal of Pediatric Psychology, 29*(4), 299–308.

Rodriguez, D., Tercyak, K. P., & Audrain-McGovern, J. (2008). Effects of inattention and hyperactivity/impulsivity symptoms on development of nicotine dependence from mid adolescence to young adulthood. *Journal of Pediatric Psychology, 33*(6), 563–575.

Sabol, S., Nelson, M., Fisher, C., Gunzerath, L., Brody, C., & Hu, S. (1999). A genetic association for cigarette smoking behavior. *Health Psychology, 18,* 7–13.

SAMHSA. (2008). *Results from the 2007 national survey on drug use and health, NSDUH: detailed tables.* Retrieved from http://www.oas.samhsa.gov/NSDUH/2k7NSDUH/tabs/Sect4peTabs10to11.pdf.

Schoedel, K. A., Hoffmann, E. B., Rao, Y., Sellers, E. M., & Tyndale, R. F. (2004). Ethnic variation in CYP2A6 and association of genetically slow nicotine metabolism and smoking in adult Caucasians. *Pharmacogenetics, 14*(9), 615–626.

Serova, L., Danailov, E., Chamas, F., & Sabban, E. L. (1999). Nicotine infusion modulates immobilization stress-triggered induction of gene expression of rat catecholamine biosynthetic enzymes. *The Journal of Pharmacology and Experimental Therapeutics, 291*(2), 884–892.

Shanahan, M. J., & Hofer, S. M. (2005). Social context in gene-environment interactions: retrospect and prospect. *The Journals of Gerontology. Series B, Psychological Sciences and Social Sciences, 60 Spec No 1,* 65–76.

Shields, A., Lerman, C., & Sullivan, P. (2004). Translating emerging research on the genetics of smoking into clinical practice: ethical and social considerations. *Nicotine and Tobacco Research, 6*(4), 675–688.

Shields, P., Lerman, C., Audrain, J., Main, D., Boyd, N., & Caporaso, N. (1998). Dopamine D4 receptors and the risk of cigarette smoking in African-Americans and Caucasians. *Cancer Epidemiology, Biomarkers & Prevention, 7,* 453–458.

Shytle, R. D., Silver, A. A., Lukas, R. J., Newman, M. B., Sheehan, D. V., & Sanberg, P. R. (2002). Nicotinic acetylcholine receptors as targets for antidepressants. *Molecular Psychiatry, 7*(6), 525–535.

Skara, S., & Sussman, S. (2003). A review of 25 long-term adolescent tobacco and other drug use prevention program evaluations. *Preventive Medicine, 37*(5), 451–474.

Skara, S., Sussman, S., & Dent, C. W. (2001). Predicting regular cigarette use among continuation high school students. *American Journal of Health and Behavior, 25*(2), 147–156.

Skowronek, M. H., Laucht, M., Hohm, E., Becker, K., & Schmidt, M. H. (2006). Interaction between the dopamine D4 receptor and the serotonin transporter promoter polymorphisms in alcohol and tobacco use among 15-year-olds. *Neurogenetics, 7*(4), 239–246.

Slomkowski, C., Rende, R., Novak, S., Lloyd-Richardson, E., & Niaura, R. (2005). Sibling effects on smoking in adolescence: evidence for social influence from a genetically informative design. *Addiction, 100*(4), 430–438.

Soldz, S., & Cui, X. (2002). Pathways through adolescent smoking: a 7-year longitudinal grouping analysis. *Health Psychology, 21*(5), 495–504.

Spear, L. P. (2000). The adolescent brain and age-related behavioral manifestations. *Neuroscience & Biobehavioral Reviews, 24*(4), 417–463.

Spear, L. P. (2002). The adolescent brain and the college drinker: biological basis of propensity to use and misuse alcohol. *Journal of Studies on Alcohol, 63*(Suppl 14), 71–81.

Spitz, M., Shi, H., Yang, F., Hudmon, K., Jiang, H., & Chanberlain, R. (1998). Case-control study of the D2 dopamine receptor gene and smoking status in lung cancer patients. *Journal of the National Cancer Institute*, *90*, 358–363.

Steptoe, A., & Butler, N. (1996). Sports participation and emotional wellbeing in adolescents. *Lancet*, *347*(9018), 1789–1792.

Storr, C. L., Ialongo, N. S., Kellam, S. G., & Anthony, J. C. (2002). A randomized controlled trial of two primary school intervention strategies to prevent early onset tobacco smoking. *Drug and Alcohol Dependence*, *66*(1), 51–60.

Swan, G. E. (1999). Implications of genetic epidemiology for the prevention of tobacco use. *Nicotine and Tobacco Research*, *1* (Suppl 1), S49–S56.

Tercyak, K. P., Peshkin, B. N., Wine, L. A., & Walker, L. R. (2006). Interest of adolescents in genetic testing for nicotine addiction susceptibility. *Preventive Medicine*, *42*(1), 60–65.

Thompson, J., Thomas, N., Singleton, A., Piggott, M., Lloyd, S., Perry, E. K., et al. (1997). D2 dopamine receptor gene (DRD2) Taq1 A polymorphism: reduced dopamine D2 receptor binding in the human striatum associated with the A1 allele. *Pharmacogenetics*, *7*(6), 479–484.

Timberlake, D. S., Rhee, S. H., Haberstick, B. C., Hopfer, C., Ehringer, M., Lessem, J. M., et al. (2006). The moderating effects of religiosity on the genetic and environmental determinants of smoking initiation. *Nicotine and Tobacco Research*, *8*(1), 123–133.

Tremblay, L. K., Naranjo, C. A., Cardenas, L., Herrmann, N., & Busto, U. E. (2002). Probing brain reward system function in major depressive disorder: altered response to dextroamphetamine. *Archives of General Psychiatry*, *59*(5), 409–416.

Tyndale, R. F., & Sellers, E. M. (2001). Variable CYP2A6-mediated nicotine metabolism alters smoking behavior and risk. *Drug Metabolism and Disposition*, *29*(4, pt. 2), 548–552.

Vallone, D., Allen, J. A., Clayton, R. R., & Xiao, H. (2007). How reliable and valid is the Brief Sensation Seeking Scale (BSSS-4) for youth of various racial/ethnic groups? *Addiction*, *102(Suppl 2)*, 71–78.

Van Tol, H. H., Wu, C. M., Guan, H. C., Ohara, K., Bunzow, J. R., Civelli, O., et al. (1992). Multiple dopamine D4 receptor variants in the human population. *Nature*, *358*(6382), 149–152.

Volkow, N. D., Fowler, J. S., Wang, G. J., & Goldstein, R. Z. (2002). Role of dopamine, the frontal cortex and memory circuits in drug addiction: insight from imaging studies. *Neurobiology of Learning and Memory*, *78*(3), 610–624.

Wahl, S. K., Turner, L. R., Mermelstein, R. J., & Flay, B. R. (2005). Adolescents' smoking expectancies: psychometric properties and prediction of behavior change. *Nicotine and Tobacco Research*, *7*(4), 613–623.

Weiss, J. W., Merrill, V., & Gritz, E. R. (2007). Ethnic variation in the association between weight concern and adolescent smoking. *Addictive Behaviors*, *32*(10), 2311–2316.

Wilfond, B. S., Geller, G., Lerman, C., Audrain-McGovern, J., & Shields, A. E. (2002). Ethical issues in conducting behavioral genetics research: the case of smoking prevention trials among adolescents. *Journal of Health Care, Law and Policy*, *6*(1), 73–88.

Wong, A. H., & Van Tol, H. H. (2003). Schizophrenia: from phenomenology to neurobiology. *Neuroscience and Biobehavioral Reviews, 27,* 269–306.

Wright, A. J., Weinman, J., & Marteau, T. M. (2003). The impact of learning of a genetic predisposition to nicotine dependence: an analogue study. *Tobacco Control, 12*(2), 227–230.

Xu, C., Goodz, S., Sellers, E. M., & Tyndale, R. F. (2002). CYP2A6 genetic variation and potential consequences. *Advanced Drug Delivery Reviews, 54*(10), 1245–1256.

14

The Dynamic Nature of Genetic and Environmental Influences on Alcohol Use and Dependence

DANIELLE M. DICK

Research in the area of alcohol dependence is particularly challenging and exciting (in the opinion of one admittedly biased researcher who has chosen to devote her energies to this area) in that understanding this debilitating disorder necessitates a dynamic perspective—one that integrates genetic, developmental, and environmental perspectives. It is a model for the kind of complex questions that form the central theme of this book. It forces us to throw out old debates about nature versus nurture. Over the course of history, widely divergent views on the "cause" of alcohol problems have been expressed, ranging from the view that alcohol problems result from deficits of moral character, to strong statements that alcohol dependence is a biological disease. Clearly, the causes of alcohol dependence are neither purely environmental nor purely genetic. Further, no one wakes up one morning to discover they have had an onset of alcohol dependence overnight; rather, the eventual diagnosis of alcohol dependence is usually preceded by a long trajectory of risk-related behaviors. Accordingly, understanding why some individuals eventually develop alcohol problems necessitates our understanding how genetic and environmental influences come together across development to contribute to different trajectories of risk.

This chapter explores the dynamic nature of genetic and environmental influences on alcohol use and alcohol problems, as they unfold across development. The chapter is divided loosely into two parts. The first reviews studies in the field of genetic epidemiology that address the dynamic developmental nature of genetic and environmental influences on alcohol dependence. This section almost exclusively discusses findings from twin studies, in part because this is the methodology that has been most used to study alcohol use and dependence, and in part because this is the method that I use in my own research. In twin studies, genetic influences are inferred, via comparisons of family members with different degrees of genetic relatedness, rather than explicitly measured. The second part of the chapter deals with studies that are aimed at gene identification, focusing on specific genes that have been associated with alcohol dependence and related traits. Although the chapter is divided in this way for the sake of presentation, a take-home message that I hope will be conferred to the reader is the need to integrate across research traditions in order to best capture the complexity of how genetic and environmental influences impact the development of alcohol dependence and to advance the science.

GENETIC EPIDEMIOLOGY

That alcohol dependence is heritable has been widely established and is largely accepted (at least among the scientific community). Heritability estimates from several large twin studies have converged on genetic estimates of 50%–60% for both men and women (Dick, Prescott, & McGue, 2009). Demonstrating the heritability of alcohol dependence was pivotal in dispelling views that alcohol dependence was simply a moral disease, or purely a result of environmental factors. Clearly and convincingly, twin studies have demonstrated that we are not all equally susceptible to developing problems associated with alcohol use. But this static estimate fails to capture the dynamic nature of genetic and environmental influences across development. What might be the genetic influences on alcohol dependence? And, since we know that the onset of alcohol dependence is usually not until the early to mid-20s, what might those genes be doing across development? We turn our attention first to research that sheds light on these questions.

Genes "for" Alcohol Dependence?

What exactly does a genetic predisposition to alcohol dependence mean? Data from several twin studies shed light on this question. In the basic twin model, comparisons of monozygotic twins (MZs; who share 100% of their genetic variation and all of their shared environment, when reared together) and dizygotic twins (DZs; who share only, on average, 50% of their genetic variation, but also share 100% of their shared environmental influences when reared together) yield information about the relative importance of genetic and environmental influences. To the extent that MZs are more alike than DZs, genetic influences are implicated. If DZs are just as similar as MZs, then shared environmental processes, such as those influences found in the shared family environment, shared peers, shared schools and neighborhoods, and the like, must predominate. If MZs are not exactly identical (as they would be if an outcome were 100% genetically influenced), then unique environmental processes must play a role. These could include environmental influences that are unique to an individual, such as a particular life event, stressor, or other influence not shared with their co-twin, and/or environmental events that differentially effect the co-twins (Turkheimer & Waldron, 2000).

Across the many twin studies conducted on alcohol dependence, an individual was far more likely to be alcohol dependent if he or she had an MZ co-twin who was an alcoholic than a DZ co-twin who was an alcoholic, thus producing the robust heritability estimates mentioned above. This basic design has been extended to test the extent to which genetic and environmental influences contribute to the covariance, or observed overlap, across multiple different disorders or outcomes. The rationale here is the same as the basic twin design, just extended to multiple variables: If MZ cross-twin cross-trait correlations are higher than are DZ cross-twin cross-trait correlations, then the correlation between the outcomes is genetically influenced. In other words, if the correlation between twin 1's alcohol use and twin 2's antisocial behavior is higher in MZs than in DZs, it suggests that genetic influences are shared between those two outcomes and that those genetic influences contribute, at least in part, to the observed overlap across alcohol use and antisocial behavior. And the logic behind comparing the cross-twin cross-trait correlations can be extended to estimate the importance of shared environmental and unique environmental influences in contributing to the correlation between the variables (in this example, alcohol use and antisocial behavior).

One of the central findings that has emerged from this kind of twin research, which has compared the genetic and environmental structure across many major psychiatric disorders, is that much of the heritability of alcohol dependence (i.e., the genetic influences on alcohol dependence) is not unique to alcohol dependence at all, but rather is shared with a number of other psychiatric conditions that fall under the general spectrum of externalizing psychopathology and includes other forms of drug dependence, adult antisocial behavior, and childhood conduct disorder. This has been robustly demonstrated across a number of independent twin samples

(Kendler, Prescott, Myers, & Neale, 2003; Krueger et al., 2002; Young, Stallings, Corley, Krauter, & Hewitt, 2000). In fact, one of the largest studies of this kind, using data from the Virginia Adult Twin Study of Psychiatric and Substance Use Disorders estimated that approximately 65% of the genetic influences on alcohol dependence were shared with these other disorders, with only about 35% of the heritability being genes that are specific to alcohol dependence (Kendler et al., 2003). These latter influences are likely to include genes that are involved in alcohol metabolism and related pathways.

Genetic variation in several of the alcohol dehydrogenase (ADH) genes, as well as in the gene *ALDH2*, which plays the primary role in converting acetaldehyde to acetate after the metabolism of ethanol to acetaldehyde by the ADH enzymes, have been associated with susceptibility to alcohol dependence (Edenberg et al., 2006; Kuo et al., 2008; Whitfield, 1997). But many of the genes that alter susceptibility to alcohol dependence must have their effects through more nonspecific pathways, such as those involved in reward dependence or behavioral disinhibition. Accordingly, they impact a number of different outcomes and contribute to the substantial comorbidity that surrounds alcohol problems. This also has important implications for understanding the developmental trajectory of risk for alcohol problems, as it suggests that a genetic predisposition toward alcohol problems can manifest as childhood behavioral problems at an earlier stage in development. So, returning to the question posed earlier, there are no genes "for" alcohol dependence—only genes that alter one's susceptibility to develop risk. Furthermore, most of these are likely to operate through nonspecific pathways that manifest as a variety of risky externalizing behaviors across development.

How Do Genetic Influences on Alcohol Use/Dependence Unfold Across Time?

Before an individual ever has the possibility of developing alcohol problems, he or she must by necessity have moved through a series of initial stages, starting with the initiation of alcohol use, the establishment of regular drinking patterns, and (for individuals who go on to develop alcohol dependence) the development of excessive drinking and related problems. Accordingly, it is important to understand how genetic and environmental influences impact these various stages of alcohol use. This is best accomplished using longitudinal research designs. Much of the longitudinal research on alcohol use has focused on the period from early adolescence to young adulthood. This period represents a critical timeframe in the developmental course for alcohol use disorders, as during this time most adolescents initiate alcohol use and subsequently move from early experimentation to more established patterns of use. Twin studies demonstrate that the importance of genetic and environmental influences changes dramatically over this developmental period.

Data from two population-based longitudinal Finnish twin studies illustrate the striking shift in the relative importance of genetic and environmental influences that occurs from early adolescence to young adulthood (Fig. 14.1): A steady increase occurs in the relevance of genetic factors across adolescence, and a corresponding and sharp decrease occurs in the relevance of common environmental influences (Rose, Dick, Viken, & Kaprio, 2001). These data demonstrate that, although alcohol initiation is largely environmentally influenced, as has also been found in numerous other twin studies (Hopfer, Crowley, & Hewitt, 2003), as drinking patterns become more regular and established across adolescence, genetic factors assume increasing importance; however, alcohol use early in adolescence is influenced largely by family, school, and neighborhood factors (Rose, Dick, Viken, Pulkkinen, & Kaprio, 2001; Rose et al., 2003). Reassuringly, a very similar pattern of results for alcohol use was obtained by a life-history method in male twin pairs from the Virginia Adult Twin Study of Psychiatric and Substance Use Disorders (Kendler, Schmitt, Aggen, & Prescott, 2008). At age 14, all twin resemblance resulted from shared environmental factors. From ages 14 to 23, shared environment became progressively less important and genetic factors more important.

Although measures of alcohol *use* clearly show genetic influence in adolescence, the more limited number of studies investigating alcohol *dependence* symptoms in early adolescence suggest a very different picture than studies of alcohol dependence in adults. Analyses of alcohol dependence symptoms at age 14 in a large sample of Finnish twins found no evidence of genetic effects in either girls or boys at this age (Rose et al., 2004).

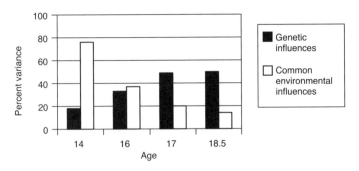

Figure 14.1. **Changing influences on alcohol use across time.**

Data from the Missouri Adolescent Female Twin Study showed a similar pattern of results, with alcohol dependence symptoms in early adolescence largely influenced by environmental factors (Knopik, 2005). These studies suggest that the factors influencing alcohol dependence symptoms that are evident very early in adolescence may differ from those that impact on the etiology of alcohol dependence in adulthood. Longitudinal studies of dependence symptoms from adolescence into adulthood will be necessary to further explore this question. Interestingly, a recent study of alcohol use problems, as assessed at five time points from ages 19 to 28 in the Dutch Twin Registry (Kendler et al., under review) found strong innovation and attenuation of genetic factors even across this age range, thus indicating that some genetic influences on alcohol problems that were evident at age 19 declined in importance across time, whereas new genetic influences became important starting at ages 21 and 23. We know that the heritability of alcohol dependence likely reflects the influence of many genes of small effect, and that these genes can be activated at different times across development (perhaps in conjunction with environmental factors, as detailed in earlier chapters in this book). These analyses suggest that different genes may be important at different developmental periods. Further, they indicate that genetic influences continue to be developmentally dynamic even beyond adolescence.

Do Genetic Influences on Alcohol Use/Dependence Change As a Function of the Environment?

For anyone reading this book in its entirety up to this point, it will come as no surprise that the answer to this question is a resounding "*yes*"! Overall heritability estimates yielded by the basic twin design essentially average across any differences that might exist by environment. If there is reason to believe that genetic influences may be more or less important under different environmental circumstances, that information should be measured and can be directly incorporated into the twin model. The degree of genetic and environmental influence can then be estimated for different environmental conditions (e.g., in urban vs. rural settings) or, in the case of environments that vary more continuously (e.g., parent–child warmth or adversity), the degree to which genetic influences change as a function of the environmental moderator can be estimated. Although the details of the modeling go beyond the scope of this chapter, they can be found in Purcell (2002). The end result is that, rather than a static heritability estimate, one can estimate changing genetic and environmental influences as a function of a measured environment.

The study of gene–environment interaction in the area of substance dependence is unique in that exposure to a particular "environment" (e.g., access to the substance) is a necessary condition for development of the disorder. At one extreme, one can imagine the case in which an individual carries a "loaded genetic predisposition" to develop alcohol problems, meaning that they carry risk alleles at every gene that is involved in altering susceptibility to develop problems. But if they are raised in an environment in which there is no access to alcohol, perhaps for geographic, political, or religious reasons, they would never become an alcoholic. In that case, the environment has "trumped" the effect of genes. That individual's genetic make-up played no role in whether or not they developed alcohol problems; it essentially never had opportunity to, due to social-environmental controls. Working from that extreme, one can then imagine a number of less extreme situations in which the environment might serve to moderate how important genetic influences are by altering opportunity to use the substance and/or the environmental or social control on the use of the substance. Indeed, as this has become a research area of increasing interest (and corresponding advances have been made in statistical modeling to test these more complex interactions), a number of such interactions have been detected.

One of the earliest illustrations of gene–environment interaction in the area of substance use research demonstrated that genetic influences on alcohol use were greater among unmarried women, whereas having a marriage-like relationship reduced the impact of genetic influences on drinking (Heath, Jardine, & Martin, 1989). Religiosity has also been shown to moderate genetic influences on alcohol use among females, with genetic factors playing a larger role among individuals without a religious upbringing (Koopmans, Slutske, van Baal, & Boomsma, 1999). Genetic influences on adolescent substance use are also enhanced in the presence of substance-using friends (Dick et al., 2007c) and in environments with lower parental monitoring (Dick et al., 2007d).

Figure 14.2 illustrates the dramatic shift in the relative importance of genetic effects that can take place across different environments using data from a Finnish twin project: At the extreme low end of parental monitoring, genetic effects assumed the greatest role in impacting adolescent smoking, whereas in homes with very high parental monitoring, genetic effects played little to no role, and common environmental factors were the most important influence (Dick et al., 2007d). Similar effects have been demonstrated for more general externalizing behavior: Genetic influences on antisocial behavior were higher in the presence of delinquent peers (Button et al., 2007) and in environments characterized by high parental negativity (Feinberg, Button, Neiderhiser, Reiss, & Hetherington, 2007), low parental

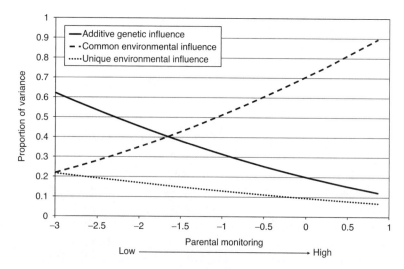

Figure 14.2. **Parental monitoring and smoking quantity.**

warmth (Feinberg et al., 2007), and high paternal punitive discipline (Button, Lau, Maughan, & Eley, 2008).

Socioregional, or neighborhood-level influences have also been shown to moderate the importance of genetic influences on substance use. Genetic influences on late adolescent alcohol use and early adolescent behavior problems are enhanced in urban environments, communities characterized by greater migration, and neighborhoods with higher percentages of slightly older adolescents/young adults (Dick, Rose, Viken, Kaprio, & Koskenvuo, 2001; Dick et al., 2009a; Rose et al., 2001). These moderation effects presumably reflect differences in availability of alcohol, a range of possible different role models, neighborhood stability, and community-level monitoring across different areas.

In looking over the list of environments for which demonstrated moderating effects on genetic influences exist, they all appear to reflect differential social control and/or opportunity, resulting in differential expression of individual predispositions (Shanahan & Hofer, 2005). Further, it is likely that the importance of different environments as moderators of genetic effects varies across developmental stage. There is some indication of this in the Finnish twin data, where parental monitoring showed significant moderating effects on substance use starting earlier in adolescence (age 14), whereas the moderating role of peer substance use was not apparent until later in adolescence (age 17) (Dick et al., 2007b), and where socioregional influences that moderated alcohol use in young adulthood did not show evidence of moderation of early adolescent alcohol use (Dick et al., 2009a). Systematic studies of the moderating role of different environmental factors across different developmental periods are needed.

Cascading Pathways of Risk in Alcohol Dependence

So far, we have discussed the dramatic developmental changes that occur in the degree of importance of genetic and environmental influences across time and on different stages of the history of alcohol use and dependence. And we have reviewed the literature suggesting widespread evidence for gene–environment interaction in alcohol use. We have not discussed evidence for gene–environment correlation, and this topic will be addressed in greater detail in a subsequent chapter (Chapter 16, by Kendler). However, here too, alcohol use represents an area in which these processes are likely to be critical, as individuals select into certain environments that alter the likelihood that they will be surrounded by, and potentially engage in, heavy drinking. And, by doing this, they are essentially altering the likelihood that they will go on to develop alcohol problems. For example, we know that individuals who are more gregarious, extroverted, and higher in sensation-seeking will be more likely to spend time in bars or perhaps to join social sororities and fraternities in college. By doing so, they are selecting into environments that pose a higher risk for heavy drinking; these environments provide an enhanced opportunity to engage in high-risk behavior and potentially go on to develop drinking problems. We also know that there is considerable genetic influence on personality traits such as extraversion and sensation-seeking (McGue & Bouchard, 1998).

So, if genes influence extraversion, which influences the probability of being in a bar, which influences the probability of heavy drinking, which influences the probability of developing alcohol dependence, are the genes the culprit? Or, is the environment of being in a bar, which the individual has selected into, the culprit? Are genes "for" extraversion then genes "for" alcohol dependence? At this point, hopefully the reader will recognize that these are all moot questions. Tallying risk as genetic or environmental is a fruitless task in this sense. The two are inexorably intertwined. What is far more important is understanding the cascading pathways of risk and the mechanisms by which risk is conferred. This will be critical if this information is ever to eventually be translated into more informed prevention and intervention programs.

From Genes to Genes–Environment–Development

As the reader has no doubt become aware by now, there has been an evolution in the kinds of questions addressed by behavioral genetic studies, moving from initial studies that were aimed simply at asking whether evidence existed for genetic effects, to much more complex study designs that ask questions about how genetic influences contribute to comorbidity across disorders, how they change across development, and how they change as a function of the environment, as well as many additional questions that are not explicitly addressed here (Neale, Boker, Xie, & Maes, 1999). Interestingly, it appears that a similar evolution is taking place in the field of statistical genetics and gene finding, although it is not nearly as developed, being a much younger research area than genetic epidemiology. Although twin studies were under way and gaining importance as a study design by the middle of the 20th century (Fuller & Thompson, 1960), it has only been in the last couple decades that advances in gene identification have made it possible to identify specific genes involved in the predispositions to complex disorders, with considerable debate still surrounding the success and potential of available methods (Altshuler, Daly, & Lander, 2008).

Several large-scale projects are under way to identify genes involved in the predisposition to alcohol dependence, including the Collaborative Study on the Genetics of Alcohol Dependence (COGA) (Begleiter et al., 1995) and the Irish Affected Sib Pair Study of Alcohol Dependence (IASPSUD) (Prescott et al., 2005b), among others. Many of these projects initially focused on gene identification using alcohol dependence diagnoses as the primary outcome of interest (Prescott et al., 2005a; Reich et al., 1998). This represents an obvious and justifiable starting place, just as the first twin studies tested simply whether any evidence existed for genetic influences on alcohol dependence as an obvious first step. However, over the past several years, there has been growing recognition among statistical geneticists and other individuals involved in gene identification projects that these efforts cannot ignore the etiological complexity inherent in alcohol dependence. Gene identification efforts have begun making progress in this respect in recent years (perhaps inspired, in part, by the lack of resounding success recognized in initial gene identification efforts focused on psychiatric outcomes). Here, we summarize some of these new developments in the area of gene identification that attempt to better capture the complexities inherent in studying alcohol dependence.

GENE IDENTIFICATION

Genes "for" Alcohol Dependence?

Large-scale efforts to identify genes involved in psychiatric disorders have not been wildly successful (Wellcome Trust Case Control Consortium, 2007). Studies in the area of alcohol dependence have met with somewhat better success, in terms of identifying replicated genetic associations, which may in part reflect the fact that we have some traction on the underlying biological processes involved in drug metabolism and drug interaction with various brain neurotransmitter receptor classes (Edenberg & Bosron, 1997). Accordingly, replicated associations with several genes involved in the breakdown of ethanol and its by-product acetaldehyde have been identified, as discussed above. Other genes identified using a candidate gene strategy that have replicated across independent samples include several genes involved in the opioid system, notably *OPRK1* and *PDYN* (Gerra et al., 2007; Williams et al., 2007; Xuei et al., 2006). In addition, the candidate gene strategy has been complemented by more systematic, atheoretical gene identification strategies. These allow one to find novel genes that might not have been identified for study as potential candidate genes a priori due to our incomplete understanding of the biological underpinnings of alcohol dependence.

Using a systematic strategy of conducting linkage analyses, which identify broad chromosomal regions likely to contain genes influencing a particular disorder, followed by association analyses, which have the goal of identifying the specific genes in the region, COGA has identified several genetic associations with alcohol dependence that have subsequently replicated in independent samples. These include *GABRA2*, a γ-aminobutyric acid (GABA) receptor gene located near a linkage peak on chromosome 4 (Covault, Gelernter, Hesselbrock, Nellissery, & Kranzler, 2004; Edenberg et al., 2004; Lappalainen et al., 2005; Soyka et al., 2008) and *CHRM2*, an acetylcholine muscarinic receptor gene located near a linkage peak on chromosome 7 (Luo et al., 2005; Wang et al., 2004). However, despite progress in identifying replicated genetic associations with alcohol dependence, it is widely believed that only a fraction of the actual genes involved in susceptibility to alcohol dependence have been identified.

More recently, the scope of study for alcohol dependence has been expanded to incorporate findings from the field of genetic epidemiology into gene identification efforts. For example, based on the converging twin evidence reviewed above—that alcohol dependence shares a genetic basis with other forms of drug dependence and with antisocial behavior— composite indexes of drug dependence (Agrawal et al., 2008; Stallings et al., 2005), or more generally, of externalizing psychopathology (Dick et al., 2007a), have been created by combining information about symptoms of alcohol dependence, illicit drug dependence, adult antisocial behavior, and childhood conduct disorder for use in gene identification projects. These analyses indicate that these combined phenotypes are useful for gene identification and can enhance the power to detect genes involved in susceptibility to alcohol dependence through more general externalizing pathways (Dick et al., 2007a). At least two of the genes originally associated with adult alcohol dependence in COGA, *GABRA2* and *CHRM2* (Edenberg et al., 2004; Wang et al., 2004) have subsequently been shown to influence externalizing psychopathology more generally (Dick, 2007).

Another strategy that has been adopted, and represents an expansion of gene identification efforts, is to analyze phenotypes related to alcohol consumption, drinking patterns, and component symptoms of alcohol dependence, which twin studies also indicate are under genetic influence, rather than just binary alcohol dependence diagnostic status. Analyzing quantitative traits can yield gains in the power to detect genes. This can happen both for both statistical reasons and/or if the quantitative trait represents an intermediary phenotype in the risk pathway from gene to outcome, rendering it "closer" to the gene action. Several bitter taste receptor genes have been associated with the maximum number of drinks an individual reports ever consuming in a 24-hour period (Wang et al., 2007). The α-synuclein gene, a candidate gene emerging from the animal literature, has been associated with craving (but not with alcohol dependence) in the COGA sample (Foroud et al., 2007). Similarly, a GABA receptor gene, *GABRA1*, was associated with several drinking behavior phenotypes, including history of blackouts, age at first drunkenness, and level of response to alcohol, although no association with alcohol dependence as defined by the *Diagnostic and Statistical Manual of Mental Disorders, Fourth Edition* (DSM-IV) was detected (Dick et al., 2005, 2006c). These initial analyses— which require replication before we should regard them as definitive— underscore the need to study unfolding risk processes in the drinking career, in addition to alcohol dependence status.

How Do Genetic Influences on Alcohol Use/Dependence Unfold Across Time?

A number of studies have begun to examine how genes identified as associated with adult alcohol dependence influence risk across development. These studies represent a first set of efforts to characterize the developmental trajectories of risk associated with specific genes. For example, consistent with the twin literature suggesting that alcohol dependence symptoms in early adolescence are more environmentally influenced, as compared to the more genetically influenced adult alcohol dependence, survival analyses of the influence of *GABRA2* on alcohol dependence in the COGA sample, as reported by participants ranging in age from 7 to 91, indicated that differences in rates of alcohol dependence by genotype were only evident after age 20 (Dick et al., 2006b). In contrast, significant genetic influences on conduct disorder were observed with *GABRA2* in children: children carrying one or more copies of the allele associated with alcohol dependence in the adult sample had nearly twice the rate of conduct problems. This is consistent with the twin studies suggesting that genes involved in general externalizing psychopathology that manifest as alcohol (or other drug) dependence in adulthood may be associated with conduct problems during childhood. In fact, two of the genes originally associated with adult alcohol dependence in COGA (both *GABRA2* and *CHRM2*) have been related to behavioral problems earlier in development, both in COGA and in independent samples (Dick et al., 2006a,b, 2009b).

The above analyses illustrate two types of developmental changes, in that the magnitude of association with a particular disorder can change across developmental stage and that specific genes can influence distinct behaviors across different developmental stages. The growing interest in incorporating a developmental component into genetic association studies is perhaps best illustrated by the fact that the focus of the data collection in the most recent funding phases of COGA, the longest running gene identification project for alcohol dependence, is on prospectively following a sample of adolescents. This will allow us to study how the risk associated with specific, identified genes unfolds across development.

Do Genetic Influences on Alcohol Use/Dependence Change As a Function of the Environment?

Paralleling the movement toward studying genetic influences across development, there has been growing interest in studying how the association between identified genes and outcome may vary as a function of environmental factors. We have used twin data on which environments moderate latent genetic risk (i.e., inferred based on twin comparisons, not measured genes) to develop hypotheses to test the moderating effects of specific identified genes. Thus, based on the twin studies reviewed above indicating that parental monitoring moderated genetic influences on substance use, we tested whether the association between *GABRA2* and adolescent externalizing behavior would vary as a function of parental monitoring. As hypothesized, and paralleling the direction of effect from twin studies, the association between *GABRA2* and externalizing behavior was stronger under conditions of lower parental monitoring and reduced under higher parental monitoring (Dick et al., 2009b). We produced similar findings for *CHRM2* and parental monitoring (Dick et al., in press). In addition, we have examined the association between *CHRM2* and externalizing behavior as a function of peer deviance (Latendresse et al., in press). Again, paralleling the direction of effect found in twin studies, which suggests that genetic influences are enhanced with increased peer deviance (Button et al., 2007; Dick et al., 2007b), the association between *CHRM2* and trajectories of externalizing behavior was stronger in the presence of a more antisocial peer group.

CONCLUSION

In summary, understanding the etiology of alcohol dependence necessitates studying how genetic and environmental factors interact across development. Arguments about the extent to which alcohol dependence is genetic versus environmental are obsolete, and must be replaced by a more informed perspective that attempts to characterize the dynamic interplay of genetic and environmental influences. The studies reviewed in this chapter illustrate how genetic influences on alcohol dependence, and related traits, can vary across development and as a function of the environment. This has been demonstrated both in twin studies, and more recently, in studies aimed at characterizing risk associated with specific genes. Integrating these two research perspectives is likely to advance our knowledge about the dynamic nature of genetic and environmental influences on alcohol dependence.

REFERENCES

Agrawal, A., Hinrichs, A. L., Dunn, G., Bertelsen, S., Dick, D. M., Saccone, S. F., et al. (2008). Linkage scan for quantitative traits identifies new regions of interest for substance dependence in the Collaborative Study on the Genetics of Alcoholism (COGA) sample. *Drug and Alcohol Dependence, 93*, 12–20.

Altshuler, D., Daly, M. J., & Lander, E. S. (2008). Genetic mapping in human disease. *Science, 322*, 881–888.

Begleiter, H., Reich, T., Hesselbrock, V., Porjesz, B., Li, T. K., Schuckit, M., et al. (1995). The collaborative study on the genetics of alcoholism. *Alcohol Health & Research World, 19*, 228–236.

Button, T. M., Corley, R. P., Rhee, S. H., Hewitt, J. K., Young, S. E., & Stallings, M. C. (2007). Delinquent peer affiliation and conduct problems: A twin study. *Journal of Abnormal Psychology, 116*, 554–564.

Button, T. M., Lau, J. Y., Maughan, B., & Eley, T. C. (2008). Parental punitive discipline, negative life events and gene-environment interplay in the development of externalizing behavior. *Psychological Medicine, 38*, 29–39.

Covault, J., Gelernter, J., Hesselbrock, V., Nellissery, M., & Kranzler, H. R. (2004). Allelic and haplotypic association of GABRA2 with alcohol dependence. *American Journal of Medical Genetics (Neuropsychiatric Genetics), 129B*, 104–109.

Dick, D. M. (2007). Identification of genes influencing a spectrum of externalizing psychopathology. *Current Directions in Psychological Science, 16,* 331–335.

Dick, D. M., Aliev, F., Wang, J. C., Grucza, R. A., Schuckit, M., Kuperman, S., et al. (2007a). Using dimensional models of externalizing psychopathology to aid in gene identification. *Archives of General Psychiatry, 65,* 310–318.

Dick, D. M., Bernard, M., Aliev, F., Viken, R., Pulkkinen, L., Kaprio, J., et al. (2009a). The role of socio-regional factors in moderating genetic influences on early adolescent behavior problems and alcohol use. *Alcoholism: Clinical and Experimental Research, 33,* 1739–1748.

Dick, D. M., Bierut, L., Hinrichs, A. L., Fox, L., Bucholz, K. K., Kramer, J. R., et al. (2006b). The role of GABRA2 in risk for conduct disorder and alcohol and drug dependence across developmental stages. *Behavior Genetics, 36,* 577–590.

Dick, D. M., Edenberg, H. J., Xuei, X., Goate, A., Hesselbrock, V., Schuckit, M., et al. (2005). No association of the GABA-A receptor genes on chromosome 5 with alcoholism in the Collaborative Study on the Genetics of Alcoholism sample. *American Journal of Medical Genetics (Neuropsychiatric Genetics), 132B,* 24–28.

Dick, D. M., Latendresse, S. J., Lansford, J. E., Budde, J., Goate, A., Dodge, K. A., et al. (2009b). The role of GABRA2 in trajectories of externalizing behavior across development and evidence of moderation by parental monitoring. *Archives of General Psychiatry, 66,* 649–657.

Dick, D. M., Meyers, J. L., Latendresse, S. J., Creemers, H. E., Lansford, J. E., Pettit, G. S., et al. (in press). CHRM2, Parental Monitoring, and Adolescent Externalizing Behavior: Evidence for Gene-Environment Interaction. *Psychological Science.*

Dick, D. M., Pagan, J. L., Viken, R., Purcell, S., Kaprio, J., Pulkkinen, L., et al. (2007b). Changing environmental influences on substance use across development. *Twin Research and Human Genetics 10,* 315–326.

Dick, D. M., Pagan, J. L., Viken, R., Purcell, S., Kaprio, J., Pulkkinen, L., et al. (2007c). Changing environmental influences on substance use across development. *Twin Research and Human Genetics, 10,* 315–326.

Dick, D. M., Plunkett, J., Wetherill, L. F., Xuei, X., Goate, A., Hesselbrock, V., et al. (2006c). Association between GABRA1 and drinking behaviors in the collaborative study on the genetics of alcoholism sample. *Alcoholism: Clinical and Experimental Research, 30,* 1101–1110.

Dick, D. M., Prescott, C., & McGue, M. (2009). *The genetics of substance use and substance use disorders.* In Y-K. Kim (Ed.), *The handbook of behavior genetics.* (pp. 433–453). New York: Springer.

Dick, D. M., Rose, R. J., Viken, R. J., Kaprio, J., & Koskenvuo, M. (2001). Exploring gene-environment interactions: Socioregional moderation of alcohol use. *Journal of Abnormal Psychology, 110,* 625–632.

Dick, D. M., Viken, R., Purcell, S., Kaprio, J., Pulkkinen, L., & Rose, R. J. (2007d). Parental monitoring moderates the importance of genetic and environmental influences on adolescent smoking. *Journal of Abnormal Psychology, 116,* 213–218.

Edenberg, H. J. & Bosron, W. F. (1997). Alcohol dehydrogenases. In F.P. Guengerich (Ed.), *Comprehensive toxicology. Vol. 3: Biotransformation* (pp. 119–131). New York: Pergamon.

Edenberg, H. J., Dick, D. M., Xuei, X., Tian, H., Almasy, L., Bauer, L. O., et al. (2004). Variations in GABRA2, encoding the 2 subunit of the GABA-A receptor are associated with alcohol dependence and with brain oscillations. *American Journal of Human Genetics, 74*, 705–714.

Edenberg, H. J., Xuei, X., Chen, H-J., Tian, H., Wetherill, L. F., Dick, D. M., et al. (2006). Association of alcohol dehydrogenase genes with alcohol dependence: A comprehensive analysis. *Human Molecular Genetics, 15*, 1539–1549.

Feinberg, M. E., Button, T. M., Neiderhiser, J. M., Reiss, D., & Hetherington, E. M. (2007). Parenting and adolescent antisocial behavior and depression: evidence of genotype x parenting environment interaction. *Archives of General Psychiatry, 64*, 457–465.

Foroud, T., Wetherill, L. F., Liang, T., Dick, D. M., Hesselbrock, V., Kramer, J., et al. (2007). Association of alcohol craving with alpha-synuclein (SNCA). *Alcoholism: Clinical and Experimental Research, 31*, 537–545.

Fuller, J. L., & Thompson, W. R. (1960). *Behavior genetics*. New York: Wiley.

Gerra, G., Leonardi, C., Cortese, E., D'Amore, A., Lucchini, A., Strepparola, G., et al. (2007). Human kappa opioid receptor gene (OPRK1) polymorphism is associated with opiate addiction. *American Journal of Medical Genetics (Neuropsychiatric Genetics), 144*, 771–775.

Heath, A. C., Jardine, R., & Martin, N. G. (1989). Interactive effects of genotype and social environment on alcohol consumption in female twins. *Journal of Studies on Alcohol, 50*, 38–48.

Hopfer, C. J., Crowley, T. J., & Hewitt, J. K. (2003). Review of twin and adoption studies of adolescent substance use. *Journal of the American Academy of Child and Adolescent Psychiatry, 42*, 710–719.

Kendler, K. S., Prescott, C., Myers, J., & Neale, M. C. (2003). The structure of genetic and environmental risk factors for common psychiatric and substance use disorders in men and women. *Archives of General Psychiatry, 60*, 929–937.

Kendler, K. S., Schmitt, J. E., Aggen, S. H., & Prescott, C. A. (2008). Genetic and environmental influences on alcohol, caffeine, cannabis, and nicotine use from adolescence to middle adulthood. *Archives of General Psychiatry, 65*, 674–682.

Koopmans, J. R., Slutske, W. S., van Baal, G. C. M., & Boomsma, D. I. (1999). The influence of religion on alcohol use initiation: Evidence for genotype x environment interaction. *Behavior Genetics, 29*, 445–453.

Krueger, R. F., Hicks, B. M., Patrick, C. J., Carlson, S. R., Iacono, W. G., & McGue, M. (2002). Etiologic connections among substance dependence, antisocial behavior, and personality: modeling the externalizing spectrum. *Journal of Abnormal Psychology, 111*, 411–424.

Kuo, P. H., Kalsi, G., Prescott, C., Hodgkinson, C. A., Goldman, D., van den Oord, E. J., et al. (2008). Association of ADH and ALDH genes with alcohol dependence in the Irish Affected Sib Pair Study of Alcohol Dependence (IASPSAD) sample. *Alcoholism, Clinical and Experimental Research, 32*(5), 785–795.

Lappalainen, J., Krupitsky, E., Remizov, M., Pchelina, S., Taraskina, A., Zvartau, E., et al. (2005). Association between alcoholism and Gamma-Amino Butyric Acid alpha2 receptor subtype in a Russian population. *Alcoholism: Clinical and Experimental Research, 29,* 493–498.

Latendresse, S. J., Bates, J. E., Goodnight, J. A., Lansford, J. E., Budde, J. P., Goate, A., et al. (in press) Differential Susceptibility to Adolescent Externalizing Trajectories: Examining the Interplay between *CHRM2* and Peer Group Antisocial Behavior. *Child Development.*

Luo, X., Kranzler, H. R., Zuo, L., Wang, S., Blumberg, H. P., & Gelernter, J. (2005). CHRM2 gene predisposes to alcohol dependence, drug dependence, and affective disorders: results from an extended case-control structured association study. *Human Molecular Genetics, 14,* 2421–2432.

McGue, M. & Bouchard, Jr. T. J. (1998). Genetic and environmental influences on human behavioral differences. *Annual Review of Neuroscience, 21,* 1–24.

Neale, M. C., Boker, S. M., Xie, G., & Maes, H. H. (1999). *Mx: Statistical Modeling (Version 5th Edition) [Computer software].* Box 126 MCV, Richmond, VA 23298: Department of Psychiatry.

Prescott, C., Sullivan, P. F., Kuo, P. H., Webb, B. T., Vittum, J., Patterson, D. G., et al. (2005a). Linkage of alcohol dependence symptoms to chromosome 4 and preliminary association with ADH-7 in the Irish affected sib pair study of alcohol dependence. *Alcoholism: Clinical and Experimental Research, 29 Suppl,* 132A.

Prescott, C., Sullivan, P. F., Myers, J., Patterson, D., Devitt, M., Halberstadt, L. J., et al. (2005b). The Irish Affected Sib Pair Study of Alcohol Dependence: study methodology and validation of diagnosis by interview and family history. *Alcoholism: Clinical and Experimental Research, 29,* 417–429.

Purcell, S. (2002). Variance components models for gene-environment interaction in twin analysis. *Twin Research, 5,* 554–571.

Reich, T., Edenberg, H., Goate, A., Williams, J., Rice, J., Eerdewegh, P. V., et al. (1998). Genome-wide search for genes affecting the risk for alcohol dependence. *American Journal of Medical Genetics, 81,* 207–215.

Rose, R. J., Dick, D. M., Viken, R. J., & Kaprio, J. (2001). Gene-environment interaction in patterns of adolescent drinking: regional residency moderates longitudinal influences on alcohol use. *Alcoholism: Clinical and Experimental Research, 25,* 637–643.

Rose, R. J., Dick, D. M., Viken, R. J., Pulkkinen, L., & Kaprio, J. (2001). Drinking or abstaining at age 14: a genetic epidemiological study. *Alcoholism: Clinical and Experimental Research, 25,* 1594–1604.

Rose, R. J., Viken, R. J., Dick, D. M., Bates, J. E., Pulkkinen, L., & Kaprio, J. (2003). It does take a village: nonfamilial environments and children's behavior. *Psychological Science, 14,* 273–277.

Shanahan, M. J., & Hofer, S. M. (2005). Social context in gene-environment interactions: retrospect and prospect. *The Journals of Gerontology. Series B, Psychological Sciences and Social Sciences, 60 (Spec. No 1),* 65–76.

Soyka, M., Preuss, U. W., Hesselbrock, V., Zill, P., Koller, G., & Bondy, B. (2008). GABA-A2 receptor subunit gene (GABRA2) polymorphisms and risk for alcohol dependence. *Journal of Psychiatric Research, 42,* 184–191.

Stallings, M. C., Corley, R. P., Dennehey, B., Hewitt, J. K., Krauter, K. S., Lessem, J. M., et al. (2005). A genome-wide search for quantitative trait loci that influence antisocial drug dependence in adolescence. *Archives of General Psychiatry, 62,* 1042–1051.

Turkheimer, E., & Waldron, M. (2000). Nonshared environment: a theoretical, methodological, and quantitative review. *Psychological Bulletin, 126,* 78–108.

Wang, J. C., Hinrichs, A. L., Stock, H., Budde, J., Allen, R., Bertelsen, S., et al. (2004). Evidence of common and specific genetic effects: association of the muscarinic acetylcholine receptor M2 (CHRM2) gene with alcohol dependence and major depressive syndrome. *Human Molecular Genetics, 13,* 1903–1911.

Wang, J. C., Hinrichs, A. L., Stock, H., Budde, J., Dick, D. M., Bucholz, K. K., et al. (2007). Functional variants in TAS2R38 and TAS2R16 influence alcohol consumption in high-risk families of African-American origin. *Alcoholism: Clinical and Experimental Research, 31,* 209–215.

Wellcome Trust Case Control Consortium (2007). Genome-wide association study of 14,000 cases of seven common diseases and 3,000 shared controls. *Nature, 447,* 661–78.

Whitfield, J. B. (1997). Meta-analysis of the effects of alcohol dehydrogenase genotype on alcohol dependence and alcoholic liver disease. *Alcohol and Alcoholism, 32,* 613–619.

Williams, T. J., LaForge, K. S., Gordon, D., Bart, G., Kellogg, S., Ott, J., et al. (2007). Prodynorphin gene promoter repeat associated with cocaine/alcohol codependence. *Addiction Biology, 12,* 496–502.

Xuei, X., Dick, D., Flury-Wetherill, L., Tian, H. J., Agrawal, A., Bierut, L., et al. (2006). Association of the kappa-opioid system with alcohol dependence. *Molecular Psychiatry, 11,* 1016–1024.

Young, S. E., Stallings, M. C., Corley, R. P., Krauter, K. S., & Hewitt, J. K. (2000). Genetic and environmental influences on behavioral disinhibition. *American Journal of Medical Genetics, 96,* 684–695.

15

The Importance of Understanding Gene–Environment Correlations in the Development of Antisocial Behavior

BRIAN M. D'ONOFRIO, PAUL J. RATHOUZ,
AND BENJAMIN B. LAHEY

Antisocial behavior is a highly prevalent and very serious public health problem. This is not only because antisocial youth harm and often traumatize their victims, but also because antisocial youth themselves are at greatly increased risk for impaired social relationships, depression, substance use disorders, incarceration, physical injury, and death from homicide and suicide (Loeber, Farrington, Stouthamer-Loeber, & Van Kammen, 1998). Furthermore, the direct financial costs to society of persistent antisocial behavior are very high, with estimates of $2 million in direct costs per antisocial individual (Cohen, 1998). Because many of the costs of antisocial behavior are indirect (e.g., the increased costs of insurance, policing, courts, medical and mental health services for victims), the true financial costs are probably much higher.

DEFINING ANTISOCIAL BEHAVIOR

A number of social and scientific constructs are used to refer to antisocial behavior across the lifespan. Although their definitions overlap, a great deal of confusion is caused by the fact that they are not identical. In criminology, the term *juvenile delinquency* is used to refer to law-breaking by children and adolescents, and *criminality* refers to law-breaking in adults. These terms broadly refer to anything from sneaking into a movie without a ticket to homicide. In the *Diagnostic and Statistical Manual of Mental Disorders, Fourth Edition* (DSM-IV) (American Psychiatric Association, 1994), the diagnosis of *conduct disorder* (CD) refers to engaging in at least three from a list of 15 antisocial behaviors within the last 12 months. The term CD only partially overlaps with delinquency in several ways. First, some symptoms of CD are not typically treated as crimes by the police and courts (e.g., bullying, staying out late without permission). Second, not all juvenile crimes are symptoms of CD (e.g., selling drugs and receiving stolen property). Third, the diagnosis of CD requires youth to engage in at least three antisocial behaviors in a relatively short time frame, whereas a youth could be considered to be delinquent on the basis of a single criminal act. From age 18 years on, the diagnosis of *antisocial personality disorder* (APD) refers to a persistent pattern of antisocial and irresponsible behavior in the presence of a number of noncriminal characteristics, including impulsivity and callous disregard for others. Thus, although nearly all individuals with APD commit crimes, most criminals do not meet criteria for APD.

In this chapter, we use the term *antisocial behavior* to refer to behaviors that violate laws or the rights of others. This term incorporates childhood conduct problems, CD, APD, delinquency, and criminality. Although there are important distinctions among these constructs, research suggests that the various measures reflect the same underlying construct to a considerable extent (e.g., Krueger, 1999).

It is important to note that complex age and sex differences exist in antisocial behavior. Aggression and conduct problems in early childhood decline in prevalence in the general population with increasing age (Moffitt, 2006; Tremblay, 2000). On the other hand, law-breaking and some covert aggressive and nonaggressive behaviors (e.g., theft with and without confrontation of the victim, running away from home overnight, and forced sex) increase in prevalence from childhood into late adolescence, then decline in prevalence into later adulthood (Farrington, 1991; Lahey et al., 2000), a developmental pattern known as the *age–crime curve* (Hirschi & Gottfredson, 1983). Thus, most serious crimes are committed by adolescents and young adults (Farrington, 1992). Although antisocial behavior is prevalent and very problematic in both sexes, it is considerably more common in males (Lahey et al., 2006; Moffitt, Caspi, Rutter, & Silva, 2001; Rutter, Caspi, & Moffitt, 2003). Because the sex difference in the prevalence of antisocial behavior is large, it will be necessary for the field to understand the causes of sex differences to fully understand the causes of antisocial behavior itself (Rutter et al., 2003).

BRIEF OVERVIEW OF CURRENT CAUSAL MODELS OF ANTISOCIAL BEHAVIOR

Several researchers have hypothesized that there are different *developmental trajectories* of antisocial behavior, with different causes (Farrington, 1991; Hinshaw, Lahey, & Hart, 1993; Loeber, 1988; Moffitt, 1993; Patterson, DeBaryshe, & Ramsey, 1989; Quay, 1987). The term *trajectory* refers to a more or less distinct temporal pattern of antisocial behavior from early childhood into adulthood. For example, two 17-year-olds arrested for shoplifting might have had very different developmental trajectories. One may have exhibited no symptoms of CD as a child and had never broken a law until skipping school and shoplifting for the first time at age 17. The other 17-year-old might have continuously met criteria for CD since early childhood, shoplifted dozens of times before, and committed violent crimes since late childhood.

Patterson and colleagues (Patterson et al., 1989; Snyder, Reid, & Patterson, 2003) distinguished between "early and late starters" in a social learning model of antisocial behavior. Early-starting delinquency begins well before school entry, in coercive interchanges with parents who sometimes model and encourage antisocial behavior. In this developmental pattern, antisocial behavior is the result of inconsistent and coercive parenting in childhood and inadequate parental supervision during adolescence. Early childhood conduct problems cause school failure and the alienation of well-behaved peers. In turn, these cause depressed mood and affiliation with antisocial peers. Patterson initially ascribed little importance to the role of individual differences in early childhood characteristics in the origins of antisocial behavior, but a more recent statement posits that early childhood cognitive deficits and greater negative emotionality promote coercive interactions with parents (Snyder et al., 2003). In contrast, in Patterson's model, late-starting delinquents have normative childhoods and become delinquent during adolescence solely because of peer influence supporting antisocial behavior.

Moffitt similarly proposed a "developmental taxonomy" of delinquency: Early-starting delinquency is thought to have its origins in neurodevelopmental deficits that are manifested in early childhood as difficult temperament, poorly developed cognitive skills, and attention-deficit hyperactivity disorder (ADHD) (Moffitt, 1993, 2006). These early characteristics are posited to give rise to early-starting delinquency, but only in the presence of maladaptive early parenting (Moffitt, 1993). In contrast, late-starting delinquency emerges in youth without histories of child conduct problems. Late-starting delinquency arises after puberty, during what is described as the vulnerable "gap" between adolescents' biological maturity and their access to adult privileges. This gap gives rise to frustrations that increase dysphoria in the youth and make them more vulnerable to peer influences to engage in antisocial behavior. As defined by Moffitt, essentially equal numbers of females and males exhibit late-onset delinquency, but males outnumber females about 3:1 in those with an early-onset trajectory (Lahey et al., 2006; Moffitt et al., 2001).

Lahey and Waldman proposed a model of antisocial behavior that is explicitly based on the developmental models of Moffitt and Patterson (Lahey & Waldman, 2003, 2005) and the criminologic model of Gottfredson and Hirschi (1990). Unlike Moffitt and Patterson, they posit that early- and late-starting delinquency are not qualitatively distinct classes, but represent an artificial dichotomization of a *continuum* of differences among antisocial trajectories. As noted by others, identifying classes corresponding to early- and late-starters in latent trajectory analyses does not imply that the groups are necessarily qualitatively distinct (Nagin & Tremblay, 2005; Sampson & Laub, 2005).

Building on the previous developmental models, Lahey and Waldman (2003, 2005) offered a set of testable hypotheses regarding individual differences in children and adolescents that make them more or less vulnerable than others to environmental influences on antisocial behavior. Specifically, Lahey and Waldman suggested that at-risk children and adolescents (a) are more likely to be exposed to environments that promote antisocial behavior, and (b) are more vulnerable to environmental influences if they are (a) lower in intelligence, (b) higher in trait negative emotionality, (c) lower in prosociality (i.e., higher in callousness as defined by Frick & White, 2008), and (d) higher in daring, which is akin to the undersocialized sensation-seeking described by Zuckerman (1996). These four enduring dispositions are posited to transact with the environment to increase risk for antisocial behavior through both gene–environment correlations (i.e., genetic factors that influence the dispositions become correlated with maladaptive environments) and gene–environment interactions (i.e., genetic factors that influence the dispositions influence vulnerability to maladaptive environments).

GENE–ENVIRONMENT CORRELATION IN THE ETIOLOGY OF ANTISOCIAL BEHAVIOR

This chapter focuses on the importance of gene–environment correlations (rGE) in understanding the etiology of antisocial behavior. Our premise is that one cannot meaningfully study either causal genetic or environmental risk factors for any form of psychopathology without fully considering rGE. This is especially true with antisocial behavior because behavior genetic research has clearly indicated that genetic factors influence such behaviors (reviewed in Rhee & Waldman, 2002).

As reviewed earlier in this volume (Chapters 1 and 4 by Kendler and Jaffee), rGE occurs when genetic variants are correlated with aspects of environmental risk (e.g., Kendler & Baker, 2007; Plomin & Bergeman, 1991). It is very likely that rGE is common, highlighting the concept of nonrandom selection into environments (i.e., individuals do not experience risky or protective environments purely by chance). Therefore, when exposure to a putative risk environment is correlated with genotype, it is possible that the observed empirical association between the environment (e.g., coercive parenting) and antisocial behavior is due in part to genetic factors that lead to both, rather than a causal effect of the environment. Indeed, genetic confounds could entirely account for the statistical association between the putative risk environment and antisocial behavior. This could lead to incorrect inferences regarding the etiologic role of the environment. It is important to note, however, that the presence of rGE does not necessarily mean that genetic factors entirely account for the association between a risk and antisocial behavior (Rutter & Silberg, 2002). Rather, the correlated genotype and environment may both be causal influences on antisocial behavior operating in an additive manner. One goal of researchers is to distinguish between these alternatives using methods like those described in this chapter.

It is essential that researchers use designs that can determine whether genetic confounding accounts for all or part of the statistical associations between putative environmental risk factors and antisocial behavior. Researchers need designs that can test multiple, alternative processes when studying associations between putative genetic and environmental risks and antisocial behavior (Moffitt, 2005; Rutter, 2007; Rutter, Pickles, Murray, & Eaves, 2001; Shadish, Cook, & Campbell, 2002). In particular, research designs must be able to help differentiate between causal mechanisms (in which exposure to the risk causes an increase in antisocial behavior), the possible role of genetic confounds, and environmental factors that influence both the risk and antisocial behavior. We emphasize throughout the current chapter how quasi-experimental behavior genetic designs have helped researchers better understand the etiology of antisocial behavior. These designs take advantage of "natural experiments" that separate genetic and environmental influences to a degree, including variations in the genetic relatedness of study participants that allow estimates of genetic confounds and relatively strong tests of causal hypotheses (Rutter et al., 2001). Next, we briefly highlight the current status of research exploring the associations between putative environmental and genetic risks across development and antisocial behavior.

rGE AND PUTATIVE PRENATAL ENVIRONMENTAL RISK FACTORS

Smoking During Pregnancy

Numerous studies have found that maternal smoking during pregnancy (SDP) is correlated with offspring antisocial behavior during childhood and adulthood. The prevailing conclusion is that SDP causes antisocial problems (Wakschlag, Pickett, Cook, Benowitz, & Leventhal, 2002), but smoking during pregnancy is also associated with numerous other risk factors for offspring antisocial behavior (e.g., Maughan, Taylor, Caspi, & Moffitt, 2004). Behavior genetic studies have found that SDP is heritable (Agrawal et al., 2008; D'Onofrio et al., 2003), which raises the possibility that shared genetic liability in the parents and offspring accounts for the statistical association between SDP and antisocial problems. That is, it is possible that genes that influence SDP in the mother also influence antisocial behavior in the offspring, resulting in rGE that gives the appearance of a causal prenatal environmental influence of SDP on the offspring's risk for antisocial behavior.

Researchers have used a number of quasi-experimental designs recently to test the environmental impact of maternal SDP on offspring behavior (Knopik, 2009). A number of studies used a sibling-comparison approach, which compared the antisocial behavior of siblings who were differentially exposed to maternal SDP (Lahey, D'Onofrio, & Waldman, 2009; Rutter, 2007). Figure 15.1 illustrates two basic approaches to studying putative risks. The between-family approach compares differentially exposed individuals (some are and are not exposed to SDP) from different families. The within-family approach, in contrast, compares siblings who are differentially exposed to SDP (i.e., a mother who varied her smoking across pregnancies). This design accounts for all shared environmental factors that influence siblings similarly and essentially rules out rGE. Through the process of meiosis, parents' genes are randomly distributed to their children, which rules out passive rGE in relatively large samples (see previous chapters for descriptions of rGE). Active and evocative rGE are essentially ruled out by the fact that genetically influenced characteristics of the offspring cannot influence their mother's prenatal smoking habits. The only exception would be the hypothetical case in which genetic variations among fetuses influence maternal smoking, such as by influencing nausea during pregnancy. Thus, in this design, if genetic or environmental confounds were responsible for the effects of SDP, there would be no difference in antisocial behavior between siblings who were differentially exposed to SDP. Recent studies have found no association between SDP and childhood conduct problems (D'Onofrio et al., 2008; Gilman, Gardener, & Buka, 2008) and

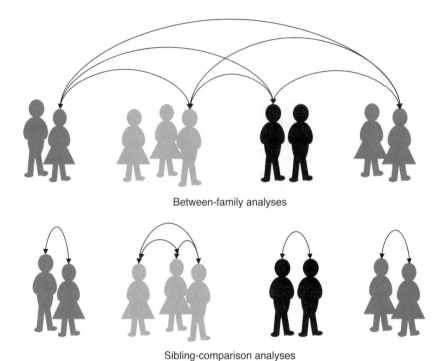

Between-family analyses

Sibling-comparison analyses

Figure 15.1. **A comparison of between- and within-family studies.** In sibling comparison studies, differences between siblings are analyzed while holding families constant. In between-family studies, differences between individuals are confounded by differences between families.

adult criminality (D'Onofrio et al., 2010a) when comparing differentially exposed siblings. The findings thus indicate that genetic confounds and family background factors, rather than the teratogenic effects of SDP, most likely account for the association between SDP and antisocial behavior.

Using a novel in vitro fertilization (IVF) cross-fostering design, Rice and colleagues (2009) reached a similar conclusion about the association between SDP and childhood conduct problems. They found no evidence of a causal environmental effect of SDP because SDP was only associated with offspring behavior problems when mother and child were genetically related, not when children were carried by a genetically unrelated surrogate. This finding, which is based on a design with different threats to internal validity than sibling-comparison studies, adds significant weight to the hypothesis that genetic confounds (rGE) were responsible for the empirical association between SDP and offspring antisocial behavior.

These findings regarding the relationship between SDP and antisocial behavior are consistent with recent quasi-experimental studies of the association between SDP and related constructs, such as academic achievement problems, lower intellectual abilities, and risk for ADHD (D'Onofrio et al., 2010b; Knopik et al., 2005; Lambe, Hultman, Torrang, MacCabe, & Cnattingius, 2006; Lundberg et al., 2010; Thapar et al., 2009). This does not mean, of course, that it is not important to reduce SDP. Apart from the health risks to the mother and to the family through passive exposure to smoke, quasi-experimental studies have found that SDP is independently associated with increased risk for preterm birth, low birth weight (Cnattingius, 2004), and infant mortality (Johansson, Dickman, Kramer, & Cnattingius, 2009), thus making the reduction of SDP an important public health goal.

Maternal Alcohol Consumption During Pregnancy

The potential causal impact of maternal alcohol consumption during pregnancy on offspring adjustment is controversial at present because of measurement difficulties and problems with accounting for the effects of covarying risks (Gray, Mukherjee, & Rutter, 2009). It is clear that heavy amounts of drinking cause fetal alcohol syndrome, but little is known about the possible causal effects of moderate levels of alcohol consumption on offspring antisocial behavior. To our knowledge, only one quasi-experimental study of prenatal alcohol exposure has been conducted. D'Onofrio and colleagues (D'Onofrio, Van Hulle, et al., 2007) compared siblings whose mothers varied their alcohol consumption across their pregnancies. The children who were exposed to more alcohol in utero had higher levels of conduct problems than did their siblings who were exposed to less in utero, which supports a tentative causal inference. Given the limited research on the topic, however, the current evidence is too sparse to conclude with any confidence that low to moderate levels of alcohol consumption causally increase risk for offspring conduct problems (Gray et al., 2009).

rGE AND PUTATIVE ENVIRONMENTAL RISK FACTORS IN EARLY CHILDHOOD

Maternal Age at Childbearing

Early maternal age at childbearing is strongly associated with a variety of adverse outcomes for both teen mothers and their children, including antisocial behavior (Coley & Chase-Lansdale, 1998; Jaffee, Caspi, Moffitt, Belsky, & Silva, 2001). Yet, the mechanisms through which early maternal age at childbearing influences offspring functioning are still unresolved. Early maternal age at childbearing is correlated with numerous putative environmental risks for offspring antisocial behavior, and early maternal age at childbearing is partly influenced by genetic factors (Rodgers, Bard, & Miller, 2007).

A number of researchers have utilized a cousin-comparison approach, in which the offspring of adult sisters discordant for having children while teenagers were compared (Geronimus, Korenman, & Hillemeier, 1994; Turley, 2003). The comparison of cousins accounts for environmental factors that all family members in an extended family share, as well as some genetic factors (cousins share 12.5% of their genetic makeup, on average). These first studies suggested that family background factors were responsible for the association between early maternal age at childbearing and offspring antisocial behavior (and related constructs) because the offspring of teen mothers had the same rates of antisocial problems as their cousins born to nonteen mothers.

Recently, however, researchers have used sibling comparisons to further explore the potential causal environment effects of early maternal age at childbearing. Sibling-comparison studies in Australia (Harden, Lynch et al., 2007) and the United States (D'Onofrio et al., 2009) found that, after controlling for birth order, offspring born when their mothers were younger had more antisocial problems than did their siblings born at older maternal ages. The study by D'Onofrio and colleagues (2009) also found the within-family association when comparing full siblings, thus controlling for paternal risk factors as well. The results support the tentative causal inference that environmental factors specifically associated with early maternal age at childbearing within a family are responsible for the increased antisocial problems found in offspring born to young mothers. Future quasi-experimental studies are needed to identify the specific environmental factors that account for this association, both to make a stronger case for environmental causation and because these environmental factors would be important targets for prevention efforts aimed at reducing antisocial behaviors.

Parenting

The belief that maladaptive parenting is a cause of antisocial behavior in children is deeply ingrained in our society. Indeed, all major theories of the etiology of antisocial problems include a major role for parenting and parent–child relationships (e.g., Lahey & Waldman, 2005; Moffitt, 2006; Snyder et al., 2003). Is there credible evidence, however, that variations in parenting are a causal environmental risk factor? As with other risk factors for antisocial behavior, parenting behaviors are highly correlated with other environmental risks and are influenced by genetic factors (review in Kendler & Baker, 2007). That is, the association between parenting and offspring antisocial behavior is confounded by rGE. Recently, however, researchers have begun to use a number of different quasi-experimental designs to test the mechanisms responsible for the associations between parenting behaviors and offspring antisocial problems.

In a longitudinal study of young children initially assessed before the age of 2 years and followed a year and a half later, Jaffee (2007) found that high-risk children who experienced improvements in sensitive and stimulating parenting had fewer behavior problems. To rule out the possibility of genetic confounds due to passive rGE, the study explored parenting in families in which the children were removed from their biological parents and were being raised in foster families. This eliminated the possible confounding due to shared genetic liability in parents and offspring, suggesting that these aspects of parenting play a causal role in the etiology of antisocial behavior.

Jaffee and her colleagues (2004) also used a twin study to explore the mechanisms through which various parenting behaviors are associated with offspring antisocial problems. Multivariate analyses of twin study data allow researchers to explore the genetic and environmental influences on individual traits, as well as the covariation between multiple traits (Neale & Cardon, 1992). The study explored the genetic and environmental influences on parenting behaviors (corporal punishment and physical maltreatment) *and* the covariation between those measures and offspring conduct problems in a sample of 5-year-old twins. The results indicated that genetic factors accounted for some of the variation in corporal punishment and that these genetic factors were largely shared with children's antisocial behavior. The results suggest that the use of corporal punishment by parents may be a response to the children's behavior, which is partially influenced by genetic factors. The finding that corporal punishment may be a contingent response to the children's antisocial behavior stresses the importance of considering child effects on parenting in the development of antisocial behavior (Bell & Harper, 1997). In contrast to the findings on corporal punishment, the association between physical maltreatment and offspring antisocial behavior were due to environmental factors associated with the maltreatment, consistent with a causal influence.

The comparison of differentially exposed identical or monozygotic (MZ) twins, is another powerful approach for studying putative environmental risk factors. The method, frequently referred to as the *co-twin control design*, is a special form of sibling comparison (see Figure 15.1). The approach accounts for all genetic confounds because MZ twins share all of their genetic makeup and environmental factors that make siblings similar (Rutter, 2007). Caspi et al. (2004) used the approach to study the importance of maternal expressed emotion in predicting antisocial problems in 7-year-old children. The authors found that the MZ twin exposed to more negative expressed emotion at age 5 years had higher rates of antisocial problems at age 7 years than his or her co-twin, even when controlling for levels of anti-social behavior at age 5 (thus reducing the likelihood of confounding child effects on parenting). The results strengthen the causal inference that exposure to negative emotions from parents increases the probability that offspring will engage in antisocial problem behaviors.

The potential causal effects of physical maltreatment also have been examined in a children-of-twins (CoT) study of harsh parenting and off-spring psychopathology (Lynch et al., 2006). The design compares differ-entially exposed offspring of MZ twins (who are as genetically similar as half siblings, although socially they are cousins) and offspring of fraternal or dizygotic (DZ) twins (who genetically *and* socially are as related as cousins) (D'Onofrio et al., 2005; Heath, Kendler, Eaves, & Markell, 1985; Silberg & Eaves, 2004). Lynch et al. (2006) found that among the offspring of MZ twins, those who were exposed to harsher parenting had greater antisocial problems than their cousins (genetic half-sibling) who were exposed to less-harsh parenting. The comparison provides a rigorous test of causality for environmental risks shared by siblings within a family because offspring of the MZ co-twins share similar genetic risk but differ in their environmental exposures.

Family Income

Low family income during childhood and adolescence has been found to be concurrently and prospectively associated with antisocial behavior (review in Conger & Donnellan, 2007). Yet, low family income is correlated with numerous putative environmental risk factors for offspring antisocial behav-ior (Evans, 2004) and may be associated with genetic factors (rGE). Indeed, some researchers have also suggested that genetic factors could partly or wholly explain the association between family income and child adjustment (Rowe & Rodgers, 1997).

A few experimental studies designed to evaluate alternative changes in the welfare system have explored whether randomly selecting families to receive additional aid results in changes in child well-being. Two studies (Gennetian & Miller, 2002; Morris & Gennetian, 2003) found that increases in family income were associated with reduced antisocial behavior in the children. These results were consistent with the results of a natural experiment that explored the changes in family functioning and child mental disorders before and after family income was increased on an Indian Reservation because of the opening of a casino (Costello, Compton, Keeler, & Angold, 2003). These studies, which reduced genetic and shared environmental confounding, also suggested that family income has a causal influence on antisocial behavior.

A number of sibling comparison studies also have tested the potential causal effects of variations in family income on offspring conduct problems. These studies (e.g., Blau, 1999; Hao & Matsueda, 2006) and a recent comparison of full siblings exposed to varying levels of family income during childhood (D'Onofrio et al., 2009), found an independent relation between family income and offspring conduct problems when many genetic and environmental confounds were controlled. The magnitude of the independent association between family income and child antisocial behavior, though, was much lower than the magnitude of the association when not comparing siblings, which is consistent with a recent review that suggests that the association is due to both causal and noncausal selection factors (Conger & Donnellan, 2007).

rGE and Putative Environmental Risk Factors in Adolescence: Parental Criticism

Adolescence is period of development in which children are becoming more independent—there is growing emotional distance between parents and their children and there is an increase in parent–child conflicts (Steinberg & Morris, 2001). Researchers are trying to understand the influences of parents on adolescent children, as well as the effects of adolescents on their parents. As noted earlier (e.g., Jaffee et al., 2004), twin studies of children can explore the importance of child effects by testing the effects of children's genetically influenced traits on parent behavior (active or evocative rGE). In addition, the CoT design (e.g., D'Onofrio et al., 2005) can help explore whether shared genetic liability passed down from parents to their offspring (passive rGE) accounts for the association between putative risks and offspring functioning. Recent studies have combined the two approaches to jointly explore the importance of genetic confounds arising from both passive rGE and active/evocative rGE (Neiderhiser et al., 2004). Using an analytical model that can examine both processes simultaneously (Narusyte et al., 2008), Narusyte and colleagues (in press) examined the mechanisms underlying the association between parental criticism and adolescent antisocial problems in Sweden. The authors found that the underlying etiological processes differed for mothers and fathers. The association between paternal criticism and offspring adjustment appeared to be due to a direct environmental influence (not shared genetic liability due to passive or evocative/active *r*GE), which is consistent with a causal inference. In contrast, the association between maternal criticism and offspring antisocial behavior was due to child effects (evocative *r*GE). That is, the results suggested that genetically influenced traits in the offspring elicit increased maternal criticism.

Parent–Child Relationships

Adolescents with poorer relationships with their parents are more likely to exhibit antisocial problems. Burt and her colleagues (2007) used an adoption study to explore the relation between parent–child relationships and offspring antisocial behavior. In principle, adoption studies allow a clean break between environmental and genetic influences because parents are genetically unrelated to their children (Plomin, DeFries, McClearn, & McGuffin, 2008). These researchers found that the association between the quality of parent–child relationships (reported by the parents) and offspring antisocial behavior (reported by the offspring) was likely due to shared and nonshared environmental influences, not to genetic confounds due to passive *r*GE. Burt and her colleagues (2006) also found that parent–child conflict was independently associated with later antisocial problems when comparing differentially exposed MZ twins in a longitudinal study. The study followed twin pairs at age 11, 14, and 17 years, which allowed the researchers to combine the benefits of a longitudinal study with the quasi-experimental controls found in the co-twin control design. The findings strongly supported a causal inference regarding the role of parent–child relationships in the etiology of antisocial behavior.

Peer Groups

Although the importance of deviant peers is important throughout childhood and adolescence (e.g., Snyder et al., 2005), deviant peer groups are a particularly strong predictor of teen antisocial behavior. Unfortunately, the direction of effects is difficult to determine, as adolescents both select and may be influenced by their peer groups. The fact that association with deviant peers is partially heritable (Hill, Emery, Harden, & Mendle, 2008; Rowe & Osgood, 1984) raises the possibility that the association between peer groups and offspring antisocial behavior may be due to active *r*GE. The causal effects of peer groups when considering *r*GE is the focus of an accompanying chapter in the current book (see Chapter 16, by Kendler).

rGE AND PUTATIVE ENVIRONMENTAL RISK FACTORS REGARDLESS OF DEVELOPMENTAL TIMING

Researchers have identified a number of risk factors for antisocial problems that are important regardless of the timing of the exposure during childhood or adolescence.

Parental Divorce

Parental separation is associated with an increased risk of antisocial behavior, but two hypothesized explanations could account for the findings (Amato, 2000). Selection factors could account for the behavior problems found in children who experienced a divorce, or parental divorce could specifically cause these problems. Because genetic factors partially explain why some couples get divorced and others do not (D'Onofrio et al., 2005; McGue & Lykken, 1992), studies must account for the possibility that genetic confounding could account for the increased rate of antisocial behavior among youth whose parents get divorced. A number of CoT studies, including studies in Australia (D'Onofrio et al., 2005, 2006) and the United States (D'Onofrio, Turkheimer et al., 2007), have found, however, that parental divorce is independently associated with offspring antisocial problems when controlling for genetic factors shared by parents and their children, as well as for measured traits of the adult twins and their spouses. Recent adoption studies (Amato & Cheadle, 2008; Burt, Barnes, McGue, & Iacono, 2008) have also found that parental divorce is associated with increased risk for antisocial problems among adopted children who are not living with genetically related parents. The CoT and adoption studies thus provide converging evidence that supports the causal inference that environmental factors associated with divorce exert a causal environmental influence on risk for antisocial behavior in the offspring.

Marital Conflict

Marital conflict has consistently been found to predict offspring antisocial problems (Cummings & Davies, 2002), with researchers calling for more process-oriented research focused on identifying the causal mechanisms involved. Causal environmental inferences are supported by a number of longitudinal studies and a recent prevention program aimed at reducing marital conflict that was associated with a decrease in child adjustment problems (Cummings, Faircloth, Mitchell, Cummings, & Schermerhorn, 2008).

Yet, two recent CoT studies suggest that confounded genetic factors may account for the association between marital conflict and offspring antisocial behavior. In a study of Australian twin families, Harden and her colleagues (Harden, Turkheimer, et al., 2007) found that offspring of MZ twins differentially exposed to varying levels of marital conflict had similar rates of antisocial problems. The antisocial behavior of the offspring of DZ twins, however, was associated with differences in marital conflict within the twin family. Schermerhorn et al. (in press) found similar results when exploring the association between measures of marital agreement and conflict with offspring antisocial problems in a sample of adult twins and their offspring in Sweden. Certainly, more research is needed to fully understand the role that marital conflict plays in causing offspring antisocial behavior. Just assuming the association is causal, in fact, may be misleading.

IMPLICATIONS OF GENE–ENVIRONMENT CORRELATIONS

We have presented arguments for why it is essential to consider *r*GE in tests of hypotheses that propose aspects of the physical and social environment exert causal environmental influences on the risk for antisocial behavior. The next decade of research on the etiology of antisocial behavior in children and adolescents will attempt to understand both the specific genetic and specific environmental causal risk factors for antisocial behavior and how they work together. To accomplish this, it is essential to consider *r*GE in theoretical models, in the design of future studies, and in the statistical analysis of data. Failure to do so will result in largely wasted research resources and opportunities because the research is at risk for claiming that noncausal factors are playing an etiologic role in antisocial behaviors.

At this point, a number of quasi-experimental studies that reduced or eliminated *r*GE in tests of causal environmental hypotheses suggest that some (but not all) of the putative environmental influences implicated in prevailing theories of the origins of antisocial behavior causally do, in fact, impact risk for antisocial behavior. It appears that a number of hypothesized environments will be accepted as causal risk factors for antisocial behavior if confirmed in future quasi-experimental studies and randomized experiments.

Therefore, one of the goals for the next phase of research on the origins of antisocial behavior must be to use designs that minimize *r*GE in screening candidate environments, to identify those that are likely to play causal roles and in replicating the findings for specific environments. This research should use longitudinal methods to consider the child's chronological age and trajectory of antisocial behavior to understand environmental risk in a developmental context. Future studies should also examine environmental risk factors in a transactional context to understand how reciprocal effects of child characteristics and environments give rise to antisocial behavior.

Just as one must rule out confounded genetic factors when attempting to identify causal environmental risk factors, one must consider the influence of confounded environmental risk factors when attempting to identify genetic variants that increase risk for antisocial behavior and other maladaptive outcomes. Although this seems obvious, many molecular genetic studies of psychopathology have ignored the effects of confounded risk environments. Because small effect sizes are commonplace in molecular genetic research, even small estimation errors of effect size estimation can lead to incorrect inferences regarding associations.

Furthermore, to fully understand the origins of antisocial behavior, it will be very advantageous to use behavior genetic designs to examine environments that may also influence antisocial behavior through gene-by-environment interaction (G × E), which may prove to be important to the etiology of antisocial behaviors (e.g., Dodge, 2009). Unfortunately, it is not feasible to ascertain samples large enough to test for interactions between very large numbers of genetic polymorphisms and all of the potentially relevant environments with adequate power. Therefore, screening for environments that are involved in G× E using nonmolecular behavior genetic designs may identify aspects of the environment for future molecular genetic tests of G × E (Moffitt, Caspi, & Rutter, 2005).

G × E has important consequences for understanding the role of rGE (Jaffee & Price, 2007), which may prove to be important to the etiology of antisocial behaviors (e.g., Dodge, 2009). When genetic and environmental risks combine in an additive manner, statistical methods for assessing rGE are available, but when the correlated genetic and environmental influences combine multiplicatively (i.e., G × E), statistical methods that model and control for G × E must be used to estimate rGE accurately (Purcell, 2002; Rathouz, Van Hulle, Rodgers, Waldman, & Lahey, 2008). This is because G × E involving the risk environment, if not properly modeled, will appear as genetic effects on the trait, which are correlated with that environment.

Conversely, as noted in earlier chapters in the book, when rGE is present, statistical methods that take rGE fully into account are needed to estimate G × E accurately (Rathouz et al., 2008). Statistical models that test for G × E must account for all genetic effects on the trait and allow all of the genetic variability to interact with the risk environment. Traditionally, models of G × E have regressed out the main effect of the risk environment. If rGE exists but is ignored, the effects of genetic variants that influence both the risk environment and the trait are masked in this approach, and the

G × E involving those genetic variants, therefore, is masked as well, which biases the estimates and tests for G × E.

Quantitative behavior genetic research alone cannot identify biological mechanisms underlying maladaptive behavior. Nonetheless, such quasi-experimental research with genetically informative samples can sometimes provide crucial information for neuroscientists. Findings from quasi-experimental designs can help tell neuroscientists where they should be searching for biological mechanisms that influence antisocial behavior. For example, when exploring the consequences of prenatal nicotine exposure on offspring aggressive behavior, recent quasi-experimental research (Knopik, 2009) suggests that neuroscientists not focus on teratogenic mechanisms. Rather, it may be more informative if neuroscience research is focused on understanding the genetic and biological factors that influence maternal nicotine dependence (Agrawal et al., 2008) and the sequelae of these factors for offspring. With the growing emphasis on translational research, scientists interested in the etiology of antisocial behavior must begin to integrate information from across various disciplines, such as quantitative behavior genetics, molecular genetics, and neuroscience (e.g., Caspi & Moffitt, 2006).

Consideration of rGE also has profound implications for the development and implementation of interventions aimed at reducing antisocial behavior. Researchers and policy makers in prevention science base their interventions on the most recent basic science that identifies true causal influences on risk behaviors (Coie, Miller-Jackson, & Bagwell, 2000). As such, quasi-experimental studies that rule out genetic confounds that arise through rGE in the test of hypothesized causal environmental risk factors can identify modifiable environmental causes that can be targeted.

CONCLUSION

The next decade of research on the etiology of antisocial behavior in children and adolescents will attempt to understand both the genetic and environmental causal risk factors for antisocial behavior and how they work together. In doing so, rGE must be considered in studies designed to test causal hypotheses for putative environmental risk factors, in studies designed to identify molecular genetic variants, in studies that provide a foundation for studies of pathophysiology, and in studies designed to test for G × E. Studies that take rGE and G × E fully into account will be in a far better position to advance understanding in all of these areas and provide the translational science database for designing and testing new methods of preventing antisocial behavior. In the future, such studies would almost certainly benefit from a developmental perspective, studying the processes through which genetic and environmental risks influence varying developmental trajectories of antisocial behavior.

ACKNOWLEDGMENTS

Preparation of this chapter was supported in part by grants R01 MH070025, R01 MH53554, and R21 MH086099 from the National Institute of Mental Health; R01 HD056354,R01 HD053550, R01 HD061384, and R01 HD061817 from the National Institute of Child Health and Human Development; and NARSAD.

REFERENCES

Agrawal, A., Knopik, V. S., Pergadia, M. L., Waldron, M., Bucholz, K. K., Martin, N. G., et al. (2008). Correlates of cigarette smoking during pregnancy and its genetic and environmental overlap with nicotine dependence. *Nicotine and Tobacco Research, 10,* 567–578.

Amato, P. R. (2000). The consequences of divorce for adults and children. *Journal of Marriage and the Family, 62,* 1269–1287.

Amato, P. R., & Cheadle, J. E. (2008). Parental divorce, marital conflict and children's behavior problems: A comparison of adopted and biological children. *Social Forces, 86,* 1139–1161.

American Psychiatric Association. (1994). *Diagnostic and statistical manual of mental disorders (Vol. 4).* Washington, DC: American Psychiatric Association.

Bell, R. Q., & Harper, L. V. (1997). *Child effects on adults.* Hillsdale, NJ: Lawrence Erlbaum.

Blau, D. M. (1999). The effect of income on child development. *The Review of Economics and Statistics, 81(2),* 261–276.

Burt, S. A., Barnes, A. R., McGue, M., & Iacono, W. G. (2008). Parental divorce and adolescent delinquency: Ruling out the impact of common genes. *Developmental Psychology, 44,* 1668–1677.

Burt, S. A., McGue, M., Iacono, W. G., & Krueger, R. F. (2006). Differential parent–child relationships and adolescent externalizing symptoms: cross-lagged analyses within a discordant twin design. *Developmental Psychology, 42,* 1289–1298.

Burt, S. A., McGue, M., Krueger, R. F., & Iacono, W. G. (2007). Environmental contributions to adolescent delinquency: A fresh look at the shared environment. *Journal of Abnormal Child Psychology,* 787–800.

Cadoret, R. J., & Cain, C. A. (1980). Sex differences in predictors of antisocial behavior in adoptees. *Archives of General Psychiatry, 37,* 1171–1175.

Caspi, A., & Moffitt, T. E. (2006). Gene-environment interactions in psychiatry: Joining forces with neuroscience. *Nature Reviews Neuroscience, 7,* 583–590.

Caspi, A., Moffitt, T. E., Morgan, J., Rutter, M., Taylor, A., Arseneault, L., et al. (2004). Maternal expressed emotion predicts childrens' antisocial behavior problems:

using monozygotic-twin differences to identify environmental effects on behavioral development. *Developmental Psychology, 50,* 149–161.

Cnattingius, S. (2004). The epidemiology of smoking during pregnancy: smoking prevalence, maternal characteristics, and pregnancy outcomes. *Nicotine and Tobacco Research, 6,* S125–S140.

Cohen, M. A. (1998). The monetary value of saving a high-risk youth. *Journal of Quantitative Criminology, 14,* 5–30.

Coie, J. D., Miller-Jackson, S., & Bagwell, C. (2000). Prevention Science. In A. J. Sameroff, M. Lewis, & S. M. Miller (Eds.), *Handbook of developmental psychopathology* (Vol. 2, pp. 94–114). New York: Spring.

Coley, R. L., & Chase-Lansdale, P. L. (1998). Adolescent pregnancy and parenthood: recent evidence and future directions. *American Psychologist, 53(2),* 152–166.

Conger, R. D., & Donnellan, M. B. (2007). An interactionist perspective on the socioeconomic context of human development. *Annual Review of Psychology, 58,* 175–199.

Costello, E. J., Compton, S. N., Keeler, G., & Angold, A. (2003). Relationships between poverty and psychopathology: a natural experiment. *Journal of the American Medical Association, 290,* 2023–2029.

Cummings, E. M., & Davies, P. T. (2002). Effects of marital conflict on children: Recent advances and emerging themes in process-oriented research. *Journal of Child Psychology and Psychiatry, 43,* 31–63.

Cummings, E. M., Faircloth, W. B., Mitchell, P. M., Cummings, J. S., & Schermerhorn, A. C. (2008). Evaluating a brief prevention program for improving marital conflict in community families. *Journal of Family Psychology, 22,* 193–202.

D'Onofrio, B. M., Goodnight, J. A., Van Hulle, C. A., Waldman, I. D., Rodgers, J. L., Rathouz, P. J., et al. (2009). A quasi-experimental analysis of the association between family income and offspring conduct problems. *Journal of Abnormal Child Psychology, 37,* 415–429.

D'Onofrio, B. M., Singh, A. L., Iliadou, A., Lambe, M., Hultman, C., Grann, M., et al. (2010a). Familial confounding of the association between maternal smoking during pregnancy and offspring criminality: A population-based study in Sweden. *Archives of General Psychiatry, 67*(5), 529–538.

D'Onofrio, B. M., Singh, A. L., Iliadou, A., Lambe, M., Hultman, C., Neiderhiser, J. M., et al. (2010b). A quasi-experimental study of maternal smoking during pregnancy and offspring academic achievement. *Child Development, 81*(1), 80–100.

D'Onofrio, B. M., Goodnight, J., Van Hulle, C. A., Waldman, I. D., Rodgers, J. L., Harden, K. P., et al. (2009). Maternal age at childbirth and offspring disruptive behaviors: testing the causal hypothesis. *Journal of Child Psychology & Psychiatry, 50,* 1018–1028.

D'Onofrio, B. M., Turkheimer, E., Eaves, L. J., Corey, L. A., Berg, K., Solaas, M. H., et al. (2003). The role of the Children of Twins design in elucidating causal relations between parent characteristics and child outcomes. *Journal of Child Psychology & Psychiatry, 44,* 1130–1144.

D'Onofrio, B. M., Turkheimer, E. N., Emery, R. E., Maes, H. H., Silberg, J., & Eaves, L. J. (2007). A Children of Twins Study of parental divorce and offspring psychopathology. *Journal of Child Psychology & Psychiatry, 48,* 667–675.

D'Onofrio, B. M., Turkheimer, E. N., Emery, R. E., Slutske, W., Heath, A., Madden, P. A. F., et al. (2005). A genetically informed study of marital instability and its association with offspring psychopathology. *Journal of Abnormal Psychology, 114,* 570–586.

D'Onofrio, B. M., Turkheimer, E. N., Emery, R. E., Slutske, W., Heath, A., Madden, P. A. F., et al. (2006). A genetically informed study of the processes underlying the association between parental marital instability and offspring adjustment. *Developmental Psychology, 42,* 486–499.

D'Onofrio, B. M., Van Hulle, C. A., Waldman, I. D., Rodgers, J. L., Harden, K. P., Rathouz, P. J., et al. (2008). Smoking during pregnancy and offspring externalizing problems: an exploration of genetic and environmental confounds. *Development and Psychopathology, 20,* 139–164.

D'Onofrio, B. M., Van Hulle, C. A., Waldman, I. D., Rodgers, J. L., Rathouz, P. J., & Lahey, B. B. (2007). Causal inferences regarding prenatal alcohol exposure and childhood externalizing problems. *Archives of General Psychiatry, 64,* 1296–1304.

Dodge, K. A. (2009). Mechanisms of gene-environment interaction effects in the development of conduct disorder. *Perspectives on Psychological Science, 4,* 408–414.

Evans, G. W. (2004). The environment of childhood poverty. *American Psychologist, 59(2),* 77–92.

Farrington, D. P. (1991). Antisocial personality from childhood to adulthood. *The Psychologist 4,* 389–394.

Farrington, D. P. (1992). Criminal career research: Lessons for crime prevention. *Studies on Crime and Crime Prevention, 1,* 7–29.

Frick, P. J., & White, S. F. (2008). Research review: The importance of callous-unemotional traits for developmental models of aggressive and antisocial behavior. *Journal of Child Psychology and Psychiatry, 49,* 359–375.

Gennetian, L. A., & Miller, C. (2002). Children and welfare reform: a view from an experimental welfare program in Minnesota. *Child Development, 73,* 601–620.

Geronimus, A. T., Korenman, S., & Hillemeier, M. M. (1994). Does young maternal age adversely affect child development? Evidence from cousin comparisons in the United States. *Population and Development Review, 20(3),* 585–609.

Gilman, S. E., Gardener, H., & Buka, S. L. (2008). Maternal smoking during pregnancy and children's cognitive and physical development: a causal risk factor? *American Journal of Epidemiology, 168,* 522–531.

Gottfredson, M. R., & Hirschi, T. (1990). *A general theory of crime.* Stanford, CA.: Stanford University Press.

Gray, R., Mukherjee, R. A. S., & Rutter, M. (2009). Alcohol consumptions during pregnancy and its effects on neurodevelopment: what is known and what remains uncertain. *Addiction, 104,* 1270–1273.

Hao, L., & Matsueda, R. L. (2006). Family dynamics through childhood: a sibling model of behavior problems. *Social Science Research, 35(2),* 500–524.

Harden, K. P., Lynch, S. K., Turkheimer, E., Emery, R. E., D'Onofrio, B. M., Slutske, W. S., et al. (2007). A behavior genetic investigation of adolescent motherhood and offspring mental health problems. *Journal of Abnormal Psychology, 116,* 667–683.

Harden, K. P., Turkheimer, E., Emery, R. E., D'Onofrio, B. M., Slutske, W. S., Heath, A. C., et al. (2007). Marital conflict and conduct disorder in children-of-twins. *Child Development, 78*, 1–18.

Heath, A. C., Kendler, K. S., Eaves, L. J., & Markell, D. (1985). The resolution of cultural and biological inheritance: informativeness of different relationships. *Behavior Genetics, 15*, 439–465.

Hill, J., Emery, R. E., Harden, K. P., & Mendle, J. (2008). Alcohol use in adolescent twins and affiliation with substance using peers. *Journal of Abnormal Child Psychology*, (36), 81–94.

Hinshaw, S. P., Lahey, B. B., & Hart, E. L. (1993). Issues of taxonomy and comorbidity in the development of conduct disorder. *Development and Psychopathology, 5*, 31–50.

Hirschi, T., & Gottfredson, M. (1983). Age and the explanation of crime. *American Journal of Sociology, 89*, 552–584.

Jaffee, S. R. (2007). Sensitive, stimulating caregiving predicts cognitive and behavioral resilience in neurodevelopmentally at-risk infants. *Development and Psychopathology, 19*, 631–647.

Jaffee, S. R., Caspi, A., Moffitt, T. E., Belsky, J., & Silva, P. (2001). Why are children born to teen mothers at risk for adverse outcomes in young adulthood? Results from a 20-yr longitudinal study. *Development and Psychopathology, 13(2)*, 377–397.

Jaffee, S. R., Caspi, A., Moffitt, T. E., Polo-Tomas, M., Price, T. S., & Taylor, A. (2004). The limits of child effects: evidence for genetically mediated child effects on corporal punishment but not on physical maltreatment. *Developmental Psychology, 40*, 1047–1058.

Jaffee, S. R., & Price, T. S. (2007). Gene-environment correlations: a review of the evidence and implications for prevention of mental illness. *Molecular Psychiatry, 12(5)*, 432–442.

Johansson, A. L. V., Dickman, P. W., Kramer, M. S., & Cnattingius, S. (2009). Maternal smoking and infant mortality: Does quitting smoking reduce the risk of infant death? *Epidemiology, 20*, 1–8.

Kendler, K. S., & Baker, J. H. (2007). Genetic influences on measures of the environment: A systematic review. *Psychological Medicine, 37*, 615–626.

Knopik, V. S. (2009). Maternal smoking during pregnancy and child outcomes: Real or spurious effect? *Developmental Neuropsychology, 34*, 1–36.

Knopik, V. S., Sparrow, E. P., Madden, P. A. F., Bucholz, K. K., Hudziak, J. J., Reich, W., et al. (2005). Contributions of parental alcoholism, prenatal substance exposure, and genetic transmission to child ADHD risk: a female twin study. *Psychological Medicine, 35*, 625–635.

Krueger, R. F. (1999). The structure of common mental disorders. *Archives of General Psychiatry, 56*, 921–926.

Lahey, B. B., D'Onofrio, B. M., & Waldman, I. D. (2009). Using epidemiologic methods to test hypotheses regarding causal influences on child and adolescent mental disorders. *Journal of Child Psychology and Psychiatry, 50*, 53–62.

Lahey, B. B., Schwab-Stone, M., Goodman, S. H., Waldman, I. D., Canino, G., Rathouz, P. J., et al. (2000). Age and gender differences in oppositional behavior and conduct problems: A cross-sectional household study of middle childhood and adolescence. *Journal of Abnormal Psychology, 109,* 488–503.

Lahey, B. B., Van Hulle, C. A., Waldman, I. D., Rodgers, J. L., D'Onofrio, B. M., Pedlow, S., et al. (2006). Testing descriptive hypotheses regarding sex differences in the development of conduct problems and delinquency. *Journal of Abnormal Child Psychology, 34,* 737–755.

Lahey, B. B., & Waldman, I. D. (2003). A developmental propensity model of the origins of conduct problems during childhood and adolescence. In B. B. Lahey, T. E. Moffitt & A. Caspi (Eds.), *Causes of conduct disorder and juvenile delinquency* (pp. 76–117). New York: Guilford Press.

Lahey, B. B., & Waldman, I. D. (2005). A developmental model of the propensity to offend during childhood and adolescence. In D. P. Farrington (Ed.), *Advances in criminological theory (Vol. 13;* pp. 15–50). Piscataway, NJ: Transaction Publishers.

Lambe, M., Hultman, C., Torrang, A., MacCabe, J., & Cnattingius, S. (2006). Maternal smoking during pregnancy and school performance at age 15. *Epidemiology, 17,* 524–530.

Loeber, R. (1988). Natural histories of conduct problems, delinquency, and associated substance abuse: Evidence for developmental progressions. In B. B. Lahey & A. E. Kazdin (Eds.), *Advances in clinical child psychology* (Vol. *11*). New York: Plenum.

Loeber, R., Farrington, D. P., Stouthamer-Loeber, M., & Van Kammen, W. B. (1998). *Antisocial behavior and mental health problems: explanatory factors in childhood and adolescence.* Mahwah, NJ: Lawrence Erlbaum Associates Publishers

Lundberg, F., Cnattingius, S., D'Onofrio, B., Altman, D., Lambe, M., Hultman, C., et al. (2010). Maternal smoking during pregnancy and intellectual performance in young adult Swedish male offspring. *Pediatric and Perinatal Epidemiology, 24*(1), 79–87.

Lynch, S. K., Turkheimer, E., Emery, R. E., D'Onofrio, B. M., Mendle, J., Slutske, W., et al. (2006). A genetically informed study of the association between harsh punishment and offspring behavioral problems. *Journal of Family Psychology, 20,* 190–198.

Maughan, B., Taylor, A., Caspi, A., & Moffitt, T. E. (2004). Prenatal smoking and early childhood conduct problems. *Archives of General Psychiatry, 61,* 836–843.

McGue, M., & Lykken, D. T. (1992). Genetic influence on risk of divorce. *Psychological Science, 6,* 368–373.

Moffitt, T. E. (1993). Adolescence-limited and life-course-persistent antisocial behavior: A developmental taxonomy. *Psychological Review, 100,* 674–701.

Moffitt, T. E. (2005). The new look of behavioral genetics in developmental psychopathology: gene-environment interplay in antisocial behaviors. *Psychological Bulletin, 131,* 533–554.

Moffitt, T. E. (2006). Life-course persistent and adolescence-limited antisocial behavior. In D. Cicchetti, & D. Cohen (Eds.), *Developmental Psychopathology, (2nd ed.),* (pp. 570–598). New York: Wiley.

Moffitt, T. E., Caspi, A., & Rutter, M. (2005). Strategy for investigating interactions between measured genes and measured environments. *Archives of General Psychiatry, 62,* 473–481.

Moffitt, T. E., Caspi, A., Rutter, M., & Silva, P. A. (2001). *Sex differences in antisocial behaviour: conduct disorder, delinquency, and violence in the Dunedin Longitudinal Study.* New York: Cambridge University Press.

Morris, P. A., & Gennetian, L. A. (2003). Identifying the effects of income on children's development using experimental data. *Journal of Marriage and Family, 65(3),* 716–729.

Nagin, D. S., & Tremblay, R. E. (2005). Developmental trajectory groups: fact or a useful statistical fiction? *Criminology: An Interdisciplinary Journal, 43,* 873–904.

Narusyte, J., Andershed, A-K., Neiderhiser, J. M., D'Onofrio, B. M., Reiss, D., Spotts, E., et al. (in press). Parental criticism and externalizing behavior problems in adolescents: the role of environment and genotype-environment correlation. *Journal of Abnormal Psychology.*

Narusyte, J., Neiderhiser, J. M., D'Onofrio, B. M., Heath, A., Reiss, D., Spotts, E., et al. (2008). Testing different types of genotype-environment correlation: an extended children-of-twins model. *Developmental Psychology, 44,* 1591–1603.

Neale, M. C., & Cardon, L. R. D. (1992). *Methodology for genetic studies of twins and families.* Dordrecht, The Netherlands: Kluwer Academic Press.

Neiderhiser, J. M., Reiss, D., Pedersen, N. L., Lichtenstein, P., Spotts, E. L., Hansson, K., et al. (2004). Genetic and environmental influences on mothering of adolescents: a comparison of two samples. *Developmental Psychology, 40,* 335–351.

Patterson, G. R., DeBaryshe, B. D., & Ramsey, E. (1989). A developmental perspective on antisocial behavior. *American Psychologist, 44,* 329–335.

Plomin, R., & Bergeman, C. S. (1991). The nature of nurture: genetic influences on "environmental" measures. *Behavioral and Brain Sciences, 10,* 1–15.

Plomin, R., DeFries, J. C., McClearn, G. E., & McGuffin, P. (Eds.). (2008). *Behavioral genetics* (Vol. 5). New York: Worth Publishers.

Purcell, S. (2002). Variance components for gene-environment interaction in twin studies. *Twin Research, 5,* 554–571.

Quay, H. (1987). Patterns of delinquent behavior. In H. Quay (Ed.), *Handbook of juvenile delinquency* (pp. 118–138). New York: Wiley.

Rathouz, P. J., Van Hulle, C. A., Rodgers, J. L., Waldman, I. D., & Lahey, B. B. (2008). Specification, testing, and interpretation of gene-by-measured-environment interaction models in the presence of gene-environment correlation. *Behavior Genetics, 38,* 301–315.

Rhee, S. H., & Waldman, I. D. (2002). Genetic and environmental influences on antisocial behavior: A meta-analysis of twin and adoption studies. *Psychological Bulletin, 128,* 490–529.

Rice, F., Harold, G. T., Boivin, J., Hay, D. F., Van den Bree, M., & Thapar, A. (2009). Disentangling prenatal and inherited influences in humans with an experimental design. *Proceedings of the National Academy of Sciences of the U.S.A., Early Edition Online.*

Rodgers, J. L., Bard, D., E., & Miller, W. B. (2007). Multivariate Cholesky models of human female fertility patterns in the NLSY. *Behavior Genetics, 37,* 345–361.

Rowe, D. C., & Osgood, D. W. (1984). Heredity and sociology theories of delinquency: a reconsideration. *American Sociological Review, 49,* 526–540.

Rowe, D. C., & Rodgers, J. L. (1997). Poverty and behavior: are environmental measures nature and nurture? *Developmental Review, 17,* 358–375.

Rutter, M. (2003). Crucial paths from risk indicator to causal mechanism. In B. B. Lahey, T. E. Moffitt, & A. Caspi (Eds.), *Causes of conduct disorder and juvenile delinquency* (pp. 3–26). New York: Guilford.

Rutter, M. (2007). Proceeding from observed correlation to causal inference: the use of natural experiments. *Perspectives on Psychological Science, 2,* 377–395.

Rutter, M., Caspi, A., & Moffitt, T. E. (2003). Using sex differences in psychopathology to study causal mechanisms: unifying issues and research strategies. *Journal of Child Psychology and Psychiatry, 44,* 1092–1115.

Rutter, M., Pickles, A., Murray, R., & Eaves, L. J. (2001). Testing hypotheses on specific environmental causal effects on behavior. *Psychological Bulletin, 127,* 291–324.

Rutter, M., & Silberg, J. (2002). Gene-environment interplay in relation to emotional and behavioral disturbance. *Annual Review of Psychology, 53,* 463–490.

Sampson, R. J., & Laub, J. H. (2005). Seductions of method: rejoinder to Nagin and Tremblay's "developmental trajectory groups: fact or fiction?" *Criminology: An Interdisciplinary Journal, 43,* 905–913.

Schermerhorn, A. C., D'Onofrio, B. M., Turkheimer, E., Ganiban, J. M., Spotts, E. L., Lichtenstein, P., et al. (in press). A genetically informed study of associations between family functioning and child psychosocial adjustment. *Developmental Psychology.*

Shadish, W. R., Cook, T. D., & Campbell, D. T. (2002). *Experimental and quasi-experimental designs for generalized causal inference.* New York: Houghton Mifflin.

Silberg, J. L., & Eaves, L. J. (2004). Analyzing the contribution of genes and parent-child interaction to childhood behavioral and emotional problems: a model for the children of twins. *Psychological Medicine, 34,* 347–356.

Snyder, J., Reid, J. B., & Patterson, G. R. (2003). A social learning model of child and adolescent antisocial behavior. In B. B. Lahey, T. E. Moffitt & A. Caspi (Eds.), *Causes of conduct disorder and juvenile delinquency* (pp. 27–48). New York: Guilford Press.

Snyder, J., Schrepferman, L., Oeser, J., Patterson, G., Stoolmiller, M., Johnson, K., et al. (2005). Deviancy training and association with deviant peers in young children: occurrence and contribution to early-onset conduct problems. *Development and Psychopathology, 17*(2), 397–413.

Steinberg, L., & Morris, A. S. (2001). Adolescent development. *Annual Review of Psychology, 52,* 83–110.

Thapar, A., Rice, F., Hay, D., Boivin, J., Langley, K., Van den Bree, M., et al. (2009). Prenatal smoking may not cause ADHD. Evidence from a novel design. *Biological Psychiatry, 66*(8), 722–727.

Tremblay, R. E. (2000). The development of aggressive behaviour during childhood: what have we learned in the past century? *International Journal of Behavioral Development, 24,* 129–141.

Turley, R. N. L. (2003). Are children of young mothers disadvantaged because of their mother's age or family background? *Child Development, 74,* 465–474.

Wakschlag, L. S., Pickett, K. E., Cook, E., Benowitz, N. L., & Leventhal, B. L. (2002). Maternal smoking during pregnancy and severe antisocial behavior in offspring: a review. *American Journal of Public Health, 92,* 966–974.

Zuckerman, M. (1996). The psychobiological model for impulsive unsocialized sensation seeking: a comparative approach. *Neuropsychobiology, 34,* 125–129.

16

Peer Group Deviance, Conduct Disorder, and Alcohol Intake

KENNETH S. KENDLER

The goal of this chapter is to illustrate the three main themes of this volume—gene × environment correlation, gene × environment interaction, and development—in research that I have done with colleagues on the concept of peer group deviance (PGD).

PEER DEVIANCE AND GENE × ENVIRONMENT CORRELATION

Peers have a broad impact on many aspects of behavior and are an important influence on a wide variety of human traits (Harris, 2002). In particular, exposure to high levels of PGD in childhood and adolescence is robustly associated with future externalizing behaviors, including substance use (Allen, Donohue, Griffin, Ryan, & Turner, 2003; Hawkins et al., 1998; Petraitis, Flay, Miller, Torpy, & Greiner, 1998). As a result, PGD plays a substantial role in most developmental models for antisocial behavior (e.g., see Coie & Miller-Johnson, 2001; Farrington, 2005; Patterson, DeBaryshe, & Ramsey, 1989). Therefore, understanding what makes individuals associate with prosocial versus antisocial friends will be critical in clarifying sources of individual differences in externalizing behaviors.

Models of person–environment interaction have traditionally emphasized the passive role of the individual (e.g., environment → person). However, prior studies of PGD have suggested important bidirectional effects (environment ↔ person) (e.g., Eiser, Morgan, Gammage, Brooks, &

Kirby, 1991; Kandel, 1979, 1996; Wills & Cleary, 1999). That is, although social pressures to conform make adolescents adopt the behaviors of their peers (via *peer influence*), adolescents also certainly seek out like-minded friends who share their own attitudes and predilections (via *peer selection*). Longitudinal studies have been key in our efforts to disentangle the effects of peer influence and selection (Dishion, Patterson, Stoolmiller, & Skinner, 1991; Fergusson, Woodward, & Horwood, 1999; Fergusson & Horwood, 1999; Kandel, 1978; Simons, Whitbeck, Conger, & Conger, 1991).

A genetic strategy offers a complementary approach to disentangling these mechanisms. As reviewed in Chapter 1 of this volume, a growing number of studies over the past decades have applied behavioral genetic models to analysis of "environmental" measures (Kendler & Baker, 2007). If the genetically influenced characteristics of an individual affect the type of friends selected, then measures of PGD should be heritable. To date, a small number of genetic studies of PGD have been done, and most (Baker & Daniels, 1990; Daniels, Dunn, Furstenberg, & Plomin, 1985; Manke, McGuire, Reiss, Hetherington, & Plomin, 1995; Pike, Nake, Reiss, & Plomin, 2000; Rose, 2002) but not all (Iervolino et al., 2002; Walden, McGue, Iacono, Burt, & Elkins, 2004) have found that PGD was substantially heritable.

In these investigations, we wanted to explore whether there would be systematic differences in the relative importance of genetic and environmental influences on PGD during different developmental periods. After all, adolescence is associated with a marked increase in autonomy (Bridges, 2003; Steinberg, 1990) and a shift in socializing influences from parents to peers as children spend less time with family and more time with friends (Larson & Richards, 1991). As children become active agents in creating their social world, genetic influence on PGD may increase.

As outlined in detail elsewhere (Kendler et al., 2007b), we examined results from the third wave of interviews conducted in adult male–male twin pairs from the Virginia Adult Twin Study of Psychiatric and Substance Use Disorders ($n = 1,802$ twins) (Kendler & Prescott, 2006). These twins come from the birth certificate–based Virginia Twin Registry and are broadly representative of white, American males with respect to age, educational status, and rates of psychopathology. We utilized a life history calendar–based interview (Freedman, Thornton, Camburn, Alwin, & Young-DeMarco, 1988) that assessed a range of constructs including PGD for five age periods: 8 to 11, 12 to 14, 15 to 17, 18 to 21, and 22 to 25. PGD was assessed by mean levels of 12 items obtained from two validated instruments (Johnston, Bachman, & O'Malley, 1982; Tarter & Hegedus, 1991) that assessed the proportion of the respondent's friends, at each particular age, who engaged in specific behaviors. These items were "smoked cigarettes," "drunk alcohol," "got drunk," "had problems with alcohol," "skipped or cut school a lot," "cheated on school tests," "stole anything or damaged property on purpose," "been in trouble with the law," "smoked marijuana," "used inhalants," "used other drugs like cocaine, downers, or LSD," and "sold or gave drugs to other kids."

The details of our statistical model are described in the footnote below.[1] Descriptive statistics revealed that both the mean and variance of PGD scores increased monotonically from ages 8–11 to 18–21, then declined at ages 22–25. The variance of PGD nearly doubles from 8–11 to 12–14, increases more slowly until age 18–21, then stabilizes. Inspection of the raw data revealed substantial individual variability in the developmental pattern of PGD. Some individuals increased in PGD across time, some decreased over time, and others remained relatively stable at either high, medium, or low levels of PGD.

Given this variation in PGD trajectories, we calculated the slope of the PGD over the five time periods for each individual. These slopes were more highly correlated in MZ (+0.60) than in DZ pairs (+0.44), indicating that genetic factors also contribute to the rate of change of PGD over time.

Our formal modeling began with a fixed-effects regression model for PGD. The best fit included a linear effect of standardized age and a spline (a change in the slope of a regression line) on the effect of age impacting solely on the oldest age group. For details, see Kendler et al. (2007b). A random effects growth curve model adding additive genetic (*A*), shared environmental (*C*), and unique (*E*) environmental effects to the best-fitting fixed effects model was fitted to the PGD data. The method we used (see McArdle [2006] for more details) parameterizes the model with unique effects of *A*, *C*, and *E* on the intercept; unique effects of *A*, *C*, and *E* on slopes; and *A*, *C*, and *E* effects common to both intercept and slope. The key results that we want to focus on here are shown in Figure 16.1, which presents the "raw" or "unstandardized" genetic and environmental variance components for PGD at each age. The pattern of results differs considerably for genetic, shared environmental, and individual-specific environmental effects. Genetic variance in PGD increases substantially and monotonically across the five age periods. Variance in PGD resulting from shared environmental effects, by

---

[1] Biometric latent growth curve analyses were performed in SAS PROC MIXED, as described by McArdle (2006). Model fit comparisons involving fixed effects were carried out using maximum likelihood, whereas model fit comparisons involving random effects were carried out using residual maximum likelihood (REML). The fit statistics used included likelihood-ratio χ^2, Akaike's information criterion (AIC) (Akaike, 1987), and the generalized coefficient of determination of Cox and Snell (Cox & Snell, 1989). Variance component models used in estimating biometric latent growth curve parameters and intraclass correlations were fit using REML in PROC MIXED. Confidence intervals for variance components were constructed using a Satterthwaite approximation (Satterthwaite, 1946).

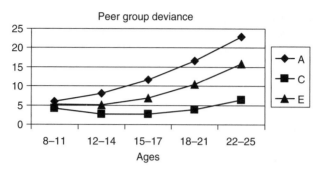

Figure 16.1. **The unstandardized additive genetic (*A*), shared or common environmental (*C*), and individual-specific environmental (*E*) variance for peer-group deviance over five time periods estimated from our best-fit growth curve model.** The *x*-axis indicates increasing age at assessment, whereas the *y*-axis reflects the unstandardized measure of variance. Figure adapted from Kendler, K. S., Jacobson, K. C., Gardner, C. O., Gillespie, N. A., Aggen, S. H., & Prescott, C. A. (2007b). Creating a social world: a developmental study of peer deviance. *Archives of General Psychiatry, 64*, 958–965, with permission.

contrast, decreases over the first three age periods and then increases modestly at ages 18–21 and 22–25. Individual specific environmental variance in PGD is relatively stable across the first three age periods but then increases sharply over the final two age periods.

We sought, in these analyses, to elucidate from a developmental perspective the sources of individual differences in PGD from mid-childhood to early adulthood. Four findings are noteworthy. First, our simple biometric growth curve model accounted for 54% of the variance of the development of PGD, supporting the explanatory power of this approach. Second, genetic factors contributed substantially to PGD, and this contribution increased as individuals aged. Third, at ages 8–11, shared family environment was an important determinant of PGD, but its importance diminished over time (but then for unknown reasons increased again in early adulthood). Fourth, individual-specific environment had an important impact on PGD that became more pronounced when twins started to live apart.

This study demonstrates the value of using behavioral genetic strategies to better understand developmental changes in key environmental exposures such as PGD. Our evidence that individual differences in PGD are heritable adds to a growing body of research demonstrating heritabilities of other putative environmental measures including parenting, social support, and stressful life events (Kendler & Baker, 2007).

Our results are best viewed as an example of active gene–environment correlation in the typology proposed by Plomin et al. (1977), which they described as follows:

Active genotype–environment correlation occurs as a result of the fact that a child is not merely the passive recipient of his environment. Rather, he contributes to his own environment and may actively seek one related to his genetic propensities. (Plomin, DeFries, & Loehlin, 1977, p. 310)

Scarr and McCartney (1983) developed this idea further, proposing a developmental theory of childhood and adolescence in which active genotype–environment correlation "becomes particularly important in later childhood and adolescence as individuals make their own environments." Our results are broadly supportive of their hypothesis. Of note, the pattern of developmental change that we saw for PGD—increase in genetic variance and decrease in the effects of shared environmental factors with aging—has also been observed in a number of genetically informative longitudinal studies of a variety of externalizing behaviors such as smoking (Koopmans, 1997; White, Hopper, Wearing, & Hill, 2003), drinking (Koopmans, 1997; Rose, Dick, Viken And, & Kaprio, 2001), and conduct disorder (Jacobson, Prescott, & Kendler, 2002; Lyons et al., 1995).

Our findings are also supportive of prior longitudinal research that suggests that a substantial proportion of the similarity between adolescents and their peer groups results from *peer selection* (Gordon et al., 2004; Kandel, 1979, 1996; van den Bree & Pickworth, 2005).

This study has implications not only for interpreting traditional studies of PGD and antisocial behavior based on nongenetic samples and for developing appropriate interventions and preventions, but also for the growing field of molecular genetics. Our field is beginning to identify individual genetic variants that impact on risk for psychiatric and substance use disorders. Along with findings reviewed above and in Chapter 1, this study suggests that the mechanisms by which some of these variants work will not be entirely reducible to "within-the-skin" physiological pathways. Rather, these genes will impact on risk through "outside-the-skin" pathways, by altering the probability of exposure to pathogenic environments (Kendler, 2001, 2005).

PEER DEVIANCE, ALCOHOL CONSUMPTION, AND GENE × ENVIRONMENT INTERACTION

The second major theme that we explore in this volume is gene × environment interaction. In the first series of analyses in this chapter, we treated PGD as the dependent variable. We were trying to understand, in a genetic and developmental context, the sources of individual differences for PGD.

In this section, we use PGD as a predictor variable and ask whether it can moderate the effect of genes on alcohol consumption. But first, a bit of background.

A range of prior studies have shown that environmental exposures can moderate the impact of genetic risk factors on alcohol-related traits (Cloninger, Bohman, & Sigvardsson, 1981; Dick, Rose, Viken, Kaprio, & Koskenvuo, 2001; Martens, Neighbors, Dams-O'Connor, Lee, & Larimer, 2007; Rose et al., 2001; Sigvardsson, Bohman, & Cloninger, 1996; Viken, Kaprio, & Rose, 2007). We want to take up this question for PGD. We want to ask whether PGD moderates the effects of two different kinds of genetic risks for alcohol use: (a) specific alcohol-related genetic risk factors, and (b) nonspecific externalizing genetic risk factors. These were indexed by psychiatric assessments conducted in the co-twin and family history information provided on the parents. Furthermore, we want to look at the moderating effects of PGD over development, from early adolescence to early adulthood. We predict that during early adolescence, at which time alcohol-use patterns are beginning to be formed, individuals at elevated genetic risk should be particularly sensitive to environments like PGD that can provide access to and support for excessive alcohol consumption (Kendler, 2001; Shanahan & Hofer, 2005).

We examined, for this investigation, the same sample of male twins described above. In the life-history calendar used in this interview, the columns were for each year of the subject's life. The first rows of this calendar, completed early in the interview, documented key changes in living situation, as well as major educational, employment, and interpersonal milestones. Toward the end of the interview, after completion of drug sections assessing standard questions about age at first use, maximal lifetime use, and symptoms of abuse and dependence, we returned to the calendar. For alcohol, we asked subjects, starting with the age at which they reported first using alcohol, the average number of times per month they consumed alcohol and the average number of drinks they consumed per day when drinking. We defined a drink as "one bottle of beer, one glass of wine, or one shot of liquor." We then moved forward in time, year by year, until reaching their present age, asking if their drinking patterns had changed, and if so, to document the new pattern. If necessary, the interviewers would use other memory prompts from the information previously recorded on the calendar to "cue" the respondent into the relevant "memory files." The test–retest reliability of our assessments of average monthly alcohol consumption (calculated as the product of the average number of days per month of drinking and the average number of drinks per day when drinking) have been presented in detail elsewhere (Kendler, Schmitt, Aggen, & Prescott, 2008, Fig. 2) and exceed +0.85 over the time period considered here. In our analyses, levels of alcohol consumption were rank-transformed to normal scores by the method of Blom (1958) for ease of interpretation, to correct for the rightward skew typically seen in such measures, and to reduce the influence of outliers. For details of our statistical methods used here.[2]

The main effects of PGD on standardized alcohol consumption (as indicated by the regression coefficient) increased over time, and all were highly statistically significant (p <0.0001): ages 12–14 (+0.22), ages 15–17 (+0.35), ages 18–21 (+0.37), and ages 22–25 (+0.41). Genetic effects specific to risk for alcohol use disorders (which were calculated from the risk for alcohol use disorders in mother, father, and co-twin) increased in their association with alcohol consumption over this time period, whereas the genetic risk for externalizing disorders (calculated from the risk for conduct disorder and antisocial behavior in the father and co-twin) maximized at ages 15–17, then declined considerably in their ability to predict alcohol use.

Turning to what we are most interested in—the interactions—we found at age 12–14 a strong and highly significant interaction (p <0.0001) between PGD and both genetic risk for alcohol use disorders and genetic risk for externalizing disorders in the prediction of alcohol consumption. We depict the second of these rather dramatic interactions in Figure 16.2. At low levels of peer deviance, genetic risk for externalizing disorders had almost no impact on the level of alcohol consumption. By contrast, at high levels of peer deviance, levels of alcohol consumption were strongly related to level of genetic risk.

At ages 15–17, a weaker interaction was seen between PGD and genetic risk for alcohol use disorders ($p < 0.01$). However, no significant interaction was seen between PGD and genetic risk for externalizing disorders. Despite the stronger main effects of PGD on alcohol use at ages 18–21 and 22–25, no interaction with either genetic effect was detected.

So, as predicted, PGD was shown to moderate genetic effects on alcohol consumption. High levels of PGD augment the impact of the genetic risks, and low levels of PGD suppress such effects. However, these effects themselves were developmentally dynamic. Environmental moderation of

---

[2] Since the residual correlation within twin pairs was substantial and higher in MZ twin pairs, regression models were run as hierarchical linear models using PROC MIXED in SAS (SAS Institute, 2008). Individual twins were treated as level 1 variables and families as level 2, and random intercepts were used to estimate family-level variance components for MZ and DZ twin pairs estimated as separate parameters. This yields parameter estimates and test statistics for regression coefficients that are adjusted for the residual intraclass correlations for MZ and DZ twins. Environmental risk factors were measured on individual twins and acted as level 1 variables, whereas genetic risks and year of birth were level 2 (family level) variables. All explanatory variables were grand mean centered.

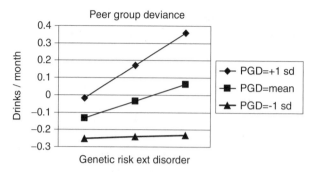

Figure 16.2. **The prediction of average standardized monthly alcohol consumption over the ages 12–14 by the genetic risk for externalizing disorders, peer group deviance (PGD), and their interaction.** The results, from the best-fit regression model, are depicted for three hypothetical individuals with a moderately high genetic risk for alcohol use disorders (values 1 standard deviation above the mean), an average genetic risk for alcohol use (mean value), and a moderately low genetic risk for alcohol use (values 1 standard deviation below the mean). Average monthly alcohol consumption is standardized, so that at the mean level of genetic risk and mean level of alcohol availability, the score is ~0. The y-axis then depicts this mean score in standard deviation units.

genetic effects was much more pronounced in early adolescence than in later periods. Genetic effects on alcohol use appear to be much more environmentally sensitive at this early developmental stage. As use patterns become more established, genetic effects appear to become less flexible and less open to environmental moderation.

These results have obvious implications for efforts to prevent the emergence of heavy drinking in adolescence. Environmental interventions that occur early in adolescence might be capable of substantially attenuating the effect of high genetic risk. Similar interventions later may be much less effective as the behavior patterns system become more "set" or developmentally canalized.

Our findings are consistent with a number of prior studies that suggest that when significant gene × environment interaction effects are found for drug-related traits, they show increased genetic effects when social constraints are minimized (e.g., low parental monitoring, low prosocial behaviors, and low parental bonding) or when the environment permits easy access to the psychoactive substance and/or encourages its use (Kendler, 2001; Shanahan & Hofer, 2005). Previous studies have demonstrated that genetic influences on adolescent substance use and externalizing behaviors more generally are magnified under conditions of lower parental monitoring (Dick et al., 2007b) and higher peer substance use/deviance (Button et al.,

2007; Dick et al., 2007a). More recently, this has also been demonstrated with respect to specific risk genes, with evidence that stronger genetic effects occur in the presence of lower parental monitoring (Dick et al., 2009) and higher peer deviance (Latendresse et al., 2009).

PEER DEVIANCE AND CONDUCT DISORDER: DEVELOPMENTALLY DYNAMIC GENE–ENVIRONMENT CORRELATION

In this last series of data analyses, we come back to an examination of gene–environment correlation, but this time in a causal and developmental context. We explore the inter-relationship over time between our old friend PGD and conduct disorder (CD). These are worth examining together because PGD and CD in children and adolescents are among the strongest predictors of future externalizing behaviors (Hawkins et al., 1998; Petraitis et al., 1998; van den Bree & Pickworth, 2005), and they typically figure prominently in developmental models for antisocial behavior (e.g., Coie &Miller-Johnson, 2001; Farrington, 2005; Patterson et al., 1989).

A central question in the literature on the development of externalizing traits and disorders has been the causal relationship between CD and PGD (Kandel, 1978, 1996). To what extent do pressures for conformity influence adolescents to adopt the behaviors of their peers via *social influence*, versus do adolescents actively seek out like-minded friends who share their attitudes and behavioral proclivities via *social selection*? Prior studies have provided mixed results regarding the relative impact of these two processes (e.g., Gordon et al., 2004; Kandel, 1978, 1996; Lacourse et al., 2006; Wills & Cleary, 1999).

Both CD (Gelhorn et al., 2005; Jacobson, Prescott, & Kendler, 2000) and PGD (Cleveland, Wiebe, & Rowe, 2005; Iervolino et al., 2002; Rose, 2002; Walden et al., 2004) have been shown to be strongly influenced by familial factors. In these analyses, we examine retrospectively reported levels of CD and PGD in the same twin sample examined above over three critical age periods: 8–11, 12–15, and 16–18. We fit developmental models to clarify the causal relationship between CD and PGD. By decomposing these developmental pathways into those resulting from genetic versus environmental factors, we hope to clarify the inter-relationship of CD and PGD over time, thereby elucidating the relative importance of social influence and social selection.

Conduct disorder was assessed by 11 items from the Semi-Structured Assessments for the Genetics of Alcoholism interview using a 4-point scale (Bucholz et al., 1994) that operationalized criteria from the *Diagnostic and*

Statistical Manual of Mental Disorders, Third Edition, Revised (DSM-III-R). However, we excluded two highly deviant and rarely endorsed items: forcing someone into sexual activity and being physically cruel to people. Furthermore, during pilot testing, several twins objected to the offensive nature of other CD items. To avert further concerns, we asked all individuals the five most commonly endorsed CD items ("physical fights," "telling lies," "playing hooky," "stealing," and "physically hurt other people.") If they responded negatively to these, we assumed they lacked CD symptoms and skipped to the next section. If they answered one or more item positively, we asked the remaining six items.

Our model fitting, performed using the Mx program (Neale, Boker, Xie, & Maes, 2003), began by attempting to discriminate between a nondevelopmental (Fig. 16.3A) and a developmental common factor model (Fig. 16.3B). The nondevelopmental model postulates that the within and cross-time correlations between CD and PGD are best understood as arising from two correlated latent variables. The developmental model (Fig. 16.3B) also assumes separate latent liabilities to CD and PGD, but also postulates within variable cross-time and cross-variable within-time transmission. *It is in the direction of this latter path—from CD to PGD or vice versa—that our model captures the predictions of the social selection versus social influence hypotheses.* We illustrate this in Figure 16.3 for genetic paths (but these models apply equally to shared and individual-specific environmental paths). In addition, we also examined the fit of a noncommon factor developmental model (depicted in Figure 16.3C), which assumes that the factors that influence CD and PGD at each age period are independent of one another. We expected that this model would likely fit best for individual-specific environmental effects that are typically occasion-specific in their impact, as well as including errors of measurement.

In longitudinal studies involving repeated measurements with the same instruments, temporal changes in variance are informative about the underlying developmental process (Eaves, Long, & Heath, 1986). Path coefficients of the model are standardized, so that the phenotypic variance is unity on the first occasion. Variances at subsequent ages are expressed relative to their initial values. Therefore, path coefficients can exceed unity, particularly when variances are increasing over time.

The models presented in Figure 16.3 contain too many nuances to be presented in detail, so I provide only summary results of the best-fitting version of each general model. We utilized the Bayesian information criterion (BIC) (Schwarz, 1978), which performs well with complex models (Markon & Krueger, 2004). By minimizing the BIC, we seek to optimize the balance of explanatory power and parsimony.

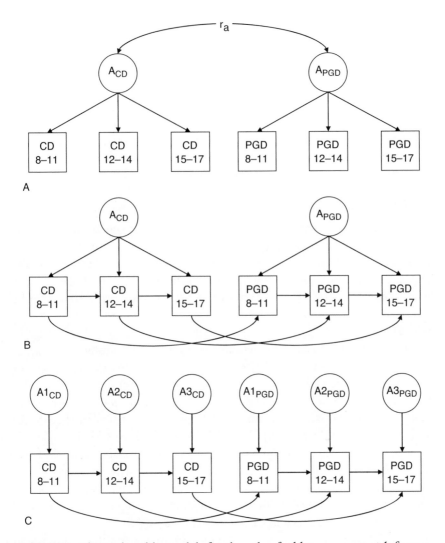

Figure 16.3. Three plausible models for the role of additive genetic risk factors (*A*) on levels of symptoms of conduct disorder (CD) and peer group deviance (PGD) for the ages of 8–11, 12–14, and 15–17. Identical models could be applied to shared or individual specific environmental effects. Model A or the *correlated factor model* is nondevelopmental in that it postulates that the within and cross-time correlations between CD and PGD result solely from two correlated latent variables with no causal paths between variables within or across time. By contrast, model B or the *casual factor model* also assumes separate latent liabilities to CD and PGD, but also postulates within-variable cross-time and cross-variable within-time transmission. The direction of this latter path—from CD to PGD or vice versa—captures the different predictions of the social selection versus social influence hypotheses. Model C

Figure 16.3. (*continued*)

or the *simple causal model* is a noncommon factor developmental model that assumes that the factors that influence CD and PGD at each age period are independent of one another. Like the causal factor model, this model also contains paths for within-variable cross-time and cross-variable within-time transmission. The subscripts $_{CD}$ and PD refer to additive genetic effects specific for conduct disorder or peer deviance, respectively. Figure adapted from Kendler, K. S., Jacobson, K., Myers, J. M., & Eaves, L. J. (2007a). A genetically informative developmental study of the relationship between conduct disorder and peer deviance in males. *Psychological Medicine,15*, 1–15, with permission.

We use as our baseline model a triple Cholesky decomposition (for *A*, *C*, and *E*), which is a saturated model of the observed genetic and environmental variances and covariances, and we use this as a basis to assess the goodness of fit of our subsequent models. We first simplified genetic factors (*A*), then individual-specific environmental factors (*E*), and finally shared environmental factors (*C*). We fitted shared environment last because its effects were known least precisely and were more modest than those seen with *A* and *E*.

I am not going to describe here the details of model-fitting (see Kendler et al., 2007a), except to say that the key features of the model—the direction of the genetic and environmental causal paths—was determined with a reasonable degree of statistical confidence. The best-fit model is depicted in Figure 16.4. It has six major features. First, forward transmission of both CD and PGD occurred from ages 8–11 to 12–14 and from ages 12–14 to 15–17. That is, levels of CD at one time period had a direct impact on levels of CD at the next time period, and prior levels of deviant behavior in peers directly influence the subsequent degree of peer deviancy. Second, genetic risk factors for CD and PGD could be best understood as a single set of common factors with similar impact at each age period. Third, the impact of individual-specific environment (and errors of measurement) could, by contrast, be best modeled as occasion-specific in nature. Fourth, more surprisingly, the influence of the shared environment was also best understood as occasion-specific in its effect. Fifth, the within-time cross-variable causal paths for genetic factors are constant over time and go from CD to PGD. The same pattern is seen for the individual-specific environment. Sixth, by contrast, the within-time cross-variable causal paths for shared environmental factors go from PGD to CD and decline sharply in magnitude over time.

Our analyses succeeded in clarifying the developmental relationship between CD (a measure of individual-level deviant behavior) and PGD (a measure of the deviance in the peer network). As we noted, prior studies had reported evidence for causal pathways both from individual to peer

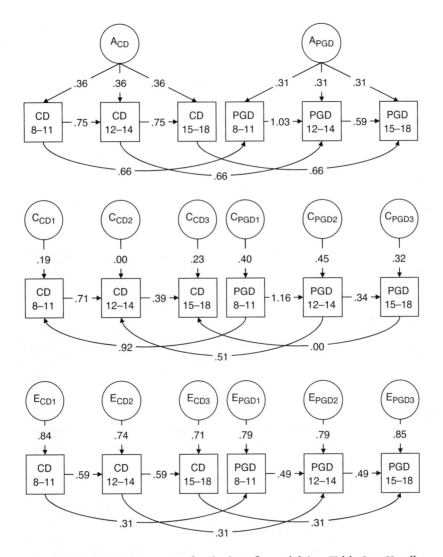

Figure 16.4. Parameter estimates for the best fit model (see Table 2 in Kendler et al. [2007a]) elucidating the impact of additive genetic effects (A), shared or common environmental factors (C), and individual-specific environmental effects (E) on levels of symptoms of conduct disorder (CD) and peer group deviance (PGD) for the ages of 8–11, 12–14, and 15–17. The subscripts ₍CD₎ and ₍PGD₎ refer to genetic or environmental effects specific for conduct disorder or peer group deviance, respectively, whereas the subscripts 1, 2, and 3 refer to factors specific to the three age periods: 8–11, 12–14, and 15–17. Figure adapted from Kendler, K. S., Jacobson, K., Myers, J. M., & Eaves, L. J. (2007a). A genetically informative developmental study of the relationship between conduct disorder and peer deviance in males. *Psychological Medicine,15*, 1–15, with permission.

deviancy via social selection (Gordon et al., 2004; Kandel, 1996; Lacourse et al., 2006) and from peer to individual deviancy via *social influence* (Gordon et al., 2004; Kandel, 1996; Wills & Cleary, 1999). We were able to show that a genetically informative longitudinal design could clarify the causal processes involved.

Of the numerous features of our findings, three are of particular developmental salience. First, our data were better explained by an active developmental model than by a static common factor model. Conduct disorder and PGD are dynamically interacting with themselves and each other over time. Second, the causal relationship between CD and PGD differed dramatically between the two sources of familial resemblance: genes and common environment. Genetic factors impacted on levels of CD, which in turn, through social selection, altered levels of PGD. Shared environmental factors acting on PGD through social influence impacted on levels of CD. When we examined our twins as an epidemiological sample, we observe a blending of these two mechanisms. Without utilizing a genetically informative design, it is unclear how these distinct causal processes could be disentangled. Third, the time course of the genetic and shared environment influences on the CD–PGD relationship differed substantially. Although the genetically driven influence of CD on PGD was constant over time, the impact of shared environment on CD mediated through PGD, very strong in later childhood (i.e., ages 8–11), declined dramatically over subsequent age periods.

These results provide further insight into the first set of findings that we presented above documenting the increasing role of genetic factors in the etiology of PGD over development. We find that CD is an important mediating variable between genes and PGD. That is, the causal pathway from genes to PGD partly flows through CD. Via processes alternatively termed *assortative friendship* (Rose, 2002), *social selection* (Patterson, Dishion, & Yoerger, 2000), or the "shopping model" (Dishion, Patterson, & Griesler, 1994),children and adolescents prone to deviant behavior actively seek out individuals who share and positively reinforce their own values, perspectives, and favored activities.

Equally interestingly, we showed that shared environmental influences on PGD in childhood and early adolescence can have strong causal impacts on CD. Other twin studies (Iervolino et al., 2002; Walden et al., 2004) and our own prior analyses in this sample (Kendler et al., 2007b), suggest important shared environmental risks for PGD. These influences surely exist at multiple levels, two of which—neighborhood and family—are of particular salience. Neighborhoods are important because antisocial boys typically find close friends within their own block (Dishion, Andrews, & Crosby, 1995). Neighborhoods differ widely in their "collective efficacy"—their level of social cohesion and tolerance of adolescent deviance (Sampson, Raudenbush, & Earls, 1997). Other neighborhood-level factors that are likely of import would include poverty levels and quality of schooling (Hawkins et al., 1998; Petraitis et al., 1998). At the family level, factors likely to impact on PGD directly (and thereby on CD indirectly) would include parental monitoring, family religious involvement, and family support for prosocial teen activities (Ary, Duncan, Duncan, & Hops, 1999; Hawkins et al., 1998; Kandel, 1996; Petraitis et al., 1998; Steinberg, Fletcher, & Darling, 1994; Walden et al., 2004).

Our findings of greater causal effect of PGD on CD at younger ages is consistent with models that predict that early association with deviant peers may be particularly potent at influencing the trajectory of future externalizing behaviors (Steinberg et al., 1994; Wills & Dishion, 2004). However, most prior observations of this association could have been a product of the child's own genetically influenced temperament causing both high level of PGD in childhood and subsequent antisocial behaviors. In our genetically informative design, we can be somewhat more confidant at inferring causal processes—that is, a true environmental effect of PGD on CD. Our results have direct relevance for intervention efforts in suggesting that alterations in levels of peer deviance, especially before age 15, can be expected to have a significant impact on levels of CD.

LIMITATIONS

The results of these three studies should be interpreted in the context of four potential methodologic limitations. First, the sample was restricted to white males born in Virginia. These results may or may not extrapolate to women or other ethnic groups. Second, the greater resemblance for PGD in MZ versus DZ twins could arise because, due to social expectations, MZ twins share more of their social network than do DZ twins. We have examined this question elsewhere (Kendler et al., 2007b) and shown that the broad pattern of results does not change when taking into account the tendency for MZ twins to co-socialize more frequently than DZ twins.

Third, given that not all eligible twins participated in this study, could our findings be unrepresentative? Using data from prior interviews, participation in this study was significantly predicted by educational status, but not by age, cannabis use, cannabis abuse/dependence, or number of DSM-IV adult antisocial symptoms. With respect to the externalizing symptoms strongly predicted by CD and PGD, this sample is likely to be broadly representative of the original twin cohort.

Last, since information on CD, PGD, and alcohol consumption were all collected retrospectively from adults, our findings could be strongly influenced by recall bias (Kandel, 1996). This is unlikely for three reasons. First, our measures of CD, PGD, and alcohol consumption all had good to excellent test–retest reliability. Second, it is difficult to construct a plausible pattern of recall bias that would produce evidence for our specific set of findings (e.g., PGD moderation of genetic effects only at earlier ages, CD → PGD genetic paths and causal PGD → CD shared environmental effects). Third, we obtained our data utilizing a life-history calendar that, by reflecting the structure of autobiographical memory and promoting sequential retrieval within memory networks, substantially improves the completeness and accuracy of retrospective recall (Belli, 1998; Freedman et al., 1988; Yoshihama, Clum, Crampton, & Gillespie., 2002).

CONCLUSION

In this chapter, we sought to illustrate the three main themes of this volume—gene–environment correlation, gene–environment interaction, and development—utilizing PGD as a central construct. We began by illustrating classical, active gene–environment correlation. An archetypal environmental construct—PGD—is substantially influenced by genetic factors (see Figure 16.1). Our genes reach out "beyond our skin" to influence key aspects of our social environment. Even this simple illustration revealed dynamic developmental processes. The role of genetic influences on PGD is not static. Rather, as an individual ages, genetic factors become progressively more important in influencing this important aspect of social environment. As we grow from childhood to young adulthood, we increasingly create our own social world, and our genes contribute substantially to that process.

Next, we shifted to consider PGD as a predictor variable, one that relates rather strongly to alcohol consumption. We showed that levels of PGD indeed do moderate genetic effects on alcohol use (see Figure 16.2). In fact, PGD interacted both with an "alcohol-specific" genetic risk factor and a more nonspecific genetic risk factor that reflected liability to externalizing disorders. However, again developmental complexities emerged. Although the impact of PGD on alcohol consumption grows with time well into early adulthood, gene × environment interaction was only observed in early to mid-adolescence. (Of note, this makes it especially unlikely that our evidence for gene–environment interaction results from a "bleeding over" of the effects of gene–environment correlation. If that were true, we should see the strongest interaction effects later in development, when the heritability of PGD maximizes). Developmental stage appears to be able to moderate gene–environment interactions. Intuitively, we would wonder whether our finding—of greater malleability of genetic influences by environments early in the developmental pathway to alcohol use—might be a more widespread phenomenon. Hopefully, this will be clarified by future research.

Our final analyses were the most explicitly developmental. Here, we wanted to see if a genetic design could "parse" the pathways from individual deviance (as assessed by symptoms of CD) to peer deviance. We succeeded, but at the price of adding yet another needed layer of complexity in our thinking and modeling. Here, we showed that the key relationship between CD and PGD could have opposite causal inter-relationships active during the same developmental time period (see Figure 16.4). Genes drive the level of CD, which influences children and adolescents to select like-minded peers. Communities and families influence tolerance for the level of PGD, which then feeds back on the individual level of CD. With yet another level of subtlety, the first effect is developmentally stable while the second quickly attenuates with age.

Beginning with the deceptively simple construct of PGD, we have been able to illustrate the importance of these three vibrant processes: gene–environment correlation, gene–environment interaction, and development. The resulting picture is dauntingly complex. Our findings, along with others presented in this volume, should—if nothing else—put definitively to rest the old idea of the static, deterministic model of gene action. If our studies here are at all representative, in its effects on the behavioral and psychiatric traits that mental health practitioners and researchers care about, our genome is highly dynamic. That is, I suspect, how the world is, and we had better start getting used to it.

ACKNOWLEDGMENTS

A number of colleagues made critical contributions to the studies reviewed in this chapter, key among whom are Charles Gardner Ph.D., Lindon Eaves D.Phil., John Myers M.S., Kristen Jacobson Ph.D., and Carol Prescott Ph.D. This work was supported in part by NIH grants DA-011287 and MH-49492. We thank Dr. Linda Corey for assistance with the ascertainment of twins from the Virginia Twin Registry, now part of the Mid-Atlantic Twin Registry (MATR), directed by Dr. Judy Silberg. The MATR has received support from the National Institutes of Health, the Carman Trust, and the WM Keck, John Templeton, and Robert Wood Johnson Foundations. Lenn Murrelle Ph.D. and Patrick Sullivan M.D. contributed to the design of this study.

REFERENCES

Akaike, H. (1987). Factor analysis and AIC. *Psychometrika, 52,* 317–332.

Allen, M., Donohue, W. A., Griffin, A., Ryan, D., & Turner, M. M. (2003). Comparing the influence of parents and peers on the choice to use drugs. *Criminal Justice and Behavior, 30,* 163–186.

Ary, D. V., Duncan, T. E., Duncan, S. C., & Hops, H. (1999). Adolescent problem behavior: the influence of parents and peers. *Behavior Research and Therapy, 37,* 217–230.

Baker, L. A., & Daniels, D. (1990). Nonshared environmental influences and personality differences in adult twins. *Journal of Personality and Social Psychology, 58,* 103–110.

Belli, R. F. (1998). The structure of autobiographical memory and the event history calendar: potential improvements in the quality of retrospective reports in surveys. *Memory, 6,* 383–406.

Blom, G. (1958). *Statistical estimates and transformed beta variables.* New York: John Wiley & Sons.

Bridges, L. J. (2003). Autonomy as an element of developmental well-being. In M. Bornstein, L. Davidson, C. L. Keyes, & K. A. Moore (Eds.) *Well-being: positive development across the life course* (pp. 167–175). Mahwah, NJ: Lawrence Erlbaum Associates, Publishers.

Bucholz, K. K., Cadoret, R., Cloninger, C. R., Dinwiddie, S. H., Hesselbrock, V. M., Nurnberger, J. I., Jr., et al. (1994). A new, semi-structured psychiatric interview for use in genetic linkage studies: a report on the reliability of the SSAGA. *Journal of Studies on Alcohol, 55,* 149–158.

Button, T. M., Corley, R. P., Rhee, S. H., Hewitt, J. K., Young, S. E., & Stallings, M. C. (2007). Delinquent peer affiliation and conduct problems: a twin study. *Journal of Abnormal Psychology, 116,* 554–564.

Cleveland, H. H., Wiebe, R. P., & Rowe, D. C. (2005). Sources of exposure to smoking and drinking friends among adolescents: a behavioral-genetic evaluation. *Journal of Genetic Psychology, 166,* 153–169.

Cloninger, C. R., Bohman, M., & Sigvardsson, S. (1981). Inheritance of alcohol abuse. Cross-fostering analysis of adopted men. *Archives of General Psychiatry, 38,* 861–868.

Coie, J. D., & Miller-Johnson, S. (2001). Peer Factors and Interventions. In R. Loeber, & D. P. Farrington (Eds.) *Child delinquents: development, intervention, and service needs* (pp. 191–209). London: Sage Publications, Inc.

Cox, D. R., & Snell, E. J. (1989). *Analysis of binary data.* London: Chapman & Hall.

Daniels, D., Dunn, J., Furstenberg, F. F., Jr., & Plomin, R. (1985). Environmental differences within the family and adjustment differences within pairs of adolescent siblings. *Child Development, 56,* 764–774.

Dick, D. M., Latendresse, S. J., Lansford, J. E., Budde, J. P., Goate, A., Dodge, K. A., et al. (2009). Role of GABRA2 in trajectories of externalizing behavior across development and evidence of moderation by parental monitoring. *Archives of General Psychiatry, 66,* 649–657.

Dick, D. M., Pagan, J. L., Viken, R., Purcell, S., Kaprio, J., Pulkkinen, L., et al. (2007a). Changing environmental influences on substance use across development. *Twin Research and Human Genetics, 10,* 315–326.

Dick, D. M., Rose, R. J., Viken, R. J., Kaprio, J., & Koskenvuo, M. (2001). Exploring gene-environment interactions: socioregional moderation of alcohol use. *Journal of Abnormal Psychology, 110,* 625–632.

Dick, D. M., Viken, R., Purcell, S., Kaprio, J., Pulkkinen, L., & Rose, R. J. (2007b). Parental monitoring moderates the importance of genetic and environmental influences on adolescent smoking. *Journal of Abnormal Psychology, 116,* 213–218.

Dishion, T. J., Andrews, D. W., & Crosby, L. (1995). Antisocial boys and their friends in early adolescence: relationship characteristics, quality, and interactional process. *Child Development, 66,* 139–151.

Dishion, T. J., Patterson, G. R., & Griesler, P. C. (1994). Peer adaptations in the development of antisocial behavior: A confluence model. In L. R. Huesmann (Ed.), *Aggressive behavior: current perspectives* (pp. 61–95). New York: Springer.

Dishion, T. J., Patterson, G. R., Stoolmiller, M., & Skinner, M. L. (1991). Family, school, and behavioral antecedents to early adolescent involvement with antisocial peers. *Developmental Psychology, 27,* 172–180.

Eaves, L. J., Long, J., & Heath, A. C. (1986). A theory of developmental change in quantitative phenotypes applied to cognitive development. *Behavior Genetics, 16,* 143–162.

Eiser, J. R., Morgan, M., Gammage, P., Brooks, N., & Kirby, R. (1991). Adolescent health behaviour and similarity-attraction: friends share smoking habits (really), but much else besides. *British Journal of Social and Clinical Psychology, 30* (Pt 4), 339–348.

Farrington, D. (2005). Childhood origins of antisocial behavior. *Clinical Psychology and Psychotherapy, 12,* 177–190.

Fergusson, D. M., & Horwood, L. J. (1999). Prospective childhood predictors of deviant peer affiliations in adolescence. *Journal of Child Psychology and Psychiatry and Allied Disciplines, 40,* 581–592.

Fergusson, D. M., Woodward, L. J., & Horwood, L. J. (1999). Childhood peer relationship problems and young people's involvement with deviant peers in adolescence. *Journal of Abnormal Child Psychology, 27,* 357–369.

Freedman, D., Thornton, A., Camburn, D., Alwin, D., & Young-DeMarco, L. (1988). The life history calendar: a technique for collecting retrospective data. *Sociological Methodology, 18,* 37–68.

Gelhorn, H. L., Stallings, M. C., Young, S. E., Corley, R. P., Rhee, S. H., & Hewitt, J. K. (2005). Genetic and environmental influences on conduct disorder: symptom, domain and full-scale analyses. *Journal of Child Psychology and Psychiatry, 46,* 580–591.

Gordon, R. A., Lahey, B. B., Kawai, E., Loeber, R., Stouthamer-Loeber, M., & Farrington, D. P. (2004). Antisocial behavior and youth gang membership: selection and socialization. *Criminology, 42,* 55–87.

Harris, J. R. (2002). *The nurture assumption: why children turn out the way they do.* New York: Touchstone/Simon & Schuster.

Hawkins, J. D., Herrenkohl, T., Farrington, D. P., Brewer, D., Catalano, R. F., & Harachi, T. W. (1998). A review of predictors of youth violence. In R. Loeber, & D. P. Farrington (Eds.), *Serious & violent juvenile offenders: risk factors and successful interventions* (pp. 106–146). London: Sage Publications, Inc.

Iervolino, A. C., Pike, A., Manke, B., Reiss, D., Hetherington, E. M., & Plomin, R. (2002). Genetic and environmental influences in adolescent peer socialization: evidence from two genetically sensitive designs. *Child Development, 73,* 162–174.

Jacobson, K. C., Prescott, C. A., & Kendler, K. S. (2000). Genetic and environmental influences on juvenile antisocial behaviour assessed on two occasions. *Psychological Medicine, 30,* 1315–1325.

Jacobson, K. C., Prescott, C. A., & Kendler, K. S. (2002). Sex differences in the genetic and environmental influences on the development of antisocial behavior. *Developmental Psychopathology, 14,* 395–416.

Johnston, L. D., Bachman, J. G., & O'Malley, P. M. (1982). *Monitoring the future: questionnaire responses from the nation's high school seniors, 1981.* Ann Arbor, MI: Institute for Social Research.

Kandel, D. (1979). Homophily, selection and socialization in adolescent friendships. *American Journal of Sociology, 84,* 427–436.

Kandel, D. B. (1978). Homophily, selection, and socialization in adolescent friendships. *American Journal of Sociology, 84,* 427–436.

Kandel, D. B. (1996). The parental and peer contexts of adolescent deviance: an algebra of interpersonal influences. *Journal of Drug Issues, 26,* 289–315.

Kendler, K. S. (2001). Twin studies of psychiatric illness: an update. *Archives of General Psychiatry, 58,* 1005–1014.

Kendler, K. S. (2005). Toward a philosophical structure for psychiatry. *American Journal of Psychiatry, 163,* 433–440.

Kendler, K. S., & Baker, J. H. (2007). Genetic influences on measures of the environment: a systematic review. *Psychological Medicine, 37,* 615–626.

Kendler, K. S., Jacobson, K., Myers, J. M., & Eaves, L. J. (2007a). A genetically informative developmental study of the relationship between conduct disorder and peer deviance in males. *Psychological Medicine, 15,* 1–15.

Kendler, K. S., Jacobson, K. C., Gardner, C. O., Gillespie, N. A., Aggen, S. H., & Prescott, C. A. (2007b). Creating a social world: a developmental study of peer deviance. *Archives of General Psychiatry, 64,* 958–965.

Kendler, K. S., & Prescott, C. A. (2006). *Genes, environment, and psychopathology: understanding the causes of psychiatric and substance use disorders.* New York: Guilford Press.

Kendler, K. S., Schmitt, J. E., Aggen, S. H., & Prescott, C. A. (2008). Genetic and environmental influences on alcohol, caffeine, cannabis, and nicotine use from adolescence to middle adulthood. *Archives of General Psychiatry, 65,* 674–682.

Koopmans, J. R. (1997). *The genetics of health-related behaviors* (Thesis/Dissertation). Vrije Universiteit.

Lacourse, E., Nagin, D. S., Vitaro, F., Cote, S., Arseneault, L., & Tremblay, R. E. (2006). Prediction of early-onset deviant peer group affiliation: a 12-year longitudinal study. *Archives of General Psychiatry, 63,* 562–568.

Larson, R., & Richards, M. H. (1991). Daily companionship in late childhood and early adolescence - changing developmental contexts. *Child Development, 62,* 284–300.

Latendresse, S. J., Bates, J., Goodnight, J. A., Dodge, K. A., Lansford, J. E., & Pettit, G. S. (2009). Differential susceptibility to discrete patterns of adolescent externalizing behavior: examining the interplay between CHRM2 and peer group deviance. Unpublished work.

Lyons, M. J., True, W. R., Eisen, S. A., Goldberg, J., Meyer, J. M., Faraone, S. V., et al. (1995). Differential heritability of adult and juvenile antisocial traits. *Archives of General Psychiatry, 52,* 906–915.

Manke, B., McGuire, S., Reiss, D., Hetherington, E. M., & Plomin, R. (1995). Genetic contributions to adolescents extrafamilial social interactions – teachers, best friends, and peers. *Social Development, 4,* 238–256.

Markon, K. E., & Krueger, R. F. (2004). An empirical comparison of information-theoretic selection criteria for multivariate behavior genetic models. *Behavior Genetics, 34,* 593–610.

Martens, M. P., Neighbors, C., Dams-O'Connor, K., Lee, C. M., & Larimer, M. E. (2007). The factor structure of a dichotomously scored Rutgers Alcohol Problem Index. *Journal of Studies on Alcohol and Drugs, 68,* 597–606.

McArdle, J. J. (2006). Latent curve analyses of longitudinal twin data using a mixed-effects biometric approach. *Twin Research and Human Genetics, 9,* 343–359.

Neale, M. C., Boker, S. M., Xie, G., & Maes, H. H. (2003). *Mx: Statistical Modeling, 6th Edition.* Dept. of Psychiatry, Virginia Commonwealth University Medical School: Box 980126, Richmond VA 23298.

Patterson, G. R., DeBaryshe, B. D., & Ramsey, E. (1989). A developmental perspective on antisocial behavior. *American Psychologist, 44,* 329–335.

Patterson, G. R., Dishion, T. J., & Yoerger, K. (2000). Adolescent growth in new forms of problem behavior: macro- and micro-peer dynamics. *Previews of Science, 1,* 3–13.

Petraitis, J., Flay, B. R., Miller, T. Q., Torpy, E. J., & Greiner, B. (1998). Illicit substance use among adolescents: a matrix of prospective predictors. *Substance Use & Misuse, 33,* 2561–2604.

Pike, A., Manke, B., Reiss, D., & Plomin, R. (2000). A genetic analysis of differential experiences of adolescent siblings across three years. *Social Development, 9,* 96–114.

Plomin, R., DeFries, J. C., & Loehlin, J. C. (1977). Genotype-environment interaction and correlation in the analysis of human behavior. *Psychological Bulletin, 84,* 309–322.

Rose, R. J. (2002). How do adolescents select their friends? A behavior-genetic perspective. In L. Pulkkinen and A. Caspi (Eds.), *Paths to successful development:*

personality in the life course (pp. 106–128). Cambridge, UK: Cambridge University Press

Rose, R. J., Dick, D. M., Viken, R. J., & Kaprio, J. (2001). Gene-environment interaction in patterns of adolescent drinking: regional residency moderates longitudinal influences on alcohol use. *Alcoholism: Clinical and Experimental Research, 25(5),* 637–643.

Sampson, R. J., Raudenbush, S. W., & Earls, F. (1997). Neighborhoods and violent crime: a multilevel study of collective efficacy. *Science, 277,* 918–924.

SAS Institute. (2008). SAS ONLINE DOC Version 9.2. SAS Institute, Inc.: Cary, NC. Available at: http://support.sas.com/documentation/cdl_main/index.html.

Satterthwaite, F. E. (1946). An approximate distribution of estimates of variance components. *Biometrics Bulletin, 2,* 110–114.

Scarr, S., & McCartney, K. (1983). How people make their own environments: a theory of genotype greater than environment effects. *Child Development, 54,* 424–435.

Schwarz, G. (1978). Estimating the dimension of a model. *Annual Statistics, 6,* 461–464.

Shanahan, M. J., & Hofer, S. M. (2005). Social context in gene-environment interactions: retrospect and prospect. *The Journals of Gerontology-Series B Psychological Sciences and Social Sciences, 60* (Spec No 1), 65–76.

Sigvardsson, S., Bohman, M., & Cloninger, C. R. (1996). Replication of the Stockholm Adoption Study of alcoholism. Confirmatory cross-fostering analysis. *Archives of General Psychiatry, 53,* 681–687.

Simons, R. L., Whitbeck, L. B., Conger, R. D., & Conger, K. J. (1991). Parenting factors, social skills, and value commitments as precursors to school failure, involvement with deviant peers, and delinquent-behavior. *Journal of Youth and Adolescence, 20,* 645–664.

Steinberg, L. (1990). Autonomy, conflict, and harmony in the family relationship. In S. S. Feldman, & G. R. Elliott (Eds.), *At the threshold: the developing adolescent* (pp. 255–276). Cambridge, MA: Harvard University Press.

Steinberg, L., Fletcher, A., & Darling, N. (1994). Parental monitoring and peer influences on adolescent substance use. *Pediatrics, 93,* 1060–1064.

Tarter, R. E., & Hegedus, A. (1991). The drug use screening inventory: it's application in the evaluation and treatment of alcohol and drug abuse. *Alcohol Health & Research World, 15,* 65–75.

van den Bree, M. B., & Pickworth, W. B. (2005). Risk factors predicting changes in marijuana involvement in teenagers. *Archives of General Psychiatry, 62,* 311–319.

Viken, R. J., Kaprio, J., & Rose, R. J. (2007). Personality at ages 16 and 17 and drinking problems at ages 18 and 25: genetic analyses of data from Finn Twin16-25. *Twin Research and Human Genetics, 10,* 25–32.

Walden, B., McGue, M., Iacono, W. G., Burt, S., & Elkins, I. (2004). Identifying shared environment contributions to early substance use: the importance of peers versus parents. *Journal of Abnormal Psychology, 113,* 440–450.

White, V. M., Hopper, J. L., Wearing, A. J., & Hill, D. J. (2003). The role of genes in tobacco smoking during adolescence and young adulthood: a multivariate behaviour genetic investigation. *Addiction, 98,* 1087–1100.

Wills, T. A., & Cleary, S. D. (1999). Peer and adolescent substance use among 6th-9th graders: latent growth analyses of influence versus selection mechanisms. *Health Psychology, 18,* 453–463.

Wills, T. A., & Dishion, T. J. (2004). Temperament and adolescent substance use: a transactional analysis of emerging self-control. *Journal of Clinical Child & Adolescent Psychiatry, 33,* 69–81.

Yoshihama, M., Clum, K., Crampton, A., & Gillespie, B. (2002). Measuring the lifetime experience of domestic violence: application of the life history calendar method. *Violence Victims, 17,* 297–317.

17

Stress Hormones, Genes, and Neuronal Signaling In Adolescence

The Perfect Storm for Vulnerability to Psychosis

ELAINE F. WALKER, HANAN D. TROTMAN, JOY BRASFIELD,
MICHELLE ESTERBERG, AND MOLLY LARSEN

Interactional models have dominated the etiological theories of psychotic disorders for many years, and the diathesis–stress model has been the chief exemplar. Early discussions of this model hypothesized an interaction between constitutional vulnerability and stress exposure that set the stage for the emergence of psychosis. However, no features of vulnerability or stress were specified, and mechanisms of the interaction between them were only vaguely described. Nonetheless, the diathesis–stress model has proven to be compatible with subsequent scientific advances in neuroscience and molecular genetics (Walker & Diforio, 1997; Walker, Mittal, & Tessner, 2008).

We now know that interactions between the environment and genetics are both complex and multifaceted, and that these interactions have significant implications for brain function. In Chapter 1, Kendler describes some of the forms these interactions can take. In addition, there are bidirectional influences of the environment and genetic factors. Kendler provides examples of how gene–environment correlations are produced by the influence of the organism's genotype on the environment. However, causal pathways exist in the opposite direction, namely the influence of the environment on the organism's genotype and on gene expression (Dolinoy, Das, Weidman, & Jirtle, 2007). The influence of the environment on mammalian genes and gene expression has been elucidated in numerous studies and is a major mechanism of interest in this chapter. In this connection, it should be noted that the environment is conceptualized broadly and includes all of the psychosocial and biological factors that impinge on the organism.

In this chapter, we examine some pivotal new findings and theoretical approaches to understanding the etiology of schizophrenia and other psychotic disorders. These highlight neurodevelopmental processes during key periods. First, the prenatal period appears to be critical for the origins of vulnerability. Numerous studies have revealed that prenatal complications are linked with risk for psychosis. Further, subtle premorbid signs of dysfunction are apparent as early as infancy in individuals who later manifest psychosis (Walker, Kestler, Bollini, & Hochman, 2004). Thus, at least for some patients, vulnerability appears to be congenital, although the clinical syndrome does not emerge until late adolescence/early adulthood. Second, research on adolescent/young adult development has become inextricably tied with the study of psychotic disorders, as our understanding of postpubertal neurodevelopment has expanded (Walker, 2002; Walker et al., 2008). Adolescence can no longer be viewed as a quiescent period with respect to brain development; instead, contemporary research demonstrates significant changes that occur at this time. These findings have provided a new framework for conceptualizing a well-established epidemiological fact: In the overwhelming majority of cases, psychotic disorders gradually unfold during adolescence and young adulthood. Rarely does clinical psychosis, especially schizophrenia, have its onset before adolescence, and it is quite unusual to observe a first episode of psychosis beyond the age of 30. Finally, we should note that our focus on the broad spectrum of psychosis, rather than narrowly defined schizophrenia, is based on the compelling evidence that many of the same genetic and environmental factors that are associated with schizophrenia can also confer risk for other forms of psychosis (St. Clair, 2009).

We begin with a discussion of prenatal neurodevelopment and the manner in which environmental and genetic factors might alter its course and set the stage for vulnerability. We then turn to the adolescent period and consider the neurodevelopmental context in which environmental and genetic factors might lead to the emergence of psychosis. Of special interest are the effects of environmental factors on gene expression, as well as their role in gene–environment interactions.

PRENATAL DEVELOPMENT

The importance of the prenatal hormonal environment in determining physical and behavioral characteristics of humans is now well established.

For example, scientific data published in the 1970s and '80s highlighted the importance of the gonadal hormone milieu to which the fetus is exposed; levels of androgens and estrogens determine both external anatomic characteristics and morphological features of the central nervous system (CNS; Einstein, 2007). These have been referred to as the "organizational" effects of hormones, because they alter prenatal development in a way that permanently impacts subsequent postnatal development. Subsequently, another realm of prenatal organizational processes has been discovered; we now know that exposure to adrenal hormones, in particular those involved in the biological stress response, can alter prenatal brain development and, as a result, have a lasting impact on the organism's behavioral characteristics (Kaiser & Sachser, 2009; Nozdrachev, 2008).

We also have greater insights into the mechanisms of these effects: numerous studies demonstrate that hormones and other factors can alter prenatal physical development via their influence on the expression of genes governing cellular changes (Dolinoy et al., 2007; Reichenberg, Mill, & MacCabe, 2009). For example, prenatal exposure to elevated levels of maternal corticosterone can alter the expression of glucocorticoid receptors (GR) in the brains of rodents, and their postnatal brain function and behavior are affected (Seckl & Meaney, 2004). Other work indicates that variations in maternal care may have similar effects on postnatal gene expression (McGowan et al., 2009). Of course, ethical considerations limit the extent to which information on these intricate processes can be obtained from human research. Most of the work illustrating these effects is from experimental studies conducted with animals. Nonetheless, there is little doubt that similar mechanisms are mediating the effects of environmental factors on human development. For example, in a recent study, we found that prenatal exposure to maternal stress was associated with altered dermal development in human offspring (King et al., 2009).

The effects just described have been referred to as *epigenetic* (i e., in addition to genetic), and they entail some complex processes that scientists are only beginning to understand (Akbarian & Huang, 2009; Mehler, 2008). Recent data show that stable levels of gene expression in cells can be epigenetically and reversibly modified, altering patterns and rates of gene expression and the phenotype of the organism. One mechanism in this process is DNA methylation. DNA that comprises the gene sequence can be modified via a biochemical process that involves the addition of a methyl group to the promoter region of the gene. The promoter region is a region of the gene that responds to factors that influence the likelihood of transcription; in other words, whether or not the DNA is expressed in messenger RNA. For the gene to influence cellular structures/functions, the enzymes that transcribe the DNA must bind to the promoter region so that the gene can be expressed.

When a methyl group binds to the promoter region, the expression of the gene is blocked, and this process is referred to as methylation. It has been shown that the reduction in GRs that results from prenatal stress, described above, are mediated by the effects of methylation on the expression of the GR gene (Seckl & Meaney, 2004).

A related mechanism in epigenesis involves histones and their modification (Chuang & Jones, 2007; Deutsch, Rosse, Mastropaolo, Long, & Gaskins, 2008). Histones are small basic proteins that also play a role in transcription. Modifications of histone proteins occur in regions of active gene expression, and these modifications can alter transcription and, thereby, gene expression. Histone acetylation promotes transcription, whereas histone deacetylation represses gene expression.

In a departure from some longstanding assumptions, research is also demonstrating how epigenetic changes can be inherited transgenerationally (Chuang & Jones, 2007; Dolinoy et al, 2007). In other words, epigenetic changes can be passed on to subsequent generations via germ cells. Further, a recent study of monozygotic (MZ) and dizygotic (DZ) twins indicates that epigenetic processes contribute to the greater phenotypic similarity of MZ over DZ pairs (Kaminsky et al., 2009). Kaminsky and colleagues examined the epigenome in samples of MZ and DZ twins and found that the intraclass correlations between the epigenetic profiles of the MZ twins was greater than for the DZ twins. In other words, the MZ pairs shared more similarities in gene expression patterns than did the DZ twins. The authors conclude that DZ twin pairs show more epigenetic differences because they arise from separate zygotes, whereas MZ twins arise from the same zygote and will therefore share epigenetic events that occur prior to the point of blastocyte splitting. This indicates that some of the phenotypic similarity previously attributed to inherited genetic factors in MZ twins is instead a consequence of environmentally induced changes in gene expression. These findings also help to explain the elevated heritability estimates derived from twin, as opposed to adoption and family, studies of mental illness (Walker, Downey, & Caspi, 1991).

It is assumed that the epigenome is particularly susceptible to environmental influence during gestation, neonatal development, puberty, and old age (Dolinoy et al., 2007). These periods are characterized by significant biological changes, many of which entail pronounced hormonal activity. Vulnerability to bioenvironmental factors may be greatest during embryogenesis, however, because the DNA synthetic rate is high, and the DNA methylation patterning required for tissue development is being established.

In addition to prenatal changes in the pattern of gene expression, it is also relevant to consider the mounting evidence that mutations occurring during embryogenesis are linked with risk for psychosis and other mental illnesses (St. Clair, 2009). These mutations can be caused by a range of factors, including both advanced parental age and prenatal exposures (Fatemi & Folson, 2009). Many of the mutations that have been identified in patients with psychotic disorders occur in genes that are known to be involved in brain development.

A review of the literature on prenatal factors in the etiology of psychosis is beyond the scope of this chapter. However, two key themes are apparent and should be noted. First, a range of prenatal complications have been linked with risk for psychosis in offspring, including maternal nutritional deficiency, viral infections, rubella, Rh incompatibility, and factors that may contribute to fetal hypoxia (Clarke, Harley, & Cannon, 2006), and many of these have also been shown to be associated with changes in gene expression (e g., Waterland & Jirtle, 2003). Second, and perhaps most surprising, maternal exposure to psychosocial stress during pregnancy has been linked with increased risk for psychosis in offspring (Beydoun & Saftlas, 2008). This has been illustrated in studies of maternal exposure to war, natural disasters, and loss of a loved one. Based on findings from animal research, described above, this relation is generally assumed to be mediated by heightened maternal glucocorticoid secretion, which compromises fetal brain development. Thus, the mothers' stress-induced cortisol secretion may have the potential to set the stage for vulnerability.

ADOLESCENCE

Within the past decade, the scientific focus on adolescence, another period characterized by important hormonal changes, has intensified. Among the most fascinating findings has been the protracted developmental processes occurring in the adolescent/young-adult brain. This has led many researchers to view adolescence as a key developmental period when hormones exert important influences on brain structure and function (Walker, 2002; Walker et al., 2008). Thus, rather than being restricted to prenatal development, organizational effects of hormones may also be critical to adolescent development.

Although the changes in gonadal hormone secretion that accompany adolescence/young adulthood are well established, it is only recently that researchers have documented developmental changes in the secretion of hormones associated with the hypothalamic-pituitary-adrenal (HPA) axis (Walker, Walder, & Reynolds, 2001). To provide a background for understanding the significance of these changes, we begin with an overview of the HPA axis, a primary system mediating the biological stress response.

The complex cascade of hormonal events and regulatory feedback to the HPA axis are discussed in recent reviews, and the reader is referred to these for more details (Charmandari, Kino, Souvatzoglou, & Chrousos, 2003; Tsigos & Chrousos, 2002; Sapolsky, 2003). Although the HPA axis is not the only neural system activated by stress exposure, it has been a major focus of attention from psychopathologists because it has been linked with a range of mental disorders, including psychoses. The HPA axis governs a hormonal cascade that has important effects on brain function, and it facilitates both physiological and behavioral responses to threats.

The major components and hormonal messengers are illustrated in Figure 17.1. In response to stressors, corticotrophin-releasing hormone (CRH) is released from the periventricular nucleus of the hypothalamus. This triggers the secretion of adrenocorticotropic hormone (ACTH) from the pituitary which, in turn, leads to the secretion of glucocorticoids (GCs)

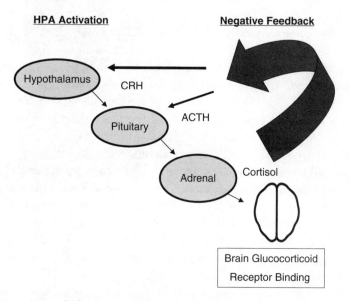

Figure 17.1. **Hypothalamus-pituitary-adrenal (HPA axis) activity** is mediated by the secretion of corticotrophin releasing hormone (CRH) from the hypothalamus, which in turn activates the secretion of adrenocorticotropin hormone (ACTH) from the pituitary, which finally stimulates the secretion of cortisol from the adrenal cortex. Cortisol interacts with its receptors in multiple target tissues including the HPA axis, where it is responsible for feedback inhibition of the secretion of ACTH from the pituitary and CRH from the hypothalamus. Image from Walker, E., Mittal, V., & Tessner, K. (2008). Stress and the hypothalamic pituitary adrenal axis in the developmental course of schizophrenia. Annual Review of Clinical Psychology, 4, 189–216, with permission.

from the adrenals. Glucocorticoids (cortisol in primates and corticosterone in rodents) have pervasive effects on organ systems, including the CNS, via the activation of GRs. Activation of these receptors triggers a feedback loop that modulates activity of the HPA axis and dampens the release of GCs. In primates, cortisol acts to synchronize other components of the stress response and alter the excitability of neuronal networks (de Kloet, 2003). These changes are typically adaptive, and have evolved to enhance physical preparedness. However, under conditions of persistent stress and concomitant cortisol elevation, the response can have adverse neurobehavioral consequences.

Findings from several lines of investigation indicate that persistently elevated GC levels can be neurotoxic, with the hippocampus being especially sensitive to these effects. Animal studies have shown that, at heightened levels, GCs can induce regression of dendritic processes, inhibit neurogenesis in the dentate gyrus, decrease neuronal survival following insults, and contribute to neuronal death (Sapolsky, 2003). Consistent with these findings, animal models (Coe et al., 2003; van der Beek et al., 2004) and neuroimaging studies of human participants, including some from our research group, have confirmed the inverse relation between GC secretion and hippocampal volume (Starkman, Giordani, Gebarski, & Schteingart, 2003; Tessner, Walker, Dhruv, Hochman, & Hamann, 2007). The inverse relation is presumed to reflect the role of the hippocampus in modulating GC secretion, as well as the effect of GCs on the hippocampus. Thus, a reduction in volume is linked with a reduction in the GCs that are needed to modulate HPA activity. Similarly, there are also significant adverse effects of elevated cortisol secretion on frontal cortex structure and function (Czeh, Perez-Cruz, Fuchs, & Flugge, 2008). These effects are presumed to be mediated by the actions of GCs on the expression of genes that govern neural plasticity. Given the effects of GCs on the brain, it is not surprising that persistently heightened GC secretion also has implications for cognitive functions (O'Brien, Lloyd, McKeith, Gholkar, & Ferrier, 2004).

Against this backdrop, we now turn to the issue of postpubertal HPA development. Recent investigations have shown that the HPA axis undergoes significant changes during adolescence, the same period when premorbid decline usually emerges in individuals subsequently diagnosed with schizophrenia. Specifically, a normative increase occurs in HPA activity that begins around puberty, with a postpubertal rise in baseline cortisol secretion that appears to be linked with pubertal stage (for a review, see Walker et al., 2008). We have collected data on cortisol levels in healthy and at-risk youth over the past 15 years in our longitudinal studies on precursors of psychosis. The data are consistent in showing a relation between age and cortisol levels for youth between the ages of 13 and 20; a clear linear increase in mean cortisol levels occurs with age, and the pattern is the same for male and female participants.

It has been suggested that this increase plays a role in the neurohormonal processes that trigger puberty and promote neuromaturation during adolescence. In particular, it appears that the HPA axis is involved in triggering adrenarche, the prelude to puberty (Rieder & Coupey, 1999; Weber, Clark, Perry, Honour, & Savage, 1997). Thus, factors that influence cortisol may alter critical chains of gonadal hormone events.

In summary, activity of the HPA axis undergoes a gradual developmental increase during the course of adolescence. Further, cortisol responses to stress increase during the course of adolescence, and cortisol levels are higher when adolescents are alone rather than with others (Adam, 2006). These, as well as other findings, have led some to conclude that adolescence is a period of increased sensitivity to stress, as well as to heightened motivation for social affiliation (Arnett, 1999; Walker, Sabuwalla, & Huot, 2004).

CORTISOL AND RISK FOR PSYCHOSIS

In addition to mounting scientific knowledge regarding normative adolescent development of the HPA axis, there is now growing evidence that heightened HPA activity may be a factor in the onset of psychosis. In the past 15 years, with a series of two cohorts of clinical and nonclinical adolescents, we have focused on the trajectories of adrenal hormone change and its relation with the progression of prodromal and psychotic symptoms during adolescence. The designation of symptoms and syndromes as "prodromal" is based on extensive evidence that individuals who are diagnosed with a psychotic disorder typically manifest a gradual onset of symptoms that entails a phase in which subclinical psychotic symptoms are observed before the first clinical episode (Klosterkötter, Hellmich, Steinmeyer, Schultze-Lutter, 2001; Miller, McGlashan, Rosen et al., 2002; Yung, Nelson, Stanford et al., 2008). These attenuated symptoms include suspiciousness, abnormal ideations, and perceptual experiences, some of the defining features of schizotypal personality disorder (SPD). Our findings indicate that, among adolescents with SPD, those who manifested a worsening of symptoms during annual follow-ups showed a more pronounced increase in cortisol release during the same period (Walker, Walder, & Reynolds, 2001). This suggests that, among those at clinical risk for developing psychotic disorders, a steeper increase in HPA activity is linked to the progression of psychopathology.

More recently, we have examined in greater detail the relation of cortisol secretion with specific prodromal symptoms in youth between the ages of 13 and 19 at baseline. The Structured Interview for Prodromal Syndromes

(SIPS) (Miller et al., 2002) was administered to participants who were clinically referred for their symptoms, and the accompanying severity scale, the Scale of Prodromal Symptoms (SOPS), was rated. The scale yields scores for four SOPS symptom dimensions: *positive* (unusual thought content, delusional ideas, suspiciousness, persecutory ideas, grandiosity, perceptual abnormalities, hallucinations, conceptual disorganization), *negative* (social isolation and withdrawal, avolition, decreased expression of emotion, decreased experience of emotion, decreased ideational richness, deterioration in role functioning), *disorganized* (odd behavior or appearance, bizarre thinking, trouble with focus and attention, impairment in personal hygiene or social attentiveness), and *general* (sleep disturbance, motor disturbance, dysphoric mood, impaired tolerance to normal stress). Individual symptoms are rated from 0 to 6, with levels 0 to 2 (non-prodromal range: none, questionable, mild), levels 3 to 5 (moderate, moderately severe, severe), and level 6 indicative of a potential psychotic state. The ratings are summed to derive a score for each of the dimensions. Individuals with positive symptom scores between 3 and 5 are considered indicative of prodromal levels.

Saliva samples for cortisol assay were obtained at least three times, on the hour, beginning at approximately 9:00 AM at each of three assessments; baseline, 7–10 months follow-up and 12–14 months follow-up. Time of day for sampling is based on evidence that, when compared to afternoon and evening values, morning values are more consistent and reliable, and their variance reflects a higher proportion (60%) of trait as opposed to state variance (Kirschbaum et al., 1990). This is assumed to reflect the cumulative effects of situational factors (e g., diet, exercise, and daily events) on variance in cortisol measured later in the day. Further, it should be noted that multiple saliva samples ($n = 3$) were obtained, so that an average could be derived, as this increases the reliability of the cortisol estimate (Li, Chiou, & Shen, 2007). (For a description of the assay procedure for salivary cortisol, see Mittal, Dhruv, Tessner, Walder, & Walker [2007]).

For those with data at all three assessments, mean cortisol values (average of the three saliva samples from each assessment) were used to calculate the area under the curve (AUC) (Fekedulegn, et al., 2007; Pruessner, Kirschbaum, Meinlschmid, & Hellhammer, 2003). The AUC is used to derive an aggregate index based on repeated measurements over time, and is often used in medication and endocrinological studies. (See Preussner et al. [2003] for more detailed information.) Following Preussner et al. (2003), *AUC with respect to the ground* (AUC_G) was calculated. AUC_G indexes the total area under the curve.

The correlations between cortisol AUC_G and symptom ratings at baseline and 1 year later are presented in Table 17.1 for the sample of 55 youth with prodromal symptoms and/or a diagnosis of SPD. As illustrated, all correlations are positive, and the correlations with baseline and follow-up ratings of positive symptoms are the largest and statistically significant. This is noteworthy, as positive symptoms are the defining symptoms of the prodrome and psychotic disorders.

To determine whether cortisol AUC was associated with changes in positive symptoms over time, regression analyses were conducted, controlling for baseline levels of positive symptoms. In this analysis, the positive symptom rating at 1-year follow-up was the dependent variable, with the baseline positive symptom score entered as a predictor in the first block, then the cortisol AUC_G in the second block. As expected, baseline positive symptoms significantly predicted follow-up symptoms ratings ($F(1,53) = 32.34, p < .01$). The results also showed a significant increment in R^2 ($F(1,52) = 3.45, p < .05$), when cortisol AUC_G was entered in the equation. Thus, consistent with prediction, the overall level of cortisol secretion as indexed through the course of the year was linked with a significant increase in positive symptoms at follow-up.

Neurochemical Mechanisms in the Relation of HPA Activity with Psychosis

Although the neurochemical processes that mediate the effects of cortisol secretion on symptom exacerbation are not known, one plausible current hypothesis concerns the effects of the HPA axis on dopamine activity (Walker et al., 2008). The relations among stress, HPA activity, and neurotransmitter activity, especially dopamine (DA), have been the subject of recent research. Although the mechanisms are not yet understood, it is clear that glucocorticoid secretion augments DA activity in certain brain regions (Dallman et al., 2004), especially the mesolimbic system (Marinelli, Rudick, Hu, & White, 2006). In a recent positron emission tomography (PET) investigation of healthy human subjects, exposure to a psychosocial stressor caused a significant release of dopamine in the ventral striatum, as indexed by a reduction in [¹¹C] raclopride binding, and the magnitude of the cortisol response was

Table 17.1. The Relation of Cortisol Secretion with Prodromal Symptom Severity Ratings in Youth at Risk for Psychosis

	Positive Symptoms		Negative Symptoms		Disorganized Symptoms		General Symptoms	
Cortisol AUC	Baseline	Follow-up	Baseline	Follow-up	Baseline	Follow-up	Baseline	Follow-up
	.24*	.34*	.13	.13	.27*	.22	.30*	.22

positively correlated with DA release in the ventral striatum ($r = 0.78$), consistent with a facilitating effect of cortisol on dopamine neuron firing (Pruessner, Champagne, Meaney, & Dagher, 2004). Conversely, animal research has shown that agents that suppress glucocorticoid secretion also reduce brain DA release in a manner similar to antipsychotics (Piazza et al., 1996). Finally, there is evidence that the administration of agents that enhance DA activity heighten HPA activity and augment cortisol secretion in humans (Philippi, Pohlenz, Grimm, Kollfer, & Schonberger, 2000). An experimental study of bupropion, a DA reuptake inhibitor, revealed that it increases cortisol release in healthy humans (Piacentini, Meeusen, Buyse, De Schutter, & De Meirleir, 2004). These and other findings document the synergistic relation between HPA activity and DA neurotransmission.

Additional research is needed to explore the possibility that HPA activation has unique effects on individuals at risk for psychosis. It has been hypothesized, for example, that vulnerability to psychosis entails a preexisting abnormality in striatal dopamine function that is triggered during adolescence when increased HPA activity augments DA release (Walker et al., 2008). But other potential mechanisms mediate the relation between HPA activity and psychosis. These entail the effects of hormones on the expression of genes that influence brain structure and function.

Hormonal Triggers of Gene Expression

At the molecular level, we have a more detailed picture of the way in which steroid hormone receptors alter neuronal function via *nongenomic* and *genomic* mechanisms (Daufeldt, Klein, Wildt, & Allera, 2006; Wintermantel, Berger, Greiner, & Schutz, 2005). Nongenomic processes are changes in neurotransmitter synthesis, release, and receptor sensitivity, and these have been implicated in the transient, activational effects of neurohormones on brain function (Rodriguez-Waitkus et al., 2003). In contrast, genomic processes are mediated by effects of hormones on the expression of genes and can therefore result in changes in brain structure and/or function that have the potential to be permanent. Genomic effects may be responsible for normal changes in brain structure and function that occur during adolescence/ early adulthood, as well as for the adverse influence of elevated GCs on cortical and hippocampal structure (Charmandari et al., 2003).

Although our understanding of the relations among hormonal changes, gene expression, and brain structure and function is limited, there is accumulating evidence that gonadal and adrenal hormones play a critical role in triggering the expression of genes that govern brain maturation (Hyman, 1996; Sisk & Zehr, 2005; Thompson et al., 2000). For example, recent studies of regional brain volumes in twins have revealed significant age-by-heritability interactions, such that white matter volume heritability estimates increased with age through adolescence, indicating that the genes governing white matter development are having an increasing influence (Wallace et al., 2006). Based on animal models, there is also evidence that the expression of some genes, as indexed by mRNA levels, increases through adolescence. Researchers have shown that mRNA levels for brain-derived neurotrophic factor (*BDNF*), which is present throughout the brain and promotes neuron survival, increase significantly during adolescence/young adulthood, then are subsequently stable. The significant increase in *BDNF* mRNA levels in the brain during the young-adult period coincides with the time when the frontal cortex matures both structurally and functionally (Webster, Weikert, Herman, & Kleinman, 2002). During the same stage, elevations in HPA activity may contribute to a heightened biological sensitivity to stress in adolescents (Walker, 2002). Animal research has demonstrated that stress exposure and stress hormones can alter the expression of mRNA for genes that control N-methyl-D-aspartate (NMDA) receptors in brain regions associated with the stress response, and this effect may be most pronounced during adolescence (Lee, Brady, & Koenig, 2003).

There are two primary modes of genomic action of GCs: transrepression and transactivation (Datson, Morsink, Meijer, & de Kloet, 2008). *Transactivation* involves the binding of GR dimers to response elements in promoter regions or introns of genes, resulting in activation or repression. *Transrepression* involves the interaction of GC monomers with other transcription factors activated via other signaling pathways. Further, some of the genes under transcriptional control of GCs, including some implicated in risk for psychosis, are involved in neuronal plasticity in the brain.

The effect of neurohormones on neuronal structure and function is assumed to mediate the relation between hormones and behavior. In addition, the potential for hormones to trigger the expression of genes suggests a mechanism through which the hormonal changes associated with puberty may serve to trigger the expression of genetic liabilities that are linked with mental disorder (Heyland, Hodin, & Reitzel, 2005; Zhang, Zhou, Li, Usrano, & Li, 2006). Consistent with this, twin studies have shown that the heritability estimate for mood disorders rises as individuals enter adolescence (Silberg et al., 1999). Kendler (Chapter 1) refers to these developmental changes as "genetic innovation"; namely, "evidence for new genetic variation that impacts on that phenotype at a later developmental period that was not evident at earlier time points." Thus, vulnerability genes for mood disorders may be expressed in response to pubertal changes.

Although we are aware of no comparable data on psychotic disorders, the modal developmental trajectory for psychosis suggests that a similar developmentally driven process may be involved. Consistent with this assumption, a postmortem study of age-related changes in the expression of schizophrenia susceptibility genes in the brain showed that the expression of three genes implicated in both brain development and risk for schizophrenia (NRG1, ERBB3, and NGFR) changed during adolescence (Colantuoni et al., 2008). Thus, changes in the expression of genes during adolescence/young adulthood may trigger brain processes that subserve the onset of psychosis.

GENE–ENVIRONMENT INTERACTIONS IN THE ETIOLOGY OF PSYCHOSIS

As noted above, the notion of a gene–environment interaction assumes that the effects of an environmental factor will vary as a function of the individual's genotype. The previous discussion focused on epigenetic effects in which environmental factors influence the expression of genes, or environmental factors alter the genotype through mutations. It is possible, however, that some epigenetic processes also entail gene–environment interactions. In other words, individuals may vary in their susceptibility to certain epigenetic processes because of their genotype. This issue is the current focus of research, and it may have relevance to the etiology of mental disorders.

To date, the research literature on gene–environment interactions in psychotic disorders is limited. Most studies suggestive of gene–stress interaction have used proxy measures for genetic vulnerability, such as a family history of psychosis. Nonetheless, a few reports indicate that some individuals are genetically susceptible to the psychogenic effects of stress and certain drugs.

Preliminary evidence suggests that polymorphisms within the catechol-O-methyltransferase (COMT), brain-derived neurotrophic factor (BDNF), and neuregulin-1 genes may interact with psychosocial stress in the development of psychosis. COMT codes an enzyme that inactivates dopamine, norepinephrine, and epinephrine. COMT augments the breakdown of dopamine in the prefrontal cortex. The COMT gene contains a functional polymorphism, involving a Met to Val substitution at codon 158, which results in two common allelic variants, the valine (Val) and the methionine (Met) allele, associated with high versus low enzyme activity, respectively. The Val allele is associated with increased COMT activity and consequent reduction of dopamine neurotransmission in the frontal cortex, but may increase levels of mesolimbic (phasic) dopamine activity.

A study of young men in military training revealed a significant interaction between COMT genotype and stress, with carriers of the Val allele showing the greatest increases in psychotic symptom severity in relation to stress exposure (Stefanis et al., 2007). A subsequent study showed that COMT Val carriers experienced more paranoia in response to event stress compared with Met carriers (Simons et al., 2009). In another report, however, researchers examined self-reported stress and symptoms in cannabis-using psychotic patients and healthy cannabis users and found that the Met/Met patients showed the largest increases in psychotic symptoms and negative affect with increased stress (van Winkel et al., 2008). Further research is needed to determine whether COMT genotype is indeed linked with stress sensitivity in individuals vulnerable to psychosis.

The gene that codes for BDNF has also been linked with stress sensitivity. BDNF is a neurotrophin that promotes the growth, differentiation, and survival of developing neurons. A common Val/Met single nucleotide polymorphism (SNP) at position 66 in BDNF was recently identified as a functional polymorphism. The Val variant is associated with higher neuronal BDNF secretory activity than is the Met variant. In the only report to date on BDNF and psychosis proneness, Met carriers showed more social-stress–induced paranoia than did individuals with the Val/Val genotype (Simons et al., 2009).

Keri et al. (2009) examined polymorphisms of the neuregulin-1 gene (SNP8NRG243177/rs6994992) in relation to psychosocial stress sensitivity because past research had shown that a polymorphism of the neuregulin-1 gene increases the risk of psychosis and affects prefrontal activation and structural connectivity in the brain. The purpose of this study was to investigate the interaction between this polymorphism and reactivity to stress. Schizophrenia patients and one of their family members participated in neutral and conflictual interactions while the relatives' critical comments and the patients' unusual thoughts were recorded. Patients with the neuregulin-1 T/T genotype, which has been associated with risk for psychosis, expressed more unusual thoughts than did C-carriers (C/T and C/C) during conflict-related interactions but not during neutral interactions.

Gene–environment interactions may also be at work in the effects of a "physical" stressor—obstetrical complications—on psychosis. In a study of obstetric complications in schizophrenia probands and family members, Nicodemus et al. (2008) examined 13 genes known to be involved in neurovascular function or to be regulated by hypoxia. Four of the genes (AKT1, with three SNPs; BDNF, with two SNPs; DTNBP1, with one SNP; and GRM3, with one SNP) showed significant genotype × obstetrical complication interaction, suggesting that genes involved in neurovascular function or regulated by hypoxia may interact with exposure to obstetric complications to increase risk for psychosis.

Perhaps the earliest gene–environment interaction reported in the literature on schizophrenia is that between COMT and the psychotogenic effects of cannabis. Caspi et al. (2005) used a longitudinal dataset to test the hypothesis that *COMT* moderated the influence of adolescent cannabis use on risk for adult psychosis. They found that carriers of the *COMT* Val allele were most likely to exhibit psychotic symptoms and to develop schizophreniform disorder if they used cannabis. Cannabis use had no such adverse influence on individuals with two copies of the Met allele. A subsequent double-blind, placebo-controlled study of tetrahydrocannabinol (THC), the active ingredient in cannabis, yielded similar results, in that patients with the *COMT* Val allele manifested more cognitive disruption in response to THC.

In summary, although there are only a handful of published reports on gene–environment interactions in relation to psychosis or psychotic symptoms, we can surely expect additional reports in the future. It will be especially important for researchers in this area to attempt replications in order to establish the reliability of published findings. Once an interaction effect is well established, the next task will be the identification of the neural mechanisms involved in the process. We may discover that many of these interactions share a final common neuropathologic pathway.

CONCLUSION

It is clear that our understanding of the genetic mechanisms involved in structural and functional brain development, both normal and abnormal, has expanded dramatically in recent years. These advances have revolutionized our views of the etiology of serious mental illnesses. As described in this chapter, the environment, broadly conceptualized, can (a) affect the expression of genes, (b) affect the genotype via the induction of mutations, and (c) interact with the genotype in determining behavioral outcomes.

It appears that vulnerability for psychosis often arises very early in development, during the prenatal period, but that the expression of this vulnerability in the form of a clinical disorder is triggered by processes that occur during adolescence/early adulthood. Although we are not yet able to identify these processes with specificity, we have some promising leads. They are likely to entail changes in the expression of genes that govern brain structure and function.

Based on these assumptions, research efforts are now being directed toward more intensive study of youth who are manifesting prodromal signs. An example of this is the North American Prodrome Longitudinal Study (NAPLs), which is a consortium of eight research sites (Emory University; Harvard University; Yale University; the University of California, Los Angeles; the University of California San Diego; the University of North Carolina; the University of Calgary; and Zucker Hillside Hospital–Albert

Einstein College of Medicine). This longitudinal study is tracking the development of youth with prodromal syndromes, and includes assessment of symptoms, environmental factors, cognition, hormones, genetics, and structural and functional brain development. Through this multisite collaboration, we hope to not only enhance our ability to identify those at greatest risk, but to also shed light on the neural mechanisms that give rise to psychosis. In particular, the NAPLs consortium will examine changes in gene expression and gene–environment interactions that may set the stage for the emergence of psychotic disorders in young adults.

REFERENCES

Adam, E. K. (2006). Transactions among adolescent trait and state emotion and diurnal and momentary cortisol activity in naturalistic settings. *Psychoneuroendocrinology*, *31*(5), 664–679.

Akbarian, S., & Huang, H. S. (2009). Epigenetic regulation in human brain-focus on histone lysine methylation. *Biological Psychiatry*, *65*(3), 198–203.

Artnett, J. (1999). Adolescent storm and stress reconsidered. *American Psychologist*, *54*, 317–326.

Beydoun, H., & Saftlas, A. F. (2008). Physical and mental health outcomes of prenatal maternal stress in human and animal studies: a review of recent evidence. *Paediatric and Perinatal Epidemiology*, *22*(5), 438–466.

Caspi, A., Moffitt, T. E., Cannon, M., McClay, J., Murray, R., Harrington, H., et al.(2005). Moderation of the effect of adolescent-onset cannabis use on adult psychosis by a functional polymorphism in the catechol-O-methyltransferase gene: longitudinal evidence of a gene-environment interaction. *Biological Psychiatry*, *57*(10), 1117–1127.

Charmandari, E., Kino, T., Souvatzoglou, E., & Chrousos, G. P. (2003). Pediatric stress: hormonal mediators and human development. *Hormone Research*, *59*(4), 161–179.

Chuang, J. C., & Jones, P. A. (2007). Epigenetics and microRNAs. *Pediatric Research*, *61*(5 Pt 2), 24R–29R.

Clarke, M. C., Harley, M., & Cannon, M. (2006). The role of obstetric events in schizophrenia. *Schizophrenia Bulletin*, *32*(1), 3–8.

Coe, C. L., Kramer, M., Czeh, B., Gould, E., Reeves, A. J., Kirschbaum, C., et al. (2003). Prenatal stress diminishes neurogenesis in the dentate gyrus of juvenile rhesus monkeys. *Biological Psychiatry*, *54*(10), 1025–1034.

Colantuoni, C., Hyde, T. M., Shruti, M., Joseph, A., Sartorius, L., Aguirre, C., et al. (2008). Age-related changes in the expression of schizophrenia susceptibility genes in the human prefrontal cortex. *Brain Structure & Function*, *213*(1-2), 255–271.

Czeh, B., Perez-Cruz, C., Fuchs, E., & Flugge, G. (2008). Chronic stress-induced cellular changes in the medial prefrontal cortex and their potential clinical implications: does hemisphere location matter? *Behavioural Brain Research*, *190*(1), 1–13.

Dallman, M. F., Akana, S. F., Strack, A. M., Scribner, K. S., Pecoraro, N., La Fleur, S. E., et al. (2004). Chronic stress-induced effects of corticosterone on brain: direct and indirect. *Annals of the New York Academy of Sciences*, *1018*, 141–150.

Datson, N. A., Morsink, M. C., Meijer, O. C., & de Kloet, E. R. (2008). Central corticosteroid actions: Search for gene targets. *European Journal of Pharmacology*, *583*(2–3), 272–289.

Daufeldt, S., Klein, R., Wildt, L., & Allera, A. (2006). Membrane initiated steroid signaling (MISS): Computational, in vitro and in vivo evidence for a plasma membrane protein initially involved in genomic steroid hormone effects. *Molecular and Cellular Endocrinology*, *246*(1–2), 42–52.

de Kloet, E. R. (2003). Hormones, brain and stress. *Endocrine Regulations*, *37*(2), 51–68.

Deutsch, S. I., Rosse, R. B., Mastropaolo, J., Long, K. D., & Gaskins, B. L. (2008). Epigenetic therapeutic strategies for the treatment of neuropsychiatric disorders: ready for prime time? *Clinical Neuropharmacology*, *31*(2), 104–119.

Dolinoy, D. C., Das, R., Weidman, J. R., & Jirtle, R. L. (2007). Metastable epialleles, imprinting, and the fetal origins of adult diseases. *Pediatric Research*, *61*(5 Pt 2), 30R–37R.

Einstein, G.[Ed]. (2007). *Sex and the brain*. Cambridge, MA: MIT Press.

Fatemi, S. H., & Folsom, T. D. (2009). The neurodevelopmental hypothesis of schizophrenia, revisited. *Schizophrenia Bulletin*, *35*(3), 528–548.

Fekedulegn, D. B., Andrew, M. E., Burchfiel, C. M., Violanti, J. M., Hartley, T. A., Charles, L.E., et al. (2007). Area under the curve and other summary indicators of repeated waking cortisol measurements. *Psychosomatic Medicine*, *69*(7), 651–659.

Heyland, A., Hodin, J., & Reitzel, A. M. (2005). Hormone signaling in evolution and development: A non-model system approach. *BioEssays*, *27*(1), 64–75.

Hyman, S. E. (1996). Relevance of gene regulation to psychiatry. *American Psychiatric Press Review of Psychiatry*, *15*, 311–330.

Kaiser, S., & Sachser, N. (2009). Effects of prenatal social stress on offspring development: Pathology or adaptation? *Current Directions in Psychological Science*, *18*(2), 118–121.

Kaminsky, Z. A., Tang, T., Wang, S. C., Ptak, C., Oh, G. H., Wong, A. H., et al. (2009). DNA methylation profiles in monozygotic and dizygotic twins. *Nature Genetics*, *41*(2), 240–245.

Keri, S., Kiss, I., Seres, I., & Kelemen, O. (2009). A polymorphism of the neuregulin 1 gene (SNP8NRG243177/rs6994992) affects reactivity to expressed emotion in schizophrenia. American Journal of Medical Genetics. *Part B, Neuropsychiatric Genetics: the Official Publication of the International Society of Psychiatric Genetics*, *150B*(3), 418–420.

King, S., Mancini-Marie, A., Brunet, A., Walker, E., Meaney, M. J. & Laplante, D. P. (2009). Prenatal maternal stress from a natural disaster predicts dermatoglyphic asymmetry in humans. *Development and Psychopathology*, *21*(2), 343–353.

Kirschbaum, C., Steyer, R., Eid, M., Patalla, U., Schwenkmezger, P., & Hellhammer, D. H. (1990). Cortisol and behavior: 2. Application of a latent state-trait model to salivary cortisol. *Psychoneuroendocrinology*, *15*(4), 297–307.

Klosterkötter, J., Hellmich, M., Steinmeyer, E. M., & Schultze-Lutter, F. (2001). Diagnosing schizophrenia in the initial prodromal phase. *Archives of General Psychiatry*, *58*(2), 158–164.

Lee, P. R., Brady, D., & Koenig, J. I. (2003). Corticosterone alters N-methyl-D-aspartate receptor subunit mRNA expression before puberty. *Molecular Brain Research*, *115*(1), 55–62.

Li, I., Chiou, H-H., & Shen, P-S. (2007). Correlations between cortisol level and internalizing disposition of young children are increased by selecting optimal sampling times and aggregating data. *Developmental Psychobiology*, *49*(6), 633–639.

Marinelli, M., Rudick, C. N., Hu, X. T., & White, F. J. (2006). Excitability of dopamine neurons: Modulation and physiological consequences. *CNS and Neurological Disorders Drug Targets*, *5*(1), 79–97.

McGowan, P. O., Sasaki, A., D'Alessio, A. C., Dymov, S., Labonte, B., Szyf, M., et al. (2009). Epigenetic regulation of the glucocorticoid receptor in human brain associates with childhood abuse. *Nature Neuroscience*, *12*(3), 342–348.

Mehler, M. F. (2008). Epigenetic principles and mechanisms underlying nervous system functions in health and disease. *Progress in Neurobiology*, *86*(4), 305–341.

Miller, T. J., McGlashan, T. H., Rosen, J. L., Somjee, L., Markovich, P. J., Stein, K., et al. (2002). Prospective diagnosis of the initial prodrome for schizophrenia based on the Structured Interview for Prodromal Syndromes: preliminary evidence of interrater reliability and predictive validity. *American Journal of Psychiatry*, *159*(5), 863–865.

Mittal, V. A., Dhruv, S., Tessner, K. D., Walder, D. J., & Walker, E. F. (2007). The relations among putative biorisk markers in schizotypal adolescents: minor physical anomalies, movement abnormalities, and salivary cortisol. *Biological Psychiatry*, *61*(10), 1179–1186.

Nicodemus, K. K., Marenco, S., Batten, A. J., Vakkalanka, R., Egan, M. F., Straub, R.E., et al. (2008). Serious obstetric complications interact with hypoxia-regulated/vascular-expression genes to influence schizophrenia risk. *Molecular Psychiatry*, *13*(9), 873–877.

Nozdrachev, A. D. (2008). Review of prenatal stress effects and the developing brain: adaptive mechanisms and immediate and delayed effects. *Human Physiology*, *34*(4), 535–536.

O'Brien, J. T., Lloyd, A., McKeith, I., Gholkar, A., & Ferrier, N. (2004). A longitudinal study of hippocampal volume, cortisol levels, and cognition in older depressed subjects. *American Journal of Psychiatry*, *161*(11), 2081–2090.

Philippi, H., Pohlenz, J., Grimm, W., Kollfer, T., & Schonberger, W. (2000). Simultaneous stimulation of growth hormone, adrenocorticotropin and cortisol with L-dopa/L-carbidopa and propranolol in children of short stature. *Acta Paediatrica*, *89*(4), 442–446.

Piacentini, M. F., Meeusen, R., Buyse, L., De Schutter, G., & De Meirleir, K. (2004). Hormonal responses during prolonged exercise are influenced by a selective DA/NA reuptake inhibitor. *British Journal of Sports Medicine, 38*(2), 129–133.

Piazza, P. V., Barrot, M., Rouge-Pont, F., Marinelli, M., Maccari, S., Abrous, D. N., et al. (1996). Suppression of glucocorticoid secretion and antipsychotic drugs have similar effects on the mesolimbic dopaminergic transmission. *Proceedings of the National Academy of Sciences of the United States of America, 93*(26), 15445–15450.

Pruessner, J. C., Champagne, F., Meaney, M. J., & Dagher, A. (2004). Dopamine release in response to a psychological stress in humans and its relationship to early life maternal care: A positron emission tomography study using [11C] raclopride. *Journal of Neuroscience, 24*(11), 2825–2831.

Pruessner, J. C., Kirschbaum, C., Meinlschmid, G., & Hellhammer, D. H. (2003). Two formulas for computation of the area under the curve represent measures of total hormone concentration versus time-dependent change. *Psychoneuroendocrinology, 28*(7), 916–931.

Reichenberg, A., Mill, J., & MacCabe, J. H. (2009). Epigenetics, genomic mutations and cognitive function. *Cognitive Neuropsychiatry, 14* (4/5), 377–390.

Rieder, J., & Coupey, S. M. (1999). Update on pubertal development. *Current Opinion in Obstetrics and Gynecology, 11*, 457–462.

Rodriguez-Waitkus, P. M., Lafollette, A. J., Ng, B. K., Zhu, T. S., Conrad, H. E., & Glaser, M. (2003). Steroid hormone signaling between Schwann cells and neurons regulates the rate of myelin synthesis. *Annals of the New York Academy of Sciences, 1007*, 340–348.

Sapolsky, R. M. (2003). Stress and plasticity in the limbic system. *Neurochemical Research, 28*(11), 1735–1742.

Seckl, J. R., & Meaney, M. J. (2004). Glucocorticoid programming. In R. Yehuda, B. McEwen (Eds.), *Biobehavioral stress response: protective and damaging effects* (pp. 63–84). New York: New York Academy of Sciences.

Silberg, J., Pickles, A., Rutter, M., Hewitt, J., Simonoff, E., Maes, H., et al. (1999). The influence of genetic factors and life stress on depression among adolescent girls. *Archives of General Psychiatry, 56*(3), 225–532.

Simons, C. J., Wichers, M., Derom, C., Thiery, E., Myin-Germeys, I., Krabbendam, L., et al. (2009). Subtle gene-environment interactions driving paranoia in daily life. *Genes, Brain, & Behavior, 8*(1), 5–12.

Sisk, C. L., & Zehr, J. L. (2005). Pubertal hormones organize the adolescent brain and behavior. *Frontiers in Neuroendocrinology, 26*(3-4), 163–174.

St. Clair, D. (2009). Copy number variation and schizophrenia. *Schizophrenia Bulletin, 35*(1), 9–12.

Starkman, M. N., Giordani, B., Gebarski, S. S., & Schteingart, D. E. (2003). Improvement in learning associated with increase in hippocampal formation volume. *Biological Psychiatry, 53*(3), 233–238.

Stefanis, N. C., Henquet, C., Avramopoulos, D., Smyrnis, N., Evdokimidis, I., Myin-Germeys, I., et al. (2007). COMT Val158Met moderation of stress-induced psychosis. *Psychological Medicine, 37*, 1651–1656.

Tessner, K. D., Walker, E. F., Dhruv, S. H., Hochman, K., & Hamann, S. (2007). The relation of cortisol levels with hippocampus volumes under baseline and challenge conditions. *Brain Research, 1179*, 70–78.

Thompson, P. M., Giedd, J. N., Woods, R. P., MacDonald, D., Evans, A. C., & Toga, A. W. (2000). Growth patterns in the developing brain detected by using continuum mechanical tensor maps. *Nature, 404*(6774), 190–193.

Tsigos, C., & Chrousos, G. P. (2002). Hypothalamic-pituitary-adrenal axis, neuroendocrine factors and stress. *Journal of Psychosomatic Research, 53*(4), 865–871.

van der Beek, E. M., Wiegant, V. M., Schouten, W. G., van Eerdenburg, F. J., Loijens, L. W., van der Plas, C., et al. (2004). Neuronal number, volume, and apoptosis of the left dentate gyrus of chronically stressed pigs correlate negatively with basal saliva cortisol levels. *Hippocampus, 14*(6), 688–700.

van Winkel, R., Henquet, C., Rosa, A., Papiol, S., Fananás, L., De Hert, M., et al. (2008). Evidence that the COMT(Val158Met) polymorphism moderates sensitivity to stress in psychosis: an experience-sampling study. *American journal of medical genetics. Part B, Neuropsychiatric genetics, 147*, 10–17.

Walker, E. F. (2002). Adolescent neurodevelopment and psychopathology. *Current Directions in Psychological Science, 11*(1), 24–28.

Walker, E. F., Walder, D. J., & Reynolds, F. (2001). Developmental changes in cortisol secretion in normal and at-risk youth. *Development and Psychopathology, 13*(3), 721–732.

Walker, E., & Diforio, D. (1997). Schizophrenia: A neural diathesis-stress model. *Psychological Review, 104*(4), 667–685.

Walker, E., Kestler, L., Bollini, A., & Hochman, K. M. (2004). Schizophrenia: etiology and course. *Annual Review of Psychology, 55*, 401–430.

Walker, E., McMillan, A., & Mittal, V. (2007). Neurohormones, neurodevelopment and the prodrome of psychosis in adolescence. In D. Romer, & E. Walker (Eds.), *Adolescent psychopathology and the developing brain: integrating brain and prevention science.* New York, NY: Oxford University Press. pp 264–284.

Walker, E., Mittal, V., & Tessner, K. (2008). Stress and the hypothalamic pituitary adrenal axis in the developmental course of schizophrenia. *Annual Review of Clinical Psychology, 4*, 189–216.

Walker, E., Downey, G., & Caspi, A. (1991). Twin studies of psychopathology: Why do the concordance rates vary? *Schizophrenia Research, 5*(3), 211–221.

Wallace, G. L., Schmitt, J. E., Lenroot, R., Viding, E., Ordaz, S., Rosenthal, M. A., et al. (2006). A pediatric twin study of brain morphometry. *Journal of Child Psychology & Psychiatry & Allied Disciplines, 47*(10), 987–993.

Waterland, R. A., & Jirtle, R. L. (2003). Transposable elements: targets for early nutritional effects on epigenetic gene regulation. *Molecular & Cellular Biology, 23*(15), 5293–5300.

Weber, A., Clark, A. J., Perry, L. A., Honour, J. W., & Savage, M. O. (1997). Diminished adrenal androgen secretion in familial glucocorticoid deficiency implicates a significant role for ACTH in the induction of adrenarche. *Clinical Endocrinology, 46*(4), 431–437.

Webster, M. J., Weikert, C. S., Herman, M. M., & Kleinman, J. E. (2002). BDNF mRNA expression during postnatal development, maturation and aging of the human prefrontal cortex. *Brain Research. Developmental Brain Research., 139*(2), 139–150.

Wintermantel, T. M., Berger, S., Greiner, E. F., & Schutz, G. (2005). Evaluation of steroid receptor function by gene targeting in mice. *Journal of Steroid Biochemistry and Molecular Biology, 93*(2-5), 107–112.

Yung, A. R., Nelson, B., Stanford, C., Simmons, M. B., Cosgrave, E. M., Killackey, E., et al. (2008). Validation of "prodromal" criteria to detect individuals at ultra high risk of psychosis: 2 year follow-up. *Schizophrenia Research, 105* (1–3), 10–17.

Zhang, L., Zhou, R., Li, X., Ursano, R. J., & Li, H. (2006). Stress-induced change of mitochondria membrane potential regulated by genomic and non-genomic GR signaling: A possible mechanism for hippocampus atrophy in PTSD. *Medical Hypotheses, 66*(6), 1205–1208.

18

The Role of Stress and Its Interaction with Genetic Predisposition in Schizophrenia

VALERIA MONDELLI, MARTA DI FORTI, CARMINE M. PARIANTE, AND ROBIN M. MURRAY

The first clinical observations on stress and mental disorders date back to the 18th century, when the idea that social stress was one of the causes of mental illness was formulated in France by Philippe Pinel, and by his student Jean Esquirol. However, it was only much later, in the 1930s, that the Swiss-American psychiatrist Adolph Meyer gave psychosocial stressors etiological prominence, putting them in the center of what he called "the psychobiology of mental disorders" (Van Praag, de Kloet, & van Os, 2004). According to Meyer's theory, mental disorders were a consequence of an unsuccessful adaptation to psychosocial adversities.

Initial formulations on the link between stress and psychosis narrowly focused on the role of psychosocial stressful events. This line of investigations provided a basis for the diathesis–stress model of schizophrenia, developed in the 1960–1970s. David Rosenthal (1970), although best known for his work in the Danish-American adoption studies that demonstrated the importance of genetic factors in schizophrenia (Rosenthal et al., 1975), played a central role in promulgating the view that the behavioral expression of the biological vulnerability for schizophrenia is influenced by exposure to stress.

However, the demonstration of brain structural abnormalities in patients with schizophrenia in the 1970s (Johnstone, Crow, Frith, Husband, & Kreel, 1976) and emphasis on neurodevelopment in the 1980s and 1990s (Murray & Lewis, 1987), led to a neglect of the role of the social environment in schizophrenia. But gradually, the wheel began to turn, and over the last decade, a renewed emphasis has been placed on the role of a variety of social stressors (Morgan & Fisher, 2007; Morgan et al., 2008). The neurodevelopmental hypothesis has also been modified to include not only neurological but also nonphysical environmental hazards operating during development (Di Forti, Lappin, & Murray, 2007; Howes et al., 2004). As part of this shift, one focus has been an attempt to elucidate the biological mechanisms mediating the effect of stressors on psychosis (Walker, Mittal, & Tessner, 2008).

In this chapter, we discuss the role of stress in schizophrenia, the types of stressors that have been implicated, the pathogenic mechanisms that appear to be involved, and finally, how stress interacts with genetic predisposition.

WHAT IS STRESS?

One of the first definitions of stress belongs to Walter Bradford Cannon (1929), who used the term "stress" to designate forces that act on the organism, disturb its homeostasis, and cause "strain." Cannon showed that both physical and psychological stimuli could evoke similar reactions. Although Cannon initiated the study of the stress response, Hans Selye (1998) has always been considered the father of stress research, describing the physical responses to stress and defining this reaction as the "general adaptation syndrome." Other researchers have focussed on the emotional response to stressors and on the psychological dimension of the stress response (Lazarus, 1966; Mason, 1971; Mikhail, 1981). In particular, Lazarus underlined the importance of how the stressor is psychologically perceived, stating that the "stress phenomena" appear only if the situation is perceived as potentially damaging or hard to cope with. Based on these new theories, stress research gradually shifted from a model looking at the physical response to stressful stimuli toward a model in which the psychological repercussions of the stressor have a central position. This shift in emphasis was then accompanied by an increasing interest in life-event research, trying to determine the risk of suffering mental problems after having being exposed to stressful life events.

Stressful Life Events in Psychosis

The first major study investigating the potential relationship between the experience of stressful events and schizophrenia was conducted at the

Institute of Psychiatry in London about 40 years ago (Brown & Birley, 1968). This study reported that patients with schizophrenia experienced more stressful events in the 3 months preceding relapse of the illness than did control subjects. Moreover, the number of stressful events was higher in the weeks just preceding the relapse; in particular, during the 3 weeks preceding the onset of the relapse, 46% of patients experienced at least one stressful event, compared with 12% in an earlier 3-month period, thus suggesting a possible relationship between life events and onset of schizophrenia (Brown & Birley, 1968). Bebbington et al. (1993) replicated these findings, reporting an excess of stressful life events over the 6 months before the relapse of psychosis, both in patients with schizophrenia and in patients with affective psychoses when compared with a psychiatrically healthy sample from the local general population.

Stressful life events have also been associated with the occurrence of relapses in psychosis by several studies (Birley & Brown, 1970; Malla, Cortese, Shaw, & Ginsberg, 1990; Ventura, Nuechterlein, Lukoff, & Hardesty, 1989). However, other studies failed to find the same results. Chung et al. (1986) reported no change in the frequency of life events before onset, and Gruen and Baron (1984) reported that only 15.4% of chronic schizophrenia patients experienced severe stressors during the 12 months before onset of illness.

Inconsistencies among studies might be due to methodological or sampling differences. For example, the retrospective design of most of these studies poses the problem of the ability of the patient to recall events over the preceding weeks to several months. This problem may be particularly relevant in psychosis, as the patient's recall may be affected by disorders of thinking, judgment, perception, language, and memory. Consistent with this possibility, longitudinal studies that examined the issue prospectively have also shown that relapses and symptom exacerbation in schizophrenia are preceded by an elevated rate of stressful life events (Hirsch et al., 1996; Pallanti, Quercioli, & Pazzagli, 1997; Ventura et al., 1989).

Another possible methodological issue concerns the type of events measured, ranging from relatively minor events (change of job) to traumatic events (tragic loss of a loved one). Indeed, as also underlined in Chapter 1, even a detailed stressful life event inventory may miss a significant stressful event for a specific individual. People may differ with respect to the experiences they find stressful. Minor interpersonal events are highly stressful for some, whereas others are relatively unperturbed by such events. These individual differences can render within-subject longitudinal studies more powerful in detecting the adverse effects of stress exposure.

Stressful events have also been described as dependent or independent, where the event dependence reflects the plausibility that the stressful life event could have resulted from the subject's own behaviour. The distinction between dependent and independent stressful events is particularly important when trying to investigate the possible role of stressful events as a cause of psychosis. Indeed, a dependent stressful event may not have a causal relationship with the psychotic disorder, but rather the psychotic disorder or its antecedents may predispose to this stressful event; an independent stressful life event is more likely to play a role in the causation of the psychiatric disorder.

Childhood Trauma and Psychosis

As well as stresses immediately proximal to the onset or relapse of psychosis, increasing attention has been paid to severe stresses in early life. Thus, a recent systematic review of the literature claimed to find an excess of childhood trauma in patients with psychotic symptoms (Read, van Os, Morrison, & Ross, 2005). However, 31 of the 51 studies considered in this review were conducted in diagnostically heterogeneous samples, in which the numbers of patients with a psychotic disorder was often unclear (Morgan & Fisher, 2007). Moreover, most of the studies focusing on psychosis have been conducted on small samples of patients with a chronic duration of illness.

Nonetheless, several large, population-based studies support the association between childhood trauma and psychosis. Working on the data concerning 8,580 participants aged 16–74 from the British National Survey of Psychiatric Morbidity, Bebbington et al. (2004) found that those who met criteria for a definite or probable psychotic disorder ($n = 60$) were over 15 times more likely to report having been sexually abused. Another study by Janssen et al. (2004) was conducted on a general population sample of 4,045 subjects in the Netherlands. This prospective study involved interviews with adults aged 18–64 years, assessed for presence of mental illness at baseline, after 1 year and again after 3 years; individuals who had experienced emotional, physical, or sexual abuse or neglect before the age of 16 were more likely to report experiencing psychotic symptoms during the 3-year follow-up (Janssen et al., 2004). The effect was strongest when psychosis was most stringently defined and remained even after a range of potential confounding factors was taken into account (adjusted odds ratio [OR] 7.3).

In both these studies, the childhood abuse was assessed retrospectively by self-report. Further support for an association between childhood trauma and psychosis derives from the study by Spauwen and colleagues (2006). These authors used data from 2,524 participants from the Early Developmental Stages of Psychopathology study and found that the experience of lifetime trauma (even if not restricted to childhood) was associated with the development of three or more psychotic symptoms during an average follow-up period of 42 months. In a large population-based study, Whitfield et al. (2005) found that respondents reporting a history of hallucinations were more likely to have been both physically (adjusted OR 1.7) and sexually (adjusted OR 1.7) abused during childhood.

Using a different approach, Spataro and colleagues (2004) studied hospital admissions rates in 1,612 subjects who, according to official records, had been sexually abused before age of 16, and compared them with admission rates in a large population-based control sample. The authors found that abused males were 1.3 times more likely, and abused females 1.5 times more likely, to have been treated for a "schizophrenic disorder." However, these differences were not statistically significant.

Interestingly, only four studies until now have investigated the effects of childhood trauma in patients with a first episode of psychosis. Studies in patients with first-episode psychosis are particularly valuable as they reduce the effects of the disorder on recall compared to previous studies that have predominantly relied on chronic patient samples. Greenfield et al. (1994) reported that 52% of 38 patients with affective and nonaffective psychoses had a history of childhood abuse; the authors did not find any difference in severity of psychotic symptoms between patients with and without history of childhood abuse. The second study, by Compton et al. (2004), found a relationship between childhood abuse and substance dependency, suggesting that childhood abuse could be an important risk factor for substance abuse comorbidity in patients with schizophrenia. The third study, by Ucok and Bikmaz (2007), reported high rates of childhood abuse (52% of their patients) and childhood neglect (43% of their patients) in first-episode schizophrenia; the presence of childhood trauma was associated with positive, but not with negative, psychotic symptoms, and the severity of childhood trauma was also correlated with the severity of positive psychotic symptoms (Ucok & Bikmaz, 2007).

In a much larger U.K. study based on the large AESOP sample of patients with their first episode of psychosis and epidemiologically based controls, Fisher et al. (2009) reported a gender difference in the association between childhood abuse and psychosis. In particular, women with first-episode psychosis were twice as likely to report either physical or sexual abuse as were women controls; when the two types of abuse were separated, there was a stronger association between physical abuse and psychosis than between sexual abuse and psychosis. In contrast, the authors found no association between physical and/or sexual childhood abuse and psychosis in men (Fisher et al., 2009).

Urbanicity and Psychosis

A possible link between stress and psychosis has been suggested also by studies showing an association between urbanicity and psychosis (Lewis, David, Andreasson, & Allebeck, 1992). Several mechanisms have been hypothesized in the past to explain this association; undeniably, it is well established that the urban environment is characterized by higher levels of deprivation and social isolation compared with the rural environment, and this could indeed represent the main mechanism mediating the association between urbanicity and schizophrenia.

Previous studies have shown a higher prevalence of stressful factors such as noise pollution, feeling unsafe, divorce, crime, and poor perceived mental and physical health in urban areas compared with rural ones, thus positing the urban environment as a model of exposure to stress (Marcelis, Takei, & van Os, 1999). Marcelis et al. investigated if the association between urbanicity and schizophrenia was mediated by an early effect (urban birth upbringing) or by a late effect (urban residence around the time of onset), and found that environmental factors associated with urbanization increase the risk for schizophrenia before rather than around the time of illness onset (Marcelis et al., 1999). These studies suggest that early and prolonged exposure to stress play a significant role in the onset of psychosis.

However, it is clear that the exposure to urban environment or to stressful events is not sufficient to cause schizophrenia and that a genetic predisposition may be needed for these environmental factors to precipitate the onset of psychosis. In the last part of this chapter, we will explore in more detail the latest studies about the interaction between stressful environments and genes.

HYPOTHALAMUS-PITUITARY-ADRENAL AXIS AND THE STRESS RESPONSE

Hypothalamus-Pituitary-Adrenal Axis

The hypothalamus-pituitary-adrenal (HPA) axis plays a fundamental role in the response to external and internal stimuli, including most notably psychological stressors. HPA axis activity is mediated by a myriad of peptides and hormones that largely involve the secretion of corticotropin-releasing hormone (CRH) and vasopressin (AVP) from the hypothalamus, which in turn activate the secretion of adrenocorticotropin hormone (ACTH) from the pituitary, which finally stimulates the secretion of the glucocorticoids (cortisol in humans and corticosterone in rodents) from the adrenal cortex (Pariante & Lightman, 2008) (see Figure 17.1 in Chapter 17).

Cortisol levels vary during the day, reaching a zenith at the time of the awakening in the morning, decreasing in the late afternoon and evening, and reaching nadir generally around midnight. Most circulating cortisol is protein-bound to the corticosteroid binding globulin, and it is only the unbound or "free" fraction that is biologically active. Although up to 95% of the cortisol secreted in the blood is bound to corticosteroid-binding globulin, the cortisol present in the saliva is all unbound cortisol (Kirschbaum & Hellhammer, 1994), and therefore can be readily measured.

Cortisol interacts with its receptors in multiple target systems, including the HPA axis, where it is responsible for feedback inhibition of the secretion of ACTH from the pituitary and CRH from the hypothalamus (Pariante & Lightman, 2008) (see Figure 18.1). Although glucocorticoids regulate the function of almost every tissue in the body, the best known physiological effect of these hormones is the regulation of energy metabolism (increased gluconeogenesis, increased lipolysis, increased protein degradation). Interestingly, the HPA axis also interacts with different neurotransmitter systems, especially the dopaminergic system; this interaction partly clarifies the underlying mechanisms linking stress and glucocorticoids to the development of psychosis. In particular, glucocorticoid secretion has been shown to increase dopaminergic activity in certain brain regions, especially the mesolimbic system (Walker et al., 2008).

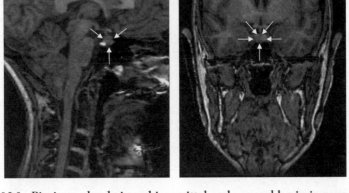
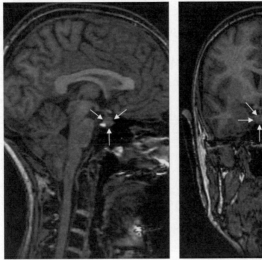

Figure 18.1. **Pituitary gland viewed in sagittal and coronal brain images.** Image from Pariante, C. M., Vassilopoulou, K., Velakoulis, D., Phillips, L., Soulsby, B., Wood, S. J., et al. (2004). Pituitary volume in psychosis. *British Journal of Psychiatry, 185,* 5–10, with permission.

Effect of Acute and Chronic Stress on the HPA Axis

Acute physical or psychological stress activates the HPA axis, increasing plasma levels of ACTH and cortisol. An adequate adaptive stress response consists of rapid cortisol secretion, followed by inhibition via a feedback system, before the HPA axis activation becomes damaging by itself. Acute time-limited rises in cortisol levels have been shown to be adaptive, but chronic stress and consequent long-term elevation in cortisol levels appear to negatively affect physical and behavioral parameters important for survival (Chrousos & Gold, 1992; McEwen & Seeman, 1999; Raison & Miller, 2003). In particular, a number of pathologies associated with stress have been attributed to glucocorticoid excess, including volumetric changes in the brain and behavioral alterations. Indeed, it has been hypothesized that prolonged activation of the HPA axis damages brain structures (especially the hippocampus) essential for HPA axis control. Such damage, in turn, has been suggested to lead to a feed-forward circuit in which ongoing stressors drive continuous glucocorticoid overproduction (the *glucocorticoid cascade hypothesis*) (Raison & Miller, 2003; Sapolsky, 2000).

Despite the popularity of the glucocorticoid cascade hypothesis, some authors have suggested that insufficient glucocorticoid signaling may play a significant role in the development and expression of pathology in stress-related disorders. Indeed, a growing literature supports the idea that glucocorticoids confer longer-term, stress-related benefits by shaping and restraining stress-related physiological processes, including inflammatory responses, activation of the sympathetic nervous system, and stimulation of CRH pathways, all of which can produce adverse health outcomes if allowed to continue unabated after crisis resolution (Raison & Miller, 2003). In view of these findings, the emphasis on HPA axis hyperactivity in psychiatric disorders is shifting to the role of insufficient glucocorticoid signaling in these conditions, possibly as a result of attenuated glucocorticoid responsiveness (Raison & Miller, 2003). A physiological view would suggest that both excessive and insufficient glucocorticoid signaling would probably have adverse consequences.

HYPOTHALAMUS-PITUITARY-ADRENAL AXIS ACTIVITY AND PSYCHOSIS

Studies on Hormonal Levels

Several studies have reported that patients in the acute phase of schizophrenia have elevated basal HPA axis activity as shown by raised cortisol and

ACTH levels and nonsuppression of cortisol secretion by dexamethasone in the dexamethasone suppression test and in the dexamethasone/CRH test (Herz, Fava, Molnar, & Edwards, 1985; Lammers et al., 1995; Ryan, Collins, & Thakore, 2003; Ryan, Sharifi, Condren, & Thakore, 2004; Sachar, Kanter, Buie, Engle, & Mehlman, 1970; Tandon et al., 1991). However, studies on patients with chronic schizophrenia have not found elevated basal cortisol levels or an increased rate of suppression in the dexamethasone suppression test, especially if the patients were medicated and clinically stable (Ismail, Murray, Wheeler, & O'Keane, 1998; Tandon et al., 1991). Unfortunately, many of the reports on baseline cortisol involve very small numbers of patients and, therefore, very low statistical power for detecting group differences (Walker & Diforio, 1997). In light of this, Walker and Diforio (1997) computed effect sizes for 11 of the previous studies reporting means and standard deviations of cortisol levels for patients with schizophrenia and healthy controls; the authors found a mean effect size across studies of 0.60, which is typically classified as moderate, supporting the notion of HPA axis hyperactivity in schizophrenia.

Five studies have investigated HPA axis activity in first-episode psychosis patients. In the first two studies, Ryan and colleagues (2003, 2004), assessed plasma cortisol levels in drug-naïve patients with a first episode of schizophrenia and age- and sex-matched controls, taking a blood sample at one time-point only during the day (at 8 AM after overnight fasting). In these studies, patients with a first episode of schizophrenia presented higher cortisol levels compared with healthy controls. However, a procedure based on a single sample for the cortisol assessment represents a limitation, since it may not provide an accurate estimate of cortisol levels and of HPA axis activity. The third study, from the same group, investigated 12 drug-naïve patients with first-episode psychosis and 12 age- and sex-matched controls, measuring plasma cortisol and ACTH levels, collecting blood samples every 20 minutes (from 1 to 4 PM) (Ryan et al., 2004). Again, patients with first-episode schizophrenia presented higher cortisol and ACTH secretion during the whole sampling period compared with controls, thus supporting the presence of HPA axis hyperactivity at this point in the illness (Ryan et al., 2004).

The fourth study on HPA axis activity in first-episode psychosis investigated the cortisol response to the dexamethasone suppression test in 56 patients with first-episode schizophrenia at the time of admission to a hospital (before starting antipsychotic treatment), at the time of discharge, and again after 1 year (Ceskova, Kasparek, Zourkova, & Prikryl, 2006). The rate of nonsuppression was 17.9% at baseline before starting the treatment, 5.3% at the time of discharge, and 16% after 1 year (Ceskova et al., 2006). In agreement with the literature in chronic schizophrenia, rates of dexamethasone nonsuppression were higher in drug-free and unmedicated patients. The increase in the rate of nonsuppression after 1 year was explained as a possible consequence of clinical deterioration and of noncompliance with treatment (Ceskova et al., 2006).

The most recent study investigated cortisol levels during the day and stressful life events in 50 patients with first-episode psychosis and 36 control subjects. In this study, patients had a trend for higher diurnal cortisol levels, with those who had received less than 2 weeks of antipsychotic treatment showing significantly higher cortisol levels than both patients with more than 2 weeks of antipsychotic treatment and healthy controls. Moreover, this study also showed that first-episode psychosis patients have a blunted cortisol awakening response compared with controls. Although the patients as a whole had experienced significantly more recent stressful events, perceived stress, and childhood trauma than did controls, the cortisol abnormalities in these patients appeared not to be driven by the excess of stressors. Indeed, diurnal cortisol levels were negatively, rather than positively, correlated with the number of recent stressful events, suggesting that factors other than the excess of psychosocial stress may also explain HPA axis abnormalities in first-episode psychosis (Mondelli et al., 2010).

Studies on Pituitary Volume

The pituitary gland plays an important role in the regulation of the HPA axis. Its volume can change in size as a consequence of both physiological and pathological alterations in the patterns of hormone secretion; in major depression, HPA axis hyperactivity has been associated with an increased volume of the pituitary gland (Axelson et al., 1992).

Studies conducted so far in psychosis confirm that the pituitary is a dynamic organ that changes according to different stages of the psychotic disorder, in response to both the disorder itself and to treatment with antipsychotics. Specifically, pituitary volume increases during the prodromal phase leading to psychosis onset (Garner et al., 2005), and it is larger (by 10%–20% compared with controls) if assessed during the first 2–6 months after psychosis onset (Pariante et al., 2004, 2005). This effect is not due to antipsychotic treatment, as it is present in antipsychotic-naïve prodromal subjects (Garner et al., 2005), as well as in neuroleptic-free patients with (first-episode) psychosis (Pariante et al., 2005), and is likely to reflect HPA axis hyperactivity. Interestingly, a recent study in drug-naïve first-episode psychosis patients has also shown that larger pituitary volume is associated with less improvement over 12 weeks in overall symptoms and positive symptoms, further supporting the importance of the HPA axis in emerging psychosis and in determining clinical outcome (Garner et al., 2009).

Following the initial enlargement in the first phases of psychosis, the pituitary tends to become smaller, as suggested by studies in patients with psychosis of at least 2 years' duration (Pariante et al., 2004). The mechanisms underlying the smaller pituitary volume are yet to be fully understood, and the explanations have ranged from being the consequence of a chronic activation of the HPA axis to being a neurodevelopmental problem (Pariante, 2008).

The Link Between Stress and HPA Axis Response in Psychosis

The increased activation of the HPA axis reported in psychosis has been hypothesized to be related to increased biological susceptibility to daily life stress, the increased level of independent stressors leading to the psychotic episode, or to the distress caused by the psychotic experience, or to all these causes (Bebbington et al., 1993; Myin-Germeys, van Os, Schwartz, Stone, & Delespaul, 2001). Only two studies have investigated the association between stressful life events and cortisol levels in psychosis. The first study reported that severity of preadmission stressors was positively correlated with cortisol levels at admission in acute psychotic patients (Mazure, Quinlan, & Bowers, 1997), thus supporting the hypothesis that stressful life events play a role in the HPA axis hyperactivity found in this condition. However, the second more recent and more surprising study, although finding both high levels of stress and of cortisol in patients with first-episode psychosis, reported an inverse relationship between recent stressful life events and diurnal cortisol levels in the same subjects (Mondelli et al., 2010).

Interestingly, previous studies in patients with chronic schizophrenia have found a blunted cortisol response to psychological and psychosocial stressors, indicating that these patients do not activate the HPA axis in response to stress like healthy controls (Albus, Ackenheil, Engel, & Muller, 1982; Breier, Wolkowitz, Doran, Bellar, & Pickar, 1988; Jansen et al., 1998). In these studies, cortisol levels were collected following different kinds of stressors, such as cold pressor test, noise, mental arithmetic (Albus et al., 1982), lumbar puncture (Breier et al., 1988), and a public-speaking task (Jansen et al., 1998). The blunted cortisol rise after these stressors seems to indicate that patients with psychosis have an impaired ability to activate the HPA axis in response to "true stressors," although we cannot exclude the possible effect of antipsychotic treatment on the activation of the HPA axis in these subjects. However, these findings suggest that the increased number of life stressors cannot explain completely the HPA axis hyperactivity reported in schizophrenia, and that other factors might better explain this abnormality.

The hypothesis that HPA axis hyperactivity is associated with the severity of psychotic symptoms has been tested only in a few studies. In cross-sectional studies, baseline cortisol levels are positively correlated with ratings of positive psychotic symptoms in nonmedicated (Rybakowski, Linka, Matkowski, & Kanarkowski, 1991) and medicated (Franzen, 1971) patients with schizophrenia. Higher post-dexamethasone cortisol levels have been found to be associated with more severe negative symptoms (Kaneko et al., 1992; Keshavan, Brar, Ganguli, & Jarrett, 1989; Newcomer, Faustman, Whiteford, Moses, & Csernansky, 1991), and positive symptoms (Kaneko et al., 1992; Walder, Walker, & Lewine, 2000), but most of the studies have shown an association of cortisol levels with negative symptoms rather than with positive symptoms (Tandon et al., 1991). In studies conducted in first-episode psychosis, Ryan and colleagues failed to find an association between cortisol levels and symptom severity (Ryan et al., 2003, 2004), whereas Ceskova and colleagues reported a positive correlation between cortisol levels after dexamethasone suppression test and negative psychotic symptoms at time of discharge (Ceskova et al., 2006).

FAMILIAL LIABILITY TO ALTERED STRESS RESPONSE IN SCHIZOPHRENIA

Several studies have described elevated HPA axis activity in the acute phases of chronic schizophrenia (Tandon et al., 1991), as well as at the time of the first psychotic episode (Pariante et al., 2004, 2005; Ryan et al., 2004), and more recently, some studies have tried to investigate if one of the possible mechanisms underlying HPA axis hyperactivity could be an increased susceptibility to daily life stress, possibly genetically mediated.

Myin-Germeys and colleagues have shown that family members of psychotic patients have an increased emotional reactivity (measured as changes in positive and negative affect) and increased intensity of subtle psychotic experiences in response to daily life stress (Myin-Germeys et al., 2001; Myin-Germeys, Delespaul, & van Os, 2005). Moreover, a recent study has reported an enlarged pituitary volume in unaffected relatives of patients with schizophrenia, suggesting that the pituitary enlargement (marker of HPA axis hyperactivity) in first-episode psychosis is at least partially explained by genetic liability (Mondelli et al., 2008). Goldstein and colleagues also reported an enlarged hypothalamic volume in both patients with schizophrenia and their unaffected relatives, further supporting the presence of a possible hyperactivity of the HPA axis in unaffected relatives of patients with schizophrenia (Goldstein et al., 2007). Interestingly, a previous study showed that first-degree relatives of patients with schizophrenia have increased striatal dopamine synthesis compared with controls (Huttunen et al., 2008), compatible with the view that a genetic liability to HPA axis hyperactivity could contribute to the pathogenesis of psychotic disorders by increasing brain dopaminergic activity.

Gene–Stress Interaction in Schizophrenia

Genetic epidemiological and molecular studies have excluded the possibility that schizophrenia might be transmitted by one single major gene. Instead, schizophrenia has come to be seen as a multifactorial disorder in which multiple small genes interact with each other and with environmental factors. Large genome-wide association (GWA) studies have revealed that the association between individual genes and psychotic disorders is weak (Purcell et al., 2009; Shi et al., 2009; Stefansson et al., 2009), and such findings have provoked considerable controversy (Collier, 2008; Crow, 2008; O'Donovan, Craddock, & Owen, 2008; Sullivan, 2008). The number of genes postulated to be involved has gradually increased from two or three to hundreds or even thousands (O'Donovan, Craddock, & Owen, 2009). This change reflects an acknowledgment that the effect sizes of those genes that have so far been putatively implicated is vanishingly small and nonspecific.

Indeed, the nonspecificity of genetic effects has been demonstrated by recent twin (Cardno, Rijsdijk, Sham, Murray, & McGuffin, 2002) and molecular studies that have found considerable genetic overlap between schizophrenia and bipolar illness, implying that some of the susceptibility to these conditions is likely to be in common (Kirov et al., 2008; Sullivan et al., 2008). Most recently, it has been discovered that variations in the number of copies of nucleotide-bases across chromosomes, known as *copy number variations* (CNVs), appear to underlie the genetic contribution to a small proportion of cases (2%–5%) of schizophrenia. Some of the same CNVs have also been associated with autism (Stefansson et al., 2008; Walsh et al., 2008; Weiss et al., 2008), implying a pleiotropic effect that may result in a spectrum of developmental psychopathology. Interestingly, CNVs have not generally been reported in bipolar disorder, a negative finding compatible with the view that bipolar disorder is less likely to be associated with developmental impairment than is schizophrenia (Murray et al., 2004).

Models implicating gene–environmental (G × E) interaction have become increasingly common in psychiatry (Caspi et al., 2005; van Os & Murray, 2008). The expression of a genetic effect (several small genes mutations or/and CNVs) on a disease may depend on environmental exposures. Although this general approach has not been without its critics (Zammit, Wiles, & Lewis, 2010), it might provide a bridge between the small main effect of putative susceptibility genes and the effect of known environmental risk factors, which by themselves leave the majority of the exposed individuals unaffected.

We noted earlier the role played by urbanicity in the etiology of psychosis. In a Dutch population study, van Os et al. used familial liability as a proxy measure of genetic risk, and tested if this moderates the effect of urbanicity (a five-level rating of population density of place of residence) in increasing the risk for psychotic disorders. They found that both familial liability and urbanicity increased the risk for psychotic disorder independently of each other, but that the effect was much greater for individuals with evidence of both. In particular, the risk of psychotic disorder in those exposed to both urbanicity and family history of psychosis was 9.72%, while it was 1.59% in those exposed only to urbanicity and 3.01% in those exposed only to family history of psychosis. These findings are suggestive of a gene × environment interaction between shared genes (familial liability) and the urban environment in the causation of psychosis (van Os, Hanssen, Bak, Bijl, & Vollebergh, 2003).

Of course, we are still very far from being able to identify which genes might moderate the effect of urbanicity and at which stage of development this interaction produces the larger impact on the risk for psychosis. It is also very uncertain which components of the urban environment play a role as risk factors for psychosis; this makes it difficult to translate the above finding into a reliable and valid experimental study design to test the biological mechanism underlying this interaction.

The first report of a specific gene-versus-environment interaction in a major psychiatric disorder came from Caspi et al. (2003), who reported that risk for depression following exposure to stressful life events is moderated by a functional polymorphism in the promoter region of the serotonin transporter (5-HTT) gene. These and subsequent similar findings have been recently challenged by a meta-analysis (Risch et al., 2009). This critical report carries its own limitations. For instance, it ignores experimental studies reporting how healthy volunteers exposed to stressors and having the 5-HTTLPR short genotype have greater stress responses (psychological, emotional, and cognitive) than do the 5-HTT long genotype (Beevers, Gibb, McGeary, & Miller, 2007; Canli & Lesch, 2007; Fox et al., 2009; Sheikh et al., 2008). These findings indicate that the verdict is still out, as also indicated by Uher and McGuffin (2008) and in Chapters 2, 3 and 12 by Uher, Shanahan and Bauldry, Perpletchikova and Kaufman in this volume.

The first attempt to explore how genes and environment might interact to initiate psychosis also came from the Dunedin study. Caspi and colleagues (2005), studying the catechol-O-methyltransferase (COMT) Val[158]Met genotype, showed that those individuals carrying the Val allele were more likely to develop psychosis following adolescent cannabis use than were those who were Met homozygotes. Subsequently, Stefanis et al. (2007) investigated how stress might precipitate psychosis in those with a specific genotype in a cohort of Greek conscripts. They found that exposure to intense and datable stress (young men at the start of a compulsory army induction program) led to an increased likelihood to experience psychotic symptoms and that carriers of the *COMT* Val allele were more susceptible to psychotic symptoms following stress than were those with the Met/Met genotype (Stefanis et al., 2007). This model of stress exposure also excludes, a priori, any gene × environment correlation, since in Greece, induction into the army is compulsory for all men and therefore not dependent upon individual, genetically influenced characteristics. In support of these findings, a study in young adult female twins reported that carriers of the Val/Val genotype displayed more feelings of paranoia in response to daily stressful events (Simons et al., 2009). In the same study, another functional polymorphism, the *BDNF* Val66Met (Simons et al., 2009) also influenced vulnerability to social-stress-induced paranoia (Simons et al., 2009).

Of course, exposure to stress itself may be significantly influenced by genetic factors. Kendler et al. (1999), in a population-based sample of 2,164 female twins, showed that genetic risk factors for major depression increase the probability of experiencing stressful life events. There is not yet evidence that the genetic factors that influence the likelihood to experience life events are positively correlated with the genetic liability to psychosis as well. Nevertheless, it is possible to speculate that genes influence the risk for psychiatric illness by causing individuals to select themselves into high-risk environments and by shaping the individual response to the latter.

Thus, the genetic susceptibility to schizophrenia is likely to be moderated by a number of genetic variants of small effect relatively common in the general population or by rare CNVs involving genes relevant to key systems and processes underlying psychosis. It is plausible that we all carry genetic variants that influence the degree and nature of our response to stress. Unfortunately, both the environmental stressors and the various small genetic variants potentially playing a role in the etiology of psychosis remain very difficult to measure.

CONCLUSION

The role of stress on the onset of psychosis was first recognized many years ago, when psychosocial stressors started to be consistently described in patients with schizophrenia and to be linked to its relapse. However, the progressive shift of psychosis research from a psychosocial to a more biological focus in the latter quarter of the 20th century led to a loss of interest in stress as a risk factor for psychosis, and to a lack of studies supporting this association. The difficulty in finding biological mechanisms explaining why some but not other subjects exposed to high levels of stress develop psychosis also contributed to this loss of interest. However, in the last decade, there has been a revival of studies examining the etiological role of the social environment in schizophrenia and the role of the HPA axis in mediating this. Furthermore, genetic epidemiological and molecular studies have facilitated the idea that genetic vulnerability may be needed for stress to precipitate the onset of psychosis. So far, only a few studies have investigated gene–stress interaction in psychosis, and we are still far from understanding which genetic factors modulate the effect of stress. However, these studies already offer a way forward to a better understanding of the etiopathogenesis of psychosis, and in particular, of the previous findings supporting the role of stress in the development of psychosis, by integrating psychosocial, biological, and genetic data.

REFERENCES

Albus, M., Ackenheil, M., Engel, R. R., & Muller, F. (1982). Situational reactivity of autonomic functions in schizophrenic patients. *Psychiatry Research, 6,* 361–370.

Axelson, D. A., Doraiswamy, P. M., Boyko, O. B., Rodrigo, E. P., McDonald, W. M., Ritchie, J. C., et al. (1992). In vivo assessment of pituitary volume with magnetic resonance imaging and systematic stereology: relationship to dexamethasone suppression test results in patients. *Psychiatry Research, 44,* 63–70.

Bebbington, P., Wilkins, S., Jones, P., Foerster, A., Murray, R., Toone, B., et al. (1993). Life events and psychosis. Initial results from the Camberwell Collaborative Psychosis Study. *British Journal of Psychiatry, 162,* 72–79.

Bebbington, P. E., Bhugra, D., Brugha, T., Singleton, N., Farrell, M., Jenkins, R., et al. (2004). Psychosis, victimisation and childhood disadvantage: evidence from the second British National Survey of Psychiatric Morbidity. *British Journal of Psychiatry, 185,* 220–226.

Beevers, C. G., Gibb, B. E., McGeary, J. E., & Miller, I. W. (2007). Serotonin transporter genetic variation and biased attention for emotional word stimuli among psychiatric in-patients. *Journal of Abnormal Psychology, 116,* 208–212.

Birley, J. L., & Brown, G. W. (1970). Crises and life changes preceding the onset or relapse of acute schizophrenia: clinical aspects. *British Journal of Psychiatry, 116,* 327–333.

Breier, A., Wolkowitz, O. M., Doran, A. R., Bellar, S., & Pickar, D. (1988). Neurobiological effects of lumbar puncture stress in psychiatric patients and healthy volunteers. *Psychiatry Research, 25,* 187–194.

Brown, G. W., & Birley, J. L. (1968). Crises and life changes and the onset of schizophrenia. *Journal of Health and Social Behavior, 9,* 203–214.

Canli, T., & Lesch, K. P. (2007). Long story short: the serotonin transporter in emotion regulation and social cognition. *Nature Neuroscience, 10,* 1103–1109.

Cannon, W. B. (1929). *Body changes in pain, hunger, fear and rage.* New York: Appleton.

Cardno, A. G., Rijsdijk, F. V., Sham, P. C., Murray R. M., & McGuffin P. (2002). A twin study of genetic relationships between psychotic symptoms. *American Journal of Psychiatry, 159 (4),* 539–545.

Caspi, A., Moffitt, T. E., Cannon, M., McClay, J., Murray, R., Harrington, H., et al. (2005). Moderation of the effect of adolescent-onset cannabis use on adult psychosis by a functional polymorphism in the catechol-O-methyltransferase gene: longitudinal evidence of a gene X environment interaction. *Biological Psychiatry, 57,* 1117–1127.

Caspi, A., Sugden, K., Moffitt, T. E., Taylor, A., Craig, I. W., Harrington, H., et al. (2003). Influence of life stress on depression: moderation by a polymorphism in the 5-HTT gene. *Science, 301*(5631), 386–389.

Ceskova, E., Kasparek, T., Zourkova, A., & Prikryl, R. (2006). Dexamethasone suppression test in first-episode schizophrenia. *Neuro Endocrinology Letters, 27,* 433–437.

Chrousos, G. P., & Gold, P. W. (1992). The concepts of stress and stress system disorders. *Overview of physical and behavioral homeostasis. Journal of the American Medical Association, 267,* 1244–1252.

Chung, R. K., Langeluddecke, P., & Tennant, C. (1986). Threatening life events in the onset of schizophrenia, schizophreniform psychosis and hypomania. *British Journal of Psychiatry, 148,* 680–685.

Collier, D. A. (2008). Schizophrenia: the polygene princess and the pea. *Psychological Medicine, 38,* 1687–1691.

Compton, M. T., Furman, A. C., & Kaslow, N. J. (2004). Preliminary evidence of an association between childhood abuse and cannabis dependence among African American first-episode schizophrenia-spectrum disorder patients. *Drug and Alcohol Dependence, 76,* 311–316.

Crow, T. J. (2008). The emperors of the schizophrenia polygene have no clothes. *Psychological Medicine, 38,* 1681–1685.

Di Forti, M., Lappin, J. M., & Murray, R. M. (2007). Risk factors for schizophrenia— all roads lead to dopamine. *European Neuropsychopharmacology, 17 (Suppl. 2),* S101–S107.

Fisher, H., Morgan, C., Dazzan, P., Craig, T. K., Morgan, K., Hutchinson, G., et al. (2009). Gender differences in the association between childhood abuse and psychosis. *British Journal of Psychiatry, 194,* 319–325.

Fox, E., Ridgewell, A., & Ashwin, C. (2009). Looking on the bright side: biased attention and the human serotonin transporter gene. *Proceedings. Biological Sciences, 276,* 1747–1751.

Franzen, G. (1971). Serum cortisol in chronic schizophrenia: a study of the adreno-cortical response to intravenously administered insulin and ACTH. *Acta Psychiatrica Scandinavica, 47,* 82–91.

Garner, B., Berger, G. E., Nicolo, J. P., Mackinnon, A., Wood, S. J., Pariante, C. M., et al. (2009). Pituitary volume and early treatment response in drug-naive first-episode psychosis patients. *Schizophrenic Research, 113,* 65–71.

Garner, B., Pariante, C. M., Wood, S. J., Velakoulis, D., Phillips, L., Soulsby, B., et al. (2005). Pituitary volume predicts future transition to psychosis in individuals at ultra-high risk of developing psychosis. *Biological Psychiatry, 58,* 417–423.

Goldstein, J. M., Seidman, L. J., Makris, N., Ahern, T., O'Brien, L. M., Caviness, V. S., Jr., et al. (2007). Hypothalamic abnormalities in schizophrenia: sex effects and genetic vulnerability. *Biological Psychiatry, 61,* 935–945.

Greenfield, S. F., Strakowski, S. M., Tohen, M., Batson, S. C., & Kolbrener, M. L. (1994). Childhood abuse in first-episode psychosis. *British Journal of Psychiatry, 164,* 831–834.

Gruen, R., & Baron, M. (1984). Stressful life events and schizophrenia. Relation to illness onset and family history. *Neuropsychobiology, 12,* 206–208.

Herz, M. I., Fava, G. A., Molnar, G., & Edwards, L. (1985). The dexamethasone suppression test in newly hospitalized schizophrenic patients. *American Journal of Psychiatry, 142,* 127–129.

Hirsch, S., Bowen, J., Emami, J., Cramer, P., Jolley, A., Haw, C., et al. (1996). A one year prospective study of the effect of life events and medication in the aetiology of schizophrenic relapse. *British Journal of Psychiatry, 168,* 49–56.

Howes, O. D., McDonald, C., Cannon, M., Arseneault, L., Boydell, J., & Murray, R. M. (2004). Pathways to schizophrenia: the impact of environmental factors. *Internation Journal of Neuropsychopharmacology, 7(Suppl 1),* S7–S13.

Huttunen, J., Heinimaa, M., Svirskis, T., Nyman, M., Kajander, J., Forsback, S., et al. (2008). Striatal Dopamine Synthesis in First-degree Relatives of Patients with Schizophrenia. *Biological Psychiatry, 63,* 114–117.

Ismail, K., Murray, R. M., Wheeler, M. J., & O'Keane, V. (1998). The dexamethasone suppression test in schizophrenia. *Psychological Medicine, 28,* 311–317.

Jansen, L. M., Gispen-de Wied, C. C., Gademan, P. J., De Jonge, R. C., van der Linden, J. A., & Kahn, R. S. (1998). Blunted cortisol response to a psychosocial stressor in schizophrenia. *Schizophrenic Research, 33,* 87–94.

Janssen, I., Krabbendam, L., Bak, M., Hanssen, M., Vollebergh, W., de Graaf, R., et al. (2004). Childhood abuse as a risk factor for psychotic experiences. *Acta Psychiatrica Scandinavica, 109,* 38–45.

Johnstone, E. C., Crow, T. J., Frith, C. D., Husband, J., & Kreel, L. (1976). Cerebral ventricular size and cognitive impairment in chronic schizophrenia. *Lancet, 2,* 924–926.

Kaneko, M., Yokoyama, F., Hoshino, Y., Takahagi, K., Murata, S., Watanabe, M., et al. (1992). Hypothalamic-pituitary-adrenal axis function in chronic schizophrenia: association with clinical features. *Neuropsychobiology, 25,* 1–7.

Kendler, K. S., Karkowski, L. M., & Prescott, C. A. (1999). Causal relationship between stressful life events and the onset of major depression. *American Journal of Psychiatry, 156,* 837–841.

Keshavan, M. S., Brar, J., Ganguli, R., & Jarrett, D. (1989). DST and schizophrenic symptomatology. *Biological Psychiatry, 26,* 856–858.

Kirov, G., Zaharieva, I., Georgieva, L., Moskvina, V., Nikolov, I., Cichon, S., et al. (2008). A genome-wide association study in 574 schizophrenia trios using DNA pooling. *Molecular Psychiatry, 14(8),* 796–803.

Kirschbaum, C., & Hellhammer, D. H. (1994). Salivary cortisol in psychoneuroendo-crine research: recent developments and applications. *Psychoneuroendocrinology, 19,* 313–333.

Lammers, C. H., Garcia-Borreguero, D., Schmider, J., Gotthardt, U., Dettling, M., Holsboer, F., et al. (1995). Combined dexamethasone/corticotropin-releasing hormone test in patients with schizophrenia and in normal controls: II. *Biological Psychiatry, 38,* 803–807.

Lazarus, R. S. (1966). *Psychological stress and the coping process.* New York: McGraw-Hill.

Lewis, G., David, A., Andreasson, S., & Allebeck, P. (1992). Schizophrenia and city life. *Lancet, 340,* 137–140.

Malla, A. K., Cortese, L., Shaw, T. S., & Ginsberg, B. (1990). Life events and relapse in schizophrenia. A one year prospective study. *Social Psychiatry and Psychiatric Epidemiology, 25,* 221–224.

Marcelis, M., Takei, N., & van Os, J. (1999). Urbanization and risk for schizophrenia: does the effect operate before or around the time of illness onset? *Psychological Medicine, 29,* 1197–1203.

Mason, J. W. (1971). A re-evaluation of the concept of "non-specificity" in stress theory. *Journal of Psychiatric Research, 8,* 323–333.

Mazure, C. M., Quinlan, D. M., & Bowers, M. B., Jr. (1997). Recent life stressors and biological markers in newly admitted psychotic patients. *Biological Psychiatry, 41,* 865–870.

McEwen, B. S., & Seeman, T. (1999). Protective and damaging effects of mediators of stress. Elaborating and testing the concepts of allostasis and allostatic load. *Annuls of the New York Academy of Sciences, 896,* 30–47.

Mikhail, A. (1981). Stress: a psychophysiological conception. *Journal of Human Stress, 7,* 9–15.

Mondelli, V., Dazzan, P., Gabilondo, A., Tournikioti, K., Walshe, M., Marshall, N., et al. (2008). Pituitary volume in unaffected relatives of patients with schizo-phrenia and bipolar disorder. *Psychoneuroendocrinology, 33,* 1004–1012.

Mondelli, V., Dazzan, P., Hepgul, N., Di, F. M., Aas, M., D'Albenzio, A., et al. (2010). Abnormal cortisol levels during the day and cortisol awakening response in first-episode psychosis: The role of stress and of antipsychotic treatment. *Schizophrenia Research, 116,* 234–242.

Morgan, C., & Fisher, H. (2007). Environment and schizophrenia: environmental factors in schizophrenia: childhood trauma—a critical review. *Schizophrenia Bulletin, 33,* 3–10.

Morgan, C., Fisher, H., Hutchinson, G., Kirkbride, J., Craig, T. K., Morgan, K., et al. (2008). Ethnicity, social disadvantage and psychotic-like experiences in a healthy population based sample. *Acta Psychiatrica Scandinavica, 119*(3), 226–235.

Murray, R. M., & Lewis, S. W. (1987). Is schizophrenia a neurodevelopmental disorder? *British Medical Journal, 295,* 681–682.

Murray, R. M., Sham, P., van Os, J., Zanelli, J., Cannon, M., & McDonald C. (2004). A developmental model for similarities and dissimilarities between schizophrenia and bipolar disorder. *Schizophrenia Research, 71,* 405–416.

Myin-Germeys, I., Delespaul, P., & van Os, J. (2005). Behavioural sensitization to daily life stress in psychosis. *Psychological Medicine, 35,* 733–741.

Myin-Germeys, I., van Os, J., Schwartz, J. E., Stone, A. A., & Delespaul, P. A. (2001). Emotional reactivity to daily life stress in psychosis. *Archives of General Psychiatry, 58,* 1137–1144.

Newcomer, J. W., Faustman, W. O., Whiteford, H. A., Moses, J. A., Jr., & Csernansky, J. G. (1991). Symptomatology and cognitive impairment associate independently with post-dexamethasone cortisol concentrations in unmedicated schizophrenic patients. *Biological Psychiatry, 29,* 855–864.

O'Donovan, M. C., Craddock, N., & Owen, M. J. (2008). Schizophrenia: complex genetics, not fairy tales. *Psychological Medicine, 38,* 1697–1699.

O'Donovan, M. C., Craddock, N. J., & Owen, M. J. (2009). Genetics of psychosis; insights from views across the genome. *Human Genetics, 126,* 3–12.

Pallanti, S., Quercioli, L., & Pazzagli, A. (1997). Relapse in young paranoid schizophrenic patients: a prospective study of stressful life events, P300 measures, and coping. *American Journal of Psychiatry, 154,* 792–798.

Pariante, C. M., Dazzan, P., Danese, A., Morgan, K. D., Brudaglio, F., Morgan, C., et al. (2005). Increased pituitary volume in antipsychotic-free and antipsychotic-treated patients of the AEsop first-onset psychosis study. *Neuropsychopharmacology, 30,* 1923–1931.

Pariante, C. M., & Lightman, S. L. (2008). The HPA axis in major depression: classical theories and new developments. *Trends in Neuroscience, 31,* 464–468.

Pariante, C. M., Vassilopoulou, K., Velakoulis, D., Phillips, L., Soulsby, B., Wood, S. J., et al. (2004). Pituitary volume in psychosis. *British Journal of Psychiatry, 185,* 5–10.

Purcell, S. M., Wray, N. R., Stone, J. L., Visscher, P. M., O'Donovan, M. C., Sullivan, P. F., et al. (2009). Common polygenic variation contributes to risk of schizophrenia and bipolar disorder. *Nature, 460,* 748–752.

Raison, C. L., & Miller, A. H. (2003). When not enough is too much: the role of insufficient glucocorticoid signaling in the pathophysiology of stress-related disorders. *American Journal of Psychiatry, 160,* 1554–1565.

Read, J., van Os, J., Morrison, A. P., & Ross, C. A. (2005). Childhood trauma, psychosis and schizophrenia: a literature review with theoretical and clinical implications. *Acta Psychiatrica Scandinavica, 112,* 330–350.

Risch, N., Herrell, R., Lehner, T., Liang, K. Y., Eaves, L., Hoh, J., et al. (2009). Interaction between the serotonin transporter gene (5-HTTLPR), stressful life events, and risk of depression. *Journal of the American Medical Association 301*(23), 2463–2471.

Rosenthal, D. (1970). *Genetic theory and abnormal behaviour.* New York: McGraw-Hill.

Rosenthal, D., Wender, P. H., Kety, S. S., Schulsinger, F., Welner, J., & Rieder, R. O. (1975). Parent-child relationships and psychopathological disorder in the child. *Archives of General Psychiatry, 32,* 466–476.

Ryan, M. C., Collins, P., & Thakore, J. H. (2003). Impaired fasting glucose tolerance in first-episode, drug-naive patients with schizophrenia. *American Journal of Psychiatry, 160,* 284–289.

Ryan, M. C., Flanagan, S., Kinsella, U., Keeling, F., & Thakore, J. H. (2004). The effects of atypical antipsychotics on visceral fat distribution in first episode, drug-naive patients with schizophrenia. *Life Science, 74,* 1999–2008.

Ryan, M. C., Sharifi, N., Condren, R., & Thakore, J. H. (2004). Evidence of basal pituitary-adrenal overactivity in first episode, drug naive patients with schizophrenia. *Psychoneuroendocrinology, 29,* 1065–1070.

Rybakowski, J., Linka, M., Matkowski, K., & Kanarkowski, R. (1991). [Dexamethasone suppression test and the positive and negative symptoms of schizophrenia]. *Psychiatria Polska, 25,* 9–15.

Sachar, E. J., Kanter, S. S., Buie, D., Engle, R., & Mehlman, R. (1970). Psychoendocrinology of ego disintegration. *American Journal of Psychiatry, 126,* 1067–1078.

Sapolsky, R. M. (2000). Glucocorticoids and hippocampal atrophy in neuropsychiatric disorders. *Archives of General Psychiatry, 57,* 925–935.

Selye, H. (1998). A syndrome produced by diverse nocuous agents. 1936. *The Journal of Neuropsychiatry and Clinical Neurosciences, 10,* 230–231.

Sheikh, H. I., Hayden, E. P., Singh, S. M., Dougherty, L. R., Olino, T. M., Durbin, C. E., et al. (2008). An examination of the association between the 5-HTT promoter region polymorphism and depressogenic attributional styles in childhood. *Personality and Individual Differences, 45(5),* 425–428.

Shi, J., Levinson, D. F., Duan, J., Sanders, A. R., Zheng, Y., Pe'er, I., et al. (2009). Common variants on chromosome 6p22.1 are associated with schizophrenia. *Nature, 460,* 753–757.

Simons, C. J., Wichers, M., Derom, C., Thiery, E., Myin-Germeys, I., Krabbendam, L., et al. (2009). Subtle gene-environment interactions driving paranoia in daily life. *Genes, Brain and Behavior, 8,* 5–12.

Spataro, J., Mullen, P. E., Burgess, P. M., Wells, D. L., & Moss, S. A. (2004). Impact of child sexual abuse on mental health: prospective study in males and females. *British Journal of Psychiatry, 184,* 416–421.

Spauwen, J., Krabbendam, L., Lieb, R., Wittchen, H. U., & van Os, J. (2006). Impact of psychological trauma on the development of psychotic symptoms: relationship with psychosis proneness. *British Journal of Psychiatry, 188,* 527–533.

Stefanis, N. C., Henquet, C., Avramopoulos, D., Smyrnis, N., Evdokimidis, I., Myin-Germeys, I., et al. (2007). COMT Val158Met moderation of stress-induced psychosis. *Psychological Medicine, 37,* 1651–1656.

Stefansson, H., Ophoff, R. A., Steinberg, S., Andreassen, O. A., Cichon, S., Rujescu, D., et al. (2009). Common variants conferring risk of schizophrenia. *Nature, 460,* 744–747.

Stefansson, H., Rujescu, D., Cichon, S., Pietiläinen, O. P., Ingason, A., Steinberg, S., et al. (2008). Large recurrent microdeletions associated with schizophrenia. *Nature, 455,* 232–236.

Sullivan, P. F. (2008). The dice are rolling for schizophrenia genetics. *Psychological Medicine, 38,* 1693–1696.

Sullivan, P. F., Lin, D., Tzeng, J. Y., van den Oord, E., Perkins, D., Stroup, T. S., et al. (2008). Genomewide association for schizophrenia in the CATIE study: results of stage 1. *Molecular Psychiatry, 13*(6), 570–584.

Tandon, R., Mazzara, C., DeQuardo, J., Craig, K. A., Meador-Woodruff, J. H., Goldman, R., et al. (1991). Dexamethasone suppression test in schizophrenia: relationship to symptomatology, ventricular enlargement, and outcome. *Biological Psychiatry, 29,* 953–964.

Ucok, A., & Bikmaz, S. (2007). The effects of childhood trauma in patients with first-episode schizophrenia. *Acta Psychiatrica Scandinavica, 116,* 371–377.

Uher, R., & McGuffin, P. (2008). The moderation by the serotonin transporter gene of environmental adversity in the aetiology of mental illness: review and methodological analysis. *Molecular Psychiatry, 13,* 131–146.

van Praag, H. M., de Kloet, E. R., & van Os, J. (2004). Traumatic life events: general issues. In *Stress, the brain and depression* (pp. 12–23). Cambridge, UK: Cambridge University Press.

van Os, J., Hanssen, M., Bak, M., Bijl, R. V., & Vollebergh, W. (2003). Do urbanicity and familial liability coparticipate in causing psychosis? American Journal of Psychiatry, *160,* 477–482.

van Os, J., & Murray, R. (2008). Gene-environment interactions in schizophrenia. *Introduction. Schizophrenia Bulletin, 34,* 1064–1065.

Ventura, J., Nuechterlein, K. H., Lukoff, D., & Hardesty, J. P. (1989). A prospective study of stressful life events and schizophrenic relapse. *Journal of Abnormal Psychology, 98,* 407–411.

Walder, D. J., Walker, E. F., & Lewine, R. J. (2000). Cognitive functioning, cortisol release, and symptom severity in patients with schizophrenia. *Biological Psychiatry, 48,* 1121–1132.

Walker, E., Mittal, V., & Tessner, K. (2008). Stress and the hypothalamic pituitary adrenal axis in the developmental course of schizophrenia. *Annual Review of Clinical Psychology, 4,* 189–216.

Walker, E. F., & Diforio, D. (1997). Schizophrenia: a neural diathesis-stress model. *Psychology Review, 104,* 667–685.

Walsh, T., McClellan, J. M., McCarthy, S. E., Addington, A. M., Pierce, S. B., Cooper, G. M., et al. (2008). Rare structural variants disrupt multiple genes in neurodevelopmental pathways in schizophrenia. *Science, 320,* 539–543.

Weiss, L. A., Shen, Y., Korn, J. M., Arking, D. E., Miller, D. T., Fossdal, R., et al. (2008). Association between microdeletion and microduplication at 16p11.2 and autism. *New England Journal of Medicine, 358,* 667–675.

Whitfield, C. L., Dube, S. R., Felitti, V. J., & Anda, R. F. (2005). Adverse childhood experiences and hallucinations. *Child Abuse and Neglect, 29,* 797–810.

Zammit, S., Wiles, N., & Lewis, G. (2010). The study of gene–environment interactions in psychiatry: limited gains at a substantial cost? *Psychological Medicine, 40,* 711–716.

Part III

Implications for the Future

19

Conclusions and Implications
for the Future

KENNETH S. KENDLER, SARA R. JAFFEE, AND DANIEL ROMER

This book has had three major themes, as applied to psychiatric and substance use syndromes and psychological traits: gene–environment interaction, gene–environment correlation, and development. Our goal in this final chapter is to reflect on this material and to provide the reader with a few take-home messages.

Most importantly, the studies presented in this book show us that the old static views about genetic and environmental contributions to behavioral and psychiatric traits do not accurately depict reality. The idea that we can study genes and environment in isolation from one another—that genes and environment just add together in acting on the behavioral traits that we care about—is unsustainable. Similarly, the idea that genes are static and stable in their effect across the lifespan—a belief which is alarmingly embedded in public consciousness—is, for many critical traits, just plain wrong. This volume illustrates an impressive and growing research literature showing how intertwined the impact of genes and environment are in the real world.

This book introduces and illustrates two quite different—but oft-confused—ways in which genes and environment interrelate: gene–environment interaction and gene–environment correlation. Gene–environment interaction can be defined in two equivalent ways, depending on whether you want to, respectively, give priority to a genetic or an environmental view of the world: (a) the impact of genes on a trait (or risk for an illness) varies substantially across different environments, *or* (b) the impact of a specific environment on a trait (or risk for an illness) varies substantially across different genotypes. Generally, the first conceptualization (that is, as a geneticist would see the world), sometimes summarized as *genetic control of sensitivity to the environment*, is the more frequently emphasized.

Gene–environment correlation, by contrast, reflects the nonrandom distribution of genetic and environmental risk factors for a disorder (or genetic and environmental influences for a trait). This can occur in several ways clearly outlined in Chapter 4, by Jaffee. One useful framework within which to think about gene–environment correlation is as an "outside-the-skin pathway for genetic effects." That is, to produce their full phenotypic effects, many genes have effects that "loop" outside the skin to impact on environmental exposures that then often feed back on the phenotype. This is appropriately contrasted with the more traditional, physiological, within-the-skin pathways of gene effects. Another useful name for gene–environment correlation is *genetic control of exposure to the environment*.

Many chapters in the book treat different aspects of gene–environment interaction. Readers of this volume will see that it is a challenging and sometimes controversial subject. As outlined in considerable detail in Chapter 1 by Kendler, "interaction" is, at its base, a statistical concept and can be rather ephemeral—hard to detect and replicate—and artifact prone. Implementing studies that have produced reliable gene–environmental interaction findings has, in some areas, proven difficult. None of these issues should take away from the fact that this paradigm has been and will continue to be deeply informative about how genes and environment jointly contribute to risk.

We also see in the volume good examples of the two different ways in which the "gene" part of gene × environment interaction is assessed. In some studies, typically utilizing twin samples, "genes" were assessed as a latent variable indexing the aggregate effect of all genes averaged across the genome. In other studies, specific genetic variants at individual genes were assessed. Not surprisingly, the former design, although uninformative about the specific biological pathways involved, tends to produce more robust and replicable findings than does the latter approach. For example, in Chapter 16, Kendler examines the interaction between peer deviance and genetic risk in predicting alcohol consumption using an aggregate genetic index measure of liability to alcohol use disorders calculated from the risk for alcohol use disorders in mother, father, and co-twin. No molecules are involved. In Chapter 14, by contrast, Dick shows that the association between variants in the receptor gene *GABRA2* and externalizing behavior was stronger with low levels of parental monitoring and reduced with higher parental monitoring. That is, she shows a molecular gene × environment interaction in the prediction of externalizing behavior.

A number of reports examined the famous (or perhaps infamous) serotonin transporter polymorphism in the context of gene–environment interactions. This literature has been one of the most controversial in recent memory (e.g., Caspi, Hariri, Holmes, Uher, & Moffitt, 2010; Risch et al., 2009), and we have no intention of trying to adjudicate this complex story here. Suffice it to say that contributors to our volume represent a spectrum from those deeply committed to (e.g., Uher) versus those rather skeptical of this specific research paradigm (e.g., Kendler).

Gene–environment correlation is a quite different thing from gene–environment interaction. Although it typically gets second billing, much less attention, and fewer headlines, it is probably more robust, less artifact prone, and at least as pervasive as gene–environmental interaction (Kendler & Baker, 2007). The concept that the skin is where genetic effects should stop is actually quite an odd one for creatures as social as humans.

This does not mean that the effects of gene–environment correlation cannot be subtle or complex. This is well illustrated in Chapter 6, by Horwitz, Marceau, and Neiderhiser, and in Chapter 15 by D'Onofrio and colleagues, which demonstrate how the nuclear family is a hotbed of interweaving gene–environment correlations. Hopefully, the results presented in this volume will convince the skeptical reader that, at least with respect to behavior, the skin is a rather arbitrary divide with respect to gene action. Put another way, we humans, in part influenced by our genes, construct and/or evoke quite a bit of our own psychosocial environment, which feeds back on our risk for illness in all sorts of ways.

Many, but not all, the chapters of this book also explore developmental effects. It is hopefully becoming self-evident to many in the field that the emergence of most psychiatric disorders and psychological traits can only be understood in a developmental context. For example, Kendler (Chapter 16) shows that alcohol use emerges from an interaction of genetic risk with key developmental experiences, particularly peer deviance. However, the degree to which genetic effects were moderated by environmental experiences itself changed over time. Genetic influences were more plastic in early adolescence (i.e., more easily moderated by environmental experiences) and became progressively less malleable as individuals passed through later adolescence and into early adulthood. The same chapter provides another illustration of how genetic and environmental influences can differentially shift in importance over development. The pathway from genes to high levels of conduct disorder to high levels of peer deviance remained strong and unchanging from childhood to late adolescence. By contrast, the path from shared environmental influences to peer deviance and back to conduct disorder attenuated rapidly with development, being quite potent in childhood and disappearing by late adolescence. The need for a developmental perspective is, if anything, more critical for those who seek to understand how genetic and environmental contributions to key disorders and traits unfold over time.

The volume also included chapters giving a glimpse of the impact of "new" genetic technologies on the fields of psychiatric and behavioral genetics. DNA methylation has revealed an entirely novel mechanism of control of gene expression. Particularly exciting from the perspective of this volume is the increasing evidence that environmental experiences can alter DNA methylation (see Chapter 7, by Mill). At least in animal models, the evidence is now convincing that one way in which the environment produces long-term changes in behavior is via an impact on DNA methylation patterns.

Gene expression, as demonstrated in Chapter 9, by Cole, provides another window into our genetic machinery. Although genomic DNA at the sequence level (the A's, T's, C's, and G's) is boringly stable over the life course (picking up now and again a rare somatic mutation), RNA levels are anything but. Cole's chapter convincingly shows how environmental experiences can alter, in a systematic and highly informative way, gene expression patterns in peripheral blood cells. Work in model organisms has similarly explored the impact of environmental effects (and drugs of abuse) on gene expression in the brain and found robust and complex effects (Kerns & Miles, 2008; Martineza, Calvo-Torrenta, & Herbert, 2002). Our increasing capacity to sequence human genes, measure methylation patterns, and assess gene expression rapidly and cheaply will surely shape the future psychiatric and behavioral genetics research in the not too distant future.

The picture of the etiology of psychiatric and substance use disorders and of key psychological traits that emerge from this book is not a simple one: gene–environment interaction and correlation are critical, and this all happens in a developmental "dance through time." The complexity can be daunting. But neither our science nor the understanding emerging therefrom will be advanced by massive oversimplification, which has sometimes characterized this field in the past. We have to embrace rather than deny the complexity.

WHAT DO WE NEED TO KNOW BEFORE RESEARCH ON GENE–ENVIRONMENT INTERPLAY CAN INFORM INTERVENTION?

Findings from studies of gene–environment interplay are already informative about which environments should be targeted for intervention. This is because some studies of gene–environment interplay are "natural experiments" that provide evidence as to whether a candidate risk environment has causal effects on psychopathology or whether the effects of the candidate environment are confounded by genotype. Prevention and intervention researchers can use the findings from these studies to make informed decisions about which aspects of the environment to target so as to ameliorate symptomatology. As reviewed by Jaffee (Chapter 4), D'Onofrio et al. (Chapter 15), and Glymour et al. (Chapter 8), several research designs are available to disentangle those risk factors (e.g., genetic and environmental) that are typically confounded. As described by Glymour et al., these designs have their unique strengths and weaknesses, and no one alone is an adequate alternative to experimental manipulation of candidate risk environments. Some make very strong assumptions that must be subject to empirical testing. Together, however, these research designs can provide convergent data. They have shown, for example, that successful interventions to reduce mothers' smoking during pregnancy are likely to reduce the prevalence of infants born small for gestational age, but are unlikely to have an effect on prevalence rates of antisocial behavior in the population (D'Onofrio et al., Chapter 15).

The next step for environmental risk research is to identify those mechanisms by which causal risk factors exert their effects. For example, there is converging evidence across studies using different research designs that parental divorce is a cause of children's antisocial behavior that operates apart from genetic confounds. Interventions to minimize the adverse impact of divorce on children would benefit from more research on *why* divorce contributes to children's antisocial behavior and why some children who experience divorce seem to be relatively unaffected. This may well be an example in which more careful delineation of the contextual and life-course components of "parental divorce" as an environmental influence needs to be applied (see Chapter 3 by Shanahan and Bauldry). Parental divorce is not simply an environmental event, and more analysis will be needed to determine its adverse elements, so that appropriate preventive action can be taken. In the case of maltreatment, the course of action is somewhat clearer in that any steps to reduce its occurrence are likely to have protective effects. Nurse visitation programs (Olds et al., 1997), for example, are one interventional strategy to prevent such influences in households defined as high risk (e.g., low parental education and income).

In some cases, it may be possible to identify biological mechanisms that underlie vulnerability to disorder. When researchers use animal models of disease, they can take full advantage of experimental design to identify causal risk factors and to probe the underlying biological mechanisms. Findings from experimental animal studies of gene–environment interplay have the potential to inform the development of novel treatments for disorder. For example, as noted by Perepletchikova and Kaufman (Chapter 12), there is evidence from studies of rats that histone deacetylase inhibitors can reverse the epigenetic changes associated with normal variations in early maternal care that produce variation in anxiety-related phenotypes in animals (Weaver, Meaney, & Szyf, 2006). Such findings have led to the investigation of histone deacetylase inhibitors as a potential treatment for depression (Covington et al., 2009). Nevertheless, it is fair to say that this research enterprise is still at a very early stage.

Tailoring of Treatment and Prevention

Findings of genotype × environment interaction have raised the prospect that interventions can be tailored to individuals with specific profiles of genetic and environmental risk. For this promise to be realized, researchers must make progress on at least four fronts. First, genotype × environment interactions must be replicated. As reviewed by Kendler (Chapter 1), Uher (Chapter 2), and Shanahan and Bauldry (Chapter 3), detecting G × E is methodologically challenging and there is considerable controversy about how best to aggregate findings from studies for the purposes of meta-analysis.

Second, researchers must identify the mechanisms by which genes and environments interact to affect behavior so as to determine how best to intervene. For example, Audrain-McGovern and Tercyak (Chapter 13) note that team sports participation protects individuals with one or two genetic risk alleles from taking up smoking or becoming more regular smokers. It is not clear, however, whether participation in team sports affords protection by increasing physical activity levels (and, thus, dopamine levels), by providing antismoking norms, or both. Interventions to prevent smoking progression would be informed by efforts to delineate those pathways. Greater understanding of such effects would enable researchers to design maximally effective interventions, so that even youth who are at high genetic risk for the adverse behavior (e.g., smoking) can be encouraged to follow a healthier behavioral trajectory. Thus, G × E research can help prevention researchers to identify interventions that are effective even for those at highest risk.

Greater understanding of how G × E effects are mediated at the biological level will also advance our ability to design effective preventive interventions. This goal will require linking genetic epidemiology with neuroscience (Caspi & Moffitt, 2006) to better identify the biological substrates that mediate environmental effects on behavior in genetically vulnerable individuals. Such mechanisms could be explored at a number of levels, including environmental alterations of DNA methylation, leading to the silencing of risk alleles (see Mill, Chapter 7), differences in cortical reactivity to aversive stimuli as a function of genotype (see Casey et al.,

Chapter 11), and differences in hormonal reactivity to stressful circumstances as a function of genotype (Gotlib, Joormann, Minor, & Hallmayer, 2007; Walker et al., Chapter 17). Identifying differences at the level of gene expression, neurotransmitter, or hormonal function among genetically identical individuals with different environmental exposures could suggest novel treatments for disorder that target those systems specifically.

Third, complex disorders are likely to be caused by multiple genes of small effect, interacting with each other and with the environment. As discussed by Audrain-McGovern and Tercyak (Chapter 13), the function of many identified gene variants is unknown. Treatments for disorder would be better informed by information about how systems of genes operate with each other and with the environment over time. It is not clear, for example, how pharmacotherapies that target the action of any one of these genes will affect the action of other genes implicated in disease. In addition, genotypes are likely to confer risk under some conditions and protection under others, which further complicates efforts to intervene at the level of the genome. At the very least, the paucity of replicated gene (or gene × environment) associations with disorder and the likely complexity of genetic effects make it currently impractical to attempt to identify at-risk individuals on the basis of genotype alone.

Fourth, researchers must grapple with the ethical implications of designing interventions for subsets of the population with specific genotypes. If this were to lead to prioritizing certain individuals for treatment on the basis of their genotype, it would have at least two consequences. It would have the potential to stigmatize the genetically "vulnerable," and it would raise the specter that others would be deemed genetically "invulnerable" and not treated even if they had been exposed to risk environments. The latter possibility is especially problematic if certain genetic variants are protective for some outcomes but not others in the face of specific environmental exposures. However, researchers rarely assess a range of outcomes in studies of gene–environment interplay, leaving it unclear, for example, whether the individual whose genotype confers protection against antisocial behavior in the face of maltreatment is also at low risk for depression and anxiety.

Take, for example, the possibility that specific genetic variants make some individuals highly susceptible to the effects of maltreatment on risk for depression or antisocial behavior. Some may argue that scarce resources for providing clinical services for maltreated children, family services to prevent maltreatment recurrence, or for monitoring families at risk of harming children should be targeted at those who carry genetic risk variants. These approaches increase the likelihood that some children will be deemed genetically "invulnerable" and not treated as quickly or comprehensively as needed. Such an approach would be ethically dubious because it would miss a substantial minority of children who would be adversely affected by maltreatment even if they did not carry risk alleles. Moreover, it would potentially stigmatize the targets of such interventions. Finally, the targeted approach to primary prevention would be practically problematic because preventative interventions are typically directed at the family level, and some children in a family will carry risk variants, but others will not.

Given these issues, researchers will have to determine whether interventions targeted to high-risk groups based on genetic characteristics are more effective than interventions delivered to all who experience adverse events, and whether the benefits of treating high-risk subgroups outweigh the costs of potentially stigmatizing those groups or failing to meet the mental health needs of those who are not deemed high risk. These issues are not unique to research on gene–environment interplay, but they will have to be adequately addressed before such strategies are adopted. Ultimately, these debates may be rendered moot. For example, the risk variants that increase susceptibility to maltreatment are common, affecting around a third of the population. Assuming that several variants interact with maltreatment exposure to increase risk for a broad range of psychiatric symptoms, most individuals will carry at least one of the risk variants, with the result being that even targeted interventions may be directed at the vast majority of those who experience maltreatment.

THE ROLE OF STRESS AND ADVERSE EVENTS IN MENTAL HEALTH

Several authors focus on the effects of early adverse events in precipitating later mental health conditions. Early adverse events have effects on depression (Chapter 12 by Perepletchiva and Kaufmann, and Chapter 2, by Uher), conduct disorder (Chapter 15, by D'Onofrio et al.), and schizophrenia (Chapter 17, by Walker et al., and Chapter 18, by Mondelli et al.) among others. Although these effects may be moderated by genetic predispositions, these events are among the most robust predictors of later mental health problems. Indeed, a recent epidemiological study found that nearly half of all child mental health disorders and about 30% of adult disorders were associated with such experiences (Green et al., 2010).

Despite these robust associations, we still lack a clear understanding of how they are mediated. Although, as suggested by Walker et al. (Chapter 17) and Mondelli et al. (Chapter 18), some may involve disruptions to the stress response system of either the parent or the offspring, it is also possible that they are mediated by epigenetic or other forms of control over gene expression (e.g., social genomics). Some may also be attributable to passive *r*GE (parental mental illness correlated with offspring mental illness). Despite these uncertainties, the path to preventing these disorders appears to lie in either reducing the rates of early adverse events or intervening with appropriate treatments to reverse a resulting dysfunctional trajectory. Kaufman's findings regarding the ameliorative effects of social support in reducing the impact of maltreatment on depression suggest that such interventions are possible and that they can overcome the genetic risks that predispose youth to adverse outcomes.

On another optimistic note, Eley's (Chapter 10) review of the effects of interventions using cognitive behavior therapy (CBT) to treat anxiety problems in children suggests that the same genetic factors that make children susceptible to such conditions may also make them responsive to interventions such as CBT. Future research should be directed at identifying the ameliorative effects of subsequent experiences, so that the burden of future disorders and maladaptive development can be prevented. It would also be interesting to determine whether programs that teach effective social problem solving and other forms of self control (i.e., preventative versions of CBT) can promote resilience in the face of harsh environments.

Adverse Environments That Prevent Healthy Youth Development

In addition to the adverse effects of early stressful environments on later disorder, it is also critical to consider the effects of such environments on the development of cognitive skills. The reciprocal effects model proposed by Dickens et al. (Chapter 5) suggests that youth growing up in resource-poor environments will not experience the same favorable evocative effects from parents and other adults that are available in more enriched environments. In addition, youth growing up in such environments will not have the same opportunities to engage in the cognitively stimulating activities available to youth in more advantaged settings. Furthermore, youth growing up in poor environments will also have greater exposure to the adverse effects of poverty and other risks for mental disorders. A range of strategies currently in place have the capability to reduce such deficits, including home visitation (Olds et al., 1997), access to preschool (Reynolds & Temple, 2008) and improved K through 12 training. Given the recent move toward reform of the U. S. healthcare system, there may also be opportunities for the primary care system to be more alert to conditions that interfere with healthy youth development. Concerns about mental health (and adequate insurance coverage for same) still lag behind attention to "physical" diseases and injuries.

LABELLING OF MENTAL ILLNESS AND RELATED CONDITIONS

Despite considerable improvement in public understanding of mental illness and its treatment, stigma associated with these conditions continues to
plague efforts at reform. Our editorial team faced this problem as we tried
to find the best terms that encompassed the many different forms of mental
health outcomes and traits that were discussed in the book. Some find the
term "mental," when applied to these problems, misleading because it suggests that the problems do not qualify as true illnesses (they are just psychological and hence "not real"). Others see the term as connoting a false
dichotomy between the mental and the physical. Our examination of
gene–environment interplay clearly indicates that behavioral and mental
health problems have a physical basis or else we would not find genetic
underpinnings to their emergence. However, labeling them as "psychiatric
disorders" can also create stigma for those so categorized because it suggests that the conditions they experience are subject to diagnosis by a specialized medical community and that the conditions are interpreted by
some to reflect personal weakness or failure. This attitude is especially evident for the addictive disorders, which are considered outside the realm of
general medicine and hence receive only limited attention by primary care
physicians (Kuehn, 2010).

The problem of stigma is especially important when considering the
first emergence of mental and behavioral health problems in childhood and
adolescence. Many of the syndromes that are eventually labeled as mental
disorders have their genesis and first appearance during these early ages
(Green et al., 2010). Although they may not qualify as full-blown conditions at this time, they do presage risk for later disorder. In addition, many
young people experience bouts of depression or substance use that fall short
of meeting full diagnostic criteria, yet are disruptive to adaptive functioning.
Indeed, estimates of lifetime experience of mental health conditions now
approach 50% when prospective methods of ascertainment are employed
(Moffitt et al., 2010). In addition, careful assessment of mental health conditions in late adolescence and early adulthood suggests similarly high rates
(Blanco et al., 2008). These estimates suggest that mental health problems
are quite common and that we risk avoiding the treatment and prevention
of these conditions due to concerns about labeling their early appearance as
signs of mental disorder. It is hoped that the findings in this book lead to a
less stigmatizing view of the mental and behavioral health problems that
emerge during childhood and adolescence and that the strong role that both
genes and the environment play in these conditions encourages readers and
the public to regard them as developmentally malleable and not reflective
of personal or moral failure.

INTERDISCIPLINARY TRAINING

For this field to move forward, scientific training must encompass a broad array of disciplines. Although the move in science has been toward increasing specialization, research on gene–environment interplay requires that we be generalists. On the one hand, it is not realistic to expect scientists to develop expertise in everything from wet lab techniques to the design of structured interviews, or to master analytic models of everything from gene expression profiles to complex survey designs. On the other hand, researchers must at least be familiar with the methodological, statistical, and conceptual issues that characterize these very different levels of analysis. One model for this sort of training is to establish centers or research units that bring together scientists interested in broadly similar phenotypes but who conduct their research at different levels of analysis, from the cellular/molecular to the social structural level. Another model employed by some graduate training programs is to require Ph.D. students to acquire minors in a related discipline.

Researchers must also be trained to convey the complexity of gene–environment interplay to other researchers and to lay audiences, including the science media. Despite an increasingly complex understanding of how genes and environments work together over time to influence development, there remains a common misperception that genotype equals destiny and that there are "genes for" complex disorders (Kendler, 2005). This responsibility to educate the media and the public will largely fall on the shoulders of those who work in this field. We hope that they and you, the reader, will take on this challenge, so that the public gains a better understanding of the myths that surround the influence of genes on behavior.

REFERENCES

Blanco, C., Okuda, M., Wright, C., Hasin, D. S., Grant, B. F., Liu, S., et al. (2008). Mental health of college students and their non-college-attending peers: results from the National Epidemiologic Study on Alcohol and Related Conditions. *Archives of General Psychiatry*, 65(12), 1429–1437.

Caspi, A., Hariri, A. R., Holmes, A., Uher, R., & Moffitt, T. E. (2010). Genetic sensitivity to the environment: the case of the serotonin transporter gene and its implications for studying complex diseases and traits. *American Journal of Psychiatry*, 167(5), 509–527.

Caspi, A., & Moffitt, T. E. (2006). Gene-environment interactions in psychiatry: Joining forces with neuroscience. *Nature Review Neuroscience*, 7, 583–590.

Covington, H. E., 3rd, Maze, I., LaPlant, Q. C., Vialou, V. F., Ohnishi, Y. N., Berton, O., et al. (2009). Antidepressant actions of histone deacetylase inhibitors. *Journal of Neuroscience*, 29(37), 11451–11460.

Gotlib, I. H., Joormann, J., Minor, K. L., & Hallmayer, J. (2007). HPA axis reactivity: a mechanism underlying the associations among 5HTTLPR, stress and depression. *Biological Psychiatry, 63*, 847–851.

Green, J. G., McLaughlin, K. A., Berglund, P. A., Gruber, M. J., Sampson, N. A., Zaslavsky, A. M., et al. (2010). Childhood adversities and adult psychiatric disorders in the National Comorbidity Survey Replication I: associations with first onset of DSM-IV disorders. *Archives of General Psychiatry, 67*(2), 113–123.

Kendler, K. S. (2005). "A gene for…": The nature of gene action in psychiatric disorders. *American Journal of Psychiatry, 162*, 1243–1252.

Kendler, K. S., & Baker, J. H. (2007). Genetic influences on measures of the environment: A systematic review. *Psychological Medicine, 37*(5), 615–626.

Kerns, R. T., & Miles, M. F. (2008). Microarray analysis of ethanol-induced changes in gene expression. *Methods in Molecular Biology, 447*, 395–410.

Kuehn, B. M. (2010). Integrated care key for patients with both addiction and mental illness. *Journal of the American Medical Association, 303*(19), 1905–1907.

Martineza, M., Calvo-Torrenta, A., & Herbert, J. (2002). Mapping brain response to social stress in rodents with c-fos expression: A review. *Stress: The International Journal on the Biology of Stress, 5*(1), 3–13.

Moffitt, T. E., Caspi, A., Taylor, A., Kokaua, J., Milne, B. J., Polanczyk, G., et al. (2010). How common are common mental disorders? Evidence that lifetime prevalence rates are doubled by prospective versus retrospective ascertainment. *Psychological Medicine, 40*(6): 899–909.

Olds, D. L., Eckenrode, J., Henderson, C. R., Kitzman, H., Powers, J., Cole, R., et al. (1997). Long-term effects of home visitation on maternal life course and child abuse and neglect. Fifteen-year follow-up of a randomized trial. *Journal of the American Medical Association, 278*(8), 637–643.

Reynolds, A. J., & Temple, J. A. (2008). Cost-effective early childhood development programs from preschool to third grade. *Annual Review of Clinical Psychology, 4*, 109–139.

Risch, N., Herrell, R., Lehner, T., Liang, K. Y., Eaves, L., Hoh, J., et al. (2009). Interaction between the serotonin transporter gene (5-HTTLPR), stressful life events, and risk of depression: A meta-analysis. *Journal of the American Medical Association, 301*(23), 2462–2471.

Weaver, I. C. G., Meaney, M. J., & Szyf, M. (2006). Maternal care effects on the hippocampal transcriptome and anxiety-mediated behaviors in the offspring that are reversible in adulthood. *Proceedings of the National Academy of Sciences of the U.S.A., 103*, 3480–3485.

Index

Note: Page numbers followed by "*f*" and "*t*" denote figures and tables, respectively.